Progress in Mathematics
Volume 134

Series Editors

H. Bass
J. Oesterlé
A. Weinstein

Algebraic Geometry and Singularities

Antonio Campillo López
Luis Narváez Macarro
Editors

Birkhäuser Verlag
Basel · Boston · Berlin

Editors:

Antonio Campillo López
Dep. de Algebra, Geometría y Topología
Fac. de Ciencias
Universidad de Valladolid
Prado de la Magdalena s/n
47005 Valladolid
Spain

Luis Narváez Macarro
Dept. of Algebra, Computación, Geometría y Topología
Fac. de Matemáticas
Universidad de Sevilla
Tarfia s/n
41012 Sevilla
Spain

A CIP catalogue record for this book is available from the Library of Congress,
Washington D.C., USA

Deutsche Bibliothek Cataloging-in-Publication Data

Algebraic geometry and singularities / Antonio Campillo López;
Luis Narváez Macarro (ed.). – Basel ; Boston ; Berlin :
Birkhäuser, 1996
  (Progress in mathematics ; Vol. 134)
  ISBN 3-7643-5334-1 (Basel ...)
  ISBN 0-8176-5334-1 (Boston)
NE: Campillo López, Antonio [Hrsg.]; GT
1991 Mathematics Subject Classification 14B05.

© 1996 Birkhäuser Verlag, P.O. Box 133, CH-4010 Basel, Switzerland
Printed on acid-free paper produced of chlorine-free pulp. TCF ∞
Printed in Germany
ISBN 3-7643-5334-1
ISBN 0-8176-5334-1

9 8 7 6 5 4 3 2 1

Table of Contents

I   Resolution of Singularities

Désingularisation en dimension 3 et caractéristique p
VINCENT COSSART

Sur l'espace des courbes tracées sur une singularité
G. GONZALEZ-SPRINBERG   M. LEJEUNE-JALABERT

Blowing up acyclic graphs and geometrical configurations
CARLOS MARIJUÁN

On moduli spaces of semiquasihomogeneous singularities
GERT-MARTIN GREUEL   GERHARD PFISTER

Stratification Properties of Constructible Sets
ZBIGNIEW HAJTO

On the linearization problem and some questions for webs in $\mathbb{C}^2$
ALAIN HÉNAUT

Globalization of Admissible Deformations
THEO DE JONG

Caractérisation géométrique de l'existence du polynôme de Bernstein relatif
J. BRIANÇON   PH. MAISONOBE

Le Polygone de Newton d'un $\mathcal{D}_X$-module
Z. MEBKHOUT

How good are real pictures?
DAVID MOND

Weighted homogeneous complete intersections
C. T. C. WALL

III   Curves and Surfaces

Degree 8 and genus 5 curves in $\mathbb{P}^3$ and the Horrocks-Mumford bundle.
M. R. GONZÁLEZ-DORREGO

Irreducible Polynomials of $k((X))[Y]$
A. GRANJA

Examples of Abelian Surfaces with Polarization type (1,3)
ISIDRO NIETO

Semigroups and Clusters at Infinity
ANA-JOSÉ REGUERA LÓPEZ

Cubic surfaces with double points in positive characteristic
MARKO ROCZEN

On the classification of reducible curve singularities
JAN STEVENS

Introduction

The volume contains both general and research papers. Among the first ones are papers showing recent and original developments or methods in subjects such as resolution of singularities, D-module theory, singularities of maps and geometry of curves. The research papers deal on topics related to, or close to, those listed above. The contributions are organized in three parts according to their contents.

Part I presents a set of papers on resolution of singularities, a topic of renewed activity. It deals with important topics of current interest, such as canonical, algorithmic, combinatorial and graphical procedures (Villamayor, Oka, Marijuán), as well as special results on desingularization in characteristic p (Cossart, Moh), and connections between resolution and structure of the space of arcs through a singularity (González-Sprinberg-Lejeune-Jalabert).

Part II contains a series of papers on the study of singularities and its connections with differential systems and deformation or perturbation theories. Two expository papers (Maisonobe-Briançon, Mebkhout) describe, in an algebro-geometric way, the interaction between singularities and D-module theory including recent progress on Bernstein polynomials and Newton polygon techniques. Geometry of foliations (Hénaut, García-Reguera), polar varieties and stratifications (Hajto) are also topics treated here. Two other papers (Wall, Greuel-Pfister) deal with quasihomogeneous singularities in the contexts of perturbations and moduli spaces. Globalization of deformations of singularities (de Jong) and determination of complex topology from the real one (Mond) complete this series of papers.

Part III consists of papers on algebraic geometry of curves and surfaces. Equisingularity of plane curve singularities over arbitrary ground fields provides irreducibility criteria (Granja) and helps the study of singularities at infinity (Reguera). Classification of space curves is the objetive of another paper (Stevens). Curves on Kummer surfaces are not appropriate to construct the Horrocks- Mumford bundle (González Dorrego). Two papers on surfaces with double points (Roczen) and abelian surfaces (Nieto) complete the volume.

The papers originated at the Third International Conference on Algebraic Geometry held at La Rábida (Spain) during the week of December 9-14, 1991. This time the main focus of the Conference was Singularity Theory, especially its algebraic aspects. This was an appropriate end of an active year in this field.

Several organizations sponsored the Conference: the European Singularity Project, the local government Junta de Andalucía, the cities of Palos de la Frontera and Huelva, and the universities of Valladolid and Sevilla. The Organizing Committee consisted of J.M. Aroca, T. Sánchez Giralda (Valladolid), F. Castro and J.L. Vicente (Sevilla). Profs. A. Campillo (Valladolid), G.M. Greuel (Kaiserslautern), Lê Dũng Tráng (Paris), I. Luengo (Madrid), L. Narváez (Sevilla) formed the Scientific Committee.

The Scientific Committee received a first version of the papers before November 30, 1992. Then the papers passed through a steady process of elaboration and refereeing. The final versions were produced before September 30, 1993. We thank the referees for their contributions. We also thank all speakers, participants, the Organizing and Scientific Committees, who helped, in one way or another, to make this book possible. We owe special thanks to J.L. Vicente for making possible again to enjoy the nice atmosphere of La Rábida. Antonio Inés organized the background tasks and Manuel Soto did an excellent work by T<small>E</small>Xing the book.

A. Campillo López

L. Narváez Macarro

July, 1994

Plenary Conferences

BRASSELET, Jean Paul
Le théorème de De Rham relatif

CASAS, Eduardo
Singularities of polar curves

COSSART, Vincent
Desingularization in dimension 3, characteristic p

LE DUNG TRANG
Lefschetz theorem and vanishing of constructible cohomologies

LEJEUNE-JALABERT, Monique
Sur l'espace des courbes des singularités de surface

MEBKHOUT, Zoghman
Le polygne de Newton du faisceau d'irrgularit

MOH, T.T.
On a Newton polygon approach to the uniformization of singularities of characteristic p

NAVARRO AZNAR, Vicente
Connexion de Gauss-Manin sur l'homotopie rationnelle

OKA, Mutsuo
On the resolution complexity of planes curves and hyperplane sections

PHAM, Frederic
Monodromy and resurgence

SAITO, Kyoji
Jacobi inversion problem and root system

SIERSMA, Dirk
Functions on singular spaces

SZPIRO, Lucien
abc and Mordell

SPIVAKOSVSKI, Marc
On the Artin approximation theorem

TEISSIER, Bernard
Local h-cobordism and the geometry of the real discriminant

WALL, Charles Terence Clagg
Weighted homogeneous complete intersections

Specialized Conferences

ALEKSANDROV, Aleksandr Grigorjevich
Deformations of zero-dimensional singularities

BALDASSARI, Francesco
Generalized Hypergeometric Functions (GHF) and variation of cohomology
(d'après Dwork and Loeser)

BIRBRAIR, Gev
The stratification of the space of homogeneous mappings

CASTELLANOS, Julio
Arf closure relative to a divisorial valuation and transversal curves

DELGADO DE LA MATA, Félix
Symmetry of semigroups and algebraic properties of curves

DU BOIS, Philippe
Forme de Seifert des Entrelacs Algébriques

GAETA, Federico
A discussion of the possible applications of the D-modules, hypergeometric
functions, etc. to the Schottky problem

GARCIA DE LA FUENTE, Julio
Polaridad relativa a una foliación en $\mathbf{P}^2$

GONZALEZ DORREGO, Rosa
Geometry and classification of Kummer surfaces in $\mathbf{P}^3$

GRANJA BARON, Angel
Irreducible polynomials of $k((X))[Y]$

GREUEL, Gert-Martin
A simple proof for the smoothness of the μ-constant stratum for curves

HAJTO, Zbigniew
Stratification Properties of Constructible Sets

HÄUSER, Herwig and MÜLLER, Gert
Automorphism groups in local analytic geometry, infinite dimensional rank
theorem and Lie groups

HENAUT, Alain
Webs of maximun rank in $\mathbb{C}^2$ wich are algebraic

HENRY, Jean-Pierre
FRONCES et DOUBLES PLIS
Towards a numerical criterion for Zariski equisingularity

JONG, Theo de
Rational surface singularities with reduced fundamental cycle and
Globalization of admisible deformations

LANDO, Sergei
Singularities of the differential forms of the highest degree and their
deformations

LAURENT, Yves
Irregular vanishing cycles for $\mathcal{D}$-modules

MAISONOBE, Philippe
Polynôme de Bernstein relatif et type topologique constant

MARIJUAN LOPEZ, Carlos
Desingularización geométrica de un grafo acíclico

MARTIN, Bernd
Moduli spaces of singularities of simplest topological type

MERLE, Michel
Multiplicités des variétés caractéristiques

MOND, David
How good are real pictures?

MORALES, Marcel
Blow-up of ideals of codimension 2

MORENO SOCIAS, Guillermo
An Ackermannian polynomial ideal

MÜLLER, Gerd
Integral varieties of Lie algebras of vector fields

NIETO, Isidro
Examples of Abelian surfaces with a level $(2, 6)$-structure

PEREZ PEREZ, Tomás and FINAT CODES, Javier
Lie Algebras preserving Tangent Spaces to R-orbits

PFISTER, Gerhard
Moduli spaces of semiquasihomogeneous singularities

PISON CASARES, Pilar
Monomial curves in $\mathbf{A}^4$

POLISCHUK, Alexander
Noncommutative projective spaces

REGUERA LOPEZ, Ana José
Plane curves associated to a semigroup

ROCZEN, Marko
Recognition of simple singularities in positive characteristic

SABBAH, Claude
Connexions mromorphes deux variables

STEIN, Harvey
Singularities, Smooth Morphisms and Lifting Lemmas

STEVENS, Jan
On the classification of reducible curve singularities

VILLAMAYOR, Orlando
On constructive or algorithmic Resolution of singularities

XAMBO, Sebastin
Rational equivalence on some families of plane curves

ZURRO MORO, M. Angeles
Abhyankar-Jung revisited

List of Participants

Prof. ALEKSANDROV, Aleksandr Grigorjevich; Moscow, Urss.
Prof. AROCA HERNÁNDEZ-ROS, J.M.; Valladolid, España.
Prof. BALDASSARI, Francesco; Padova, Italia.
Prof. BERMEJO DÍAZ, María Isabel; Sta. Cruz de Tenerife, España.
Prof. BIRBRAIR, Gev; Jerusalem, Israel.
Prof. BRASSELET, Jean Paul; Luminy (Marsella), Francia.
Prof. CABRERO VELASCO, Rafael; Valladolid, España.
Prof. CAMPILLO LÓPEZ, A.; Valladolid, España.
Prof. CANO TORRES, Felipe; Valladolid, España.
Prof. CANO TORRES, José; Valladolid, España.
Prof. CARNICER, Manuel; Valladolid, España.
Prof. CASAS ALVERO, E.; Barcelona, España.
Prof. CASSOU-NOGUES, Pierrete; Talence, France.
Prof. CASTELLANOS, Julio; Tenerife, España.
Prof. CASTRO JIMÉNEZ, Francisco; Sevilla, España.
Prof. CHARDIN, Marc; Palaiseau, France.
Prof. COSSART, Vincent; Jouy en Josas, France.
Prof. DELGADO DE LA MATA, Félix; Valladolid, España.
Prof. DU BOIS, Philippe; Angers, France.
Prof. ENCINAS CARRIÓN, Santiago; Valladolid, España.
Prof. FERNÁNDEZ DOMÍNGUEZ, Jesús; ??, ??.
Prof. FERNÁNDEZ GUTIÉRREZ, Diego; Valladolid, España.
Prof. FINAT CODES, Javier; Valladolid, España.
Prof. GAETA, Federico; Madrid, España.
Prof. GALINDO PASTOR, Carlos; Valladolid, España.
Prof. GARCÍA BARROSO, Evelia; Santa Cruz de Tenerife, España.
Prof. GARCÍA DE LA FUENTE, Julio; Valladolid, España.
Prof. GIMÉNEZ, Philippe; Saint Martin d'Hères, France.
Prof. GONZÁLEZ DORREGO, Rosa; Toronto, Canada.
Prof. GONZÁLEZ SPRINBERG, Gerardo; Saint Martin d'Hères, France.
Prof. GRANGER, Jean Michel; Angers, France.
Prof. GRANJA BARÓN, Ángel; León, España.
Prof. GREUEL, Gert-Martin; Kaiserlautern, Germany.
Prof. GUDIEL RODRÍGUEZ, Félix; Sevilla, España.
Prof. GUEMES ALZAGA, María Belén; Valladolid, España.
Prof. GUILLÉN, F.; Barcelona, España.
Prof. HAJTO, Zbigniew; Valladolid, España.
Prof. HAUSER, Herwig; Innsbruck, Austria.
Prof. HÉNAUT, Alain; Talence, Francia.
Prof. HENRY, Jean-Pierre; Palaiseau, France.
Prof. HERNANDO MARTÍN, María del Carmen; Barcelona, España.
Prof. HIRONAKA, E.; Boon, Germany.

INÉS CALZÓN, Antonio;   Sevilla,   España.

Prof. JONG, Theo de.;   Kaiserlautern,   Germany.

Prof. KIYEK, Karl Heinz;   Paderborn,   Germany.

Prof. LANDO, Sergei;   Moscow,   Urss.

Prof. LAURENT, Yves;   Saint Martin d'Hères,   France.

Prof. LE DUNG TRANG;   Paris,   France.

Prof. LEJEUNE-JALABERT, Monique;   Saint Martin d'Heres,   France.

Prof. LUENGO VELASCO, Ignacio;   Madrid,   España.

Prof. MAHAMMED, Norreddine;   Villeneuve d'Ascq.,   Francia.

Prof. MAISONOBE, Philippe;   Nice,   France.

Prof. MARIJUÁN LOPEZ, Carlos;   Valladolid,   España.

Prof. MARTIN, Bernd;   Berlin,   Germany.

Prof. MARTÍNEZ MARTÍNEZ, María del Carmen;   Valladolid,   España.

Prof. MEBKHOUT, Zoghman;   Paris,   Francia.

Prof. MERLE, Michel;   Nice,   France.

Prof. MOH, T.T.;   Purdue,   U.S.A.

Prof. MOND, David;   Coventry,   Great Britain.

Prof. MORALES, Marcel;   Saint Martin d'Heres,   France.

Prof. MORENO SOCÍAS, Guillermo;   Paris,   France.

Prof. MOZO FERNÁNDEZ, Jorge;   Valladolid,   España.

Prof. MULLER, Gerd;   Mainz,   Germany.

Prof. NARVÁEZ MACARRO, Luis;   Sevilla,   España.

Prof. NATANZON, Sergej;   Moscow,   URSS.

Prof. NAVARRO AZNAR, Vicente;   Barcelona,   España.

Prof. NETO, Orlando;   Lisboa,   Portugal.

Prof. NIETO, Isidro;   México D.F.,   México.

Prof. NUÑEZ JIMÉNEZ, Carolina Ana;   Valladolid,   España.

Prof. NUSS, Philippe;   Strasbourg,   France.

Prof. OKA, Mutsuo;   Tokyo,   Japan.

Prof. OLIVEIRA, Bruno;   Lisboa,   Portugal.

Prof. PASCUAL GAINZA, Pere;   Barcelona,   España.

Prof. PERAIRE DURBA, Rosa;   Barcelona,   España.

Prof. PÉREZ PÉREZ, Tomás;   Valladolid,   España.

Prof. PHAM, F.;   Nice,   France.

Prof. PFISTER, Gerhard;   Belin,   Germany.

Prof. PIEDRA SÁNCHEZ, Ramón;   Sevilla,   España.

Prof. PISÓN CASARES, Pilar;   Sevilla,   España.

Prof. POLISCHUK, Alexander;   Moscow,   URSS.

Prof. POSISZELSKY, Leonid;   Moscou,   URSS.

Prof. REGUERA LÓPEZ, Ana José;   Valladolid,   España.

Prof. REY, Jérôme;   Toulouse,   France.

Prof. RIVERO ÁLVAREZ, Margarita;   La Laguna (Tenerife),   España.

Prof. ROCZEN, Mark;   Berlin,   Germany.

Prof. RODRÍGUEZ SÁNCHEZ, María Cristina;   León,   España.

Prof. SABBAH, Claude;   Palaiseau,   France.

Prof. SAITO, K.;   Kyoto,   Japan.

Prof. SÁNCHEZ GIRALDA, Tomás;   Valladolid,   España.

Prof. SAULOY, Jacques;   Toulouse,   France.

Prof. SERRANO, F.;   Barcelona,   España.

Prof. SIERSMA, Dirk;   Utrecht,   Pays-Bas.

Prof. SOLEEV, Ahmadjon;   Samarkand,   URSS.

Prof. SPIVAKOSVSKI, M.;   Toronto,   Canada.

Prof. STASICA,;   Warsaw,   Polland.

Prof. STEIN, Harvey;   México D.F.,   México.

Prof. STEVENS, Jan;   Hamburg,   Germany.

Prof. SZPIRGLAS, Aviva;   Letaneuse,   France.

Prof. SZPIRO, Lucien;   Orsay,   France.

Prof. TEISSIER, Bernard;   Paris,   France.

Prof. TIEP, Pham Huu;   Moscow,   URSS.

Prof. TRAN, Hoi Ngoc;   Ho Chi Mihn City,   Vietnam.

Prof. VICENTE CÓRDOBA, José Luis;   Sevilla,   España.

Prof. VILLAMAYOR, Orlando;   Madrid,   España.

Prof. WALL, Charles Terence Clagg;   Liverpool,   Great Britain.

Prof. XAMBO, S.;   Madrid,   España.

Prof. XUAN HAI, Bui;   Ho Chi Mihn City,   Vietnam.

Prof. ZURRO MORO, M. Ángeles;   Valladolid,   España.

Part I

Resolution of Singularities

Désingularisation en dimension 3 et caractéristique p

Vincent Cossart

Soit Y une variété projective singulière de dimension 3 sur un corps k algébriquement clos de caractéristique p différente de 2, 3 et 5. Nous voulons convaincre le lecteur qu'il existe une construction courte et claire d'une désingularisation de Y. Ce qui signifie qu'il existe une variété projective régulière Y_1 et un morphisme projectif $p_1 \colon Y_1 \to Y$ qui est un isomorphisme au-dessus de l'ouvert de régularité de Y.

Nous rappelons d'abord les définitions classiques des différentes notions de désingularisation, ensuite, nous montrons comment relier trois articles différents pour obtenir notre désingularisation.

1 Différentes notions de désingularisation

Dans toute cette conférence, Y est une variété projective singulière sur un corps k algébriquement clos de caractéristique p.

1) Un modèle projectif régulier W de Y est une variété projective W régulière munie d'un morphisme birationnel

$$p \colon W \longrightarrow Y.$$

2) Une désingularisation de Y est un morphisme projectif

$$\pi \colon X \longrightarrow Y,$$

où X est une variété projective régulière et π est un isomorphisme au-dessus Y_{reg}, l'ouvert de régularité de Y.

3) En général, notre variété Y est un fermé d'une variété projective régulière Z. On appelle alors *désingularisation plongée* de Y un morphisme birationnel projectif

$$q \colon Z' \longrightarrow Z,$$

tel que Z' est une variété projective régulière, q est un isomorphisme en dehors du lieu singulier de Y et, si on désigne par Y' le transformé strict de Y, c'est à dire l'adhérence de l'image inverse de Y_{reg}, q restreint à Y' est une désingularisation de Y au sens de 2).

Remarquons qu'on obtient généralement les désingularisations plongées comme des composés d'éclatements à centres réguliers inclus dans les lieux singuliers des transformés stricts de Y.

Remarquons enfin que la désingularisation que nous allons construire est une désingularisation au sens de 2) et pas de 3), mais que sa construction nécessite des théorèmes de désingularisations plongées en dimension 2, ce qui ne donne aucun espoir de faire une récurrence éventuelle sur la dimension de Y.

Progress in Mathematics, Vol. 134
© 1996 Birkhäuser Verlag Basel/Switzerland

2   Première reduction

Théorème 2-1. (Abhyankar) *Soit Y une variété projective sur un corps k algébriquement clos, alors il existe une variété projective W et un morphisme birationnel p: W → Y avec tous les points de W de multiplicité inférieure ou égale à* dim(*Y*) !

On trouvera la preuve de ce Théorème dans [1] et une version simplifiée dans [9]. L'idée est de plonger *Y* dans un $\mathbb{P}^N$ avec *N* très grand, puis de faire des projections stéréographiques successives depuis des points de multiplicités maximales de *Y* et de ses projetées, le lieu des points de grande multiplicité finit par se projeter sur un point, puis disparait. La difficulté essentielle est d'éviter le cas où *Y* ou un de ses projetés est un cône, difficulté que l'on résoud en augmentant *N* si nécessaire.

Remarquons que ces changements de *N* et ces projections successives nous enlève tout espoir d'obtenir une désingularisation plongée.

3   Deuxième réduction, construction d'un modèle projectif

Théorème 3-1. (Abhyankar) *Soit W une variété projective de dimension 3 sur un corps k algébriquement clos et telle que tous ses points sont de multiplicité strictement inférieure à la caractéristique de k, alors il existe une désingularisation*

$$\pi: X \longrightarrow W$$

de W au sens de 2).

Bien sûr, on trouve la preuve de ce Théorème dans [1]. Remarquons simplement que, d'après [2], sous les hypothèses du Théorème, on a le *contact maximal* en tout point de *W*, ce qui signifie qu'on peut employer l'argument récurrent utilisé en caractéristique 0 par Hironaka, c'est à dire que le problème se réduit localement à la désingularisation d'un *exposant idéaliste* [8] porté par un espace régulier de dimension 3. Bref, on se ramène à la désingularisation d'un idéal plongé dans un espace régulier de dimension 3. C'est un problème que l'on sait résoudre simplement [3, 5].

De toute façon, la preuve de ce Théorème nécessite des résultats très fins de désingularisations plongées en dimension 2, c'est donc ici que tout espoir de récurrence sur la dimension de *W* s'effondre.

(3-2) Constatons cependant que, partant de *Y*, nous avons obtenu *W* birationnellement équivalente à *Y* et avec ses points de multiplicité inférieure ou égale à 6 = dim(*Y*) ! et qu'ensuite, si la caractéristique de *k* est strictement plus grande que 6, c'est à dire différente de 2, 3 et 5, nous obtenons *X* régulière birationnellement équivalente à *W* et donc à *Y*, bref, *X* est un modèle projectif régulier de *Y*.

4 Troisième réduction, birationnel devient projectif

Théorème 4-1. *Avec les hypothèses de 3-2, il existe un entier n et une suite d'éclatements*

$$\pi(i)\colon X(i+1) \longrightarrow X(i), \qquad 0 \le i \le n-1, \qquad X(0) = X$$

à centres réguliers et un morphisme projectif $\pi\colon X(n) \to Y$ tels que, si on pose

$$e = \pi(0) \circ \cdots \circ \pi(n-1),$$

π se factorise par e.

C'est un résultat bien classique dont on trouvera une preuve dans [4].

Bien sûr, $X(n)$ est birationnellement équivalente à X et donc à Y et, les centres d'éclatements étant réguliers, $X(n)$ est régulière.

Ce Théorème améliore la situation puisque nous avons maintenant une variété régulière $X(n)$ et un morphisme projectif birationnel

$$q\colon X(n) \longrightarrow Y,$$

mais $X(n)$ n'est en général pas une désingularisation de Y au sens de 2), car q n'est pas toujours un isomorphisme au-dessus de Y_{reg}.

Pour simplifier les notations, $X(n)$ sera désormais notée Z.

5 Final: Morphisme projectif birationnel devient désingularisation

Théorème 5-1. *Avec les hypothèses et notations de 4, il existe un carré com-mutatif de morphismes projectifs*

où $p\colon Y' \to Y$ est composé d'éclatements à centres réguliers se projetant sur $\operatorname{Sing}(Y)$ et $p'\colon Z' \to Y'$ projectif birationnel avec Z' régulière tel que le lieu fondamental de p'^{-1} est union de deux fermés disjoints F_1 et F_2 avec $p(F_1) \subset \operatorname{Sing}(Y)$ et $\operatorname{Sing}(Y') \subset F_1$.

L'idée de la preuve est la suivante : le lieu fondamental de $Y \to Z$ contient $\operatorname{Sing}(Y)$, en dehors de $\operatorname{Sing}(Y)$, ce lieu fondamental F est de codimension au plus 2 c'est à dire de dimension au plus 1, s'il contient des points isolés et des courbes disjointes de $\operatorname{Sing}(Y)$, on prend $Y' = Y$ et $Z' = Z$. Sinon, en éclatant les points d'intersection de l'ensemble G des courbes de F non incluses dans $\operatorname{Sing}(Y)$ et de $\operatorname{Sing}(Y)$ et un nombre fini de points se projetant sur $G \cap \operatorname{Sing}(Y)$, on sépare les transformées strictes de ces derniéres du lieu singulier du transformé Y_0 de Y. On peut facilement construire un morphisme $Z_0 \to Z$ se factorisant par Y_0 et composé d'éclatements à centres réguliers se projettent sur les points de $G \cap \mathrm{Sing}(Y)$ et de plus tel que

est commutatif.

On peut avoir créé des courbes appartenant au lieu fondamental de $Y_0 \to Z_0$ non incluses dans $\mathrm{Sing}(Y_0)$ et intersectant $\mathrm{Sing}(Y_0)$, ces courbes se projettent sur les centres d'éclatements composant $Y_0 \to Y$, donc sur $\mathrm{Sing}(Y)$, pour s'en débarasser, il suffit de les éclater un certain nombre de fois et de modifier Z_0 en conséquence. Pour celà, on utilise le lemme de Zariski [12, p.538]:

Lemme 5-2. *Soient X' et Y' deux variétés projectives de dimension 3 telles qu'il existe $\pi': X' \to Y'$ un morphisme projectif birationnel. Soit Γ une courbe de Y' non contenue dans le lieu singulier de Y'. Alors, au-dessus du point générique η de Γ, π' est composé d'éclatements de courbes se projetant horizontalement sur Γ.*

Nous n'en dirons pas plus sur la construction de $p': Z' \to Y'$. Le lecteur peut se référer à [4].

Montrons simplement que nous en avons pratiquement terminé. Soit X_0 le recollement de $Z' - p'^{-1}(F_2)$ et de $Y' - F_1$ le long de $Y' - (F_1 \cup F_2) = Z' - p'^{-1}(F_1 \cup F_2)$ et soit $p_0: X_1 \to Y'$, le morphisme donné par le recollement, p_0 est projectif. En effet soit i l'immersion ouverte de $Y' - F_2$ dans Y' et soit I' un idéal dont l'éclatement est p'. Alors le faisceau $I_1 = i_* i^*(I)$ est un idéal de $\mathcal{O}_{Y'}$. En effet, $i^*(I)$ est la restriction de I à $Y' - F_2$ et, comme la codimension de F_2 est supérieure ou égale à 2 et que Y' est régulier en tout point de F_2, les sections de I_1 sont en fait des sections de $\mathcal{O}_{Y'}$. Cet idéal I_1 coïncide avec I' sur $Y' - (F_1 \cup F_2)$, p' étant un isomorphisme au-dessus de $Y' - (F_1 \cup F_2)$, I_1 est inversible sur $Y' - (F_1 \cup F_2)$ et, par le même argument que ci-dessus, en tout point de F_2, un générateur local se prolonge en un générateur local. Ainsi, I_1 est inversible sur $Y' - F_1$ et coïncide avec I sur un voisinage de F_1. Il est clair que p_0 est l'éclatement de I_1.

Maintenant, il suffit de remarquer que $p_1 = p \circ p_0: X_0 \to Y$ est notre désingularisation.

References

[1] ABHYANKAR S.S. : Resolution of singularities of embedded algebraic surfaces New-York , London. Academic Press 1966.

[2] COSSART V. : Contact maximal en caractéristique positive et petite multiplicité Duke Math. J. 63 n 1 June 1991.

[3] COSSART V. : Desingularization of embedded excellent surfaces . Tôhoku Math. J. 33 n 1, pp. 25–33, 1981.

[4] COSSART V. : Modèle projectif et désingularisation. Mathematische Annalen 1992

[5] COSSART V., GIRAUD J., ORBANZ U. : Resolution of surface singularities . Lect. Notes Math., vol. 1101, 1984.

[6] GIRAUD J. : Etude locale des singularités. Publ. Math. Orsay. 1972.

[7] GIRAUD J. : Contact maximal en caractéristique positive. Ann. Sci. Ec. Norm. Sup., IV, Série 8, fasc.2, 1975.

[8] HIRONAKA H. : Idealistic Exponents of Singularity. In : IGUSA (ed.) Alg. Geom. pp. 52–125, J.J. Sylvester Symp., John HOPKINS Univ., Baltimore, Maryland 1976.

[9] LIPMAN J. : Introduction to resolution of singularities. Proc. Symp. Pure Math. Humboldt State Univ. California 1974.

[10] SÁNCHEZ-GIRALDA T. : Teoría de singularidades de superficias algebroides sumergidas. Monogr. Mem. Mat. IX, 1976.

[11] ZARISKI O. : A simplified proof for the resolution of singularities of an algebraic surface. Ann. Math., 43, pp. 583–592, 1942.

[12] ZARISKI O. : Reduction of the singularities of algebraic three dimensional varieties. Ann. Math. 45, pp. 472–542, 1944.

Address of author:

Université de VERSAILLES, 45 avenue des Etats-Unis, F78035 VERSAILLES Cedex
ou LABORATOIRE 213 Université PARIS 6, 4 place Jussieu, 75252 PARIS Cedex
France
Email: cossart@ariana.polytechnique.fr

Sur l'espace des courbes tracées sur une singularité

G. Gonzalez-Sprinberg and M. Lejeune-Jalabert

1 Introduction

Dans un preprint non publié vers les années 65, J. Nash commençait l'étude de l'espace $\mathcal{H}$ des courbes (formelles) tracées sur un germe de variété singulière (V, O). La question générale est d'élucider les correspondances entre les propriétés algébro-géométriques de $\mathcal{H}$ d'une part et les propriétés de divers modèles birationnels propres au-dessus de (V, O), en particulier ses désingularisations, d'autre part.

Nous présentons ici quelques uns des développements récents du sujet. Grâce à un théorème de M. Artin, le calcul des jets de courbes à un ordre donné fait intervenir un système d'équations en un nombre fini de variables mesuré par la fonction β d'Artin. Si (V, O) est une singularité d'hypersurface, on a obtenu des majorations de cette dernière en terme de la géométrie de l'éclatement de l'idéal jacobien (section 2).

J. Nash avait introduit une subdivision de l'espace $\mathcal{H}$ en familles ne dépendant que de l'anneau local $\mathcal{O}_{V,O}$ de V en O ([16]). Il posait la question de la relation entre ces familles et les composantes de la fibre exceptionnelle dites essentielles d'une désingularisation de (V, O), *i.e.*, apparaissant, à équivalence birationnelle près, dans toute désingularisation. On a déterminé les diviseurs essentiels de certaines singularités de dimension trois (section 3).

Dans le cas d'une singularité isolée d'hypersurface, on a aussi obtenu des équations et inéquations de l'espace des jets à un ordre donné par une méthode algorithmique. La géométrie de l'éclatement jacobien intervient encore dans son test d'arrêt. Ces calculs ont conduit à définir une autre subdivision de $\mathcal{H}$ en familles que l'on est parvenu à relier à la géométrie des désingularisations plongées dans quelques exemples (section 4 et section 6).

La non nullité de l'espace des jets à l'ordre un de courbes se traduit géométriquement par l'existence de courbes lisses en O. Lorsque V est une surface, on a obtenu des critères géométrique d'existence de courbes lisses et de branches lisses d'une section hyperplane générale faisant intervenir, ou bien des désingularisations, ou bien des chaînes de points infiniment voisins satisfaisant des conditions de transversalité, celles-ci en un nombre fini. Les éclatements définissant chaque chaîne aboutissent à des singularités dont la section hyperplane générale possède des branches lisses. On retrouve géométriquement le caractère fini du problème de la détermination des tangentes aux courbes lisses en O sur la surface, résultant de la traduction précédente et du section 2.

Deux exemples élémentaires développés au section 6 illustrent les différentes notions de familles introduites précédemment .

Progress in Mathematics, Vol. 134
© 1996 Birkhäuser Verlag Basel/Switzerland

2   Structure pro-algébrique de l'espace des courbes et la fonction de M. Artin d'une singularité

Dans toute la suite, nous désignerons par singularité le germe (V, O) en un point fermé singulier O d'un schéma algébrique ou formel V, non nécessairement réduit, défini sur un corps algébriquement clos $\mathbf{k}$.

Suivant J. Nash [16], nous désignerons par courbe (ou arc) tracée sur la singularité (V, O), une courbe paramétrée formelle, *i.e.*, $\mathbf{k}[\![t]\!]$ désignant l'anneau des séries formelles à une variable à coefficients dans $\mathbf{k}$, un $\mathbf{k}$-morphisme :

$$h : (T, o) := \operatorname{Spec} \mathbf{k}[\![t]\!] \longrightarrow (V, O).$$

La donnée d'un tel morphisme est équivalente à celle d'une paramétrisation *i.e.*, d'un $\mathbf{k}$-morphisme local (continu) de l'anneau local $\mathcal{O}_{V,O}$ de V en O dans $\mathbf{k}[\![t]\!]$.

Dans toute la suite, $\mathcal{H}$ désignera l'ensemble des courbes tracées sur (V, O). L'espace $\mathcal{H}$ a une structure pro-algébrique naturelle. En effet, pour tout entier $i \geq 0$, soit $\mathcal{H}_i$ l'ensemble des courbes à l'ordre i sur (V, O), *i.e.*, des k-morphismes locaux $\mathcal{O}_{V,O} \to \mathbf{k}[\![t]\!]/(t)^{i+1}$. Les projections canoniques :

$$\mathbf{k}[\![t]\!] \longrightarrow \mathbf{k}[\![t]\!]/(t)^{i+1} \longrightarrow \mathbf{k}[\![t]\!]/(t)^{j+1}, i \geq j$$

définissent par composition des morphismes :

$$\mathcal{H} \xrightarrow{\rho_i} \mathcal{H}_i \xrightarrow{\rho_{ji}} \mathcal{H}_j, i \geq j$$

formant un système projectif et on a $\mathcal{H} = \varprojlim \mathcal{H}_i$. (Si $h \in \mathcal{H}$, nous dirons que $\rho_i(h)$ est son i-ème jet).

On vérifie facilement que $\mathcal{H}_i$ est un ensemble algébrique affine et que chaque ρ_{ji} est un morphisme algébrique, d'où la structure pro-algébrique de $\mathcal{H}$.

Puisque O est un point singulier de V, en général le morphisme ρ_i n'est pas surjectif. Nash considère donc son image $\rho_i(\mathcal{H})$ qu'il désigne par $\mathrm{Tr}(i)$. C'est l'ensemble des i-jets des courbes tracées sur (V, O).

Théorème 2-1. (Nash) $\mathrm{Tr}(i)$ *est un sous-ensemble constructible de* $\mathcal{H}_i$.

Rappelons que ceci signifie que c'est la réunion finie de parties localement fermées (*i.e.*, fermées dans un ouvert) de $\mathcal{H}_i$.

Cette propriété ne résulte pas directement de la définition de $\mathrm{Tr}(i)$, mais c'est, comme nous allons le voir, une conséquence immédiate du théorème d'approximation d'Artin sur les solutions des équations algébriques et formelles [1]. Soit $\widehat{\mathcal{O}_{V,O}}$ le complété de $\mathcal{O}_{V,O}$ pour la topologie définie par son idéal maximal ; $\widehat{\mathcal{O}_{V,O}}$ est $\mathbf{k}$-isomorphe au quotient d'un anneau de séries formelles $\mathbf{k}[\![X_1, \dots, X_n]\!] = \mathbf{k}[\![\underline{X}]\!]$ par un idéal $(f_1, \dots, f_r)$. Les courbes tracées sur (V, O) ont pour représentation paramétrique les solutions $\underline{X}(t) \in \mathbf{k}[\![t]\!]^n$ s'annulant en

0 du système d'équations implicites indépendantes de t, $f_s(\underline{X}) = 0$, $1 \leq s \leq r$. Le théorème d'Artin appliqué à ce système assure l'existence d'une fonction numérique $\beta : \mathbb{N} \to \mathbb{N}$ ayant la propriété suivante : Pour tout $i \in \mathbb{N}$, si $f_s(\underline{X}(t)) \equiv 0 \bmod(t)^{\beta(i)+1}$, $1 \leq s \leq r$, il existe $\underline{\widetilde{X}}(t) \in \mathbf{k}[\![t]\!]^n$ tel que

1. $f_s(\underline{\widetilde{X}}(t)) = 0$, $1 \leq s \leq r$

2. $\underline{X}(t) - \underline{\widetilde{X}}(t) \in (t)^{i+1}$.

Autrement dit :

$$\mathrm{Tr}(i) = \rho_{i,\beta(i)}(\mathcal{H}_{\beta(i)}).$$

Or, par un résultat classique de Chevalley, l'image d'un ensemble algébrique par un morphisme est un sous-ensemble constructible. Ainsi, la détermination de $\mathrm{Tr}(i)$ se ramène à la résolution d'un système d'équations et d'inéquations en un nombre fini de variables.

Définition 2-2. *On appelle fonction d'Artin de la singularité (V, O) la fonction $\beta \colon \mathbb{N} \to \mathbb{N}$ qui fait correspondre à l'entier i le plus petit entier $\beta(i)$ ayant les propriétés 1 et 2 ci-dessus.*

C'est un invariant formel de la singularité. On vérifie en effet facilement que β ne dépend ni du système de coordonnées $(X_1, \ldots, X_n)$ ni des équations locales $f_s(\underline{X}) = 0$ choisies pour représenter (V, O).

On sait en fait peu de choses sur la fonction d'Artin β d'une singularité. Pour une singularité isolée d'hypersurface, on dispose d'une majoration effective de β par une fonction linéaire [10]. Cette majoration a été améliorée depuis par M. Hickel. Dans [8], également consacré à l'étude des singularités d'hypersurfaces, Hickel compare de plus la fonction d'Artin de (V, O) et celle de $(\mathrm{Sing}\, V, O)$ où $\mathrm{Sing}\, V$ désigne le lieu singulier de (V, O). L'énoncé précis est le suivant :

Théorème 2-3. (Hickel) *Soit (V, O) une singularité d'hypersurface complexe formelle définie dans $(\mathbb{C}^n, O)$ par l'équation $f(X_1, \ldots, X_n) = 0$. Soit $(\mathrm{Sing}\, V, O)$ le sous-schéma (en général non réduit) de (V, O) défini dans $(\mathbb{C}^n, O)$ par l'idéal*

$$J(f) = \Big(\frac{\partial f}{\partial X_1}, \ldots, \frac{\partial f}{\partial X_n} \Big)$$

de $\mathbb{C}[\![X_1, \ldots, X_n]\!] = \mathbb{C}[\![\underline{X}]\!]$. Enfin, soit β (resp. β_{Sing}) la fonction d'Artin de (V, O) (resp. $(\mathrm{Sing}\, V, O)$). On a :

$$\beta(i) \leq \beta_{\mathrm{Sing}}(i) + i.$$

Il en résulte :

Corollaire 2-4. *Soit (V, O) une singularité isolée d'hypersurface. On a :*

$$\beta(i) \leq \theta i + i$$

où θ est l'exposant de Lojasiewicz de la singularité.

(2-5) Rappelons que l'exposant de Lojasiewicz d'une singularité isolée d'hypersurface est un nombre rationnel positif. On trouvera par exemple dans [11] les diverses caractérisations suivantes de θ provenant des divers avatars de la clôture intégrale d'un idéal :

1. M étant l'idéal maximal de $\mathbb{C}[\![X]\!]$, soit

$$\nu_{J(f)}(M^k) := \{\, \sup n \mid n \in \mathbb{N}, M^k \subset J(f)^n \,\}$$

et soit

$$\overline{\nu}_{J(f)}(M) := \lim_{k \to \infty} \nu_{J(f)}(M^k)/k.$$

On a :

$$\theta = 1/\overline{\nu}_{J(f)}(M).$$

2. Soit $\mathcal{C}$ l'ensemble des courbes tracées sur $(\mathbb{C}^n, O)$

$$1/\theta = \inf_{h \in \mathcal{C}} \frac{\operatorname{ord}_t M \circ h}{\operatorname{ord}_t J(f) \circ h}.$$

Ici, I étant un idéal quelconque de $\mathbb{C}[\![X]\!]$ et h la paramétrisation d'une courbe

$$\operatorname{ord}_t I \circ h = \sup\{n \mid n \in \mathbb{N}, h(I) \subset (t)^n\}.$$

3. Enfin, si f est une série convergente, il existe un voisinage U de O dans $\mathbb{C}^n$ et une constante $c \in \mathbb{R}_{>0}$ tels que l'on ait ;

$$\forall x \in U, \sup_i \left| \frac{\partial f}{\partial X_i}(x) \right| \geq c|x|^\theta$$

et θ est la borne inférieure des exposants pour lesquels une telle inégalité est satisfaite.

Enfin, sous les mêmes hypothèses, Teissier exprime θ en fonction du contact en O de l'hypersurface avec une courbe polaire générale [20].

Précisément, soit P_Λ la courbe polaire d'équations

$$\left(\frac{1}{\lambda_1}\right)\frac{\partial f}{\partial X_1} = \cdots = \left(\frac{1}{\lambda_n}\right)\frac{\partial f}{\partial X_n}$$

correspondant à l'hyperplan Λ de $\mathbb{C}^n$ d'équation $\lambda_1 X_1 + \cdots + \lambda_n X_n = 0$ et soit $P_\Lambda = \cup_q \Gamma_q$ sa décomposition en composantes irréductibles. Si Λ est un hyperplan assez général, le nombre de composantes irréductibles de P_Λ est un entier ℓ indépendant de Λ. De plus, si $1 \leq q \leq \ell$, la multiplicité m_q de Γ_q en O et sa multiplicité d'intersection avec V en O, $(V \cdot \Gamma_q)_O := m_q + e_q$ sont indépendantes de Λ. On a

$$\theta = \sup_q \frac{e_q}{m_q}.$$

Il en résulte facilement que la majoration du 2-4 ci-dessus de la fonction d'Artin d'une singularité isolée d'hypersurface est la meilleure que l'on puisse obtenir puisque si (V, O) est un germe de courbe plane analytiquement irréductible

1. $\sup_i \beta(i)/i = \theta + 1$

2. il existe un entier i_0 tel que $\beta(i_0) = (\theta + 1)i_0$.

Il suffit, en effet, de considérer une des courbes Γ_q telle que $\theta = e_q/m_q$. En la normalisant, on obtient une paramétrisation

$$h : \mathbb{C}[\![\underline{X}]\!] \longrightarrow \mathbb{C}[\![t]\!]$$

de Γ_q telle que $\mathrm{ord}_t\, M \circ h = m_q$ et que $\mathrm{ord}_t\, f \circ h = (V \cdot \Gamma_q)_O = m_q(\theta + 1)$. Posons $i_0 = m_q$ et supposons que $\beta(i_0) < i_0(\theta+1)$. Par définition de β, il existe alors $\widetilde{h} \in \mathbb{C}[\![t]\!]^n$ tel que $f(\widetilde{h}) = 0$ et que $h - \widetilde{h} \in (t)^{i_0+1}$.

Puisque (V, O) est une branche plane, $\sum_q m_q = m - 1$ où m est la multiplicité de V en O. D'où $\mathrm{ord}_t\, M \circ h = m_q = i_0 < m$. D'autre part, $\widetilde{h}$ est une représentation paramétrique de (V, O). D'où $\mathrm{ord}_t\, M \circ \widetilde{h}$ est un multiple de m ; en particulier $\mathrm{ord}_t\, M \circ \widetilde{h} \geq m \geq i_0 + 1$. Ces inégalités contredisent le fait que $h - \widetilde{h} \in (t)^{i_0+1}$. C'est donc que $\beta(i_0) \geq i_0(\theta + 1)$ d'où l'égalité au vu de 2-4.

3 Familles de courbes (selon J. Nash) et désingularisations

J. Nash ne démontre pas le théorème 2-1 en utilisant le théorème d'Artin mais une désingularisation[1] de (V, O) lorsqu'on sait que celle-ci existe, par exemple si $\mathbf{k}$ est un corps de caractéristique 0 ou si $\dim(V, O) \leq 3$. Il en déduit un résultat plus précis.

Pour simplifier et jusqu'à la fin de cette section, nous supposerons que O est un point singulier isolé.

Théorème 3-1. (Nash) *Soit $\pi : X \to (V, O)$ une désingularisation d'une singularité isolée. Pour tout $i \in \mathbb{N}$, le nombre de composantes irréductibles $r(i)$ de l'adhérence de Zariski de $\mathrm{Tr}(i)$ dans $\mathcal{H}_i$ est inférieur ou égal au nombre de composantes irréductibles e_π du lieu exceptionnel $\pi^{-1}(O)$ de π.*

Les étapes essentielles de la démonstration de Nash sont les suivantes :

Soit $i \in \mathbb{N}$; la variété X étant non singulière, pour tout point fermé Q de $E := \pi^{-1}(O)$, l'ensemble des courbes à l'ordre i sur (X, Q) coïncide avec celui des i-jets de courbes tracées sur (X, Q). Soit $\mathcal{H}_i^E$ leur réunion lorsque Q décrit E. En utilisant le critère valuatif de propreté et la remarque précédente, on montre que $\mathrm{Tr}(i)$ est l'image de la projection naturelle $\mathcal{H}_i^E \to \mathcal{H}_i$ induite par π. Le nombre de composantes irréductibles $r(i)$ de $\mathrm{Tr}(i)$ est donc au plus celui de $\mathcal{H}_i^E$.

---

[1] Une désingularisation de (V, O) est un morphisme propre et birationnel $\pi : X \to (V, O)$ tel que X soit non singulier et qui induit un isomorphisme de $X \setminus \pi^{-1}(\mathrm{Sing}\,V)$ sur $X\,\mathrm{Sing}\,V$.

L'ensemble $\mathcal{H}_i^E$ a une structure naturelle de cône algébrique lisse au-dessus de E. (Ce n'est pas un fibré vectoriel si $i \neq 1$). Le nombre de composantes irréductibles de $\mathcal{H}_i^E$ coïncide donc avec celui de E.

Comme par ailleurs, si $j \geq i$, $\rho_{ij}(\mathrm{Tr}(j)) = \mathrm{Tr}(i)$, $r(i)$ est une fonction croissante de i et puisqu'elle est bornée, il existe $r \geq 1$ tel que si $i \gg 0$, $r(i) = r$ et, bien sûr, $r \leq e_\pi$ pour toute désingularisation π.

Problème 3-2. Si (V, O) est une singularité isolée de surface et si π est sa désingularisation minimale, a-t-on $r = e_\pi$? Ce problème posé par Nash dans [16] reste entièrement ouvert. On trouve dans [9] une réponse positive partielle à une question qui en dérive.

On suppose ici que (V, O) est normal. Les composantes irréductibles E_1, ..., E_s de $E = \pi^{-1}(O)$ sont alors des courbes projectives. On appellera "générique" toute courbe $h \in \mathcal{H}$ telle que $h(T) \neq O$ et dont l'unique relèvement $\widehat{h} : (T, o) \to X$ soit une courbe non singulière, transverse à l'un des E_i, $1 \leq i \leq s$, en point "général" de celui-ci, *i.e.*, appartienne à un ouvert de Zariski de E_i précisé dans chaque énoncé. La condition de transversalité exigée ici est algébrique. Elle signifie que $\widehat{O} = \widehat{h}(o)$ est un point non singulier de E_i et que si P est l'idéal de $\mathcal{O}_{X,\widehat{O}}$ définissant E_i, $\mathrm{ord}_t \, P \circ \widehat{h} = 1$.

Problème 3-3. Soit (V, O) une singularité de surface normale et soit $\varphi : (\mathbf{k}^2, o) \to (V, O)$ un germe de morphisme formel. On suppose qu'il existe une courbe non singulière $h : (T, o) \to (\mathbf{k}^2, o)$ telle que la courbe $\varphi \circ h : (T, o) \to (V, O)$ soit "générique" au sens précédent. Le morphisme φ se factorise-t-il par la désingularisation minimale de (V, O) ?

Il en est ainsi si V a en O une singularité torique, c'est-à-dire s'il existe une variété torique affine W (voir définition ci-dessous), un point w de W et un $\mathbf{k}$-isomorphisme entre les complétés de $\mathcal{O}_{V,O}$ et $\mathcal{O}_{W,w}$ ([9], Prop. 3.1.1).

On se ramène à ce cas si V est une surface définie dans $\mathbf{k}^3$ par $f \in \mathbf{k}[[x_1, x_2, x_3]]$ non dégénérée relativement à son polyèdre de Newton [2] et si la courbe exceptionnelle de la transformée stricte V_η de V dans l'éclatement normalisé de l'idéal engendré par les monômes de f (ou éclatement de Newton) ne contient pas de courbes rationnelles non singulières. L'argument de [9] Section 2 2.2 s'applique puisque si V_η est singulière, ses singularités sont toriques (cf. [6]).

La relation entre les deux problèmes est expliquée dans l'introduction de [9].

En toutes dimensions, le théorème 3-1 permet de définir une subdivision de $\mathcal{H}$ en familles non nécessairement disjointes. En effet, puisque $r(i) = r$ si $i \gg 0$, pour $j \geq i \gg 0$, l'image par ρ_{ij} de chaque composante irréductible

---

[2]Soit $f = \sum c_r x^r$ et soit $\varepsilon(f) := \{r \mid c_r \neq 0\}$. Le polyèdre de Newton de f est l'enveloppe convexe de $\{r + \mathbb{R}_{\geq 0}^3 \mid r \in \varepsilon(f)\}$. On dit que f est non dégénérée si pour toute face compacte γ du polyèdre de Newton de f, le lieu singulier de la surface d'équation $f_\gamma = \sum_{r \in \gamma} c_r x^r$ ne rencontre pas le tore $\mathbf{k}^{*3}$ de $\mathbf{k}^3$.

de $\overline{\mathrm{Tr}(j)}$ est dense dans une composante irréductible de $\overline{\mathrm{Tr}(i)}$. Si $i \gg 0$, on peut donc indexer les composantes irréductibles $C_1(i), \ldots, C_r(i)$ de $\overline{\mathrm{Tr}(i)}$ de sorte que, si $j \geq i$, $\rho_{ij}\big(C_\lambda(j)\big)$ soit dense dans $C_\lambda(i)$, $1 \leq \lambda \leq r$. Pour tout λ, $1 \leq \lambda \leq r$, on peut alors considérer $F_\lambda := \big\{h \in \mathcal{H} \mid \rho_i(h) \in C_\lambda(i)$ si $i \gg 0\big\}$. Le sous-ensemble F_λ de $\mathcal{H}$ est une famille de courbes au sens de Nash et $\mathcal{H} = \cup_\lambda F_\lambda$.

Pour toute désingularisation π de (V, O), on peut montrer qu'il existe $i \in \mathbb{N}$, un ouvert de Zariski $\Omega_\lambda(i)$ de $C_\lambda(i)$ et une composante irréductible E_λ de $\pi^{-1}(O)$ tels que si $h \in \mathcal{H}$ et $\rho_i(h) \in \Omega_\lambda(i)$, alors le point exceptionnel $\widehat{h}(o)$ de la courbe $\widehat{h} \colon (T, o) \to X$ qui relève h appartient à E_λ et ne dépend que de $\rho_i(h)$.

Chaque famille F_λ repère donc une composante irréductible bien précise de l'ensemble exceptionnel de chaque désingularisation de (V, O). On peut se ramener au cas où celle-ci est un diviseur en la faisant éclater, puis en désingularisant l'espace ainsi obtenu. Deux quelconques de ces diviseurs sont birationnellement équivalents entre eux relativement à V i.e., l'application birationnelle entre les deux désingularisations de (V, O) les contenant respectivement les met en correspondance (ou encore ils définissent la même valuation divisorielle du corps des fonctions rationnelles de (V, O)).

Les familles de courbes ainsi définies déterminent donc des composantes "essentielles" qui doivent apparaître, à équivalence birationnelle près, dans l'ensemble exceptionnel de toute désingularisation de (V, O). La représentation des composantes essentielles par des familles de courbes est un problème largement ouvert.

Si (V, O) est un germe de surface normale, les diviseurs essentiels pour V, c'est-à-dire les diviseurs irréductibles définis à équivalence birationnelle relative à V près, qui possèdent un représentant dans toute désingularisation de V sont évidemment les composantes irréductibles du lieu exceptionnel de la désingularisation minimale (cf. Problème 3-2). Si V_σ est la variété torique de dimension 3 associée à un cône rationnel polyédral fortement convexe de $\mathbb{R}^3$, C. Bouvier et G. Gonzalez-Sprinberg caractérisent les diviseurs essentiels de V_σ ([3], Prop. 6). Ici, on ne suppose plus la singularité isolée.

Avant de pouvoir énoncer leur résultat, il est nécessaire de faire quelques brefs rappels relatifs au dictionnaire variété torique, éventail (voir par exemple [21]).

Étant donné un tore algébrique $T \sim (\mathbf{k}^*)^d$, une variété torique est une variété algébrique normale contenant T comme ouvert de Zariski dense et munie d'une action de T, $T \times V \to V$, prolongeant $T \times T \to T$. Étant donné un réseau $N \sim \mathbb{Z}^d$, on appelle éventail une famille finie Σ de cônes rationnels polyédraux fortement convexes satisfaisant :

1. si τ est une face de $\sigma \in \Sigma$, alors $\tau \in \Sigma$

2. si σ_1, $\sigma_2 \in \Sigma$, alors $\sigma_1 \cap \sigma_2$ est une face de σ_1 et de σ_2.

Nous noterons V_Σ la variété torique associée à Σ ; V_Σ est affine si et seulement si Σ est l'ensemble des faces d'un cône σ comme ci-dessus. Dans ce cas, σ désignera à la fois le cône et l'éventail de ses faces. Il existe une correspondance bijective entre les T-orbites de X_Σ et les cônes de Σ. La dimension de l'orbite $\mathbb{O}_\tau$ correspondant au cône $\tau \in \Sigma$ est la codimension de τ.

Une subdivision de Σ est un éventail Σ' tel que pour chaque $\sigma' \in \Sigma'$, il existe $\sigma \in \Sigma$, $\sigma' \subset \sigma$ et dont les supports $|\Sigma| = \cup_{\sigma \in \Sigma}\, \sigma$ et $|\Sigma'| = \cup_{\sigma' \in \Sigma'}\, \sigma'$ coïncident. Elle détermine un morphisme propre, birationnel et équivariant de $V_{\Sigma'}$ sur V_Σ. Le lieu exceptionnel de ce morphisme est la réunion des orbites $\mathbb{O}_{\sigma'}$ de T dans $V_{\Sigma'}$ pour $\sigma' \in \Sigma'$ et $\sigma' \notin \Sigma$. D'où la correspondance bijective entre les composantes de codimension 1 du lieu exceptionnel de π et les arêtes de Σ' qui ne sont pas des arêtes de Σ. Enfin chaque désingularisation équivariante de la variété V_Σ correspond à une subdivision régulière Σ' de Σ (i.e., chaque cône $\sigma' \in \Sigma'$ est engendré par un sous-ensemble d'une base de N).

Soit donc V_σ la variété torique affine associée au cône σ de N. On désigne par G_σ (ou G) le système générateur minimal du semi-groupe $\sigma \cap N \setminus \{O\}$. C'est l'ensemble des éléments minimaux de $\sigma \cap N \setminus \{O\}$ pour la relation d'ordre $r \geq r'$ s'il existe $r'' \in \sigma \cap N$ tel que $r = r' + r''$.

Théorème 3-4. ([3]) *Soit V_σ une variété torique affine. Alors le système générateur minimal G de σ est l'ensemble des vecteurs extrémaux communs à toutes les subdivisions régulières de σ.*

On a donc une correspondance bijective entre les éléments de G qui ne sont pas situés sur des arêtes de σ et les diviseurs contenus (à équivalence birationnelle près) dans le lieu exceptionnel de toute désingularisation équivariante de V_σ.

Si $\mathbf{k} = \mathbb{C}$ et si $\dim V_\sigma = 3$, ces mêmes diviseurs sont essentiels pour toute désingularisation de V.

La preuve de la dernière assertion de 3-4 utilise la théorie des modèles minimaux ([15]).

Par exemple, il n'y a pas de diviseur essentiel exceptionnel pour le cône C ayant pour base la quadrique non singulière $\mathbb{P}^1_{\mathbf{k}} \times \mathbb{P}^1_{\mathbf{k}}$. Celui-ci correspond au cône σ de $\mathbb{R}^3$ engendré par $e_1 = (1,0,0)$, $e_2 = (0,0,1)$, $u = (0,1,0)$, $v = (1,-1,1)$. Le système générateur minimal de σ se réduit à ces 4 vecteurs. Les subdivisions I et I' de σ représentées dans la figure 3.1, définissent deux petites désingularisations de C ayant chacune un $\mathbb{P}^1$ pour lieu exceptionnel.

Géométriquement, on a éclaté le cône de base une droite de l'une ou l'autre des familles de droites sur $\mathbb{P}^1 \times \mathbb{P}^1$.

L'éclatement du sommet de C domine ces deux petites désingularisations. Il correspond à la subdivision II. On obtient II à partir de I (resp. I') en éclatant le $\mathbb{P}^1$ exceptionnel.

Il y a donc une seule famille de courbes sur C et une seule composante essentielle, à équivalence birationnelle et ce n'est pas un diviseur.

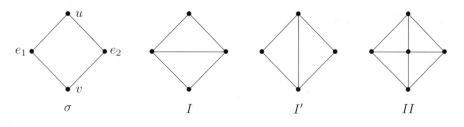

Figure 3.1: Subdivisions de σ.

4   Courbes sur une singularité isolée d'hypersurface

Si (V, O) est une singularité isolée d'hypersurface, la constructibilité des ensembles $\mathrm{Tr}(i)$ définis au section 2 est obtenue dans [10] en adaptant au cas singulier la méthode des approximations successives ([15] I.2 Section 2 pp. 54–57 ou "De la réduction des équations affectées" XXXVI).

Nous conservons ici les notations introduites au section 2, notamment celles de 2-5. Pour toute suite décroissante d'entiers positifs $m_1, \ldots, m_i$ majorés par la multiplicité m_0 de V en O, un algorithme fournit, à partir d'une équation $f(X_1, \ldots, X_n) = 0$ de (V, O), les équations et inéquations d'un sous-ensemble localement fermé $\mathcal{H}_{m_0, \ldots, m_i}$ de $\mathcal{H}_i$. Chacun des ensembles $\mathrm{Tr}(i)$, $i \geq 1$, se trouve être la réunion des images par ρ_{ij}, $j \geq i$, d'un nombre fini de sous-ensembles $\mathcal{H}_{m_0, \ldots, m_j}$.

Au lieu de l'exposant de Łojasiewicz θ de la singularité, l'exposant qui intervient ici pour borner j est l'exposant intrinsèque φ tel que :

$$1/\varphi = \bar{\nu}_{j(f)}(\mathfrak{m}) = \inf_{h \in \mathcal{H}} \frac{\mathrm{ord}_t\, \mathfrak{m} \circ h}{\mathrm{ord}_t\, j(f) \circ h} = \inf_{h \in \mathcal{H}} \frac{\mathrm{ord}_t\, M \circ h}{\mathrm{ord}_t\, J(f) \circ h}$$

où $\mathfrak{m}$ est l'idéal maximal de $\mathcal{O}_{V,O}$ et $j(f) := J(f)\mathcal{O}_{V,O}$ (comparer avec 2-5 ; au lieu de l'éclatement normalisé de $J(f)$ dans $(\mathbf{k}^n, O)$, on considère l'éclatement normalisé de $j(f)$ dans $\mathcal{O}_{V,O}$, c'est-à-dire la modification de Nash normalisée de (V, O)) ; φ est aussi un nombre rationnel. Puisque l'ensemble $\mathcal{H}$ des courbes tracées sur (V, O) est inclus dans l'ensemble $\mathcal{C}$ des courbes tracées sur $(\mathbf{k}^n, O)$, on a $\varphi \leq \theta$. Par exemple si (V, O) est une branche plane de multiplicité m, ayant β_g pour plus grand exposant caractéristique (cf. 4-6 ci-dessous) et $\mu = \mathrm{rg}_{\mathbf{k}}\, \mathbf{k}[\![x_1, x_2]\!]/J(f)$ pour nombre de Milnor, on a $\theta = \frac{\mu + \beta_g - 1}{m}$ et $\varphi = \frac{\mu + m - 1}{m}$ ([13], [20]).

Théorème 4-1. *Soit (V, O) une singularité isolée d'hypersurface de multiplicité m_0. Pour tout entier $i \geq 1$, soit :*

$$\mathcal{M}_i := \left\{ (m_1, \ldots, m_i) \in \mathbb{N}^i \,\middle|\, m_0 \geq m_1 \geq \cdots \geq m_i = 1 \right\}.$$

Alors

$$\mathrm{Tr}(i) \setminus (0) = \left[\bigcup_{(m_1,\ldots,1)\in\mathcal{M}_{[\varphi i]}} \rho_{i,[\varphi i]}(\mathcal{H}_{m_0,m_1,\ldots,1}) \right] \setminus (0)$$

où $[\varphi i]$ *désigne la partie entière de* φi.

D'où la constructibilité de $\mathrm{Tr}(i)$, puisque $\mathcal{M}_{[\varphi i]}$ est un ensemble fini. On a de plus une inclusion :

$$\mathrm{Tr}(i) \subset \bigcup_{m_0\geq\cdots\geq m_i\geq 1} \left[\bigcap_{1\leq j\leq i} \rho_{ji}^{-1}(\mathcal{H}_{m_0,\ldots,m_j}) \right]$$

induisant sur $\mathrm{Tr}(i)$ une partition compatible avec le système projectif ρ_{ji}. Pour toute suite d'entiers $m_0 \geq m_1 \geq \cdots \geq m_i \geq \cdots \geq 1$, on peut donc considérer

$$\mathcal{H}_{m_0,\ldots,m_i,\ldots} := \left\{ h\in\mathcal{H} \mid \rho_i(h)\in\mathcal{H}_{m_0,\ldots,m_i}, i\geq 1 \right\}.$$

On obtient ainsi une partition de $\mathcal{H}$ en familles.

(4-2) La suite d'entiers $\underline{m} = (m_0,\ldots,m_i,\ldots)$ ainsi associée à $h\in\mathcal{H}$ est appelée sa suite des multiplicités de Nash. Si $h(T) = O$, c'est la suite constante $(m_0,\ldots,m_0,\ldots)$. Sinon, soit $e := \inf\{i \mid \rho_i(h)\neq 0\}$ la multiplicité de h et soit $L := \inf\{i \mid m_i = 1\}$. D'après 4-1, si O est un point singulier isolé de V on a $1 \leq L \leq e\varphi$. Pour chaque entier $e \geq 1$, les courbes de multiplicité e se répartissent donc dans un nombre fini de familles. L'entier L est aussi l'ordre du jet de h qui suffit à déterminer la limite (lorsque t tend vers zéro) de l'hyperplan tangent à V en $h(t)$.

La suite des multiplicités de Nash $\underline{m} = (m_0,\ldots,m_i,\ldots)$ de $h\in\mathcal{H}$ se calcule facilement par la formule récurrente suivante :

$$m_0 + \cdots + m_i = \min_{\widetilde{h}\in\mathcal{C}(h,i)} \mathrm{ord}_t f\circ\widetilde{h}, i\geq 0, \qquad (4.1)$$

où $\mathcal{C}(h,i) := \{ \widetilde{h}\in\mathcal{C} \mid \rho_i(\widetilde{h}) = \rho_i(h) \}$.

Autrement dit, c'est le minimum des multiplicités d'intersection locale en O de l'hypersurface V avec une courbe tracée sur son espace ambiant $(\mathbf{k}^n, O)$ dont le i-ème jet coïncide avec celui de h.

Il est facile de vérifier à partir de (4.1) qu'on a aussi :

Proposition 4-3. *Soit* $\Gamma_h\colon (T,o) \to (V\times T, O\times o) =: (W_0, O_0)$ *le graphe de* $h\in\mathcal{H}$ *et soit :*

$$\cdots \longrightarrow W_i \xrightarrow{\pi_i} W_{i-1} \longrightarrow \cdots \longrightarrow W_1 \xrightarrow{\pi_1} W_0$$

la suite de morphismes telle que

1. π_i *soit l'éclatement de centre* O_{i-1}, $i\geq 1$

2. O_i *soit le point exceptionnel de la transformée stricte de* Γ_h *sur* W_i, $i\geq 1$. *Alors* m_i *est la multiplicité de* W_i *en* O_i, $i\geq 0$.

C'est sous cette forme qu'apparaît la suite $\underline{m}$ dans [8].

Si la multiplicité de h est 1, i.e., si c'est une courbe non singulière, on a un énoncé similaire en remplaçant (W_0, O_0) par (V, O) et en éclatant les points exceptionnels des transformées strictes successives de h.

Si $\mathbf{k}$ est un corps de caractéristique 0, on peut aussi retrouver la suite des multiplicités de Nash de $h \in \mathcal{H}$ à partir de l'idéal jacobien $J(f)$ de f. En effet, on a :

Proposition 4-4. *Soit* $\mathbf{k}$ *un corps de caractéristique* 0 ,

$$m_0 + \cdots + m_i - (i+1) = \min_{\widetilde{h} \in \mathcal{C}(h,i)} \operatorname{ord}_t J(f) \circ \widetilde{h}, i \geq 0.$$

Remarque 4-5. *Plus généralement, soit* $\widetilde{h} \in \mathcal{C}$ *tel que* $\rho_i(\widetilde{h}) \in \operatorname{Tr}(i)$. *On détermine la suite* $(m_0, \ldots, m_i)$ *telle que* $\rho_j(\widetilde{h}) \in \mathcal{H}_{m_0, \ldots, m_j}$, $0 \leq j \leq i$, *par les formules (4.1) appliquées pour tout entier* $j \leq i$.

(4-6) Si (V, O) est une branche plane (i.e., un germe de courbe plane analytiquement irréductible) définie sur un corps de caractéristique 0 et si $h \colon (T, o) \to (V, O)$ est sa normalisation, la connaissance de la suite $\underline{m}(h)$ est équivalente à celle de la multiplicité m et des exposants caractéristiques de Puiseux $\beta_1, \ldots, \beta_g$ de (V, O), donc aussi de son type topologique. Rappelons que dans un système de coordonnées convenable, h est donné par

$$\begin{cases} x & = t^m \\ y & = \displaystyle\sum_{0 \leq j \leq \ell_1} a_{1j} t^{\beta_1 + j d_1} + \cdots + \sum_{0 \leq j} a_{gj} t^{\beta_g + j d_g} \end{cases}$$

où posant $d_0 = m$, on a $d_i = pgcd(\beta_i, d_{i-1})$, $1 \leq i \leq g$, $d_g = 1$ et $a_{i0} \neq 0$, $1 \leq i \leq g$.

Proposition 4-7. *En écrivant la suite* $\underline{m}(h)$ *sous la forme*

$$(\mu_1^{r_1}, \ldots, \mu_{t-1}^{r_{t-1}}, \mu_t, \ldots)$$

où $\mu_i^{r_i}$ *représente* μ_i *répété* r_i *fois et* $\mu_i \neq \mu_{i+1}$, $1 \leq i < t$, *alors*

$$\underline{m}(h) = (d_0^{\beta_1}, \ldots, d_i^{\beta_{i+1} - \beta_i}, \ldots, d_g, \ldots).$$

D'où $L = \inf\{i \mid m_i = 1\} = \beta_g$.

Si $i \geq \beta_g$ (i.e., si $m_i = 1$), on détermine les points infiniment voisins de O communs à (V, O) et aux courbes définies paramétriquement par $\widetilde{h} \colon (T, o) \longrightarrow (\mathbf{k}^2, O)$ telles que $\rho_j(\widetilde{h}) \in \mathcal{H}_{m_0, \ldots, m_j}$, $1 \leq j \leq i$, où $\underline{m}$ est la suite des multiplicités de Nash de la normalisation h de (V, O) déterminée ci-dessus. Un point Q est infiniment voisin de O, si c'est un point d'une variété Z obtenue à partir de $(\mathbf{k}^2, O)$ par une suite finie d'éclatements de points. On dit qu'une courbe tracée sur $(\mathbf{k}^2, O)$ passe par Q, si Q appartient à sa transformée stricte dans Z. La notion de point infiniment voisin a été introduite par M. Nœther dans [18] pour résoudre les singularités des courbes planes.

L'énoncé fait apparaître des désingularisations plongées de (V, O). Soit $\pi \colon Z \longrightarrow (\mathbf{k}^2, O)$ un morphisme propre induisant un isomorphisme de $Z \setminus \pi^{-1}(O)$ sur $\mathbf{k}^2 \setminus \{O\}$. On dit que c'est une désingularisation plongée de (V, O) si Z est non singulier et si $\pi^{-1}(V)$ est un diviseur à croisements normaux. Soit

$$\cdots \longrightarrow Z_i \xrightarrow{\pi_i} Z_{i-1} \longrightarrow \cdots \longrightarrow Z_1 \xrightarrow{\pi_1} Z_0 = (\mathbf{k}^2, O)$$

la suite de morphismes telle que

1. π_i soit l'éclatement de centre O_{i-1}, $i \geq 1$

2. O_i soit le point exceptionnel de la transformée stricte de V sur Z_i, $i \geq 1$ et $O_0 = O$.

(On dit aussi que O_i est le point de V dans le i-ème voisinage infinitésimal de O.)

Il existe un entier r tel que, si $i \geq r$, le morphisme $\pi_1 \circ \cdots \circ \pi_i$ soit une désingularisation plongée de (V, O) ; $\pi_1 \circ \cdots \circ \pi_r$ est la désingularisation plongée minimale de (V, O).

Proposition 4-8. *Soit $p := \pi_1 \circ \cdots \circ \pi_r$ la désingularisation plongée minimale d'une branche plane (V, O) comme ci-dessus et soit $\underline{m} = (m_0, \ldots, m_i, \ldots)$ la suite des multiplicités de Nash de sa normalisation.*

Si $\widetilde{V} \in \mathcal{C}$ est une branche définie paramétriquement par $\widetilde{h} \colon (T, o) \longrightarrow (\mathbf{k}^2, O)$ telle que $\rho_j(\widetilde{h}) \in \mathcal{H}_{m_0, \ldots, m_j}$, $0 \leq j \leq i$ (cf. Remarque 4-5), pour un entier $i \geq \beta_g = \inf\{i \mid m_i = 1\}$, alors

1. *p est aussi la désingularisation plongée minimale de $\widetilde{V}$*

2. *$\widetilde{V}$ passe par le point de V situé dans le $(i - \beta_g + r)$-ème voisinage infinitésimal de O.*

Si $\rho_{i+1}(\widetilde{h}) \notin \mathrm{Tr}(i+1)$, alors $\widetilde{V}$ et V n'ont pas de point commun dans un voisinage infinitésimal de O d'ordre $> i - \beta_g + r$.

Pour montrer 4-8, on utilise la formule de Nœther qui exprime la multiplicité d'intersection $(V \cdot \widetilde{V})_O$ de V et $\widetilde{V}$ en O en fonction des multiplicités de V et $\widetilde{V}$, *i.e.*, de leurs transformées strictes respectives, aux points infiniment voisins Q de O qu'elles ont en commun

$$(V \cdot \widetilde{V})_O = \sum_Q \mathrm{mult}_Q V \cdot \mathrm{mult}_Q \widetilde{V}.$$

Les hypothèses entraînent que :

$$(V \cdot \widetilde{V})_O \geq m_0 + \cdots + m_i = \beta_1 d_0 + \cdots + (\beta_g - \beta_{g-1})d_{g-1} + (i - \beta_g + 1)$$

et que, quitte à effectuer un changement de paramétrisation, $\widetilde{V}$ admet une représentation paramétrique $\widetilde{h}$ qui coïncide jusqu'à l'ordre i avec l'expression de h donnée par 4-6.

Exemple 4-9. Soit V la courbe définie paramétriquement par $x = t^4$, $y = t^6 + t^7$; alors on a $m = 4$, $\beta_1 = 6$, $\beta_2 = 7$ et $\underline{m} = (4^6, 2, 1, \ldots)$. Si $\widetilde{h}$ est donné par $\widetilde{h} \circ x = \sum_{i \geq 1} a_i t^i$, $\widetilde{h} \circ y = \sum_{i \geq 1} b_i t^i$, on a $\rho_j(\widetilde{h}) \in \mathcal{H}_{m_0,\ldots,m_j}$, $0 \leq j \leq 7$ si et seulement si :

$$\begin{cases} a_1 = a_2 = a_3 = b_1 = \cdots = b_5 = 0, \\ b_6^2 - a_4^3 = 0, \\ 4a_4 b_7^2 - 12a_5 b_6 b_7 - 4a_4^3 b_6 + 9a_4^2 a_5^2 = 0, \\ a_4 \neq 0, \\ 2a_4 b_7 - 3a_5 b_6 \neq 0. \end{cases}$$

Étant donné $h \colon (T, o) \longrightarrow (\mathbf{k}^2, O)$, les formules de (4.1) définissent, pour tout $i \geq 0$ une $\mathbf{k}$-valuation discrète de rang 1, v_i, du corps des fractions K de $R := \mathcal{O}_{\mathbf{k}^2, O}$, i.e., un homomorphisme du groupe multiplicatif $K \setminus \{0\}$ dans $\mathbb{Z}$, tel que $v_i(\mathbf{k}) = 0$ et $v_i(z_1 + z_2) \geq \min\big(v_i(z_1), v_i(z_2)\big)$.

Comme $v_i(R) \geq 0$ et $v_i(\operatorname{Max} R) > 0$, on dit que v_i domine (ou est centrée dans) R. La théorie des valuations est développée au chapitre VI de [23]. Pour chaque $i \geq 0$, la valuation v_i est divisorielle, i.e., il existe un modèle normal Z dominant birationnellement $(\mathbf{k}^2, O)$ et un diviseur irréductible E de Z tel que $v_i(z)$ soit l'ordre, $v_E(z)$, du zéro ou du pôle de $z \in K$, le long de E.

Il résulte immédiatement de 4-8 qu'on a

Remarque 4-10. *Soit E_i le diviseur exceptionnel créé par l'éclatement du point O_i de V dans le i-ème voisinage infinitésimal de O, $i \geq 1$. Si $i \geq \beta_g$, on a*

$$v_i(z) = \min_{\widetilde{h} \in \mathcal{C}(h,i)} \operatorname{ord}_t z \circ \widetilde{h} = v_{E_{i-\beta_g+r}}(z), z \in K.$$

En effet, si $z \in R$ et si W est la courbe d'équation $z = 0$, $(W \cdot \widetilde{V})_O$ est aussi le nombre d'intersection de la transformée totale de W dans $Z_{i-\beta_g+r+1}$ avec la transformée stricte de $\widetilde{V}$. Or celle-ci est non singulière et transverse à $E_{i-\beta_g+r}$ et lorsque $\widetilde{h}$ varie dans $\mathcal{C}(h,i)$, son point exceptionnel est mobile sur $E_{i-\beta_g+r}$.

Enfin un point infiniment voisin Q de O détermine un idéal complet (ou intégralement clos) simple I_Q de R primaire pour $M := \operatorname{Max} R$ (voir [23] Appendice 5). La courbe définie par un élément général de I_Q est analytiquement irréductible. Ces courbes ont pour seuls points base dans les voisinages infinitésimaux successifs de O le point Q et ses images par la suite d'éclatements qui lui donne naissance. Leurs transformées strictes sur l'espace obtenu en éclatant Q sont non singulières et transverses à E_Q, le diviseur créé par cet éclatement.

Zariski a montré dans [22] que tout idéal complet dans un anneau local régulier de dimension 2 admet une factorisation unique en produit d'idéaux complets simples, factorisation qui correspond à la factorisation d'un élément général de I en produit de facteurs analytiquement irréductibles.

Les notations étant toujours celles utilisées tout au long de 4-6, on vérifie que si $i \geq r$

$$I_{O_i} = \big\{ z \in R \mid v_{i-r+\beta_g}(z) \geq v_{i-r+\beta_g}(f) \big\},$$

où

$$v_{i-r+\beta_g}(f) = \beta_1 d_0 + \cdots + (\beta_g - \beta_{g-1}) d_{g-1}(i - r + 1) d_g.$$

La normalisation de la courbe définie par un élément général de I_{O_i} est un élément général de $\mathcal{C}(h, i-r+\beta_g)$. Comparer avec le théorème 11.2 de [22], voir aussi 3.11 de [7] et [28], section 7.

5   Courbes lisses sur une singularité de surface

En appliquant l'algorithme des approximations successives discuté au section 4 à une singularité isolée d'hypersurface (V, O), on détermine d'abord si l'ensemble Tr(1) des 1-jets de courbes tracées sur (V, O) est réduit ou non à $\{0\}$ autrement dit si (V, O) contient des courbes lisses. Les résultats des calculs effectués à partir de la donnée d'une équation de V ne dépendent que de l'anneau local $\mathcal{O}_{V,O}$.

Pour une singularité de surface (S, O), on dispose aussi de critères d'existence de courbes lisses tracées sur (S, O) qui font intervenir une désingularisation de S et son cycle maximal (voir définition 5-1 ci-dessous). Ici on suppose seulement la surface S réduite et équidimensionnelle.

Dans toute la suite, $\mathcal{L}$ désignera l'ensemble des courbes $h \colon (T, o) \to (S, O)$ de multiplicité 1 dont l'image $\Gamma := h(T)$ (nécessairement lisse) n'est pas continue dans le lieu singulier de S. Comme le complété de $\mathcal{O}_{\Gamma,O}$ est $\mathbf{k}$-isomorphe à l'anneau de séries formelles $\mathbf{k}[\![t]\!]$, on dira plutôt que $\Gamma \in \mathcal{L}$. Si $\mathcal{L} \neq \emptyset$, on dira que S possède la propriété (cl) en O. Les résultats énoncés dans ce section sont démontrés dans [6].

Définition 5-1. *Soit (S, O) une singularité d'une surface réduite, équidimensionnelle et soit $\pi \colon X \to (S, O)$ un morphisme propre et birationnel. On appelle cycle maximal (de π) la composante $\mathcal{Z}_X$ de codimension (ou dimension) 1 du cycle sous-jacent au sous-schéma de X défini par $\mathfrak{m}\mathcal{O}_X$ où $\mathfrak{m}$ est l'idéal maximal de $\mathcal{O}_{S,O}$.*

On a $\mathcal{Z}_X = 0$ si et seulement si π est un morphisme fini. Si E est une composante irréductible de codimension 1 de $\pi^{-1}(O)$, la multiplicité m_E de E dans $\mathcal{Z}_X$ est par définition la longueur de l'anneau artinien $\mathcal{O}_{X,E}/\mathfrak{m}\mathcal{O}_{X,E}$.

On remarque aussi que si π est une désingularisation de (S, O), $\mathcal{Z}_X$ est le cycle défini par la partie divisorielle de $\mathfrak{m}\mathcal{O}_X$; enfin, si le morphisme π se factorise par l'éclatement de S de centre O, $\mathcal{Z}_X$ provient d'un diviseur de Cartier sur X et pour toute courbe $E \subset \pi^{-1}(O)$, le nombre d'intersection $(\mathcal{Z}_X \cdot E)$ est un entier ≤ 0.

Théorème 5-2. *Soit (S, O) une singularité de surface comme dans 5-1 et soit $\pi\colon X \to (S, O)$ une désingularisation.*

Pour que S possède la propriété (cl) en O, il faut et il suffit que :

- *ou bien $\mathcal{Z}_X \neq 0$ et il existe une courbe irréductible $E \subset \pi^{-1}(O)$ telle que $m_E = 1$.*

- *ou bien il existe un point isolé Q dans $\pi^{-1}(O)$ et un entier $m \geq 1$ tel que $\mathfrak{m}\mathcal{O}_{X,Q} = (u, v^m)$ pour un système régulier de paramètres convenable (u, v) de $\mathcal{O}_{X,Q}$.*

De plus, si $\Gamma \in \mathcal{L}$ et si le point exceptionnel $F_X(\Gamma)$ de sa transformée stricte sur X n'est pas un point isolé de $\pi^{-1}(O)$, alors :

1. *$F_X(\Gamma)$ est un point non singulier de $\pi^{-1}(O)$.*

2. *Si E désigne l'unique composante irréductible de $\pi^{-1}(O)$ qui contient à $F_X(\Gamma)$, on a $m_E = 1$.*

3. *Au voisinage de $F_X(\Gamma)$, $\mathfrak{m}\mathcal{O}_X$ est un $\mathcal{O}_X$-module inversible.*

Enfin, si $Q \in X$ possède les propriétés 1, 2 et 3 ci-dessus, il existe une courbe (formelle) $\Gamma \in \mathcal{L}$ telle que $Q = F_X(\Gamma)$.

Ce critère s'applique en particulier aux singularités rationnelles de surface en connaissant seulement la matrice d'intersection des composantes irréductibles E_i (ou encore le graphe dual pondéré) du diviseur exceptionnel de leurs désingularisations minimales. En effet, le cycle maximal coïncide avec le cycle fondamental, *i.e.*, le plus petit cycle $\mathcal{Z}$ à support exceptionnel tel que $(\mathcal{Z} \cdot E_i) \leq 0$, pour tout i ([2]). On constate que la singularité de type $\mathbb{E}_8$ de la surface d'équation $x^2 + y^3 + z^5 = 0$ est le seul point double rationnel qui ne possède pas la propriété (cl). Par contre, toutes les singularités toriques de surface possèdent la propriété (cl), puisque leur cycle fondamental est réduit.

On peut aussi reconnaître si une section hyperplane générale de (S, O) possède une branche lisse. Le critère fait intervenir une condition numérique d'intersection pour le cycle maximal d'une désingularisation qui domine l'éclatement $\sigma_1\colon S_1 \to (S, O)$ de centre O. A cause de la propriété universelle de la normalisation, une telle désingularisation domine aussi l'éclatement normalisé $\bar{\sigma}_1\colon \overline{S}_1 \to (S, O)$. Le morphisme $\bar{\sigma}_1$ est la composition $\sigma_1 \circ n_1$ où $n_1\colon \overline{S}_1 \to S_1$ est la normalisation de S_1.

Avant de préciser la notion de section hyperplane générale, il est peut-être utile de rappeler que si (Z, O) est une variété formelle non singulière de dimension minimale contenant (S, O), son espace tangent $T_{Z,O}$ s'identifie à l'espace tangent de Zariski $T_{S,O} := \operatorname{Spec} \operatorname{Sym} m/m^2$ de S en O. D'autre part, la courbe exceptionnelle $\sigma_1^{-1}(O)$ est la courbe projective réduite de $\operatorname{Proj} T_{S,O}$ sous-jacente à $\operatorname{Proj} C_{S,O}$ où $C_{S,O} := \operatorname{Spec} \otimes_{n \geq 0} \mathfrak{m}^n/\mathfrak{m}^{n+1}$ est le cône tangent à S en O. La courbe $\bar{\sigma}_1^{-1}(O)$ est considérée de même comme une courbe réduite.

Définition 5-3. *Une section hyperplane de* (S, O) *est une courbe (non nécessairement réduite) sur* (S, O) *admettant pour équation locale* $h = 0$ *où* $h \in \mathfrak{m} \setminus \mathfrak{m}^2$.

On dit qu'elle est générale si l'hyperplan H de $\operatorname{Proj} T_{S,O}$ ayant pour équation $h \bmod \mathfrak{m}^2 = 0$ coupe transversalement $\operatorname{Proj} |C_{S,O}| = \sigma^{-1}(O)$ en évitant :

- *les points singuliers de* $\sigma_1^{-1}(O)$

- *les points exceptionnels de la transformée stricte du lieu singulier de* S, *si* O *n'est pas un point singulier isolé*

- *les images par* n_1

  – *des points singuliers (isolés) de* $\overline{S}_1$

  – *des points singuliers de* $\bar{\sigma}_1^{-1}(O)$ *et des points de ramification de la restriction* $n_{1|\bar{\sigma}_1^{-1}(O)} \colon \bar{\sigma}_1^{-1}(O) \longrightarrow \sigma_1^{-1}(O)$.

Les génératrices de $C_{S,O}$ correspondant aux points de $\sigma_1^{-1}(O)$ énumérés ci-dessus seront dites spéciales.

Si $(S, O) \subset (\mathbb{C}^3, O)$, une génératrice qui n'est pas tangente au lieu singulier de S est une génératrice spéciale si et seulement si c'est une tangente exceptionnelle au sens de [12].

Définition 5-4. *On dit que* S *possède la propriété* (clh) *en* O *s'il existe une section hyperplane générale de* (S, O) *qui possède une branche lisse.*

En fait, s'il en est ainsi, toute section hyperplane générale possède une branche lisse. En effet, si $\overline{\mathcal{Z}}_1 = \Sigma m_i E_i$ est le cycle maximal de $\bar{\sigma}_1$ et si $|\overline{\mathcal{Z}}_1|$ est la courbe réduite sous-jacente, toute section hyperplane générale de (S, O) possède $-(\overline{\mathcal{Z}}_1 \cdot |\overline{\mathcal{Z}}_1|)$ branches dont $-(\overline{\mathcal{Z}}_1 \cdot E_i)$ de multiplicité m_i.

Théorème 5-5. *Soit* (S, O) *une singularité de surface comme dans 5-1 et soit* $\pi \colon X \to (S, O)$ *une désingularisation qui domine l'éclatement de centre* O. *Pour que* S *possède la propriété* (clh), *il faut et il suffit qu'il existe une courbe irréductible* $E \subset \pi^{-1}(O)$ *telle que la multiplicité* m_E *de* E *dans le cycle maximal* $\mathcal{Z}_X$ *soit 1 et que* $(\mathcal{Z}_X \cdot E) < 0$.

(5-6) Le théorème 5-2 permet de définir une subdivision de $\mathcal{L}$ en familles disjointes associées à certaines composantes irréductibles de la fibre exceptionnelle $\pi^{-1}(O)$ de la désingularisation minimale $\pi \colon X \to (S, O)$. En effet, pour toute composante irréductible, courbe ou point isolé, E de $\pi^{-1}(O)$, on peut considérer $\mathcal{L}_E := \{\Gamma \in \mathcal{L} \mid F_X(\Gamma) \in E\}$ où comme dans 5-2, $F_X(\Gamma)$ est le point exceptionnel de la transformée stricte de Γ sur X.

On a $\mathcal{L} = \cup_E \mathcal{L}_E$ et c'est une partition. Ici, une famille de courbes lisses désignera un sous-ensemble non vide de $\mathcal{L}$ apparaissant dans cette partition. Par extension, si $\mathcal{L}_E \neq \emptyset$, on dira que S possède la propriété (cl) en O relativement à E. Pour qu'il en soit ainsi, il faut et il suffit que E possède les propriétés énoncées dans la 1-ère partie du théorème 5-2. De même

Définition 5-7. *On dira que S possède la propriété (clh) en O relativement à E si $\mathcal{L}_E$ contient une branche lisse d'une section hyperplane générale.*

Nous verrons sur un exemple au section 6, que s'il en est ainsi, il peut exister des courbes $\Gamma \in \mathcal{L}_E$ qui ne sont pas contenues dans une section hyperplane générale de (S, O).

Ici, encore on peut caractériser les composantes irréductibles de $\pi^{-1}(O)$ concernées. On considère d'abord le cas où E est un point isolé de $\pi^{-1}(O)$.

Le morphisme π étant la désingularisation minimale de (S, O), $\overline{S}$ sa normalisation, le morphisme $\overline{\pi} \colon X \to \overline{S}$ qui factorise π est la désingularisation minimale de $\overline{S}$. Par suite, $\overline{\pi}$ est un isomorphisme local au voisinage de E. La normalisation sépare les composantes analytiquement irréductibles ou feuillets de (S, O). La singularité (S, O) possède donc un unique feuillet $\mathcal{F}$ dont la normalisation passe par $\overline{E} := \overline{\pi}(E)$.

Proposition 5-8. *Soit E un point isolé de la fibre exceptionnelle de la désingularisation minimale de (S, O) et soit $\mathcal{F}$ le feuillet de (S, O) qu'il détermine. Pour que S possède la propriété (clh) relativement à E, il faut et il suffit que $\mathcal{F}$ soit non singulier en O.*

Considérons maintenant le cas où E est une courbe.

Remarquons d'abord que si $\pi_1 \colon X_1 \to S_1$ est la désingularisation minimale de l'éclaté S_1 de S en O et $n_1 \colon \overline{S}_1 \to S_1$ sa normalisation, alors il existe un diagramme commutatif

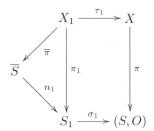

car la désingularisation $\sigma_1 \circ \pi_1$ de (S, O) se factorise par la désingularisation minimale π et par l'éclatement normalisé $\overline{\sigma}_1 = \sigma_1 \circ n_1$, et $\overline{\pi}_1$ est la désingularisation minimale de $\overline{S}_1$. De plus, τ_1 étant un morphisme propre et birationnel entre surfaces lisses, il est la composition d'une suite finie d'éclatements de points. On vérifie que c'est la suite de longueur minimale pour laquelle l'image réciproque de $\mathfrak{m}\mathcal{O}_X$ soit inversible.

Proposition 5-9. *Soit E une composante irréductible de dimension 1 de la fibre exceptionnelle de la désingularisation minimale π de (S, O) et soit m_E sa multiplicité dans le cycle maximal de π. Pour que S possède la propriété (clh) en O relativement à E, il faut et il suffit que $m_E = 1$ et que l'image par π_1 (ou $\overline{\pi}_1$) de la transformée stricte E_1 de E par τ_1 soit une courbe.*

Dans ce cas, le cône sur la courbe projective exceptionnelle $\pi_1(E_1)$ de σ_1 est une composante irréductible de $C_{S,O}$ qu'on désignera par C_E. On en conclut aussi que si S possède la propriété (cl) mais pas la propriété (clh) relativement à une courbe E contractée par π, le morphisme $\overline{\pi}_1$ contracte E en un point $\overline{O}_1$ de $\overline{S}_1$ nécessairement singulier, car $\overline{S}_1$ est normale et $\overline{\pi}_1$ est sa désingularisation minimale. La génératrice de $C_{S,O}$ correspondant au point $O_1 := n_1(\overline{O}_1) = \pi_1(E_1)$ de $\sigma_1^{-1}(O)$ est donc une génératrice spéciale (cf. 5-3). Ceci suggère de distinguer 2 types de génératrices spéciales.

Définition 5-10. *Soit L une génératrice spéciale de $C_{S,O}$ et soit $O_1 := \operatorname{Proj} L$ le point de $\sigma_1^{-1}(O)$ correspondant à sa direction. On dira que L est une génératrice spéciale ordinaire si $\overline{S}_1$ est non singulière en tout point de $n_1^{-1}(O_1)$, singulière dans le cas contraire.*

Il résulte facilement de 5-2 3 que la singularité (S_1, O_1) possède la propriété (cl) relativement à la courbe E_1 contractée en O_1 par la désingularisation minimale π_1 de S_1. Si (S_1, O_1) ne possède pas la propriété (clh) relativement à E_1, on peut itérer la construction précédente. Ce processus s'arrête au bout d'un nombre fini de pas.

Théorème 5-11. *Soit $\pi: X \to (S,O)$ la désingularisation minimale d'une singularité de surface. Si S possède la propriété (cl) en O relativement à une composante irréductible E de dimension 1 de $\pi^{-1}(O)$, il existe un unique diagramme commutatif*

$$
\begin{array}{ccccccccc}
X_{\ell+1} & \xrightarrow{\tau_{\ell+1}} & X_\ell & \xrightarrow{\tau_\ell} & \cdots & \longrightarrow & X_1 & \xrightarrow{\tau_1} & X_0 = X \\
\downarrow{\scriptstyle \pi_{\ell+1}} & & \downarrow{\scriptstyle \pi_\ell} & & & & \downarrow{\scriptstyle \pi_1} & & \downarrow{\scriptstyle \pi_0 = \pi} \\
S_{\ell+1} & \xrightarrow[\sigma_{\ell+1}]{} & S_\ell & \xrightarrow[\sigma_\ell]{} & \cdots & \longrightarrow & S_1 & \xrightarrow[\sigma_1]{} & S_0 = (S,0)
\end{array}
$$

où :

1. *π_i est la désingularisation minimale de S_i, $0 \le i \le \ell + 1$.*

2. *σ_i est l'éclatement de centre O_{i-1}, $1 \le i \le \ell + 1$, et $O_0 = O$.*

3. *$\pi_i(E_i) = O_i$, $0 \le i \le \ell$, et $\dim \pi_{\ell+1}(E_{\ell+1}) = 1$ où E_i désigne la transformée stricte de E dans X_i, $1 \le i \le \ell + 1$.*

4. *Si $0 \le i < \ell$, la génératrice L_i de C_{S_i, O_i} qui correspond à O_{i+1} est une génératrice spéciale singulière.*

5. *S_i possède la propriété (cl) en O_i relativement à E_i, $0 \le i \le \ell$ et ℓ est le plus petit entier i pour lequel S_i possède la propriété (clh) en O_i relativement à E_i.*

6. *Si F_{S_i} (resp. $F_{X_{\ell+1}}$) désigne l'application qui à $\Gamma \in \mathcal{L}$ fait correspondre le point exceptionnel de sa transformée stricte sur S_i, $1 \le i \le \ell + 1$, (resp. $X_{\ell+1}$), alors $F_{S_i}(\mathcal{L}_E) = O_i$, $1 \le i \le \ell$, $F_{S_{\ell+1}}(\mathcal{L}_E)$ est un ouvert dense de $\pi_{\ell+1}(E)$ et $F_{X_{\ell+1}}(\mathcal{L}_E) = E_{\ell+1} \setminus \operatorname{Sing}(\pi \circ \tau_1 \circ \cdots \circ \tau_{\ell+1})^{-1}(O)$.*

La *désingularisation minimale* $\overline{\pi}_{\ell+1}$ *de la normalisation* $\overline{S}_{\ell+1}$ *de* $S_{\ell+1}$ *et* $\tau_1 \circ \cdots \circ \tau_{\ell+1}$ *sont des isomorphismes au voisinage de* $F_{X_{\ell+1}}(\mathcal{L}_E)$.

Si (S,O) est une singularité de surface normale, on retrouve géométriquement le caractère fini du problème de la détermination de Tr(1).

Dans ce cas, la fibre exceptionnelle de la désingularisation minimale n'a pas de points isolés.

On observe d'abord que si $(Z,O) = Z_0$ est une variété formelle non singulière de dimension minimale contenant (S,O) et si D_i désigne le diviseur exceptionnel de l'éclatement $Z_{i+1} \to Z_i$ de centre O_i, alors :

(t1) la génératrice L_i de C_{S_i,O_i} de 5-11 (iv) est transverse à D_{i-1} en O_i, $1 \leq i < \ell$.

(t2) la composante irréductible C_{E_ℓ} de C_{S_ℓ,O_ℓ} déterminée par E_ℓ—voir 5-9, S_ℓ possède la propriété (clh) en O_ℓ relativement à E_ℓ—n'est pas contenue dans l'hyperplan tangent à $D_{\ell-1}$ en O_ℓ.

Définition 5-12. *On désigne par chaîne de points infiniment voisins de O sur S une suite finie $\{O_i\}_{i\geq 0}$ où $O_0 = O$ et O_i est un point de la fibre exceptionnelle de l'éclatement $\sigma_i\colon S_i \to S_{i-1}$ de centre O_{i-1}, $i \geq 1$.*

Théorème 5-13. *Soit (S,O) une singularité de surface normale. Il n'existe qu'un nombre fini de chaînes de points infiniment voisins $\{O_i\}_{0\leq i\leq \ell}$ de O sur S telle que pour $0 \leq i < \ell$, la génératrice L_i de C_{S_i,O_i}, qui correspond à O_{i+1} soit une génératrice spéciale singulière avec la propriété de transversalité (t1) ci-dessus si $i \neq 0$.*

Si $\{O_i\}_{0\leq i\leq \ell}$ est une telle chaîne, toute courbe irréductible E_ℓ telle que S_ℓ possède la propriété (clh) en O_ℓ relativement à E_ℓ et telle que la composante irréductible C_{E_ℓ} de C_{S_ℓ,O_ℓ} associée aît la propriété de transversalité (t2) ci-dessus provient, par la construction du théorème 5-11, d'une courbe irréductible E telle que S possède la propriété (cl) en O relativement à E.

Remarque 5-14. *L'énoncé précédent permet une construction algorithmique des familles de courbes lisses.*

6 Deux exemples

Nous comparons sur deux exemples de points double de surfaces plongées dans $\mathbb{C}^3$ les diverses notions de famille introduites précédemment.

6.1 Singularité de type $\mathbb{A}_2$

La surface S d'équation $x^2 + y^2 + z^3 = 0$ possède à l'origine une singularité de type $\mathbb{A}_2$. C'est à la fois un point double rationnel et une singularité torique. Sa désingularisation minimale est l'éclatement $\sigma_1\colon S_1 \to (S,O)$ de centre O.

La fibre exceptionnelle $\sigma_1^{-1}(O)$ est la réunion de deux $\mathbb{P}^1$, d_1 et d'_1, d'auto-intersection -2, se coupant transversalement.

Le cône tangent $C_{S,O}$ est réduit ; c'est la réunion de deux plans se coupant transversalement. On vérifie que $\mathrm{Tr}(1) \setminus (0) \simeq \mathrm{Reg}\, C_{S,O}$. Son adhérence de Zariski $\overline{\mathrm{Tr}(1)} \simeq C_{S,O}$ a donc deux composantes irréductibles et il y a donc deux familles de courbes au sens de Nash (section 3).

D'autre part, S possède la propriété (cl) en O relativement à d_1 et d'_1 car le cycle maximal est $\mathcal{Z}_1 = d_1 + d'_1$. Il y a donc aussi deux familles de courbes lisses $\mathcal{L}_{d_1}$ et $\mathcal{L}_{d'_1}$ (section 5). De plus, S possède la propriété (clh) en O relativement à d_1 et d'_1 d'après 5-9. Toute section hyperplane générale possède deux branches lisses. Dans cet exemple, toute courbe lisse sur (S,O) est une branche lisse d'une section hyperplane générale. La seule génératrice spéciale de $C_{S,O}$ est son lieu singulier et c'est une génératrice spéciale ordinaire.

En appliquant l'algorithme du section 4, on montre que les courbes tracées sur (S,O) de multiplicité au plus 3 se répartissent dans les 4 familles $\mathcal{H}_{2^i\bar{1}}$, $1 \leq i \leq 4$, où 2^i signifie 2 répété i fois et $\bar{1}$ est la suite (infinie) constante de valeur 1. On a $\mathcal{L} = \mathcal{L}_d \cup \mathcal{L}_{d'_1} = \mathcal{H}_{2\bar{1}}$. Si $x = \sum a_i t^i$, $y = \sum b_i t^i$, $z = \sum c_i t^i$ est la paramétrisation d'une courbe tracée sur (S,O) et e est sa multiplicité, les conditions suivantes caractérisent son appartenance à $\mathcal{H}_{2^i\bar{1}}$, $1 \leq i \leq 4$:

$\mathcal{H}_{2\bar{1}}$	$\mathcal{H}_{2^2\bar{1}}$	$\mathcal{H}_{2^3\bar{1}}$	$\mathcal{H}_{2^4\bar{1}}$
$a_1^2 + b_1^2 = 0$	$e = 2$, $a_2^2 + b_2^2 = 0$	$e \geq 2$, $a_3^2 + b_3^2 + c_2^3 = 0$	$e \geq 3$, $a_4^2 + b_4^2 = 0$
$a_1 \neq 0$	$a_2 \neq 0$	$a_3 \neq 0$	$a_4 \neq 0$

La transformée stricte sur S_1 d'une courbe générale tracée sur (S,O) appartenant à $\mathcal{H}_{2^i\bar{1}}$, $1 \leq i \leq 4$, est une courbe lisse et transverse à d_1 ou d'_1 en $Q \neq d_1 \cap d'_1$ si $i = 1$, ayant un contact d'ordre 2 avec d_1 ou d'_1 en $Q \neq d_1 \cap d'_1$ si $i = 2$, transverse à d_1 et d'_1 en $O_1 = d_1 \cap d'_1$ si $i = 3$, ayant un contact d'ordre 2 avec d_1 ou d'_1 en $O_1 = d_1 \cap d'_1$ si $i = 4$ (voir figure 6.2).

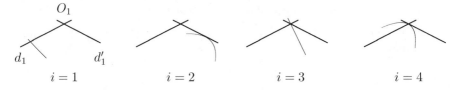

$i = 1$ $i = 2$ $i = 3$ $i = 4$

Figure 6.2: Transformée stricte sur S_1 d'un courbe tracée sur (S,O).

Enfin, on observe une généralisation partielle de la proposition 4-8. L'éclatement $p_1 \colon Z_1 \to (\mathbb{C}^3, O)$ de centre O n'est pas une désingularisation plongée de S car au voisinage de O_1 la transformée totale de S n'est pas un diviseur à croisements normaux. On en obtient une en considérant $p = p_1 \circ p_2 \circ p_3$ où $p_2 \colon Z_2 \to Z_1$ est l'éclatement de centre O_1 et $p_3 \colon Z \to Z_2$ celui de centre la courbe intersection des 2 composantes irréductibles exceptionnelles de $p_1 \circ p_2$. On désignera par D_ℓ la transformée stricte sur Z du diviseur exceptionnel du $(\ell+1)$-ème éclatement, $0 \le i \le 2$. (On remarque qu'en éclatant d_1 ou d'_1, on obtient d'autres désingularisations plongées de S).

On observe que si $\underline{m} = (2^i, \bar{1})$ avec $i = 1, 3, 4$ et si $\widetilde{\Gamma}$ est une branche définie paramétriquement par $\widetilde{h} : (T, o) \to (\mathbb{C}^3, O)$ telle que $\rho_j(\widetilde{h}) \in \mathcal{H}_{m_0, \dots, m_j}$, $1 \le j \le i$, alors

1. La transformée stricte $\widetilde{\Gamma}$, sur Z est lisse et transverse à tous les diviseurs D_ℓ qui contiennent son point exceptionnel $\widetilde{O}$, i.e., D_0 si $i = 1$, D_2 si $i = 3$, D_0 et D_2 si $i = 4$.

2. La transformée stricte $\widehat{S}$ de S sur Z passe par $\widetilde{O}$.

Si $\rho_{i+1}(\widetilde{h}) \notin \mathrm{Tr}(i+1)$, alors $\widehat{S}$ et $\widetilde{\Gamma}_Z$ sont transverses en $\widetilde{O}$.

6.2 Une singularité elliptique simple

La surface S d'équation $x^2 + y^3 + z^6 = 0$ possède à l'origine une singularité elliptique simple. La fibre exceptionnelle de sa désingularisation minimale $\pi : X \to (S, O)$ est une courbe elliptique E. Il y a donc une seule famille de courbes au sens de Nash (section 3).

D'autre part, S possède la propriété (cl) en O relativement à E et il y a donc aussi une seule famille de courbes lisses $\mathcal{L}_E$ (section 5). Mais S n'a pas la propriété (clh) en O relativement à E. Le diagramme commutatif de 5-11 relatif à E est le suivant :

$$
\begin{array}{ccccc}
X_2 & = & X_1 & \xrightarrow{\tau_1} & X \\
\downarrow{\scriptstyle \pi_2} & & \downarrow{\scriptstyle \pi_1} & & \downarrow{\scriptstyle \pi} \\
S_2 & \xrightarrow{\sigma_2} & S_1 & \xrightarrow{\sigma_1} & S
\end{array}
$$

où σ_1 est l'éclatement de centre O, σ_2 celui de centre le point $O_1 := \mathrm{Sing}\, S_1$, la désingularisation minimale π_1 de S_1 est l'éclatement normalisé de O_1 et celle de S_2, π_2, est sa normalisation. (En effet, $\mathrm{Sing}\, S_2 = \sigma_2^{-1}(O_1) := d_2$ est un $\mathbb{P}^1$ le long duquel S_2 a une singularité ordinaire sauf en 4 points du type "parapluie de Whitney". L'un d'eux est le point d'intersection de d_2 avec la transformée stricte $d_1 \simeq \mathbb{P}^1$ de $\sigma_1^{-1}(O)$; d_1 et d_2 se coupent transversalement). Enfin τ_1 est la contraction de la transformée stricte de d_1 sur X_2 et si E_1 est la transformée stricte de E sur $X_1 = X_2$, on a $\pi_2(E_1) = d_2$ et $\pi_1(E_1) = O_1$.

Ainsi, S_1 vérifie la propriété (clh) en O relativement à E_1. Une section hyperplane générale quelconque de (S_1, O_1) a 2 branches lisses ayant un contact d'ordre 2. Le cône C_{S_1, O_1} a 4 génératrices spéciales, toutes ordinaires, G_i, $1 \le i \le 3$, et la tangente G_0 en O_1 à $\sigma_1^{-1}(O)$, correspondant respectivement aux 4 "parapluies de Whitney" sur S_2.

Puisque $\pi_1^{-1}(O_1) = E_1$, alors (S_1, O_1) possède une seule famille de courbes lisses $\mathcal{L}_{E_1}$. De plus, comme E_1 est une courbe non singulière et π_1 domine l'éclatement de O_1, le théorème 5-2 assure que toute génératrice G de C_{S_1, O_1} est la tangente à une branche lisse Γ de cette famille. Si $G = G_i$, $0 \leq i \leq 3$, une telle Γ ne peut être contenue dans une section hyperplane générale de (S_1, O_1). La singularité (S_1, O_1) et la courbe E_1 illustrent la remarque suivant la définition 5-7.

Il résulte immédiatement de ce qui précède et de 5-11 (vi) que Tr(1) est la génératrice L_0 de $C_{S,O}$ correspondant à O_1. C'est l'unique génératrice spéciale de $C_{S,O}$ et elle est singulière.

Les courbes lisses sur (S, O) se répartissent dans les 2 familles $\mathcal{H}_{2^3\bar{1}}$ et $\mathcal{H}_{2^4\bar{1}}$. On vérifie que le jet à l'ordre 3 de la paramétrisation d'une courbe lisse Γ dans $\mathcal{H}_{2^3\bar{1}}$ est :

$$\left\{ \begin{array}{l} x = a_3 t^3, \\ y = b_2 t^2 + b_3 t^3, \\ z = c_1 t + c_2 t^2 + c_3 t^3, \end{array} \right. \quad \text{avec} \quad \left\{ \begin{array}{l} a_3^2 + b_2^3 + c_1^6 = 0, \\ a_3 \neq 0, \\ c_1 \neq 0, \end{array} \right.$$

et dans $\mathcal{H}_{2^4\bar{1}}$ est :

$$\left\{ \begin{array}{l} x = 0, \\ y = b_2 t^2 + b_3 t^3, \\ z = c_1 t + c_2 t^2 + c_3 t^3 \end{array} \right. \quad \text{avec} \quad \left\{ \begin{array}{l} b_2^3 + c_1^6 = 0, \\ b_2^3 b_3 + 2c_1^5 c_2 = 0, \\ c_1 \neq 0. \end{array} \right.$$

Dans le premier cas la tangente à la transformée stricte de Γ sur S_1 n'est pas une génératrice spéciale de C_{S_1, O_1}, dans le deuxième cas, c'est l'une des G_i, $1 \leq i \leq 3$.

Y a t-il une relation entre le fait que la section hyperplane de (S, O) appartienne aussi à $\mathcal{H}_{2^3\bar{1}}$ et le fait que (S, O) possède une désingularisation plongée induisant sa désingularisation minimale ?

Bibliographie

[1] Artin M. Algebraic approximation of structures over complete local rings. Publ. Math. IHES 36 1969 pp23–58

[2] Artin M. On isolated rational singularities of surfaces. Ann. of Math.88 1966 pp129–136

[3] Bouvier C. et Gonzalez-Sprinberg G. *G*-désingularisations de variétés toriques. C. R. Acad. Sci. Paris Sér. I Math. à paraître.

[4] Campillo A. et Castellanos J. Arf closure relative to a divisorial valuation and transversal curves. Prepubl. Dep. de Algebra, Geometría y Topología, Univ. Valladolid, n 11 à paraître.

[5] Gonzalez-Sprinberg G. et Lejeune-Jalabert M. Modèles canoniques plongés. Kodai Math. J. 14 n 2 1991 pp194–209.

[6] Gonzalez-Sprinberg G. et Lejeune-Jalabert M. Courbes lisses, cycle maximal et points infiniment voisins des singularités de surface. prepub196 1992 , à paraître.

[7] Gonzalez-Sprinberg G. Désingularisation des surfaces par des modifications de Nash normalisées. Séminaire Bourbaki 661 1986 pp85–86

[8] Hickel M. Fonction de Artin et germes de courbes tracées sur un germe d'espace analytique à paraître Amer. J. Math.

[9] Lejeune-Jalabert M. Arcs analytiques et résolution minimale des singularités des surfaces quasi-homogènes. Springer Lecture Notes in Math. 777 1980 pp303–336.

[10] Lejeune-Jalabert M. Courbes tracées sur un germe d'hypersurface. Amer. J. Math. 112 1990 pp525–568.

[11] Lejeune-Jalabert M. et Teissier B. Clôture intégrale des idéaux et équisingularité. Séminaire École Polytechnique (1974) Publ. Inst. Fourier.

[12] L.T. Lê D.T. et Teissier B. Sur la géométrie des surfaces complexes, I. Tangentes exceptionnelles. Amer. J. Math. 101 1979 pp420–452

[13] Milnor J. Singular points of complex hypersurfaces. Princeton Univ. Press and Univ. of Tokyo Press. 1968

[14] Mori S. Flip theorem and the existence of minimal models for 3-folds. Journal of the A.M.S. 1 1988 pp117–253

[15] Newton I. De Methodis Serierum et Fluxionum. 1670, The Mathematical papers of I. Newton, Vol. III, Cambridge Univ. Press, Traduction française de Buffon-Debure 1740, Blanchard 1966

[16] Nash J. Arc structure of singularities. Preprint non publié

[17] Nobile A. On Nash theory of arc structure of singularities. Annali di Mat. Pura ed Applic. 160 1991 129–146

[18] Nœther. M Ueber die singulären Werthsysteme einer algebraischen Function und die singulären Punkte einer algebraische Curve. Math. Ann 1875, 166–182

[19] Spivakovsky M. Valuations in function fields of surfaces. Amer. J. Math. 112 1990 107–156

[20] Teissier B. Variétés polaires, I Inv. Math. 40 1977 267–292

[21] Kempf G., Knudsen F., Mumford D. et Saint Donat B. Toroïdal Embeddings, I Lecture Notes in Math., **339**, Springer 1973

[22] Zariski O. Polynomial ideals defined by infinitely near base points. Amer. J. Math. 60 1938 151–204

[23] Zariski O., Samuel P. Commutative algebra, II Appendice 5, Van Nostrand 1960

Addresses of authors:

Institut Fourier, Univ. Grenoble I
B.P. 74
38402 Saint-Martin d'Hères Cedex
France
Email: gonsprin@fourier.grenet.fr

Email: lejeune@fourier.grenet.fr,

BLOWING UP ACYCLIC GRAPHS AND GEOMETRICAL CONFIGURATIONS

Carlos Marijuán[1]

1 Introduction

Blowing up is a useful technique in algebraic and analytic geometry. In particular, it is the main tool for proving resolution of singularities. Hironaka [2] proved in 1964 that every algebraic variety over a field of characteristic zero admits a resolution of singularities which is obtained by successive blowing ups of certain regular centers. Moreover, he proves the stronger version of embedded resolution of singularities, i.e., for every (singular) subvariety X of a smooth variety Z there exists a sequence of birational morphisms

$$Z_N \longrightarrow Z_{N-1} \longrightarrow \cdots \longrightarrow Z_1 \longrightarrow Z_0 = Z, \qquad (1.1)$$

such that π_i is the blowing up of Z_{i-1} at a regular center C_i which is transversal to the exceptional divisor E_{i-1} of $\pi_{i-1} \circ \cdots \circ \pi_1$, and such that the strict transform X_N of X at Z_N is smooth and transversal (normal crossing) to the exceptional divisor E_N.

The embedded resolution of singularities was extended to analytical spaces in 1974 in [1]. Recently Villamayor [4] proved a canonical theorem of embedded resolution of singularities, i.e., a procedure to define a concrete sequence such as (1.1) for each variety X. This leads us to consider an algorithmic point of view (see [5]).

Looking for such an algorithmic viewpoint, our initial motivation, in this paper, is to try to understand the behaviour, after blowing up π_{i+1}, of the set (or a subset) of subvarieties E_i created at Z_i by the sequence of transformations.

More precisely, we are interested in the following question: if $\pi \colon Z' \to Z$ is a blowing up of a smooth variety at a regular center C, and if $\mathbf{C}$ is a smooth geometrical configuration of subvarieties of Z, try to describe the transform $\mathbf{C}'$ of $\mathbf{C}$ in Z', i.e., the set of strict transforms of subvarieties in $\mathbf{C}$ not included in C and the fibers of those contained in C, as well as the irreducible components of the intersections of such objects. Here, by a smooth configuration, we mean a subset of smooth subvarieties meeting pairwise transversally, i.e., such that at each point in the intersection, both subvarieties are locally described by the vanishing of elements in a common system of parameters.

To focus this question in a combinatorial way, we associate an incidence graph to each configuration, i.e., the acyclic (oriented) graph given by the inclusion ordering on the subvarieties of the configuration, weighted by the

[1]Supported by DIGICYT PB91 0210 C0201

dimensionality of those subvarieties. Then we introduce, in a purely combinatorial way (section 3), the blowing up of a weighted acyclic graph with center at one of its points. The main result in the paper (section 4) proves that the graph associated to the blow up configuration $\mathbf{C}'$ is the transitivization of the blow up of the graph associated to the configuration $\mathbf{C}$.

To complete the paper, we give a canonical procedure to "desingularize" (in some sense) a general acyclic graph. Namely, from a given acyclic graph we obtain another one, which we call total or geometric blow up, by blowing up successively the levels of the points of the original graph. The geometric blow up has a nice structure which is made up from hypercubes and it is also provided of a nice weight function, as it is determined by the weight of a maximal dominant element. The cubic structure on the total blow up is described in the autor's Ph. Thesis ([3]).

2   Basic concepts and notations

By a graph we will mean a couple (X, G) where X is a finite set and $G \subset X \times X - \{(x,x) : x \in X\}$. Elements in X and G are called points and arcs respectively. A labeling of X by the label set E is a bijective map $\tilde{x} \colon E \to X$; $\tilde{x}(e)$ being denoted by x_e for any $e \in E$.

A path joining the point x_1 with x_q in (X, G) of length $q-1$ is an injective map $s \colon \{1, 2, \dots, q\} \to X$, $q \geq 2$, such that $\big(s(i), s(i+1)\big) \in G$, and it will be denoted alternatively as $s(1)s(2) \dots s(q)$ or $x_1 x_2 \dots x_q$. If $x_1 x_2 \dots x_q$ is a path and $(x_q, x_1) \in G$ then the sequence $x_1 x_2 \dots x_q x_1$ is said to be a cycle. A graph is said to be acyclic if it has no cycles. Finally for a semipath in (X, G) we mean an injective sequence such that for any $i, 1 \leq i \leq q - 1$, one has either $\big(s(i), s(i+1)\big) \in G$ or $\big(s(i+1), s(i)\big) \in G$. Semicycles can be defined in the same way.

A subgraph of (X, G) is a graph (Y, H) with $Y \subset X$ and $H \subset G$. For the induced subgraph on Y we will mean the graph $(Y, G/Y)$ where $G/Y = G \cap (Y \times Y)$.

Given a graph (X, G), the equivalence relationship "$x \equiv y$ if and only if x and y are in a semipath" gives rise to the partition of X into connected components $X = X_1 \cup \dots \cup X_r$. Roughly speaking the connected components are the induced subgraphs $(X_i, G/X_i)$. If $r = 1$ the graph is connected.

For a point $x \in X$ we will consider the sets $\overline{x}$ (resp. $\underline{x}$) consisting of x and those points $y \in X$ such that there exists a path from y to x (resp. x to y). A point x is said to be dominant if $(y, x) \in G$ for any $y \in \overline{x}$.

Acyclic graphs have some nice properties. First, for an acyclic graph there exists at least one point x (resp. y) such that $\overline{x} = \{x\}$ (resp. $\underline{y} = \{y\}$). Such a point will be said to be a minimal (resp. maximal) in the graph. Moreover, the points in an acyclic graph (X, G) can be distributed by levels $N_0, N_1, \dots$

as follows: $N_0 = \{x \in X : x \text{ is minimal in } (X, G)\}$, and recursively for $p \geq 1$,

$$N_p = \{x \in X - \cup_{i=0}^{p-1} N_i : x \text{ is minimal in } (X - \cup_{i=0}^{p-1} N_i, G/X - \cup_{i=0}^{p-1} N_i)\}.$$

Thus one has a partition of X, $X = \cup_{i=0}^{k} N_i$, k being the height of the graph, i.e., the greatest index such that $N_k \neq \emptyset$.

Second, we have the following characterization of acyclic graphs (see [3] for details): A graph (X, G), with $\text{card}(X) = n$, is acyclic if, and only if, there exists a labeling of X by the label set $E = \{1, \dots n\}$ such that if $(x_i, x_j) \in G$ then $i < j$. In this case, we will say that (X, G) is naturally ordered.

Given two graphs $(X, G), (X', G')$, by a graph morphism from (X, G) to (X', G') we mean a mapping $\Delta \colon X \to X'$ such that for every arc $(x, y) \in G$, one has either $\Delta(x) = \Delta(y)$ or $(\Delta(x), \Delta(y)) \in G'$. The morphism is an isomorphism when it has an inverse, i.e., if Δ is bijective and $(x, y) \in G$ if and only if $(\Delta(x), \Delta(y)) \in G'$. The morphisms which take maximal points to maximal points will be called dominant.

A graph (X, G) is said to be transitive if for any pair of arcs (x, y), (y, z) with $x \neq z$, then (x, z) is also an arc. The graph is antisymmetric if it has no pair of symmetric arcs, i.e., if (x, y) and (y, x) are not both in G for $x \neq y$. A transitive graph is acyclic if and only if it is antisymmetric, and an acyclic transitive graph (X, G) has a partial ordering $(X, \leq)$ where $x < y$ if and only if $(x, y) \in G$. For an acyclic graph (X, G) the transitivized graph is the graph (X, G^t) where $G^t = G \cup \{(x, y) : x \neq y \text{ and there exists a path } x_1 \dots x_q, q \geq 2,$ with $x_1 = x$ and $x_q = y\}$.

Example 2-1. If $X = \{x_i : 1 \leq i \leq 12\}$ and

$$\begin{aligned}
G = \{&(x_1, x_5), (x_5, x_7), (x_7, x_9), (x_1, x_7),\\
&(x_1, x_3), (x_3, x_9), (x_3, x_{12}), (x_{11}, x_{12}),\\
&(x_2, x_6), (x_6, x_8), (x_8, x_{10}), (x_2, x_8),\\
&(x_2, x_{10}), (x_6, x_{10}), (x_2, x_4), (x_4, x_{10})\}
\end{aligned}$$

then (X, G) is a naturally ordered acyclic graph with two connected components (X_1, G_1) and (X_2, G_2).

Points with equal height in the figure 2.1 belong to the same level in the graph. The connected component (X_2, G_2) is the induced subgraph on $\overline{x}_{10} = \{x_2, x_4, x_6, x_8, x_{10}\}$ and x_{10} is the unique dominant maximal element of the graph (X, G).

The map $\Delta \colon X_1 \to X_2$ given by $\Delta(x_{12}) = x_4, (x_{11}) = x_2$ and $\Delta(x_i) = x_{i+1}$, for any $i \leq 9$, is a non-dominant graph morphism. The transitivized graph of the induced subgraph on $\overline{x}_9 = \{x_1, x_3, x_5, x_7, x_9\}$ is $(\overline{x}_9, G/\overline{x}_9{}^t)$ with $G/\overline{x}_9{}^t = G/\overline{x}_9 \cup \{(x_1, x_9), (x_5, x_9)\}$. This graph is isomorphic to the connected component (X_2, G_2) and the restriction of the map Δ to $\overline{x}_9$ is now a dominant isomorphism between both of them.

A weight map on an acyclic graph (X, G) is an increasing map $w \colon X \to \mathbb{N}$, i.e., such that $w(x) < w(y)$ for any $(x, y) \in G$. The triplet (X, G, w) will be

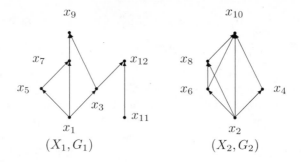

Figure 2.1: Graph of example 2-1.

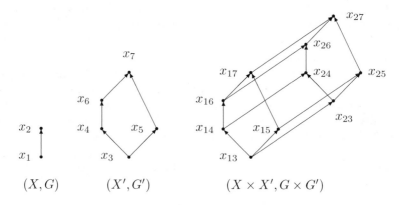

Figure 2.2: Graphs of example 2-2.

called a weighted acyclic graph and the weight map will be said to be transversal on (X, G) if for every pair of points x, y one has $w(x) + w(y) \leq w(z) + w(z')$ for any z, $z' \in G$ such that $z \in \max(\overline{x} \cap \overline{y})$ and $z' \in \min(\underline{x} \cap \underline{y})$ where $\max(-)$ (resp. $\min(-)$) stands for the set of maximals (resp. minimals) for the induced subgraph. A path $x_1 \ldots x_q$ is said to be complete for the weight w if for any i, $2 \leq i \leq q$, one has $w(x_i) = w(x_{i-1}) + 1$. A weighted acyclic graph (X, G, w) is said to be completely transversal for w if all paths in (X, G) are complete for w. Two given weighted acyclic graphs (X, G, w), (Y, H, v) are equivalent if there exists an isomorphism of graphs $\Phi: (X, G) \to (Y, H)$ such that $w(x) - v\big(\Phi(x)\big) = k$ for k fixed and any $x \in X$.

Finally, for two given graphs (X, G), (X', G') we will denote by $(X \times X', G \times G')$ the product graph on $X \times X'$ where $G \times G'$ denotes the set of arcs $\big\{\big((x, x'), (y, y')\big) \in (X \times X') \times (X \times X') : \big(x = y \text{ and } (x', y') \in G'\big) \text{ or } \big((x, y) \in G \text{ and } x' = y'\big)\big\}$.

Example 2-2. The figure 2.2 represents a product of graphs where each point $(x_i, x_j) \in X \times X'$ has been labeled x_{ij}.

The map $w \colon X' \to \mathbb{N}$ given by $w(x_i) = i$ for any $x_i \in X'$ is a weight map on the graph (X', G') but w is not transversal since $x_3 \in \max(\overline{x}_5 \cap \overline{x}_6)$, $x_7 \in \min(\underline{x}_5 \cap \underline{x}_6)$ and $w(x_5)+w(x_6) \not\leq w(X_3)+w(x_7)$. However, w is transversal on the weighted graph (X', G'', w) where $G'' = G' \cup \{(x_4, x_5)\}$.

In the graph (X', G') there is no complete path for the weight map w and the weighted acyclic graph (X', G', w') is not completely transversal for any weight map w' on (X', G') as there are two paths of different length joining the point x_3 with x_7.

Throughout the paper several graphs will be constructed from original ones. The point sets X' for such graphs will usually be denoted by $X' = \{x_\alpha : \alpha \in I'\}$ where I' is an explicitly described set. Moreover, I' will usually be a subset of sequences $i_1 \ldots i_p$ of elements of some known label set (the label set of some known graph) for various integers p. Sometimes p will be 1; thus, elements in various graphs will usually be denoted as x_i for the same subindex i, which will not be a loss of generality.

3 Blowing up acyclic graphs

In this section we give the main concept of this paper, the blowing up at a point of an acyclic graph with transversal weight map.

Let $X = \{x_i : i \in I\}$ be a finite set of points and let (X, G, w) be a connected acyclic graph, with a dominant maximal element x_z and with transversal weight map $w \colon X \to \mathbb{N}$. (If the acyclic graph (X, G) does not have a dominant maximal element, we can consider the acyclic graph obtained from (X, G) connecting every point $x_i \in X$ with a new point x_z by means of the arc (x_i, x_z)).

Having fixed a point $x_c \in X$, which we shall call blowing up center, let us consider the following parts in the set of points X:

1. the set of points connected to the blowing up center by a path: $\overline{x}_c$,

2. the set of points connected from the blowing up center by an arc: $x_c{\uparrow} = \{x_j : (x_c, x_j) \in G\}$,

3. the set of points connected from the blowing up center by a path: $x_c^* = \underline{x}_c - \{x_c\}$,

4. the set of paths connecting $\overline{x}_c$ with $x_c{\uparrow}$ without passing through $\overline{x}_c \cup x_c{\uparrow}$, which we shall call transversal paths to the blowing up center, i.e., the set A_c given by

$$A_c = \big\{ s \colon \{1, \ldots, q\} \to X :$$
$$q > 2, s(1) \in \overline{x}_c, s(i) \notin \overline{x}_c \cup x_c{\uparrow} \text{ for } 1 < i < q, s(q) \in x_c{\uparrow} \big\},$$

5. the set of points which are not joined by a path to the blowing up center through which some transversal paths passes and which we shall call transversal points to the blowing up center:

$$B_c = \{x_t \in X - (\overline{x}_c \cup x_c\uparrow) :$$
$$\text{there exists a path } s \in A_c \text{ with } s(i) = x_t, 1 < i < q\}.$$

For each transversal point $x_t \in B_c$, we will consider the set $A_c(x_t)$ of transversal paths $s \in A_c$ passing through x_t, its first term being $s(1)$, maximal in (X, G) of the set of the first terms of the transversal paths passing through x_t and its last term being, $s(q)$, minimal in (X, G) of the set of the last terms of the transversal paths passing through x_t: $A_c(x_t) = \{s \in A_c : s(i) = x_t \in B_c$ for any $1 < i < q$, $s(1) \in \max\{s'(1) : s' \in A_c, s'(j) = x_t$ for any $1 < j < q'\}, s(q) \in \min\{s'(q') : s' \in A_c, s'(k) = x_t$ for any $1 < k < q'\}\}$.

Finally, set $B'_c = \{x_t \in B_c : \text{there exists a path } s \in A_c(x_t) \text{ with } s(2) = x_t$ and $w(s(1)) + w(s(q)) - w(x_c) - 1 \geq w(x_t)\}$.

Definition 3-1. *Let (X, G, w) be a connected acyclic graph with dominant maximal element and with transversal weight map w, let x_c be a point of G and B_c, B'_c the sets introduced before. We define the blow up of (X, G, w) with center at x_c and relative to B_c to be the graph (X_c, G_c, w_c) where*

$$X_c = (X - \overline{x}_c) \cup \{x_{ij} : x_i \in \overline{x}_c, x_j \in x_c\uparrow\}$$
$$\cup \{x_{ij} : (x_i, x_j) \in G, x_i \in \overline{x}_c - \{x_c\}, x_j \in x_c^* - x_c\uparrow\} \cup T_c,$$

where $T_c = \{x_{ht} : x_t \in B'_c, s \in A_c(x_t), s(1) = x_h\}$, and $G_c = (G - G/\overline{x}_c - \{(x_i, x_j) \in G : x_i \in \overline{x}_c, x_j \in x_c^*\} - \{(x_h, x_t) \in G : x_t \in B_c, s \in A_c(x_t), s(1) = x_h\}) \cup (G/\overline{x}_c \times G/x_c\uparrow) \cup \{(x_{ht}, x_{hk}), (x_{ht}, x_t) : x_t \in B'_c, s \in A_c(x_t), s(1) = x_h, s(2) = x_t, s(q) = x_k\} \cup \{(x_{ij}, x_j) : (x_i, x_j) \in G, x_i \in \overline{x}_c, x_j \in x_c^*\} \cup \{(x_{hk}, x_t) : x_t \in B_c - B'_c, s \in A_c(x_t), s(1) = x_h, s(2) = x_t, s(q) = x_k\} \cup \{(x_{ht_i}, x_{ht_j}) : x_{t_i}, x_{t_j} \in B'_c, s \in A_c(x_{t_i}), s' \in A_c(x_{t_j}), (x_{t_i}, x_{t_j}) \in G, s(1) = s'(1) = x_h\}$.

The weight map $w_c : X_c \to \mathbb{N}$ is defined by:

$$\begin{cases} w_c(x_j) = w(x_j) & \text{if } x_j \in X_c \cap X, \\ w_c(x_{ij}) = w(x_i) + w(x_j) - w(x_c) - 1 & \text{if } x_{ij} \in X_c - X - T_c, \\ w_c(x_{ht}) = w(x_t) - 1 & \text{if } x_{ht} \in T_c. \end{cases}$$

Notice that the new graph is defined by preserving some labels i and adding new labels ij, ht, etc. to the labeling set. In other words, the point set X_c can be obtained from X by replacing the points of $\overline{x}_c$ with those of $\overline{x}_c \times x_c\uparrow$, adding a point x_{ij} for each arc (x_i, x_j) connecting $x_i \in \overline{x}_c - \{x_c\}$ with $x_j \in x_c^* - x_c\uparrow$; and also adding the points of the set T_c. The induced subgraph by (X, G) on $X - \overline{x}_c$ is preserved and the induced subgraph on $\overline{x}_c$ is replaced by $G/\overline{x}_c \times G/x_c\uparrow$ which is connected to $G/x_c\uparrow$ from $\{x_c\} \times G/x_c\uparrow$.

The points $(x_i, x_j) \in \overline{x}_c \times x_c\uparrow$ are denoted by x_{ij} and then the arcs $(x_i, x_j) \in G$ connecting $\overline{x}_c$ with x_c^* becoming arcs $(x_{ij}, x_j) \in G_c$. The arcs $(x_h, x_t) \in G$ disappear and each point $x_{ht} \in T_c$, associated to the transversal path s passing through $x_t \in B_c'$, with $s(1) = x_h$, $s(2) = x_t$, $s(q) = x_k$, is linked to the subgraph $G/\overline{x}_c \times G/x_c\uparrow$ in the point x_{hk} and to the induced subgraph on $X - \overline{x}_c$ in the point x_t. If, on the contrary, $s(2) = x_t \in B_c - B_c'$, then the arc (x_h, x_t) of G is replaced by the arc (x_{hk}, x_t) in G_c; moreover, some of the points $x_{ht} \in T_c$ can be connected to each other.

Proposition 3-2. *The blow up graph (X_c, G_c, w_c) is a connected acyclic graph with dominant maximal element and with transversal weight map. The sets B_c' and T_c are empty if and only if for each transversal point $x_t = s(2)$, with $s \in A_c(x_t)$, one has*

$$w\big(s(1)\big) + w\big(s(q)\big) = w(x_c) + w(x_t).$$

PROOF. Obviously, the blow up graph is acyclic, connected and with dominant maximal element x_z. The mapping w_c is increasing in the induced subgraphs on $X_c \cap X = X - \overline{x}_c$ and on T_c. It is also increasing on $\overline{x}_c \times x_c\uparrow$ since, if (x_{ij}, x_{ik}) is an arc of the product graph $G/\overline{x}_c \times G/x_c\uparrow$, then $(x_j, x_k) \in G/x_c\uparrow$ with $w(x_j) < w(x_k)$ and, therefore, $w_c(x_{ij}) = w(x_i) + w(x_j) - w(x_c) - 1 < w(x_i) + w(x_k) - w(x_c) - 1 = w_c(x_{ik})$. Analogously, the same holds for the arcs (x_{hj}, x_{ij}) with $(x_h, x_i) \in G/\overline{x}_c$. Moreover, keeping the notations as above, one has for (x_{ij}, x_j),

$$w_c(x_{ij}) = w(x_i) + w(x_j) - w(x_c) - 1 < w(x_j) = w_c(x_j).$$

We also have for (x_{hk}, x_t),

$$w_c(x_{hk}) = w(x_h) + w(x_k) - w(x_c) - 1 < w(x_t) = w_c(x_t),$$

as $x_t \notin B_c'$, and for (x_{ht}, x_{hk})

$$w_c(x_{ht}) = w(x_t) - 1 \le w(x_h) + w(x_k) - w(x_c) - 2 < w_c(x_{hk}),$$

as $x_t \in B_c'$. Thus, for (x_{ht}, x_t),

$$w_c(x_{ht}) = w(x_t) - 1 < w(x_t) = w_c(x_t),$$

and then w_c is a weight map on X_c.

It is also clear that w_c is transversal over the induced subgraphs on $X_c \cap X = X - \overline{x}_c$ and on $(X_c \cap X) \cup T_c$ and over the join of $G/\overline{x}_c \times G/x_c\uparrow$ with the induced subgraph on $X_c \cap X$. It is so also over $G/\overline{x}_c \times G/x_c\uparrow$ since, for each semicycle consisting of the paths $x_{hj}x_{ij}x_{ik}$ and $x_{hj}x_{hk}x_{ik}$ one has $w_c(x_{ij}) + w_c(x_{hk}) = w_c(x_{hj}) + w_c(x_{ik})$.

Finally, to obtain the equality to characterize when B'_c and T_c are empty sets, it is sufficient to bear in mind the inequality between weights defining the point set $B_c - B'_c$ and the inequality giving the transversal condition for the weight map w.   □

Definition 3-3. *Under the conditions of definition 3-1, the blowing up of a connected acyclic graph (X, G, w) at a point x_c is defined to be the morphism $\pi_c \colon (X_c, G_c, w_c) \to (X, G, w)$ given by*

$$\begin{cases} \pi_c(x_j) = x_j & \text{if } x_j \in X_c \cap X, \\ \pi_c(x_{ij}) = x_i & \text{if } x_{ij} \in X_c - X - T_c, \\ \pi_c(x_{ht}) = x_h & \text{if } x_{ht} \in T_c. \end{cases}$$

It is clear that if $(x, y) \in G_c$, then either $\pi_c(x) = \pi_c(y)$ or $\big(\pi_c(x), \pi_c(y)\big) \in G$ since for the arcs $(x_{ht}, x_t) \in G_c$, with $x_{ht} \in T_c$, one has $x_h = s(1)$ and $x_t = s(2)$, and for the other arcs it is obvious. In consequence, π_c is a dominant morphism of acyclic graphs.

We will call exceptional divisor to the maximal point x_{cz} of the induced subgraph on $X_c - X$ and each point x_{iz}, with $x_i \in \overline{x}_c - x_c$, will be said to be an exceptional fiber of x_i in the exceptional divisor; so that the induced subgraph on $\{x_{iz} \colon x_i \in \overline{x}_c\}$ will be refered to as the exceptional subgraph of the blow up graph. The points $x_j \in X_c \cap X$, preserving the label by the blowing up, shall be called strict transforms (of the points $x_j \in X - \overline{x}_c$). The points x_{cj} with $x_j \in x_c\!\uparrow$ shall be called the cuts of the exceptional divisor x_{cz} with the strict transform x_j and the points x_{ij}, with $x_i \in \overline{x}_c - x_c$, $x_j \in x_c\!\uparrow$, the cuts of the exceptional fiber x_{iz} with the strict transform x_j. Finally, the points $x_{ht} \in T_c$ will be called transversal cuts of the exceptional fiber x_{hz} with the strict transform x_t.

Theorem 3-4. *If two given weighted acyclic graphs $(X, G, w), (Y, H, v)$ are equivalent by means of an isomorphism Φ and if $\pi_c \colon (X_c, G_c, w_c) \to (X, G, w)$, $\pi'_c \colon (Y_c, H_c, v_c) \to (Y, H, v)$ are their blowing up morphisms with center at x_c, $y_c = \Phi(x_c)$, respectively, then the blow up acyclic graphs (X_c, G_c, w_c), (Y_c, H_c, v_c) are equivalent for an isomorphism Δ such that $\pi'_c \circ \Phi = \Delta \circ \pi_c$ and $w_c(x) - v_c\big(\Delta(x)\big) = w(y) - v\big(\Phi(y)\big) = k$, where $y = \pi_c(x)$ and k is a constant integer.*

PROOF. If the graph (X, G) is labeled by means of the bijection $\tilde{x} \colon E \to X$ with $\tilde{x}(e) = x_e$, we can consider the graph (Y, H) to be labeled by means of the bijection $\tilde{y} = \Phi \circ \tilde{x} \colon E \to Y$, thus we have $\tilde{y}(e) = y_e = \Phi(x_e)$ which is coherent with the notation given to the centers of the blowing ups and, in this way, one has $\pi'_c \circ \Phi = \Delta \circ \pi_c$ if Δ is the isomorphism given by

$$\begin{cases} \Delta(x_j) = y_j & \text{if } x_j \in X_c \cap X, \\ \Delta(x_{ij}) = y_{ij} & \text{if } x_{ij} \in X_c - X - T_c, \\ \Delta(x_{ht}) = y_{ht} & \text{if } x_{ht} \in T_c. \end{cases}$$

Moreover, if $x_j \in X_c \cap X$ we have $w_c(x_j) - v_c(\Delta(x_j)) = w(x_j) - v(y_j) = w(x_j) - v(\Phi(x_j)) = k$, if $x_{ij} \in X_c - X - T_c$, then $w_c(x_{ij}) - v_c(\Delta(x_{ij})) = w(x_i) + w(x_j) - w(x_c) - 1 - (v(y_i) + v(y_j) - v(y_c) - 1) = k$, and if $x_{ht} \in T_c$, then $w_c(x_{ht}) - v_c(\Delta(x_{ht})) = w(x_t) - 1 - (v(y_t) - 1) = k$. $\qquad\square$

Remark 3-5.

1. If the blowing up center is a minimal point x_o of the weighted acyclic graph (X, G, w), then there is no transversal point and so A_o, B_o and T_o are empty sets. In this case, we will denote the blow up graph with center in x_o by (X_o, G_o, w_o) where:

$$X_o = (X - \{x_o\}) \cup \{x_{oj} : x_j \in x_o\uparrow\}$$

and

$$G_o = G - \{(x_o, x_j) : x_j \in x_o\uparrow\} \cup$$
$$\cup \{(x_{oj}, x_{ok}) : (x_j, x_k) \in G/x_o\uparrow\} \cup \{(x_{oj}, x_j) : x_j \in x_o\uparrow\}.$$

The transversal weight map $w_o: X_o \to \mathbb{N}$ is given by $w_o(x_j) = x_j$ if $x_j \in X_o \cap X = X - \{x_o\}$ and by $w_o(x_{oj}) = w(x_j) - 1$ if $x_{oj} \in X_o - X$. The π_o-blowing up of (X, G) is the acyclic graph morphism $\pi_o: (X_o, G_o, w_o) \to (X, G, w)$ given by $\pi_o(x_j) = x_j$ if $x_j \in X_o \cap X$ and by $\pi_o(x_{oj}) = x_o$ otherwise.

2. Acyclic transitive graphs (that represent partial orderings) are especially interesting from the point of view of blowing ups. If (X^t, G^t, w^t) is a graph of this kind, under the conditions of definition 3-1 one has, in general, that its blow up graph (X_c^t, G_c^t, w_c^t) is not transitive. More precisely, in (X^t, G^t, w^t) we have $x_c\uparrow = x_c^*$ and so the point set of the blow up graph is

$$X_c^t = (X^t - \overline{x}_c) \cup \{x_{ij} : x_i \in \overline{x}_c, x_j \in x_c^*\} \cup T_c.$$

Now, the arcs of the set $\{(x_{ij}, x_j) : (x_i, x_j) \in G^t, x_i \in \overline{x}_c, x_j \in x_c^*\}$ are redundant by transitivity over the arcs $(x_{ij}, x_{cj}) \in G^t/\overline{x}_c \times G^t/x_c^*$ and (x_{cj}, x_j). For every path $x_{t_1} x_{t_2} \ldots x_{t_{q-1}} x_{t_q} \in A_c(x_{t_2})$, with $x_{t_1} = x_h$ and $x_{t_q} = x_k$, since $w_c^t(x_{hk}) < w_c^t(x_k)$, there exists i, with $2 \le i \le q - 1$, such that $w_c^t(x_{t_{i-1}}) \le w_c^t(x_{hk}) < w_c^t(x_{t_i})$, i.e., $x_{t_{i-1}} \in B_c'$ and $x_{t_i} \in B_c - B_c'$. Also the arcs (x_{hk}, x_{t_j}) are redundant for $j > i$ and the arcs (x_{ht_j}, x_{hk}) are redundant for $j < i - 1$. Thus, one has, on the one hand, the arc (x_{hk}, x_{t_i}) and, on the other, the path $x_{ht_2} \ldots x_{ht_{i-1}}$ connected to the product graph $G^t/\overline{x}_c \times G^t/x_c^*$ by the arc $(x_{ht_{i-1}}, x_{hk})$ and to the path $x_{t_2} \ldots x_{t_{i-1}}$ by the arcs (x_{ht_j}, x_{t_j}), $2 \le j \le i - 1$.

On the contrary, the transitivity fails on $G^t/\overline{x}_c \times G^t/x_c^*$, on the connection of this with G^t/x_c^* and on the connection, between them, of the paths $x_{ht_2} \ldots x_{ht_{i-1}}$ and $x_{t_2} \ldots x_{t_{i-1}}$.

4 Graphic representation of the blowing up for a geometric configuration

Throughout this section, for a variety we will mean a connected and smooth algebraic or complex analytic variety, and all the subvarieties considered will be connected and smooth. Two subvarieties are said to be transversal if at each point of their intersection there exists a system of local coordinates $x_1, \ldots, x_n$ such that both varieties are locally defined by $x_i = 0, i \in A$ and $x_i = 0, i \in B$, A and B being non-empty subsets of $\{1, 2, \ldots, n\}$, $A \not\subset B$ and $B \not\subset A$.

Let us consider a smooth geometric configuration $\mathbf{C}$, i.e., a set consisting of a variety Z and a finite subset of subvarieties such that

1. For each couple of subvarieties in $\mathbf{C}$, either there is an inclusion, their intersection is empty or they are transversal.

2. Each component of the intersection of two subvarieties in $\mathbf{C}$, is a subvariety in $\mathbf{C}$.

If $C \subset Z$ is a subvariety of dimension r and Z' the blow up of Z with center C, then the points in the inverse image by the blowing up morphism $p_C \colon Z' \to Z$, of a point $P \in C$ correspond biunivocally to the sets of $(r + 1)$-vector subspaces of the tangent space to the ambient variety, $T_P Z$, containing the tangent space to the blowing up center, $T_P C$. That is to say, to each point $P \in C$ corresponds in Z' the fiber $F_P = \{$Vector subspaces of $T_P Z$ generated by $T_P C$ and a vector $v \notin T_P C\}$.

On the other hand, the inverse image of the blowing up center C is the exceptional divisor $E_{cz} = \cup_{P \in C} F_P$. One has $\mathrm{codim}_Z F_P = r + 1$, $\mathrm{codim}_Z E_{cz} = 1$ and a bijection between Z' and $(Z - C) \cup E_{cz}$.

If the subvariety J contains the blowing up center C, we will denote the strict transform of J in Z' by J', i.e., J' is the closure in Z' of the inverse image of $J - C$ by the blowing up morphism,

$$\overline{p_C^{-1}(J - C)} = J'.$$

The intersection of the exceptional divisor E_{cz} with the strict transforms J' will be denoted by CJ'. One has $\dim J' = \dim J$ and $\dim CJ' = \dim J - 1$.

If I is a subvariety contained in C, we will denote the fiber of I in the exceptional divisor E_{cz} by $F_{iz} = \cup_{P \in I} F_P$. Let us denote the cuts of each fiber F_{iz} with each strict transform J' of a variety J as above by IJ'. One has $\dim F_{iz} = \dim I + \dim Z - \dim C - 1$ and $\dim IJ' = \dim I + \dim J - \dim C - 1$.

If T represents a transversal subvariety to the blowing up center C in a subvariety H, then each vector tangent to T in a point $P \in H$ has its corresponding vector subspace $V \in F_P$ and the strict transform T' is a subvariety of Z' for which the inverse image of P by the blowing up morphism is the collection of subspaces $V \in F_P$ corresponding to the vectors $v \in T_P T$, so that

$$\overline{p_C^{-1}(T - C)} = T'.$$

The intersection of the strict transform T' with the exceptional divisor E_{cz} will be denoted by HT'. One has $\dim T' = \dim T$ and $\dim HT' = \dim T - 1$.

If L is a disjoint subvariety with C, we will denote its corresponding transform in Z' by L'. One has $\dim L' = \dim L$.

Let $\mathbf{C'}$ be the blow up configuration, i.e., the configuration of the subvarieties of Z' consisting of the objects Z', E_{cz}, J' and CJ' for $J \supset C, F_{iz}$ and IJ' for $I \subset C, T'$ and HT' for T transversal to C, and L' for $L \cap C = \emptyset$, and J, I, T, L in the configuration $\mathbf{C}$. Note that all the above objects are (connected and smooth) subvarieties of Z'. Then the blowing up morphism gives rise to the map $p_c\colon \mathbf{C'} \to \mathbf{C}$ given, with notations as above, by $p_c(Z') = Z, p_c(E_{cz}) = C, p_c(J') = J, p_c(CJ') = C, p_c(F_{iz}) = p_c(IJ') = I, p_c(T') = T, p_c(HT') = H$ and $p(L') = L$.

We can associate a weighted directed graph (X, G, w) to each geometric configuration $\mathbf{C}$ where points in X represent the subvarieties in $\mathbf{C}$, the arcs represent the inclusion relations between them, so that $(H, K) \in G$ if and only if $H \subset K$, and the weights are the dimensions of the corresponding subvarieties. Thus in the above situation, both the smooth geometric configuration $\mathbf{C}$ formed by the variety Z and their subvarieties, and its blow up $\mathbf{C'}$ have associated weighted graphs (X, G, w) and (X', G', w') which clearly are connected transitive acyclic and have only one maximal element (Z and Z') which dominates the points in its respective graph. Note that the weight maps w and w' are transversal because of the condition 1 of transversality in the geometric case. We can associate a dominant graph morphism $\tilde{p}_c\colon (X', G') \to (X, G)$ to the map $p_c\colon \mathbf{C'} \to \mathbf{C}$ in a natural way.

We will prove that the concept of blowing up for a weighted acyclic graph given in the preceding section supports the blowing up of the associated geometric configuration; more precisely, we are going to prove that the blow up of the graph (X, G, w), associated to the geometric configuration $\mathbf{C}$, at the blowing up center C, is an acyclic graph (X_c, G_c, w_c) whose transitivized is the graph (X', G', w') associated to the configuration $\mathbf{C'}$.

Lemma 4-1. *If $T_2 \subset T_3 \subset \ldots \subset T_{q-1}$ is a chain in the geometric configuration $\mathbf{C}$, of transversal subvarieties to the blowing up center C, $T_1 \subset T_2$ a component of the intersection of T_{q-1} and C, and T_q a minimal subvariety of those that contain T_{q-1} and C, then for every i, with $2 \leq i \leq q-1$, and with notations as above, one has:*

1. $T_1 T_i' \subseteq T_1 T_q'$.

2. $\dim T_1 T_i' = \dim T_1 T_q'$ *if and only if* $T_1 T_i' = T_1 T_q' \subset T_i'$. *And, in this case,* $\dim T_1 + \dim T_q = \dim C + \dim T_i$.

3. $\dim T_1 T_i' < \dim T_1 T_q'$ *if and only if* $T_1 T_i' = T_1 T_q' \cap T_i'$. *And, in this case,* $\dim T_1 + \dim T_q > \dim C + \dim T_i$.

4. *Either* $T_1 T_q' \subset T_i'$ *for any* $i \geq 2$, *or there exists an* i, *with* $2 < i \leq q$, *such that* $T_1 T_q' \subset T_j'$ *for* $i \leq j \leq q$ *and* $T_1 T_j' = T_1 T_q' \subset T_j'$ *for* $2 \leq j < i$.

5. If $T_1 \subset T_2 \subset \ldots \subset T_q$ is a complete chain (i.e., if $\dim T_{i+1} = \dim T_i + 1$, $i = 1, \ldots, q-1$) and i is the smaller integer for which $T_1 T_q' \subset T_i'$, then $\dim T_q = \dim C + i - 1$.

PROOF. 1. Since $T_1 = T_i' \cap C$ and $T_1 \subset T_q'$, one has $T_1 T_i' = E_{cz} \cap T_i' = F_{t_1 z} \cap T_i' \subseteq F_{t_1 z} \cap T_q' = T_1 T_q'$.

2. The equivalence is a consequence of 1. Since $\dim T_1 T_q' = \dim T_1 + \dim T_q - \dim C - 1 < \dim T_i' = \dim T_i$, one has $\dim T_1 + \dim T_q \leq \dim C + \dim T_i$. The opposed inequality follows from the condition of transversality on the subvarieties.

3. The equivalence is a consequence of the following facts: $T_1 T_i' \subseteq T_1 T_q'$, $T_1 T_i' \subseteq T_i'$, $T_1 T_q'$ does not contain T_i ($T_i - T_1$ is not empty, as T_i is transversal) and $\dim T_1 T_i' = \dim T_i - 1$. Now, the relation between dimensions follows taking into account that $\dim T_i \leq \dim T_1 T_q'$.

4. If $\dim T_1 T_2' = \dim T_1 T_q'$ from 2, one has $T_1 T_q' \subset T_2' \subset \cdots \subset T'q$ and $\dim T_1 + \dim T_q = \dim C + \dim T_2$. If otherwise, from 3 one has, $T_1 T_2' = T_1 T_q' \cap T_2'$ and so $T_1 T_q'$ is not contained in T_2'. Moreover, since $T_1 \cap T_q' \subset T_q'$, it is clear that there exists a minimum i, with $2 < i \leq q$, such that $T_1 T_q' \subset T_i' \subset \cdots \subset T_q'$ and $T_1 T_q'$ is not contained in T_{i-1}' which, by 3, means that $T_1 T_i' = T_1 T_q' \cap T_j'$, for $2 \leq j < i$. Note that, in general, one has not $T_1 T_i' = T_1 T_q'$.

5. If the chain is complete, one has $T_1 T_i' = T_1 T_q'$ by 4, and $\dim T_1 + \dim T_q = \dim C + \dim T_i = \dim C + \dim T_1 + i - 1$ by 2, therefore $\dim T_q = \dim C + i - 1$. In particular, if $i = 2$, $\dim C = \dim T_q - 1$. $\qquad\square$

Theorem 4-2. Let $\mathbf{C}$ be a geometric configuration of subvarieties of Z and let (X, G, w) be the associated graph labeled by the own configuration $\mathbf{C}$ by means of the bijection $\tilde{x} \colon \mathbf{C} \to X$ which takes each subvariety H in a point of X denoted by x_h, the weight being given by $w(x_h) = \dim H$. Let $\mathbf{C}'$ be the blow up configuration of $\mathbf{C}$ with center at $C \in \mathbf{C}$ and (X_c', G_c', w_c') its associated weighted graph. The map $p_c \colon \mathbf{C}' \to \mathbf{C}$ induces a dominant acyclic transitive graph morphism $\sigma_c \colon (X_c', G_c', w_c') \to (X, G, w)$ given by $\sigma_c(K') = x_h$ where $H = p_c(K')$ for $K' \in \mathbf{C}'$. Let (X_c, G_c, w_c) be the weight acyclic transitive blow up graph of (X, G, w) on the blowing up center x_c and let $\pi_c \colon (X_c, G_c, w_c) \to (X, G, w)$ be the corresponding blowing up.

Then, the weighted graph (X_c', G_c', w_c') is isomorphic to the transitivized of (X_c, G_c, w_c) by an isomorphism Δ_c preserving the weights and such that $\sigma_c = \pi_c \circ \Delta_c$.

PROOF. Is clear that the graphs (X, G, w) and (X', G', w') are connected, transitive and acyclic with dominant maximal element and with transversal weight map, thus there exists the acyclic blow up graph (X_c, G_c, w_c) with dominant maximal element and with transversal weight map.

The subvarieties J containing the blowing up center C provide strict transforms J' and cuts CJ' with the exceptional divisor E_{cz}. These subvarieties

$J \supset C$ are represented in the acyclic transitive graph (X, G, w) by the points of the set $x_c \!\uparrow\, = x_c^* = \{x_j : (x_c, x_j) \in G\}$.

The subvarieties I contained in the blowing up center C have fibers F_{iz} in the exceptional divisor cutting strict transforms J' at IJ'. These subvarieties $I \subset C$ are represented in the acyclic transitive graph (X, G, w) by the points of $\bar{x}_c = \{x_i : (x_i, x_c) \in G\} \cup \{x_c\}$ and, in the blow up graph, it is the points of $\pi_c^{-1}(\bar{x}_c) = \bar{x}_c \times x_c \!\uparrow$ that correspond to them.

With the notation as above, define the mapping $\Delta_c : X_c' \to X_c$ as follows:

$$\Delta_c(E_{cz}) = x_{cz}, \qquad \Delta_c(F_{iz}) = x_{iz}, \qquad \Delta_c(J') = x_j,$$
$$\Delta_c(CJ') = x_{cj}, \qquad \Delta_c(IJ') = x_{ij}.$$

By construction, Δ_c carries the exceptional configuration on the exceptional subgraph, the strict transforms of the subvarieties $J \supset C$ on the strict transform points of the points $x_j \in x_c^*$ and the cuts of the exceptional configuration with the strict transforms on the cuts of the exceptional subgraph with the strict transform points.

Also by construction, Δ_c is an isomorphism between the part of the blow up configuration consisting of the subvarieties related to the blowing up center and the induced subgraph by (X_c, G_c, w_c) on $\bar{x}_c \times x_c \!\uparrow$. We can obviously extend this isomorphism to the enlargement obtained by adding the arc set $\{(CJ', J') : J \supset C\} \subset G_c'$. Neither the induced subgraph on $\bar{x}_c \times x_c \!\uparrow$ nor the corresponding join between $\bar{x}_c \times x_c \!\uparrow$ and $x_c \!\uparrow$ contain the redundant arcs by transitivity and, so, the isomorphism must be, in fact, established between the transitivized graphs of these.

The subvarieties T transversal to the blowing up center provide strict transforms T' and cuts HT' with the exceptional divisor for the components H of the intersection $T \cap C$. These transversal subvarieties T are represented in the graph (X, G, w) by the points $x_t \in B_c$ and, after lemma 4-1, we can consider the set $B_c' = \{x_t \in B_c : \text{there exists a path } s \in A_c(x_t) \text{ with } s(2) = x_t \text{ and } w(s(1)) + w(s(q)) - w(x_c) - 1 \geq w(x_t)\}$.

Thus, if T and H are as above and $K \in \mathbf{C}$ is minimal between those that contain T and C, we can apply the lemma 4-1 to the chain $H \subset T \subset K$, and then, with notation as at the beginning, we define:

$$\Delta_c(T') = x_t, \qquad \Delta_c(HT') = \begin{cases} x_{hk} & \text{if } \dim HK' < \dim T' \\ x_{ht} & \text{if } \dim HK' \geq \dim T' \end{cases}$$

In this way, interpreting the results 2 and 3 of the lemma 4-1, one has, on the one hand, that the arc $(x_{hk}, x_t) \in G_c$ if $x_t \in B_c - B_c'$ corresponds to the relation $HK' \subset T'$ and, on the other, that the equality $HT' = HK' \cap T'$ is represented in the blow up graph by the arcs $(x_{ht}, x_{hk}) \in G_c$ and $(x_{ht}, x_t) \in G_c$ if $x_t \in B_c'$.

If T_i and T_j are transversal to the blowing up center at the maximal subvariety H, with $T_i \subset T_j$ and $\dim HK' \geq \dim T'_j$, then $HT'_i \subset HT'_j$ (which justifies the presence of the arcs (x_{ht_i}, x_{ht_j}) in the graphic blow up).

The induced subgraph on the join of the path $x_{ht_i} \ldots x_{ht_j}$ with the transversal path $x_{t_i} \ldots x_{t_j}$ is not transitive, and so the isomorphism can be established with the transitivized of this subgraph.

Finally, if L is a disjoint subvariety with the blowing up center C, then define

$$\Delta_c(L) = x_l.$$

Δ_c is, by construction, an acyclic transitive graph isomorphism and a simple verification allows us to obtain $\sigma_c = \pi_c \circ \Delta_c$. To prove $w'_c = w_c \circ \Delta_c$, it is sufficient to bear in mind lemma 4-1 for the transversal subvarieties (the other cases are trivial). In fact one has

1. if $\dim HK' < \dim T'$, then $w_c \circ \Delta_c(HT') = w_c(x_{hk})$. On the other hand, $\dim HK' = \dim H + \dim K - \dim C - 1 < \dim T$ or $\dim H + \dim K \leq \dim C + \dim T$ and, by the transversality condition, one has $\dim H + \dim K = \dim C + \dim T$, so $w_c(x_{hk}) = \dim T - 1 = w'_c(HT')$;

2. if $\dim HK' \geq \dim T'$, then $w_c \circ \Delta_c(HT') = w_c(x_{ht}) = \dim T - 1 = w'_c(HT')$. $\square$

5   Geometric modification for acyclic graphs

If (X, G, w) is a connected acyclic graph with dominant maximal element and with transversal weight map, its blow up (X_c, G_c, w_c) is a graph of the same kind, so it can be blown up again at any of its points. If the graph (X, G, w) is labeled by the label set $E = \{1, 2, \ldots, n\}$, n being the card(X), so that the graph is naturally ordered by levels, i.e., if $x_i \in N_p, x_j \in N_q$ and $p < q$, then $i < j$, we can blow up successively with center on x_i, from $i = 1$ to $i = n$.

Definition 5-1. *With conditions as above, if the blow up of (X, G, w) with center on x_1 is denoted (X_1, G_1, w_1) and if, for every $i = 2, \ldots, n$, the blow up of $(X_{i-1}, G_{i-1}, w_{i-1})$ with center on x_i is denoted (X_i, G_i, w_i), then the graph (X_n, G_n, w_n) will be called total or geometric blow up of (X, G, w). Each blowing up on x_i has associated the graph morphism $\pi_i \colon (X_i, G_i, w_i) \to (X_{i-1}, G_{i-1}, w_{i-1})$ defined at 3-3; then the morphism*

$$\pi \colon (X_n, G_n, w_n) \to (X, G, w)$$

will be called geometric modification of (X, G, w), where $\pi = \pi_1 \circ \pi_2 \circ \ldots \circ \pi_n$. We will also consider, for every i, with $1 \leq i \leq n$, the partial geometric modification $\pi_1 \circ \pi_2 \circ \ldots \circ \pi_i \colon (X_i, G_i, w_i) \to (X, G, w)$ and we will say that (X_i, G_i, w_i) is the i-th partial geometric blow up of (X, G, w).

These concepts are relative to the ordering of the elements of X. If $x_1, \ldots, x_q$ are consecutive points of level N_0 and $(x_{s(1)}, \ldots, x_{s(q)})$ is a permutation of these points, the qth partial geometric blow up (X_q, G_q, w_q) follows also as the successive blowing ups of the points $x_{s(1)}, \ldots, x_{s(q)}$. Note that, in these blowing ups there is no transversality since B_c is empty.

Once a blowing up with center on a point x_c, with $1 \leq c \leq q$, has been realized the induced subgraph by the graph (X_c, G_c, w_c) on the point set $\{x_{cj} : x_j \in x_c\uparrow\}$ is a transversal structure at the next blowing up of the graph (X_q, G_q, w_q) with center on a point x_{j_1} such that $(x_c, x_{j_1}) \in G$. In this transversal structure, every transversal path $x_{cj_1} \ldots x_{cj_k}$ is connected to the path $x_{j_1} \ldots x_{j_k}$ by the arcs (x_{cj_i}, x_{j_i}), for $1 \leq i \leq k$, and $w(x_{cj_i}) = w(x_{j_i}) - 1$, and then, for every transversal point x_{cj_i} of this path, one has $x_{cj_i} \notin B_c'$. Moreover, for each semicycle formed by the paths $x_{cj_i} x_{cj_{i+1}} x_{j_{i+1}}$ and $x_{cj_i} x_{j_i} x_{j_{i+1}}$, one has $w(x_{cj_{i+1}}) + w(x_{j_i}) = w(x_{cj_i}) + w(x_{j_{i+1}})$. Therefore, at the blowing up of (X_q, G_q, w_q) with center at a point $x_{j_1} \in N_1$, such that $(x_c, x_{j_1}) \in G$, there exist transversal points but both B_c' and T_c are empty.

If $x_{r+1}, \ldots, x_u$ are consecutive points of level N_p and $(x_{s(r+1)}, \ldots, x_{s(u)})$ is a permutation of these points, then, the uth partial geometric blow up (X_u, G_u, w_u) also results from the successive blowing ups at the points

$$x_1, \ldots, x_r, x_{s(r+1)}, \ldots, x_{s(u)}.$$

In consequence, if X is classified by levels, $X = \cup_{p=0}^k N_p$, with $k = \text{height}(X, G)$, the geometric blow up (X_n, G_n, w_n) also results from the successive blowing ups at the points $x_{s(1)}, \ldots, x_{s(n)}$, $(x_{s(1)}, \ldots, x_{s(n)})$ being a permutation of $x_1, \ldots, x_n$ preserving the order of the levels, i.e., if $x_{s(i)} \in N_p, x_{s(j)} \in N_q$ and $p < q$, then $s(i) < s(j)$. In all steps the transversal sets B_c' and T_c are empty and so, after proposition 2.2, we can state:

Theorem 5-2. *If (X, G, w), with $\text{card}(X) = n$, is a connected acyclic graph naturally ordered by levels with dominant maximal element x_z and with transversal weight map, then its geometric blow up (X_n, G_n, w_n) is a connected weighted graph with dominant maximal element x_z and with completely transversal weight map. The weight $w(x_z) = w_n(x_z)$ of the dominant maximal point determines the weight map w_n as follows: if $x \in X_n$, then $w_n(x)$ is equal to $w(x_z)$ minus the length of any path connecting x with x_z. The graphic structure (X_n, G_n, w_n) is, therefore, independent of the weight map w given for the original graph (X, G, w).*

The preceding results present nice applications to geometry in the case that one blows up points succesively by levels in the graph associated to a geometric configuration. In general, one obtains a "cubic" structure on the geometric blow up graph, i.e., one can describe the set of subgraphs which are hypercubes (see [3] for details). Such relationship with geometry and the study of the cubic structure for acyclic graphs will be the subject of a forthcoming paper.

References

[1] Aroca, Hironaka, Vicente. "Desingularization theorems", Mem. Inst. Jorge Juan
     CSIC, vol 30, Madrid, 1974.

[2] Hironaka, H. "Resolution of singularities of an algebraic variety over a field of
     characteristic zero", Ann. Math. vol 79, pp 109–326, 1964.

[3] Marijuán, C. "Una teoría birracional para los grafos acíclicos", Ph.D. Thesis,
     Universidad de Valladolid, 1988.

[4] Villamayor, O. "Constructiveness of Hironaka's resolution", Ann. Sc. Ecole Nor-
     male Superieure 4, serie, t 22, pp 1–32, 1989.

[5] Villamayor, O. "Introduction to the algorithm of resolution", These proceedings.

Address of author:

Escuela Universitaria Politécnica
Francisco Mendizábal, 1
47014 Valladollid
España
Email: marijuan@cpd.uva.es

On a Newton polygon approach to the uniformization of singularities of characteristic p

T. T. Moh

1 Introduction

1.1 Resolution and Uniformization

This paper is a report about some results of uniformization of singularities in characteristic p under the guidance of Prof H. Hironaka.

The problem of resolution of singularities is well known among mathematicians. Briefly, the simplest form is the following:

Basic Problem of Resolution: Given a finitely generated function field $k(x_1, \ldots, x_n)$ over k, does there exist a nonsingular projective model for it?

This existence-type question is very hard to answer in general. Usually we may begin with some singular projective model X for it. Now a refined problem may be formulated as:

Problem of Resolution of Singularities: Assume that k is algebraically closed. Does there exist a non-singular projective model Y and a proper map $F: Y \to X$ such that F is an isomorphism over some open dense subset U of X?

Historically, for curves over the complex numbers $\mathbb{C}$, Kronecker, Max Noether and others have solved the problem of resolution of singularities.

For surfaces over $\mathbb{C}$, the first correct solution to the problem of resolution of singularities was due to R. Walker in 1936.

The great mathematician O. Zariski established a profound way of attacking this problem. Let us introduce his concept of uniformization. As is well known, a Riemann Surface is nothing but the collection of all valuations, with a certain natural topology, of a one-dimensional function field. Zariski generalized the above concept and defined the "Zariski-Riemann Surface" of any function field as the collection of all valuations with a natural topology. Given any projective model X of a function field K, then the Zariski-Riemann Surface of K is quasi-compact and dominates X, i.e., every valuation has a center at X. If X is non-singular, then certainly every valuation has a non-singular center at X. Zariski raised the following infinitesimal problem:

Basic Problem of Uniformization: Given a finitely generated function field K over k and a valuation V of K, does there exist a projective model X, for which the center of V is non-singular?

From 1939 to 1944, Zariski published a sequence of papers [28]–[32] which established among other results the following,

Progress in Mathematics, Vol. 134
© 1996 Birkhäuser Verlag Basel/Switzerland

1. For dimensions 1 or 2, uniformization $\Leftrightarrow$ resolution.

2. If the ground field is of characteristic 0, then uniformization is true for any dimension.

3. If the ground field is algebraically closed and of characteristic 0, then resolution is true for dimension 3.

Zariski's work was obviously a summit of mathematics in the 40's and will be analyzed forever by future mathematicians. We would like to remark an important point in his arguments about uniformization. Firstly, by his reduction to hypersurfaces arguments, we may consider an equation of the form

$$y^m + \sum f_i(x_1, \ldots, x_n) y^{m-i} = 0.$$

Furthermore, by a translation of the form

$$y \longmapsto y + \frac{1}{m} f_1(x_1, \ldots, x_n),$$

we may assume that $f_1 = 0$. We may call this the "killing of f_1" after Abhyankar. Note that from time immemorial this technique has been used to solve a quadratic equation. With it Zariski easily showed that either all coefficients $f_i(x_1, \ldots, x_n)$ are isolated with respect to blow-ups, hence there is a reduction of the number of variables, or there is a reduction of multiplicity m which is certainly an improvement on the singularity.

The above-mentioned technique cannot be used mechanically in uniformization for characteristic $p > 0$. The "killing of f_1" simply does not exist in characteristic p, if the number m in the above translation is zero, or even worse, if f_1 accidentally is zero, Zariski's results cannot be reproduced. We believe that the concept of maximal clean variable (cf. subsections 2.4, 3.1) is the essence in any characteristic of the method of "killing of f_1" of Zariski.

Among Zariski's outstanding students, S. S. Abhyankar and H. Hironaka made prominent contributions to this problem.

In his Ph. D. thesis of 1956, S. Abhyankar used a sequence of going-up and coming-down theorems to reduce the surface uniformization problem to a set of Artin-Schreier type equations

$$z^m + f(x, y) = 0, \qquad p \nmid m,$$
$$z^p + f(x, y) = 0,$$
$$z^p + f(x, y)z + q(x, y) = 0.$$

The middle equation can be considered fundamental. S. Abhyankar found some control terms and proved a descending regularity for them. Thus S. Abhyankar established uniformization for surfaces, and by a previous result of Zariski, resolution for surfaces.

Ten years later Abhyankar sharpened his uniformization algorithm for surfaces and established that for a 3-fold, if the multiplicity is coprime to the characteristic of the ground field, then there is a uniformization. Thus if the multiplicity < the characteristic of the field, the singularity can be resolved. By adopting an Albanese map, which guarantees all multiplicities ≤ (dim)!, Abhyankar established uniformization and then globalized it to a resolution of 3-folds with the exception of $p = 2, 3, 5$ ($\leq 3!$) of course.

In 1964, H. Hironaka published a celebrated paper [12] and solved completely the problem of resolution for any dimensional algebraic variety for characteristic zero. Later on Hironaka, J. M. Aroca and J. U. Vicente completed the resolution for complex analytic spaces. Grothendick once claimed orally that Hironaka's work is the most complicated mathematical work. Only recently Hironaka's work has been analyzed and simplified by O. E. Villamayor [25], E. Bierstone-P.Milman [6] and B. Youssin [27] independently in more constructive ways. Especially, the Newton polygon assume a prominent position in B. Youssin's work.

In Hironaka's method the uniformization is never considered as the first step in the resolution. However, Hironaka, and later Spivakovsky, remarked that Zariski might be right and there should be a way to do uniformization first.

1.2 Uniformization in characteristic p

Let us recall the results of Abhyankar. For the uniformization of 3-folds in characteristic p, after Abhyankar, we have to finish the following cases,

Case	p	multiplicity
(1)	5	$5(= p)$
(2)	3	$3(= p)$
(3)	2	$2(= p)$
(4)	2	$4(= p^2)$

In 1984 at the conference of La Rabida, V. Cossart gave a report on the uniformization of the following equation (his Ph. D. thesis, Orsay 1987),

$$T^p - f(x, y, z) = 0.$$

Although in the above cases (1), (2) and (3), the defining equations may not be purely inseparable equations as above, however, the above equations are indeed the kernel of the problem (see below). Moreover, Cossart's method can be globalized to a resolution. Thus there are resolutions for 3-folds in all characteristic except 2.

In May, 1992, M. Spivakovsky announced a proof of canonical uniformization, and hence a resolution, for any dimension in characteristic p.

1.3   A Newton polygon approach to uniformization

Given a formal power sequence $F(x_1, \ldots, x_n) = \sum a_{i_1 \cdots i_n} x^{i_1} \cdots x^{i_n} = 0$, vanishing at the origin, we define the Newton polygon of F as the convex closure of the set $\mathbf{S} = \{(m_1, \ldots, m_n)$, where $x^{m_1} \cdots x^{m_n}$ is a monomial with non-zero coefficient in the expression of $f(x_1, \ldots, x_n)\}$.

Clearly the above defined Newton polygon depends on the variables x_1, $\ldots$, x_n chosen. We shall select them such that the Newton polygon is one of the best (cf. section 2.1, maximal clean). Then we shall use the particular Newton polygons to deduce a sequence of non-negative integers, the numerical characters, $\mathbf{NC}(F)$. They should satisfy the requirements that,

1. The point is smooth if and only if $\mathbf{NC}(F) = (1)$.

2. $\mathbf{NC}(F)$ should drop lexicograghically under at least one permissible blow-up,

where a blow-up is permissible if and only if the center I satisfies the following requirements:

1. It can be extended to a coordinate system.

2. $\operatorname{ord}_I(F) \geq \operatorname{ord}(F)$.

We may even use numerical characters of different lengths: for numerical characters of different lengths, we fill with zeroes at the end to make them equal length and then order them lexicographically. We shall name the above as the lexicographic order. Traditionally, we start our numerical character with $\operatorname{ord}(F)$ and then add more terms if necessary. Note that the number $\operatorname{ord}(F)$ is stable under a permissible blow-up in the sense that it will not increase under a permissible blow-up. It is easy to find examples to show that $\operatorname{ord}(F)$ might be stationary after a blow-up. So we have to find additional terms for the numerical characters. Suppose we look at a partial numerical character sequence $(a_1, a_2, \ldots, a_m)$, following Hironaka, we wish firstly to show that it is stable under a permissible blow-up in the sense that it will not increase under a permissible blow-up. However, in the characteristic p, the most natural ones might increase (jumping in our terminology, cf. section 3) while the increments will be bounded (cf. 3.2).

Moreover, after every jumping, the resulting equations have some special forms, thus define some special singularities. Henceforth we may substract a number (i.e., the adjustor, cf. 3.4) for those special singularities to eliminate the jumping, or we simply treat the new singularities as different types. Since the increments are bounded, then there will be only finitely many new types if classified by the increments. If after a sequence of permissible blow-ups, the singularity type returned to the original type, and the numerical character will not increase, then we may view the partial numerical characters as stable under the permissible blow-ups. Secondly, we will see if those sequence will drop after the singularities return to their original types under some permissible blow-ups. If they do, then we are done. Otherwise, we have to look for the next terms in the numerical characters.

The above program works for the case $\tau_1 \geq n - 2$ (for the definition of τ_1 the reader is referred to 2.1 and the purely inseparable hypersurface singularities of degree p in 1.2).

2 Newton polygon and uniformization for $\tau_1 \geq n - 1$

2.1 Newton polygons and unformization for $\tau_1 = n$

We will use the elementary method of the Newton polygon to explore the uniformization problems in characteristic p.

Given any power sequence $F(x_1, \ldots, x_n) = 0$, let

$$d_1 = \operatorname{ord} F(x_1, \ldots, x_n).$$

We have the following well known proposition.

Proposition 2-1. *The number d_1 is stable with respect to all permissible blow-ups.*

Let the Newton polygon of F at the origin (associated with the variables $x_1, \ldots, x_n$) be the convex closure of the set $\mathbf{S} = \{(m_1, \ldots, m_n)$, where $x^{m_1} \cdots x^{m_n}$ is a monomial in the expression of $f(x_1, \ldots, x_n)$ with non-zero coefficient$\}$.

After H. Hironaka, we will modify a little the above definition. Let $\mathbf{U}$ be the first quadrant of the real space, $\mathbf{U} = \mathbb{R}^+ \oplus \cdots \oplus \mathbb{R}^+$, and let the Newton polygon $\mathbf{ND}(F, x)$ be the convex closure of the set $\mathbf{S}' = \{(m_1, \ldots, m_n) + \mathbf{U}$, where $x^{m_1} \cdots x^{m_n}$ is a monomial with non-zero coefficient in the expression of $f(x_1, \ldots, x_n)\}$

We should select the variables x such that the leading form of $F(x)$ depends on the least number of variables. In other words, the face which corresponding to the leading form of $F(x)$ of the Newton polygon should lie in the smallest dimensional linear subspace. After Hironaka, this dimension will be called $\tau_1(F) = \tau_1$. If $\tau_1(F) = n$, then we will define the numerical character $\mathbf{NC}(F)$ to be (d_1). It is easy to see that any permissible blow-up will reduce the numerical character $\mathbf{NC}(F)$ of the power sequence $F(x)$. Thus we have the following proposition.

Proposition 2-2. *If $\tau_1(F) = n$, then (d_1) will drop at all resulting points after blowing up the origin.*

Incidentally, in characteristic zero, the linear span of a projective variety can be computed by taking derivatives. For the computation of the linear span (or directrice in the terminology of Giraud) in characteristic p, the reader is refered to two aricles [10], [11] of Giraud.

For the elegance of the numerical characters introduced later, let us introduce the following number of free variable λ_1,

$$\lambda_1 = \tau_1.$$

2.2 The stability of $(d_1, n - \lambda_1, n - \tau_1)$

The stability can be deduced from the following lemma.

Lemma 2-3. (Hironaka) *The vector variable y_2 may not be one dimensional. Let $g(y_2, z_1)$ be homogeneous. Then we have: the minimal number of variables of $g(y_2, z_1) \geq$ the minimal number of variables of $g(0, z_1)$. If they are equal, then all y_2 can be removed from g by a linear transformation $z_1 \mapsto z_1 + ay_2$.*

PROOF. Hironaka proved the above lemma for the characteristic zero case. A modified version of it will establish the general case. □

We have the following proposition,

Proposition 2-4. *The sequence $(d_1, n - \lambda_1, n - \tau_1)$ is stable under any permissible blow-up.*

PROOF. If d_1 decreases, then we are done. Otherwise we use the preceding lemma with $g(y - 2, z_1) =$ the leading form of F. □

2.3 Notations

Let us divide the variables $(x_n, \dots, x_1)$ into blocks in reverse order as follows:

$$\overbrace{x_n, \dots, \underbrace{\overbrace{x_{n_r}, \dots, x_{n_{r-1}+1}}^{y_3} \cdots}_{z_r}, \underbrace{x_{n_2}, \dots, x_{n_1+1}}_{z_2}, \underbrace{x_{n_1}, \dots, x_1}_{z_1}}^{y_2},$$

namely, we let y_1, y_r and z_r be the following vectors of variables.

$$
\begin{aligned}
y_1 &= (x_n, \dots, x_1), \\
y_r &= (x_n, \dots, x_{n_{r-1}+1}), \\
z_r &= (x_{n_r}, \dots, x_{n_{r-1}+1}).
\end{aligned}
$$

We will abuse the above notations and write

$$
\begin{aligned}
(y_2, z_1) &= (x_n, \dots, x_1), \\
y_2 &= (y_3, z_2), \\
&\quad \dots\dots\dots \\
y_r &= (y_{r+1}, z_r).
\end{aligned}
$$

2.4 The presentation of the equation

Let us assume that the vector variable z_1 generates the leading form of $F(y_1)$ minimally. Then we have the following expression with $y_2 = (x_{n_1+1}, \ldots, x_n)$,

$$\begin{aligned} f(y_1) &= \sum f_{j_1}(y_2) z_1^{j_1} \\ &= F^\ell(z_1) + F^s(y_2, z_1) + F^r(y_2, z_1), \end{aligned} \tag{2.1}$$

where

1. $F^\ell(z_1)$ is the leading form of $F(x_n, \ldots, x_1)$,

2. $F^s(y_2, z_1) \in k[\![y_2]\!][z_1]$ with $\deg_{z_1} F^s(y_2, z_1) < d_1$,

3. $F^r(y_2, z_1) \in k[\![y_2, z_1]\!]$ with $d_1 \leq \operatorname{ord}_{z_1} F^r(y_2, z_1)$ and $d_1 < \operatorname{ord} F^r(y_2, z_1)$.

Note that the above decomposition is unique. Moreover the above decomposition can be explicitly written as

$$F^\ell(z_1) = \sum_{|j_1|=d_1} a_{j_1} z_1^{j_1},$$

$$F^s(y_2, z_1) = \sum_{|j_1|<d_1} \widetilde{f}_{j_1}(y_2) z_1^{j_1}, \tag{2.2}$$

$$F^r(y_2, z_1) = \sum_{d_1 \leq |j_1|} \widetilde{f}_{j_1}(y_2) z_1^{j_1}, \tag{2.3}$$

where in (2.3), $|j_1| = d_1 \Rightarrow 0 < \operatorname{ord} \widetilde{f}_{j_1}(y_2)$.

In the above expression (2.2), the coefficients $\widetilde{f}_{j_1}(y_2)$ might have some common monomial factor in y_2. We will only factor out monomials in weight. Then we are forced to factor out monomials in fractional powers. We have

$$F = F^\ell + \sum (y_2^{\alpha_2})^{d_1 - |j_1|} f_{j_1}(y_2) z_1^{j_1} + F^r. \tag{2.4}$$

where (i) α_2 is a non-negative rational vector, (ii) $f_{j_1}(y_2)$ belongs to $k\langle y_2 \rangle$, the Puisseux field of fractional power sequences, (iii) $(y_2^{\alpha_2})^{d_1-|j_1|} f_{j_1}(y_2)$ belongs to $k[\![y_2]\!]$.

2.5 The number d_2 and uniformization for $d_2 = 0$ or ∞

With the variable z_1 selected above, we define the weighted second order d_2 of F as follows,

$$d_2 = \min\left\{ \frac{\operatorname{ord} f_i}{d_1 - |i|} \right\} = \text{ weighted second order of } F.$$

We plan to use the number d_2 in our numerical characters. Note that d_2 are positive rational numbers with denominators bounded by $d_1!$. We may allow such numbers in our numerical characters. We tempt to use $(d_1, n - \lambda_1, n - \tau_1, d_2)$ as the next extension of our numerical characters. However, there are two problems: (i) only the vector space of the leading forms of the vector variable z_1 is determined, we may modify the vector variable z_1 by higher terms, thus the number $|\alpha_2| + d_2$ will be changed accordingly. Let us consider the simple case that $\tau_1 = n - 1$. We shall identify all variables in z_1, i.e., we project the point $(a_1, \ldots, a_{n-1}, a_n)$ in the Newton polygon in $\mathbb{R}^n$ to $(a_1 + \cdots + a_{n-1}, a_n)$ in $\mathbb{R}^2$, we shall select the variable z_1 with the first slope of the resulting Newton polygon to be the maximal one. In fact for a plane algebroid curve, the above requirements will locate the first characteristic pair. In general, if $\tau_1 \leq n - 2$, then the resulting Newton polygon will be of dimension 3 or higher. There will be certainly no first Newton slope to be studied.

However, we may view the requirements on the first Newton polygon differently. Note that the point $(0, d_1)$ is in the image, if we further project with this point as the center to $\mathbb{R}$, then our requirement of the first slope of the resulting Newton polygon to be maximal is equivalent to the requirement that the image of the Newton polygon under the above mentioned projection is minimal. We shall define a vector variable z_1 to be a maximal clean variable for the equation $F = 0$, if after identifying all variables in z_1 and the projection with the point $(0, \ldots, 0, d_1)$ as the center, the image of the Newton polygon of F is set-theoretic minimal. The relation of our maximal clean variable z_1 to Zariski's method of "killing f_1" is as follows, in the case that z_1 is 1-dimensional, "killing f_1" will achieve the maximal clean in one step in characteristic zero. We shall proclaim the following proposition without proof.

Proposition 2-5. *For any power sequence F, there is a maximal clean vector variable z_1.*

(ii) if we allow α_2 to be arbitrary, then the above sequence $(d_1, n - \lambda_1, n - \tau_1, d_2)$ may increase after a permissible blow-up even if the characteristic is 0. In fact in any case we should at least allow $y_2^{\alpha_2}$ to be the exceptional hyperplanes in the weighted sense. In our present paper we will assume that the vector variable y_2 includes all exceptional hyperplanes and furthermore in the above eq. (2.4) we assume that α_2 is the maximal one with the variables y_2 given. Let us remark that in characteristic 0, the sequence $(d_1, n - \lambda_1, n - \tau_1, d_2)$ is stable, while in characteristic p, it is not (cf. section 3).

Let us treat the two simplest cases: $d_2 = 0$ or ∞.

Case 1: Suppose that $d_2 = 0$. In characteristic 0, we have quasi-ordinary singularities. We define the numerical character $\mathbf{NC}(F)$ to be $(d_1, n - \lambda_1, n - \tau_1, 0, |\alpha_2|)$ and a permissible center $\mathbf{CC}(F)$ to be $(z_1, [y_2])$ where $[y_2]$ is a set of variables $(x_{k1}, \ldots, x_{ks})$ such that with the vector $\alpha_2 = (a_i)$, we have (1) $a_{k1} + \cdots + a_{ks} \geq 1$ and (2) s is the smallest number with property (1). Then it is routine to see that the numerical character will drop, if a permissible center is blown up.

Case 2: Suppose that $d_2 = \infty$. Then we have $F^s = 0$ in Eq (2.1), it is easy to see that the numerical character $\mathbf{NC}(F) = (d_1, n - \lambda_1, n - \tau_1)$ will drop if the ideal (z_1) is blown up.

2.6   Uniformization for $\tau_1 = n - 1$

It suffices to observe that in the above eq. (2.1), if $F^s(y_2, z_1) \neq 0$, i.e., $d_2 < \infty$, then $d_2 = 0$, due to the maximality of α_2.

3   Jumping lemma and Uniformization for $\tau_1 = n - 2$

3.1   Presentation of the equation

Recall the following defining equation,

$$F = F^\ell(z_1) + \sum (y_2^{\alpha_2})^{d_1 - |j_1|} f_{j_1}(y_2) z_1^{j_1} + F^r(y_2, z_1), \qquad (3.5)$$

where (1) z_1 is supposed to be a minimal set of generators for the leading form F^ℓ of F and as a vector variable of dimension $n-2$, and (2) y_2 is a vector variable of dimension 2. We will abuse our notation and write $z_1 = z$, $y_1 = (x, y, z)$ and $y_2 = (x, y)$. We shall rewrite the above eq. (2.4) in the following form,

$$F = F^\ell(z) + \sum (x^\alpha y^\beta)^{d_1 - |i|} f_i(x, y) z^i + F^r(x, y, z). \qquad (3.6)$$

In the above equation we shall factor out x and y as much as possible in the weighted sense. Furthermore, we shall modify the vector variable z by

$$z \mapsto z + x^\alpha y^\beta h(x, y) \in k[\![x, y, z]\!],$$

so that the image of the Newton polygon $\mathbf{ND}(F, y_1)$ is minimal under the map σ which will (1) identify all variables in z_1, (2) project the resulting Newton polygon with center $(0, 0, d_1)$ to $\mathbb{R}^2$. Indeed such vector variable z_1 exists and will be named a maximal clean variable. If the characteristic is zero and z_1 is one dimensional, then Zariski's method of "killing f_1" (see 1.1) will achieve a maximal clean variable in one step.

We shall assume that the vector variable z_1 is a maximal clean variable. Let us work on the second tier of Newton polygon. From the above eq. (3.6), we shall consider the union of the Newton polygons of $f_i(x, y)$ in the weighted sense: let

$$S_2 = \bigcup \frac{1}{d_1 - |i|} \mathbf{ND}(f_i, (x, y)),$$

and define $\mathbf{ND_2}(F, y_1)$ to be the convex closure of $\mathbf{S_2}$. Naturally we have

$$
\begin{aligned}
d_1 &= \operatorname{ord} F, \\
d_2 &= \min\left\{\frac{\operatorname{ord} f_i}{d_1 - |i|}\right\} = \text{weighted second order of } F, \\
c_i &= d_2(d_1 - |i|), \\
J &= \text{the leading indices of } F = \{i : \operatorname{ord} f_i = c_i\}.
\end{aligned}
$$

It follows from 2.3 that we shall only consider $0 < d_2 < \infty$. The weighted second leading forms are the following set of homogeneous polynomials,

$$\ell f_2(F, y_1) = \{\text{leading form of } f_i : \text{order of } f_i = c_i\}.$$

Now we may look for a minimal set of variables to generate the weighted second leading form $\ell f_2(F, y_1)$ of F. However, we must respect the outside term $x^\alpha y^\beta$ and proclaim the following fixed variable condition: if α is non-zero, then x is not allowed to be modified, and will be called a fixed variable, if β is non-zero, then y is not allowed to be modified, and will be called a fixed variable. A variable will be called a free variable if it is not fixed.

Let z_2 be a minimal set of variables for $\ell f_2(F, y_1)$ under the above fixed variable condition. Let

$$
\begin{aligned}
\tau_2 &= \tau_1 + \text{ the dimension of } z_2, \\
\lambda_2 &= \lambda_1 + \text{ the number of free variables in } z_2.
\end{aligned}
$$

We may separate the discussions into two cases according to $\tau_2 = n$ or $n - 1$, due to our assumption that $\tau_1 = n - 2$:

Case 1, $\tau_2 = n$. Then $y_1 = (x, y, x_{n-2}, \ldots, x_1) = (x, y, z)$, and our selection of variables is completed. We will complete the presentation as

$$F = F^\ell(z) + \sum (x^\alpha y^\beta)^{d_1 - |i|} f_i(x, y) z^i + F^r(x, y, z), \qquad (3.7)$$

$$f_i(x, y) = f_i^\ell(x, y) + f_i^r(x, y), \qquad (3.8)$$

where

$$
f_i^\ell(x, y) = \begin{cases} \text{the leading form of } f_i(x, y) & \text{if } i \in J, \\ 0 & \text{otherwise.} \end{cases}
$$

The uniformization data in this case will be $\{d_1, \lambda_1, \tau_1, d_2, \lambda_2, \tau_2, \alpha, \beta\}$.

Case 2, $\tau_2 = n - 1$. Then z_2 is one dimensional. Let $z_2 = y$. Then we have $y_1 = (x, y, z)$, $y_2 = (x, y)$, $y_3 = (x)$. Let the presentation be written as

$$F = F^\ell(z) + \sum (x^\alpha y^\beta)^{d_1 - |i|} f_i(x, y) z^i + F^r(x, y, z), \qquad (3.9)$$

$$f_i(x, y) = a_i y^{c_i} + \sum_{j < c_i} (x^{\alpha_3})^{c_i - j} f_{ij}(x) y^j + f_i^r(y). \qquad (3.10)$$

where

$$a_i \quad \begin{cases} \neq 0 & \text{if } i \in J, \\ = 0 & \text{if } i \notin J. \end{cases}$$

Here as usual, we require that α_3 is as large as possible. Due to the fixed variable condition, the variable y is either fixed or free. If it is fixed, then we do not have to do anything and simply declare the variable y to be a maximal clean variable. If it is free, then we may modify y by a transformation $\sigma: y \mapsto y + x^{\alpha_3} h(x) \in k[\![x, y]\!]$ so that the projection with center $(0, d_2)$ of the $\mathbf{ND}_2$ is minimal in $\mathbb{R}$. In this case the variable y with the minimal property will be called a maximal clean variable. We have the following proposition.

Proposition 3-1. *In the above situation, we can select variables z_1 and y such that both variables are maximal clean.*

Let us assume that it is done. Then we will require that α_3 is maximal (cf. 2.5). Then we define d_3 as

$$d_3 = \min\left\{ \frac{\operatorname{ord} f_{ij}}{c_i - j} \right\}.$$

It is easy to see that $d_3 = \infty$ or 0.

Subcase (A): If $d_3 = \infty$, we can further conclude that

$$f_i(x, y) = \delta_i(x, y) y^{c_i} \quad \forall\, i, \qquad \text{where } \delta_i \text{ is a unit for some i.}$$

Due to the maximal property of β we must have $c_i = 0$, i.e., $d_2 = 0$. We are finished by quoting 2.3.

Subcase (B): Let us consider the case that $d_3 = 0$. Then α_3 is finite. Due to the maximality of α_3, we may conclude that

$$f_{ij}(x) = \delta_i(x) \quad \forall\, i, \qquad \text{where } \delta_i \text{ is a unit for some i.}$$

In this subcase our uniformization data will consist of $\{d_1, \lambda_1, \tau_1, d_2, \lambda_2, \tau_2, \alpha, \beta, \alpha_3\}$. Furthermore, it follows from the fact that $a_i y^{c_i}$ must be the sole leading form for some i that we always have $\alpha_3 > 1$.

3.2 The jumping phenomena

As pointed out by the Proposition in 2.5, the sequence $(d_1, n - \lambda_1, n - \tau_1)$ will not increase after a blow-up, if $\tau_1 = n - 1$. In fact, the preceding is true even if $\tau_1 \neq n - 1$. In our present situation, we may consider the extended sequence $(d_1, n - \lambda_1, n - \tau_1, d_2)$. The following example illustrates that if the characteristic is positive, this extended sequence might increase.

Example 3-2. Given an equation

$$F = z^p + x^{3p-2} y(y - 2x + x^3) = 0,$$

it is easy to see that z is a maximal clean variable and the weighted second order $d_2 = 1/p$. After blowing up the origin, let us factor out x, the proper transform at $p_1 = [0, 1, 0]$ is

$$\overline{F} = z^p + x^{2p}\big[(y+1)(y-1) + (y+1)x^2\big] = 0.$$

The variable z is not maximal clean for the above equation. We shall replace z by $z - x^2$, then we get

$$F' = z^p + x^{2p}(y^2 + x^2 + x^2 y).$$

Note that the weighted second order becomes $d_2' = 2/p$ and there is an increment of d_2 by $1/p$.

Let us examine the above example closely. We observe that this jumping of d_2 happens when both x and y are fixed, i.e., both appears in the outside monomial $x^\alpha y^\beta$. After the jumping, the form of the equation becomes better, which can be quantified in the following way,

$$
\begin{aligned}
e_2 &= \min \frac{\operatorname{ord} f_i(0, y)}{d_1 - |i|}, \\
\beta &= 0, \\
e_2 &\leq \frac{d_2 + 1}{p}.
\end{aligned}
$$

The blow-up may be viewed as a transformation between categories of singularities. When some number jumps unexpectedly, the singularity is transformed from the general category to an improved category.

3.2.1 Jumping lemma

For the general situation we have the following jumping lemma ([18], [9]).

Lemma 3-3. Let k be a field of characteristic p, and e be given by $y^m f(y) \in k[y^{p^{e-1}}] \backslash k[y^{p^e}]$ and $0 \neq \lambda \in k$. Let $\deg f(y) = d_2$, $(y+\lambda)^m f(y+\lambda) = \sum a_i y^i$, $c = \min\{i : a_i \neq 0, p^e \nmid i\}$. Then $c \leq d_2 + p^{e-1}$.

PROOF. Let $f(y) = y^n \cdot f^*(y)$ with $f^*(0) \neq 0$. Then we have $p^{e-1} \mid m + n$. Without losing generality we may assume $n = 0$, $p^{e-1} \mid m$, and $f(y) \in k[y^{p^{e-1}}]$. Let us use the p^{e-1}th Hasse derivative $d_y^{(p^{e-1})}$ which is a derivative on $k[y^{p^{e-1}}]$. Thus we get

$$d_y^{(p^{e-1})} = d_{y+\lambda}^{(p^{e-1})}$$

and

$$(y+\lambda)^{m-p^{e-1}}, \; y^{c-p^{e-1}} \mid d_y^{(p^{e-1})}\big((y+\lambda)^m f(y+\lambda)\big).$$

Moreover, the right hand side is a polynomial of degree $\leq d_2 + m - p^{e-1}$. Since $\lambda \neq 0$, $(y+\lambda)^{m-p^{e-1}}$ and $y^{c-p^{e-1}}$ are coprime, the left hand side of the above is divisible by the product of the right hand side of the above. After comparing the degrees of both sides, we have

$$m - p^{e-1} + c - p^{e-1} \le d_2 + m - p^{e-1},$$

or $c \le d_2 + p^{e-1}$. □

For later references we will list the following simple corollary.

Corollary 3-4. *If* $m = 0$ *in the above lemma, then there is no jump, i.e., the inequality of the lemma is replaced by* $c \le \deg f$.

3.2.2 Weak stability for d_2

The example 3-2 indicates a general phenomenon: the jumping happens only if $\tau_2 = n$ and $\lambda_2 = n - 2$, and after the jumping, we always have $e_2 \le d_2 + (1/p)$. Note that $d_2 \le e_2$. In fact we have the following proposition.

Proposition 3-5. *Suppose that we have* $\tau_1 = n - 2$. *Then after blowing up a permissible center, we always have*

$$\left(d_1, n - \lambda_1, n - \tau_1, d_2 + \frac{1}{p} \right) \ge (d_1', n - \lambda_2', n - \tau_2', d_2')$$

where ' *indicates the data for any resulting point after the blow-up.*

We will use the above proposition to define weak stability. We may separate all singularities with $\tau_1 = n - 2$ into two types: the general type and the improved type after a jumping. Let us consider the case that d_2 is large, i.e., $d_2 > (1/p)$. Then clearly we can not factor out y, i.e., $v(y) > v(x)$, after we factor out x a finitely many times. It is not hard to see that the number d_2 must be restored at that moment. If the new leading forms depend on y only, we may still consider it as of improved type. Otherwise a factorization of x will decrease the number d_2, while factoring out y will at least cut $d_2 + (1/p)$ in half which is less than d_2.

3.3 Five types of singularities

To complete a detailed uniformization, for the case $\tau_1 = n - 2$, we have the five types of singularities in figure 3.1.

The types N_1 and E_1 are considered to be general, while the types N_2 and E_2 are considered to be improved and better. A blow-up will not only change the numerical data $(d_1, n - \lambda_1, n - \tau_1, d_2, n - \lambda_2, n - \tau_2, \ldots)$, but also change the types of the singularities. To smooth out our arguments, let us introduce the adjustor in the following subsection.

$N \quad \left(d_2 > \frac{1}{p} \right)$	
Type	Conditions
N_1	if (1) $\beta \neq 0$ or (2) $\beta = 0$ and $e_2 > d_2 + \frac{1}{p}$
N_2	if (3) $\beta = 0$ and $e_2 \leq d_2 + \frac{1}{p}$

$E \quad \left(d_2 \leq \frac{1}{p} \right)$	
Type	Conditions
E_0	if $d_2 = 0$
E_1	if (1) $0 < d_2$ and $\beta \neq 0$ or (2) $0 < d_2 < e_2$ and $\beta = 0$
E_2	if (3) $0 < d_2 = e_2$ and $\beta = 0$

Figure 3.1: Singularities for the case $\tau_1 = n - 2$.

3.4 Adjustor Δ_2

We will assign numbers to quantify all singularity types as follows. The name adjustor was given by H. Hironaka. We define the adjustor $\Delta_2(F)$ to be

$$\Delta_2(F) = \min \begin{cases} \delta + \varepsilon & \text{for type } N_1, \\ \varepsilon & \text{for type } E_1, \\ -d_2 & \text{if } d_2 < 1 \text{ and } \alpha, \beta \in \mathbb{Z}^+, \\ 0 & \text{otherwise}, \end{cases} \tag{3.11}$$

where $\delta = 1/p$, $\varepsilon = 1/(2d_1!)$ in the present section.

Remark 3-6. *We will explain the selection of the value for ε. Note that $d_1!$ serves as a common denominator for all rational values d_2, e_2 etc. Hence we have, for instance*

$$d_2 < e_2 \Rightarrow d_2 + \varepsilon < e_2.$$

Remark 3-7. *The numbers λ, ε may take different values for later uniformizations.*

3.5 Modified numerical character $\overline{\mathbf{NC}}(F)$

Given an equation $F = 0$, let z and y, if possible, be maximal clean variables; we define the modified numerical character $\overline{\mathbf{NC}}(F)$, to be

$$\overline{\mathbf{NC}}(F) = \begin{cases} (d_1) & \text{if } \tau_1 = n, \\ (d_1) & \text{if } \tau_1 = n - 1, \ d_2 = \infty, \\ (d_1, n - \lambda_1, n - \tau_1, 0, \alpha_2) & \text{if } \tau_1 = n - 1, \ d_2 = 0, \\ (d_1, n - \lambda_1, n - \tau_1, 0, \alpha + \beta) & \text{if } \tau_1 = n - 2, \ d_2 = 0, \\ (d_1, n - \lambda_1, n - \tau_1, d_2 + \Delta_2, \\ \quad n - \lambda_2, n - \tau_2, 0, \alpha_3, \alpha + \beta) & \text{otherwise}. \end{cases}$$

Condition	$\overline{\mathbf{CC}}(F)$
$\tau_1 = n$	(z_1)
$\tau_1 = n-1,\ \alpha_2 = \infty$	(z_1)
$d_2 \leq 1,\ \alpha \geq 1$	(x, z)
$d_2 \leq 1,\ \alpha < 1$ and $\beta \geq 1$	(y, z)
otherwise	(x, y, z)

Figure 3.2: Modified centers.

A presentation of F is said to be minimal if the corresponding $\overline{\mathbf{NC}}(F)$ is minimal lexicographically among all possible modified numerical characters. We will call the corresponding modified numerical character the minimal modified numerical character of the singularity.

3.6 Modified centers $\overline{\mathbf{CC}}(F)$

See figure 3.2 for a list of modified centers.

3.7 Uniformization for $\tau_1 \geq n-2$

We will establish that the modified numerical character will drop after blowing up the modified canonical center. In fact, we have the following proposition.

Proposition 3-8. *Present a hypersurface singularity $F = 0$ by any maximal clean variable y_1. Suppose that $\tau_1 \geq n-2$. If we blow up the modified center, then the modified numerical character $\overline{\mathbf{NC}}(F)$ drop strictly at all resulting points.*

PROOF. It is a case by case direct computation of some ten pages. □

Applying the above proposition to the minimal modified numerical character of a hypersurface singularity, we have the following proposition.

Proposition 3-9. *If the leading form of a power series F requires at least $n-2$ variables, then there is a sequence of blow-ups such that after a finite steps all resulting points have smaller multiplicities.*

Note that the above proposition implies an uniformization for surface singularities for any characteristic.

4   The classification of 3-dimensional singularities and uniformization for $\lambda_2 \geq 3$ or $\lambda_2 = 2, \pi_2^* \geq 2$

4.1   Introduction

The singularities with $\tau_1 \leq n-3$ are very complicate. Note that the dimension of the hypersurface must be bigger than or equal to three. We shall restrict our attention to the simplest case of dimension 3. On the other hand, due to the previous discussions, for the hypersurface singularities of dimension 3, we shall only consider the case that $\tau_1 \leq n-3$, in other words, $\tau_1 = 1$. Let the variable for the leading form be T. Then our equation $F = 0$ can be written in one of the following forms,

$$F = T^p - x^{\ell p}y^{mp}z^{np}f_0(x,y,z), \qquad (4.12)$$

$$F = T^{p^r} - x^{\ell p^r}y^{mp^r}z^{np^r}f_0(x,y,z), \qquad (4.13)$$

$$F = T^{d_1} + \sum (x^\alpha y^\beta z^\gamma)^{d_1-j_1}f_{j_1}(x,y,z)T^{j_1} + F^r. \qquad (4.14)$$

From our experiences, the above eq. (4.12) is fundamental. An thoroughly understanding of this case will help us to give a solution to eq. (4.13) which will yield a solution to the general eq. (4.14). The rest of this article will be devoted to a description of a general solution to eq. (4.12), and sometimes with explicit references, the solutions to some special cases of eq. (4.14), using the Newton polygons.

To smooth out our notations, we shall replace eq. (4.12) by the following equation

$$F = T^p - x^{\ell p}y^{mp}z^{np}\big(f_{0pd_2}(x,y,z) + \cdots\big), \qquad (4.15)$$

where T is a clean variable, i.e., the leading form of $x^{\ell p}y^{mp}z^{np}f_0(x,y,z)$ and all front of the Newton polygon are not pth powers, pd_2 is the order of f_0, and f_{0pd_2} is the leading form of f_0. Note that if the number $d_2 = 0$ or ∞, then the uniformization has been established (cf. 2.5). Thus we may assume that $0 < d_2 < \infty$.

Recall that the number λ_2 is defined as follows (see 3.1),

$$\lambda_2 = \big(\text{minimal number of free variables of } x^{\ell p}y^{mp}z^{np}f_{0pd2}(x,y,z)\big) + 1,$$

where the term free variable means a variable which appears in $f_{0pd_2}(x,y,z)$ while absent from the monomial part. The very first classification of the singularities is by the numbers τ_2 and λ_2. In this section we will establish the uniformization for the case $\tau_2 \geq \lambda_2 \geq 3$ or $\tau_2 \geq \lambda_2 = 2, \pi_2^* \geq 2$ (for the definition of π_2^*, the reader is refered to 4.3). In section 5, we will establish the same proposition for the case $\lambda_2 = 2, \pi_2^* = 1$. In section 6 we will establish the same proposition for the case $\lambda_2 = 1$.

4.1.1 Leading form jumping or no jumping

After blowing up, let us select the variable T to remove all pth power terms, i.e., let the variable T be a maximal clean variable. If the old leading form f_{0pd_2} still contribute some terms of degree $\leq pd_2$, then we say that there is no leading form jumping. In fact all discussions in section 4 are about no leading form jumping. Otherwise there is a leading form jumping, which will the main topics in sections 5 and 6. In the leading form no jumping cases, the number d_2 is of course stable. Although in higher dimensional cases, even if there is no leading form jumping, the sequence $(d_1, n - \lambda_1, n - \tau_1, d_2, n - \lambda_2)$ is not stable, however under the restriction of this section, we will realize that the above sequence is indeed stable.

4.2 Uniformization for the general Eq. (4.14) for $\lambda_2 \geq 3$

We have the following easy proposition.

Proposition 4-1. Given a general eq. (4.14), if $\lambda_2 = 4$, then after any blow-up, the numerical character $(d_1, n - \lambda_1, n - \tau_1, d_2)$ will drop.

PROOF. Note that under the present condition, the outside monomial $x^\alpha y^\beta z^\gamma$ is void, and the set of leading form $\{f_{j_1} : j_1 \epsilon J\}$ alwaly requires 3 variables x, y, z. We will be allowed to form linear transformations among x, y, z. Thus we may assume that the values of x, y, z under the given valuation are all distinct. If T assumes the minimal value, then the new equation will be a monomial times a unit and the numerical character will be (d_1'). If not, it is easy to see the numerical character will drop. □

The case $\lambda_2 = 3$ is only slightly more complicated.

Proposition 4-2. Given a general eq. (4.14), if $\lambda_2 = 3$ and $\tau_2 = 4$, then after any blow-up, the numerical character $(d_1, n - \lambda_1, n - \tau_1, d_2)$ will drop.

PROOF. In this case we may assume that $\alpha \neq 0, \beta = \gamma = 0$. We always factor out x if its value is among the smallest with respect to the valuation. Then we may even assume that there is no translation. Our proposition follows routinely. □

If $\lambda_2 = 3$ and $\tau_2 = 3$, then the above eq. (4.14) may be rewritten as follows,

$$F = T^p - \sum (x^\alpha)^{d_1 - j_1}(f_{j_1}(x,y,z)T^{j_1} + F^r, \tag{4.16}$$

$$f_{j_1} = f_{j_1}^\ell(y,z) + \sum (x^{\alpha_3})^{c_i - |j_2|} f_{j_2}(y,z) + f_{j_1}^r(x,y,z), \tag{4.17}$$

with T maximal clean. Moreover, the variables y, z will be maximal clean for Eq (4.17), i.e., the number α_3 is the maximum possible one under all transformations of the following type,

$$(y,z) \mapsto (y,z) + x^\alpha\big(h_1(x,y,z), h_2(x,y,z)\big).$$

We have the following proposition.

Proposition 4-3. *Given a general eq. (4.14), if $\lambda_2 = 3$ and $\tau_2 = 3$, then after any blow-up, the numerical character $(d_1, n - \lambda_1, n - \tau_1, d_2, n - \lambda_2, n - \tau_2, 0, \alpha_3)$ will drop.*

PROOF. If x does not assume the minimal value for the valuation, then it is easy to see that the sequence $(d_1, n - \lambda_1, n - \tau_1, d_2)$ will drop. If x assume the minimal value, then it is not hard to see the sequence $(d_1, n - \lambda_1, n - \tau_1, d_2, n - \lambda_2, n - \tau_2)$ will not increase. Assume that it is stable after the blow-up, then we may prove that α_3 will decrease.   □

Remark 4-4. *For the proofs of the preceding three propositions, we first establish that the corresponding sequences are stable.*

4.3   A classification for the purely inseparable equation. (4.12) for $\lambda_2 = 2$ by π_2^*

In many cases, the following is fulfilled for the purely inseparable eq. (4.15) for the free variable z,

$$F \;=\; T^p - x^{\ell p} y^{mp} \left(f_{0 p d_2}(x, y, z) \right.$$
$$\left. + \sum (x^{\alpha_3})^{p d_2 - j_2} f_{0 j_2}(y, z) + f_0^r \right), \quad (4.18)$$

$$f_{0 p d_2}^{\ell}(x, y, z) \;\notin\; k[x, y, z^p]. \quad (4.19)$$

In general we have the following definition.

Definition 4-5. *Let us consider eq. (4.18) with $\lambda_2 = 2$. Let the numerical datum ρ, and π_2^* be defined with respect to the free variable z as follows (with the convention $\deg 0 = -1$):*

$$\rho = \deg_z \left(\frac{\partial}{\partial z} f_0^{\ell}(x, y, z) \right) + 1,$$

and $\pi_2^* = 1 +$ minimal number of variables of $\frac{\partial}{\partial z} f_0^{\ell}(x, y, z)$ subject to the fixed variable conditions of $x^{\ell} y^m z^n$.

Remark 4-6.

1. $\rho = 0 \Leftrightarrow \pi_2^* = 1$.

2. *the condition of eq. (4.19) is satisfied $\Leftrightarrow 1 \le \rho \le p d_2 \Leftrightarrow \pi_2^* \ge 2$.*

3. $\pi_2^* = 2 \Leftrightarrow$ *for some $c \ne 0$ either (a) $\rho = p d_2 (\not\equiv 0 \pmod{p})$ and $f^{\ell} - c z^{p d_2} \in k[x, y, z^p]$ or (b) $f^{\ell} - c x^{p d_2 - 1} z \in k[x, y, z^p]$ or (c) $f^{\ell} - c y^{p d_2 - 1} z \in k[x, y, z^p]$.*

4. $0 < \rho < p d_2 \Rightarrow \pi_2^* \ge 3$.

5. ρ is invariant under all free variable transformation $\sigma: z \to z + h(x, y)$.

6. $\tau_2 \geq \pi_2^*$.

The following example is enlightening about the possible role played by π_2^*.

Example 4-7. Let F be given as

$$F = T^p + x^{p-2}y(z^{p+1} + (y - 2x)z^p + x^{3p+2}). \tag{4.20}$$

Let us blow up (x, y, z, T) and consider a valuation v with $v(x) < v(y + x)$, $v(z)$, $v(T)$. At the new point p_1, we have then

$$F' = T^p + x^p\big[(y + 1)z^{p+1} + (y + 1)(y - 1)z^p + x^{2p+1})\big],$$

and we have to use a transformation σ of the following form to find a maximal clean variable T,

$$\sigma: T \to T - xz.$$

Then we have the following equation

$$\sigma F' = T^p + x^p[z^{p+1} + y^2 z^p + x^{2p+1} + yz^{p+1}]. \tag{4.21}$$

Observe the following transformation of the numerical characters. Original:

$$(d_1, n - \lambda_1, n - \tau_1, d_2, n - \lambda_2, n - \tau_2) = \left(p, 3, 3, \frac{p+1}{p}, 2, 0\right).$$

New:

$$(d_1, n - \lambda_1, n - \tau_1, d_2, n - \lambda_2, n - \tau_2) = \left(p, 3, 3, \frac{p+1}{p}, 2, 2\right).$$

It is not hard to conclude that the coefficient $(y - 2x)$ of z^p destroys the non-increasing property of $(d_1, n - \lambda_1, n - \tau_1, d_2, n - \lambda_2, n - \tau_2)$. The interesting thing to observe is that although, $n - \tau_2$ does increase, $n - \pi_2^*$ ($= 2$) is stable.

We will further classify the equations by the constant π_2^*, its value will indicate how the free variable z appears in the leading form of f_0. If $\pi_2^* = 1$, then all terms involve z to some pth powers. If $\pi_2^* = 2, 3, 4$, then some term will involve z to a non-pth power. It turns out to be fairly easy for the later case, which we will discuss in this section.

4.4   Uniformization for $\lambda_2 = 2$, $\tau_2 \geq 3$ and $\pi_2^* \geq 2$

We have the following two cases: $\tau_2 = 3, 4$.

4.4.1   Uniformization for $\lambda_2 = 2$, $\tau_2 = 4$ and $\pi_2^* \geq 2$

In fact we have the following propositions.

Proposition 4-8. *Assume that $\lambda_2 = 2$, τ_2 and $\pi_2^* \geq 2$. After blowing up the permissible center (x, y, z, T), let the new equation be F' and let us denote the new uniformization data with $'$. We have*

$$(d_1, n - \lambda_1, n - \tau_1, d_2, n - \lambda_2) > (d_1', n - \lambda_1', n - \tau_1', d_2', n - \lambda_2').$$

PROOF. Let us rewrite eq. (4.12) below,

$$F = T^p + (x^\ell y^m)^p f_0(x, y, z), \tag{4.22}$$
$$f_0 = f_0^\ell(x, y, z) + f_0^r. \tag{4.23}$$

For the valuations with $v(T)$, $v(z) < v(y)$, $v(x)$ or $v(T)$ minimal, the arguments are routine. Let us assume that $v(x)$ or $v(y) =$ minimal value. We may assume that $v(x) =$ minimal. There will be a number μ_2 such that $v(z - \mu_2 x) > v(x)$. If $\mu_2 \neq 0$, then a free variable transformation σ of the form $z \mapsto z + \mu_2 x$ will reduce μ_2 to 0, and maximal clean stays. It will be thus assumed. Moreover there is a number μ_3 such that $v(y - \mu_3 x) > v(x)$. If $\mu_3 = 0$, then this is a "blow-up without translation", and it is easy to see that we must have a drop of $(d_1, n - \lambda_1, n - \tau_1, d_2)$. We are finished in this case. Suppose that $\mu_3 \neq 0$. Since $\pi_2^* \geq 3$, then in the following expression of f_0^ℓ,

$$f_0^\ell(x, y, z) = \sum f_{0j}^\ell(x, y) z^j,$$

there must be a term z^j with non-zero coefficient and $p \nmid j$. Under the blow-up, the particular term will be transformed to

$$(y + \mu_3)^{mp} f_{0j}^\ell(1, y + \mu_3) z^j.$$

Due to the non-pth power of the term z^j, this term can not be cleaned away by changing T. Either this term has an order $< pd_2$ or y turns to a free variable and thus $\lambda_2' \geq 3$. Hence we are done in this case. $\qquad\square$

4.4.2   Uniformization for $\lambda_2 = 2$, $\tau_2 = 3$ and $\pi_2^* \geq 2$

As we observe that $\tau_2 \geq \pi_2^*$, we must have $\pi_2^* = 3$. Let z be the free variable and y be the fixed variable. We have the following equation with j vectors,

$$F = T^p + (x^\ell y^m)^p f_0(x, y, z), \tag{4.24}$$
$$f_0 = f_0^\ell(y, z) + \sum (x^{\alpha_3})^{pd_2 - |j|} f_{0j}(x)(y, z)^j + f_0^r. \tag{4.25}$$

Let us define d_3 as follows,

$$d_3 = \min\left\{ \frac{\text{ord } f_{0j_2}}{pd_2 - j_2} \right\}.$$

As usual we require that the variables y, z are selected with $\alpha_3 + d_3$ maximum possible and the variable T is selected with d_2 maximum possible. We name the variables y, z, T maximal clean. We proclaim the following proposition,

Proposition 4-9. *The maximal clean variables* y, z, T *exist.*

Furthermore we require that α_3 is maximum possible. As usual we have either some f_{0j} is a unit (i.e., $d_3 = 0$) or all f_{0j} are zeros (i.e., $d_3 = \infty$ and we define $\alpha_3 = \infty$). Then we define the ideal $\mathbf{CC}(F)$ as follows,

$$\mathbf{CC}(F) = \begin{cases} (x, y, z, T) & \text{if } d_3 = 0, \\ (y, z, T) & \text{if } d_3 = \infty \text{ and } p \leq (m + d_2)p, \\ (x, y, z, T) & \text{if } d_3 = \infty \text{ but } p \not\leq (m + d_2)p. \end{cases}$$

We have the following proposition.

Proposition 4-10. *Assume that* $\lambda_2 = 2$, $\pi_2^* \geq 3$ *and* $\tau_2 = 3$. *Then after blowing up the ideal* $\mathbf{CC}(F)$ *as defined above, we always have,*

$$(d_1, n - \lambda_1, n - \tau_1, d_2, n - \lambda_2, n - \tau_2, 0, \alpha_3) > \\ (d_1', n - \lambda_1', n - \tau_1', d_2', n - \lambda_2', n - \tau_2', 0, \alpha_3')$$

PROOF. It is a detailed analysis as in the proof of the above proposition and is omitted. □

Remark 4-11. *Since we always have* $\tau_2 \geq \pi_2^*$, *then the dicussions in 4.4.1 and 4.4.2 establish the unformization for the case* $\pi_2^* \geq 3$.

4.5 Uniformization for $\lambda_2 = 2$ and $\pi_2^* = 2$

By the materials in 4.4.1 and 4.4.2, we shall only consider the case $\tau_2 = 2$. It is not hard to see (cf. Remark 4-6, 3 in 4.3) that $\rho = pd_2 \neq 0 \pmod{p}$. We have the following expression of our defining equation,

$$F = T^p + (x^\ell y^m)^p f_0(x, y, z), \tag{4.26}$$

$$f_0 = z^{pd_2} + \sum (x^\alpha y^\beta)^{pd_2 - j_2} f_{0j_2}(x, y) z^{j_2} + f_0^r(x, y, z). \tag{4.27}$$

We will require that the variable z is maximal clean for eq. (4.27) ,i.e., the projection of $\mathbb{R}^3$ to $\mathbb{R}^2$ with center $(0, 0, pd_2)$ of the Newton polygon of f_0 is the minimal one. Indeed, such maximal clean variable z exists. Or we may follow Zariski and simply require f_{0pd_2-1} to be zero to find a maximal clean variable.

Because $\rho = pd_2 \neq 0 \pmod{p}$, it is not hard to see that there is no leading form jumping, especially no jumping of d_2. However there are two further complications: one involves the concept of resolvers (cf. 4.5.1) and another involves the concept of double cleaning (cf. 4.5.3).

We separate our discussions into two cases:

1. $\rho = pd_2 \neq 0, 1 \pmod{p}$.

2. $\rho = pd_2 \equiv 1 \pmod{p}$.

4.5.1   Resolvers

Let us consider the case $\rho = pd_2 \not\equiv 0, 1 \pmod{p}$. Then the uniformization for equations (4.26) and (4.27) is very much like the uniformization for algebraic surface (cf. section 3). However, there is a complication when $d_3 = 0$. Let us use the following example to illustrate the concept of resolvers.

Example 4-12. Let the equation $F = 0$ be as follows with $p = 5$,

$$F = T^5 + x^3 y^2 (z^2 + xy^4 \delta).$$

It is easy to see that the ideals (x, z, T) and (y, z, T) are not permissible for the blow-up. We have to blow up the ideal (x, y, z, T). Let us consider the valuation with $v(x) < v(y), v(z), v(T)$. Then the proper transform F' is

$$F' = T^5 + x^2 y^2 (z^2 + x^3 y^4 \delta').$$

There is a decrease of ℓ from 3 to 2 and an increase of α from 1 to 3. It is not hard to construct an example to show the opposite phenomenon.

The couples (ℓ, m) and (α, β) should be balanced against each other in the theory of uniformization of singularities. We will introduce the concept of resolvers as follows.

In general,

$$
\begin{aligned}
y_1^{\alpha_1} &= 1, \\
y_i^{\alpha_i} &= y_{r+1}^{\beta_i} z_r^{\beta_{i,r}} \dots z_i^{\beta_{ii}} \qquad \text{for } i = 2, \dots, r, \\
y_{r+1}^{\alpha_{r+1}} &= y_{r+1}^{\beta_{r+1}}.
\end{aligned}
$$

In our special case with $d_3 = 0$, we have:

$$
\begin{aligned}
y_1^{\alpha_1} &= 1, \\
y_2^{\alpha_2} &= x^\ell y^m z^0 \text{ with } \beta_2 = (\ell, m), \\
y_3^{\alpha_3} &= x^\alpha y^\beta \text{ with } \beta_3 = (\alpha, \beta).
\end{aligned}
$$

We define the type of the system of equations (4.26) and (4.27) as a sequence (t_1, t_2, t_3) where

$$
t_i = \begin{cases} I & \text{if } d_i + |\alpha_i| - |\beta_i| \geq 1, \\ II & \text{if } d_i + |\alpha_i| - |\beta_i| < 1. \end{cases}
$$

We define two sets $T^\circ(I)$ and $T^\circ(II)$ by

$$
\begin{aligned}
T^\circ(I) &= \{i : t_i = I\}, \\
T^\circ(II) &= \{i : t_i = II\}.
\end{aligned}
$$

Note that $1 \in T^{\circ}(I)$, and $3 \in T^{\circ}(II)$ if $d_3 = 0$. We use the following notation for $i \in T(II)$

$$\bar{\beta}_i = \frac{1}{1 - d_i - |\alpha_i| + |\beta_i|} \beta_i.$$

Note that we always have

$$1 < d_i + |\alpha_i|, \qquad 1 < |\bar{\beta}_i|.$$

For our purposes we define the first resolver $R_1(F)$ as the following equation

$$R_1(F) = \sum_{i \in T^{\circ}(II)} y_{r+1}^{\bar{\beta}_i} = 0. \tag{4.28}$$

Note that $1 < |\bar{\beta}_i|$. We may rewrite

$$R_1(F) = y_{r+1}^{\alpha_1^{(1)}} F^{(1)}, \tag{4.29}$$

where $\alpha_1^{(1)}$ is taken to be maximal. Let us consider the equation $F^{(1)} = 0$ with all variables fixed. In our present case, $F^{(1)}$ is either a unit or a binomial with positive order. Let us define the numerical character $\mathbf{NC}(F)$ as follows. If $F^{(1)}$ is a unit, then we add a tail part $(0, |\alpha_1^{(1)}|)$ to the numerical character $(d_1, n - \lambda_1, n - \tau_1, d_2, n - \lambda_2, n - \tau_2)$. If it is a binomial with positive order, then we add a tail part, the numerical character of $F^{(1)}$, to the above numerical character. The permissible center $\mathbf{CC}(F)$, will be (T, z) if $F^{(1)}$ is a unit, and (T, z, y, x) otherwise. Let us look at our previous example again.

Example 4-13. (Example 4-7 continued) In the example above we will compute the resolver $R_1(F)$, the associate equation $F^{(1)}$ and $\mathbf{NC}(F)$,

$$\begin{aligned}
R_1(F) &= x^5 y^{\frac{10}{3}} + xy^4 = xy^{\frac{10}{3}}(x^4 + y^{\frac{2}{3}}) \\
F^{(1)} &= x^4 + y^{\frac{2}{3}}, \\
\mathbf{NC}(F) &= \left(5, 3, 3, \frac{2}{5}, 2, 0, \frac{2}{3}, 1, 1, 6\right).
\end{aligned}$$

A computation for F' will produce the following data,

$$\begin{aligned}
F' &= T^5 + x^2 y^2 (z^2 + x^3 y^4 \delta'), \\
R_1(F') &= x^{\frac{10}{3}} y^{\frac{10}{3}} + x^3 y^4 = x^3 y^{\frac{10}{3}}(x^{\frac{1}{3}} + y^{\frac{2}{3}}), \\
F'^{(1)} &= x^{\frac{1}{3}} + y^{\frac{2}{3}}, \\
\mathbf{NC}(F') &= \left(5, 3, 3, \frac{2}{5}, 2, 0, \frac{1}{3}, 1, 1, 2\right).
\end{aligned}$$

Indeed there is a drop of the numerical character $\mathbf{NC}$.

4.5.2 Uniformization for $pd_2 \not\equiv 0,1 \pmod p$

The uniformization for this case is very much like the process of section 3. We will select z such that the coefficient f_{0pd_2-1} is zero, i.e., we use Zariski's method of "killing f_1" to achieve a maximal clean variable z. We will specify the numerical character $\mathbf{NC}(F)$ and the permissible center $\mathbf{CC}(F)$ as follows.

$$\mathbf{NC}(F) = \begin{cases} cf.4.5.1 & \text{if } d_3 = 0, \\ (d_1, n - \lambda_1, n - \tau_1, \overline{\mathbf{NC}}(f_0)) & \text{otherwise,} \end{cases}$$

$$\mathbf{CC}(F) = \begin{cases} cf.4.5.1 & \text{if } d_3 = 0, \\ (T, \overline{\mathbf{CC}}(f_0)) & \text{otherwise.} \end{cases}$$

For the definitions of $\overline{\mathbf{NC}}(f_0)$ and $\overline{\mathbf{CC}}(f_0)$, the reader is referred to 3.5 and 3.6. We will state the following proposition without proof.

Proposition 4-14. If $\lambda_2 = 2$, $\pi_2^* = 2$ and $\rho = pd_2 \not\equiv 0,1 \pmod p$, then after we blow up the permissible center $\mathbf{CC}$, the numerical character $\mathbf{NC}$ will drop at every resulting point.

4.5.3 Double cleaning

Note that we always define the concept of maximal clean by the minimal property of some projections of the Newton polygon of an equation. For practical purposes as in the preceding subsection, we simply change one set of variables to achieve that maximal clean property. However, in general we have to change several sets of variables to achieve the maximal clean property. The following example is illustrative.

Example 4-15. Let an equation $F = 0$ be defined as

$$\begin{aligned} F &= T^p + x^{p-1}f_0, \\ f_0 &= z^{p+1} + 0z^p + x^p z + x^{p+1} + x^{p+2} + y^{p+2}. \end{aligned}$$

Each of the above equations appears to be cleaned. Let a transformation σ be given as

$$\sigma: \begin{cases} T &\longrightarrow T - xz, \\ z &\longrightarrow z + x. \end{cases}$$

Then we have

$$\sigma F = T^p + x^{p-1}(z^{p+1} + 0z^p + 0z + x^{p+2} + y^{p+2}).$$

We notice that not only the coefficient of z^p may be zero, but also sometimes the leading coefficient of z, if p-th power, can be removed.

The minimal property of the projection of the Newton polygon is achieved by a double clean of changing both T and z. Note that we have the following simple lemma,

Lemma 4-16. *If* $pd_2 \equiv 1 \pmod{p^e}$, $pd_2 \not\equiv 1 \pmod{p^{e+1}}$, *then the expansion of* $(z+h)^{pd_2}$ *is of the following form*

$$(z+h)^{pd_2} = z^{pd_2} + hz^{pd_2-1} + ch^{p^e}z^{pd_2-p^e} + \text{ lower terms,}$$

where $c = (pd_2 - 1)/(p^e) \not\equiv o \pmod{p}$.

PROOF. Trivial. □

Recall the following presentation of $F = 0$,

$$F = T^p + (x^\ell y^m)^p f_0(x,y,z), \tag{4.30}$$

$$f_0 = z^{pd_2} + \sum (x^\alpha y^\beta)^{j_2} f_{0j_2}(x,y) z^{pd_2-j_2} + f_0^r. \tag{4.31}$$

Using the property $pd_2 \neq 0$, we may assume that

$$f_{0pd_2-1} = 0. \tag{4.32}$$

However as indicated by previous Example, the above requirement (4.32) may not be the best possible one. One of the significations of the above Lemma is that the terms of the coefficient of $z^{pd_2-p^e}$ may be cleaned out piecewise. Thus we may request that if a term $x^{rp^e}y^{sp^e}$ of the coefficient of $z^{pd_2-p^e}$ satisfies the following conditions be cleaned away.

1. $r + \alpha, s + \beta, \ell, m$ are integers.

2. If $x^{rp^e}y^{sp^e}$ corresponds to a vertex of the projection with center $(0,0,0,d_1)$ of the Newton diagram of F and there is no term from the coefficient of z^{pd_2-j} for some $j < p^e$ corresponding to the same vertex.

We have the following understanding of the term "maximal clean" in this case,

Definition 4-17. *Given presentations (4.25) and (4.26), with* $\rho = pd_2 \equiv 1 \pmod{p}$. *If* $f_{0pd_2-1} = 0$, *moreover, there are no terms in* f_{0p^e} *that satisfy the above conditions 1 and 2. Then the equation is said to be maximal cleaned.*

We can show that that any equation can be maximal cleaned. In a process of blow-up, before the type of the equation is improved to a previous type, we have to double clean the equation. The following lemma will put a bound on the increment of d_3.

Lemma 4-18. (Jumping Lemma) *Suppose that* $y^\beta g(y) \notin y^{m'}k[y^{p^r}]$ *where* $1 \leq r$. *Then we always have, for arbitrary* $h(y)$ *and* $\mu \neq 0$, *the following*

$$\text{ord}\left\{(y+\mu)^\beta g(y+\mu) - (y+\mu)^{m'}h\left((y+\mu)^{p^r}\right)\right\} \leq \deg(g(y) + p^{r-1}).$$

PROOF. Let m^* be a number such that $m^* + m' \equiv 0 \pmod{p^r}$. Then we have

$$y^\beta g(y) \notin y^{m'}k[y^{p^r}] \Leftrightarrow y^{m^*+\beta}g(y) \notin k[y^{p^r}].$$

After multiplying the formula in the statement by y^*, the lemma follows from the Jumping Lemma of 3.2.1. □

4.5.4 Uniformization for $\lambda_2 = 2, \pi_2^* = 2$ and $\rho = pd_2 \equiv 1 \pmod{p}$

If $pd_2 = 1$, then we may take $f_0 = z$ and $d_2 = 0$. Recall that the case $d_2 = 0$ has been solved in 3.1. Thus we may assume that $p < pd_2$ or $1 < d_2$. The signification of the above remark is that we do not have to worry about the complicate theory of resolvers used in the previous subsections 4.5.1, 4.5.2. Recall the definition of maximal clean from 4.5.3. Using the jumping lemma of 4.5.3 we may prove the following lemma.

Lemma 4-19. Let F' be a resulting equation from F by a permissible blow-up. Let d_3' be the corresponding number for F'. Then we always have that $d_3' \leq d_3 + 1/p$.

There is a seesaw procedure of increasing d_3 by an amount of $1/p$ and an improvement of the form as indicated by $m = \beta = 0$. After that if the improved form is reverted, there will be a reduction of d_3 by half at least. As usual we may reduce to the case that $d_3 \leq (1/p)$. If $\alpha \geq 1$ (or resp. $\beta \geq 1$), then we may blow up (x, T) (or resp (T, y)) to achieve $\alpha, \beta < 1$. The rest will be routine. There might be an increment which occurs only if $\alpha' = \alpha + \beta + d_3 - 1$ is an integer, thus it must be 1. then we may reduce it to 0 by blowing up (T, X). We may summarize this subsection and the above discussions to claim the following proposition:

Proposition 4-20. If $\lambda_2 = 2, \tau_2 = \pi_2^* = 2$, and $\rho = pd_2 \equiv 1 \pmod{p}$, then there is a tree of blow-ups such that after finitely many times, there is a reduction of $(d_1, n - \lambda_1, n - \tau_1, d_2, n - \lambda_2, n - \tau_2)$.

Proposition 4-21. It is not hard to see that the materials in this section can be generalized to the general eq. (4.14).

5 Uniformization for $\lambda_2 = 2$ and $\pi_2^* = 1$

5.1 Further classifications of the singularities, and the number σ

Explicitly we have the following maximal clean equations (5.33) and (5.34) satisfying conditions 1 and 2;

$$F = T^p + x^{\ell p} y^{mp} f_0(x, y, z), \tag{5.33}$$
$$f_0 = f_0^\ell(x, y, z) + \cdots. \tag{5.34}$$

1. $f_0^\ell(x, y, 0) \neq f_0^\ell(x, y, z)$, i.e., z does appears in $f_0^\ell(x, y, z)$ and $\lambda_2 \geq 2$.

2. $f_0^\ell(x, y, z) \in k[x, y, z^p]$, i.e., $\pi_2^* = 1$.

The following example will be illustrative for the blow-up process.

Example 5-1. Let the singularity be defined by the following equation (cf. Example 3-2 in 3.2),

$$F = T^p + x^{3p-2} y\big((y - 2x)z^p + x^{3p+1}\big).$$

Let us blow up the origin (T, z, y, x) and factor out x. Its proper transform at the point $p_1 = [0, 1, 0, 0]$ is

$$\overline{F} = T^p + x^{3p}\big((y+1)(y-1)z^p + (y+1)x^{2p}\big).$$

The variable T is not maximal clean for the equation. We shall replace T by $T - x^3 z - x^5$, then we get

$$F' = T^p + x^{3p}(y^2 z^p + y x^{2p}).$$

Note that the second weighted order d_2 of F increases from $(1+p)/p$ to $(2+p)/p$.

Note that in the above example the variable z serves as a dummy, and we may view it as a trivial generalization of the jumping phenomenon of 3.2. For the convenience of our discussions, we will name a singularity defined by our original equations (5.33) and (5.34) satisfying conditions 1 and 2 above as a singularity of α-type. After blowing up, let us select the variable T to remove all pth power terms, i.e., let the variable T be a maximal clean variable. If the old leading form f_{0pd_2} still contribute some terms of degree $\leq pd_2$, then we say that there is no leading form jumping (in fact all discussions in section 4 are about no leading form jumping). Otherwise there is a leading form jumping (cf. Example 5-1). In the leading form jumping cases, the emphasis is on the form f_{0pd_2+1}.

Definition 5-2. *The number $\overline{\lambda}_2$ is defined as*

$$\overline{\lambda}_2 = (\text{minimal number of free variables of } f_{0pd_2+1}) + 1.$$

Definition 5-3. *The number λ_2^* is defined as*

$$\lambda_2^* = \big(\text{minimal number of variables of } f_{0pd_2+1} \bmod \ (\text{fixed variables})\big) + 1.$$

Proposition 5-4. *If we have $\lambda_2 = 2$ and there is a leading form jumping, then we always have*

$$f'_{0pd_2+1} \quad (\text{mod fixed variables}) \notin k[x^p, y^p, z^p],$$

and $\lambda_2^{'*} \geq \lambda_2 + 1 = 3$.

PROOF. It follows from the jumping lemma.   □

Given an α-type singularity. Suppose that after the blow-up, there is a leading form jumping, then it will become a β-type singularity as follows.

Definition 5-5. *A β-type singulqrity is a singularity defined by the following equations*

$$\begin{aligned}
F &= T^p + x^{\ell' p} f'_0(x, y, z), & (5.35) \\
f'_0 &= f'_{0pd_2}(x, y, z) + g(y, z) + x h(x, y, z), & (5.36)
\end{aligned}$$

with the conditions as follows.

1. $\lambda_2^* \geq 3$ and f'_{0pd_2+1} mod (fixed variables) $\notin k[x^p, y^p, z^p]$.

2. ord $h(x, y, z) \geq pd_2$.

For the singularities with $\lambda_2^* \leq 2$, the reader is refered to section 6 for the so called δ-type, i.e., $\lambda_2^* = 2$ and γ-type, i.e., $\lambda_2^* = 1$, singularities. Further notice that for a β-type singularity, there is no condition on the term f'_{0pd_2}, which may or may not be zero. If it is zero, then we call it a special β-type singularity. Otherwise, we say that it is non-special. For a special β-type singularity, ord $f'_0 = pd_2 + 1$, i.e., $f'_{0pd_2} = 0$, we may consider it as an α-type singularity with $pd_2' = pd_2 + 1$ and $\lambda_2 \geq 3$. By the results of our preceding sections, we conclude that its second weighted order d_2' will drop back after finitely many blow-ups. In fact, we have to separate our discussions into the following cases,

$$\begin{cases} f_{0pd_2} \neq 0 \begin{cases} f_{0pd_2} \in (x), \\ f_{0pd_2} \in (x, y, z) \setminus (x), \end{cases} \\ f_{0pd_2} = 0. \end{cases}$$

As indicated by our example, we wish to show that if we blow up a α-type singularity, then it may (1) become a better singularity which has been covered in the previous discussions and of a smaller numerical character, or (2) stays in α-type with a smaller numerical character, or (3) become a β-type singularity and with a smaller numerical character. Although the above can be done in its whole detail, we will only discuss the difficult cases in the present article.

To compare and connect the α-type and the β-type singularities, we introduce the following new number σ,

1. for α-type, $\sigma = pd_2 - \deg_z f_{0pd_2}(x, y, z)$.

2. for β-type, $\sigma = \ell\big(f_{0pd_2+1}(0, y, z)\big) - 1$.

where $\ell\big(g(y, z)\big) = $ minimal multiplicity of factors of $g(y, z)$. In the previous example, the types of the singularities changed from α to β, while the number σ stays. This is an useful information.

5.2 Leading form jumping or no jumping

We shall state the following proposition for reference.

Proposition 5-6. If there is no leading form jumping, then there is a droping of the partial numerical character $(d_1, n - \lambda_1, n - \tau_1, d_2, n - \lambda_2)$ after blowing up of (T, z, y, x).

PROOF. Omitted. □

$\lambda_2 = 3, 4$				see section 4
$\lambda_2 = 2$	$\pi_2^* \geq 1$			see section 4
	$\pi_2^* = 1$	leading form no-jumping		see above
		leading form jumping	$\sigma \neq 0$	see 5.3
			$\sigma = 0$	see 5.4, 5.5
$\lambda_2 = 1$				see section 6

Figure 5.3: Leading form jumping diagram.

The diagram in figure 5.3 will be helpful for our understanding.

5.3 Reduction of the numerical data σ to 0

Let us consider the case $\sigma > 1$. The following example is illustrative.

Example 5-7. Let the singularity be defined by the following equation (cf. Example in 3.2),

$$F = T^p + x^{p-1}y\big((y^p - xy^{p-1} - x^p)z^p + xz^{2p} + x^{7p+1}\big).$$

Let us blow up the origin (T, z, y, x) and factor out x. Its proper transform at the point $p_1 = [0, 1, 0, 0]$ is

$$\overline{F} = T^p + x^{2p}\big((y^{p+1} - 1)z^p + (y+1)xz^{2p} + x^{5p+1}\big).$$

The variable T is not maximal clean for the equation. We shall replace T by $T + x^2 z$, then we get

$$F' = T^p + x^{2p}(y^{p+1}z^p + xz^{2p} + xyz^{2p} + x^{5p+1})$$

We should consider the above singularity β-type. Let us blow up the origin (T, z, y, x). Then we have four possibilities: (i) $v(T) < v(z), v(y), v(x)$, there will be a reduction of d_1. (ii) $v(z) < v(y), v(x)$ and $v(z) \leq v(T)$, there will be a reduction of d_2 as α-type. (iii) $v(y) \leq v(x)$ and $v(y) \leq v(z), v(T)$, there will be a reduction of d_2 as α-type. (iv) $v(x) = $ minimal, after factoring out x. Its proper transform at all points as α-type will have a smaller d_2 except at the point $p_1 = [0, 0, 0, 0]$. At that point the proper transform is

$$F' = T^p + x^{2p+1}(z^{2p} + y^{p+1}z^p + xyz^{2p} + x^{3p}).$$

The singularities go through α-type $\mapsto$ β-type $\mapsto$ α-type, and the numbers σ go through $p \mapsto p \mapsto 0$.

In fact the above example indicates a proof of the following proposition.

Proposition 5-8. Let us consider a singularity of α-type. Then we have (1) If the number $\sigma \not\equiv 0 \pmod{p}$, then there is a number n such that along any valuation, within n steps of blow-ups, we have one of the following two:

(i) *There is a reduction of the partial numerical character $(d_1, n - \lambda_1, n - \tau_1, d_2)$.*

(ii) *The preceding partial numerical character stays, and there is an increase of λ_2.*

(2) *If the number $\sigma \equiv 0 \pmod{p}$, then there is a number n such that along any valuation, within n steps of blow-ups, we have one of the following three:*

(i) *There is a reduction of the partial numerical character $(d_1, n - \lambda_1, n - \tau_1, d_2)$.*

(ii) *The preceding partial numerical character stays, then there is an increase of λ_2*

(iii) *The preceding partial numerical character and λ_2 stay, and σ becomes 0.*

PROOF. We have to separate into two cases: $\sigma > 1$ and $\sigma = 1$. The above proposition simply proclaims that the difficult case is $\sigma = 0$. The rest of the proof is omitted. □

5.4 Uniformization for $\sigma = 0$, the non-cycling case

Let us start with a singularity of α-type. Recall that we have $\lambda_2 = 2$ and $\pi_2^* = 1$ and $pd_2 \equiv 0 \pmod{p}$, i.e., d_2 is an integer. It follows from the assumption $\pi_2^* = 1$ that equations (5.33) and (5.34) of 5.1 may be rewritten as

$$F = T^p + (x^\ell y^m)^p f_0(x, y, z), \tag{5.37}$$
$$f_0 = cz^{pd_2} + \cdots, \qquad\qquad c \neq 0. \tag{5.38}$$

Note that the so called "leading form no jumping" cases has been settled in 5.2. We may only consider the "leading form jumping case". We may thus assume that the singularity always ends up in the β-type as follows,

$$F' = T^p + (x^{\ell'})^p f_0'(x, y, z), \tag{5.39}$$
$$f_0' = f_{0pd_2}'(x, y, z) + yg(y, z) + xh(x, y, z), \tag{5.40}$$

where g(y,z) is a homogeneous polynomial of degree pd_2, $g(0, z) \neq 0$ and ord $h(x, y, z) \geq pd_2$.

At the beginning, the number ℓ' is an integer, otherwise there is no leading form jumping. However, this property of being an integer is not stable with respect to permissible blow-ups. We will drop this restriction for β-type singularities.

If we start with a β-type singularity and blow up successively a suitable sequence of permissible centers, then there are two possibilities: (i) there is an obvious reduction of the singularity in the process without reverting to the original system of equations for α-type singularity, we refer this as a non-cycling case, or (ii) the system of defining equations could return to the forms of eq. (5.37) and (5.38), we refer this as a cycling case. In fact, we have the following diagram for a β- type singularity,

$$\sigma = \begin{cases} f_{0pd_2} \neq 0 & \begin{cases} f_{0pd_2} \in (x) & \text{non-cycling,} \\ f_{0pd_2} \notin (x) & \text{cycling,} \end{cases} \\ f_{0pd_2} = 0. & \text{cycling.} \end{cases}$$

Let us study the non-cycling case. We shall consider the following system of equations which generalizes Equations (5.35) and (5.36),

$$F' = T^p + (x^{\ell'} y^{m'})^p f_0'(x, y, z), \qquad (5.41)$$

$$f_0' = f_{0pd_2}'(x, y, z) + yg(y, z) + xh(x, y, z), \qquad (5.42)$$

where m' may not be zero and satisfying $f_{0pd_2}' \neq 0 \in (x)$, $\deg g(y, z) = pd_2$ and $g(0, z) \neq 0$. Clearly if $v(T)$ or $v(z) < v(y)$, $v(x)$, then we will have a reduction. Moreover if $v(x)$ is minimal, then after we factor out x, there will be a reduction of pd_2 by at least one, moreover a changing variable T may increase it by one, then a detailed analysis will show that there must be an increase of λ_2. So we are left with the study of $v(y)$ minimal and smaller than $v(x)$. It is not hard to see that either there is a reduction, if f_{0pd_2}' depends on y, or the above equations (5.41) and (5.42) will be stable (while equations (5.35) and (5.36) will not be stable), if f_{0pd_2}' is independent of y. We may treat x and z as variables and y as coefficient, and define a numeral α_3 (i.e., the first Newton slope) for the above equations which will drop after every blow-up. Then we are done. The above is a sketch of the following proposition,

Proposition 5-9. *After blowing up a singularity of the α-type, suppose it is transformed to β-type. If $f_{0pd_2}'(x, y, z) \neq 0 \in (x)$, then by further blowing up finitely many times, we always have either a reduction of the partial numerical character $(d_1, n - \lambda_1, n - \tau_1, d_2, n - \lambda_2)$ or it stays and we have an increase of π_2^*.*

5.5 Uniformization for $\sigma = 0$, the cycling case

If we blow up a permissible center for a hypersurface $F = 0$ of α-type, suppose that there is a leading form jumping, then it is transformed to a β-type singularity. In 5.4, we have treated the simple case $f_{0pd_2}' \in (x) \setminus (0)$. However, it may be 0 or not divisible by x. In the present section we will give a refined classification. Suppose that $f_{0pd_2}'(0, y, z) \neq 0$. If it requires two variables to be expressed, then we must have $\lambda_2 = 3$ and we are done. So we may assume that it requires only one variable, let it be z (the proof is the same if it is y). For the sake of uniformization, we will restrict the β-type to the $\beta^{(1)}$-type (see below) and further separate the α-type into $\alpha^{(1)}$-type and $\alpha^{(2)}$-type as follows.

For $\alpha^{(2)}$ types with $\ell \neq 0$, $m \neq 0$,

$$
\begin{align}
F &= T^p + (x^\ell y^m)^p f_0(x, y, z), \tag{5.43}\\
f_0 &= z^{pd_2} + \Sigma (x^\alpha y^\beta)^{pd_2 - i} f_{0i}(x, y) z^i + f_0^r(x, y, z). \tag{5.44}
\end{align}
$$

For $\beta^{(1)}$ types with ord $g'_{0pd_2}(0, y) = 1$,

$$
\begin{align}
F' &= T^p + x^{\ell' p} f_0'(x, y, z), \tag{5.45}\\
f_0' &= g'_{0pd_2}(x, y) z^{pd_2} + \Sigma x^{\alpha'(pd_2 - i)} f_{0i}'(x, y) z^i + f_0'^r(x, y, z). \tag{5.46}
\end{align}
$$

For $\alpha^{(1)}$ types

$$
\begin{align}
F'' &= T^p + x^{\ell'' p} f_0''(x, y, z), \tag{5.47}\\
f_0'' &= z^{pd_2} + \Sigma x^{\alpha''(pd_2 - i)} f_{0i}''(x, y) z^i + f_0''^r(x, y, z). \tag{5.48}
\end{align}
$$

5.6 A classification of cyclings

These cycling phenomena can be further separated into six cases (i) $\alpha^{(1)} \rightleftarrows \alpha^{(1)}$, (ii) $\alpha^{(2)} \rightleftarrows \alpha^{(2)}$, (iii) $\beta^{(1)} \rightleftarrows \beta^{(1)}$, (iv) $\alpha^{(2)} \rightleftarrows \alpha^{(1)}$, (v) $\alpha^{(2)} \rightleftarrows \beta^{(1)}$ and (vi) $\alpha^{(2)} \rightarrow \beta^{(1)} \rightarrow \alpha^{(1)} \rightarrow \alpha^{(2)}$. We may draw the following diagram.

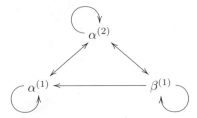

The first three cases are with the types stable and can be uniformized mechanically: we may define some numerical characters for each case, such that after every blow-up, the numerical character will drop (strictly). For the last three cases, we notice the pivot role played by $\alpha^{(2)}$-type, i.e., every cycle of them must passes through a $\alpha^{(2)}$-type singularity. Essentially, we have to show that starting with a singularity of $\alpha^{(2)}$-type, with a loop of blow-ups which produces another singularity of $\alpha^{(2)}$-type, then there will be a drop of the numerical character. There we have to prove a jumping lemma, define the numerical characters for each type, and to smooth out the passages of types, defining the adjustors (cf. section 3).

For instance, for the passage from $\alpha^{(2)}$-type to $\beta^{(1)}$-type, we need the following lemmas for "double-cleaning".

Lemma 5-10. If $p = 2, m \not\equiv 0 \pmod{p}$, then

$$
y^{m+\beta} g(y) \in k[y^2] + y^m k[y^2].
$$

PROOF. Trivial.   □

Lemma 5-11. *Suppose that $m \not\equiv 0 \pmod{p}, \beta = up^r$ and $p \nmid u$ where $r \geq 1$. Let $\mu \neq 0$ and ϕ be the projection of $(\mu + y)^n$ to k^3 as follows*

$$\phi(\mu + y)^n = \left[\mu^n, \binom{n}{1}\mu^{n-1}, \binom{n}{p^r+1}\mu^{n-p^r-1} \right].$$

Then $\phi(\mu + y)^{m+\beta}$ is not a linear combination of $\sum a_i \phi(\mu + y)^{r_i p}, \sum b_i \phi(\mu + y)^{m+s_i p^e}$.

PROOF. Omitted. □

Lemma 5-12. *Given $y^{m+\beta} g(y)$ with $m \not\equiv 0 \pmod{p}$ such that (1) its order is maximal $\mathrm{mod}(k[y^p] + y^m k[y^{p^e}])$ i.e.,*

$$\mathrm{ord}\left(y^{m+\beta} g(y) + r(y^p) + y^m s(g^{p^e}) \right) \leq \mathrm{ord}\, y^{m+\beta} g(y) < \infty,$$

for all polynomials $r(y^p)$ and $s(y^{p^e})$. (2) Let $y^\beta g(y) \in k[y^{p^r}] \setminus k[y^{r-1}]$ and r is maximal under the restriction of (1). Then for any $\mu \neq 0$, we have

$$\mathrm{ord}\left((y + \mu)^{m+\beta} g(y + \mu) + r(y^p) + (y + \mu)^m s(y^{p^e}) \right) \leq \deg g + p^r + 1.$$

for all polynomials $r(y^p)$ and $s(y^{p^e})$.

PROOF. Omitted. □

5.7 Uniformization for $\lambda_2 = 2$ and $\pi_2^* = 1$

Following the preceding subsection 5.6, there are some 25 more pages of detailed arguments to establish the following proposition.

Proposition 5-13. *Given a singularity with $\lambda_2 = 2$ and $\pi_2^* = 1$. Then there is a number n such that along any valuation v, after at most n blow-ups, there will be either a reduction of the partial numerical character $(d_1, n - \lambda_1, n - \tau_1, d_2, n - \lambda_2)$, or it stays and an increase of π_2^*.*

PROOF. Omitted. □

6 Uniformization for $\lambda_2 = 1$

6.1 Introduction

Explicitly we have the following maximal clean equations (6.49) and (6.50) satisfying conditions 1 and 2,

$$F = T^p + (x^\ell y^m z^n)^p f_0(x, y, z), \tag{6.49}$$

$$f_0 = f_0^\ell(x, y, z) + \cdots. \tag{6.50}$$

1. There is no free variable in f_0^ℓ, i.e., $\lambda_2 = 1$.

2. $f_0^\ell(x, y, 0) \neq f_0^\ell(x, y, z)$, i.e., z does appear in the leading form.

Let us blow up the origin (T, z, y, x). If $v(T)$ is minimal or $v(z)$ is minimal and $< v(y)$, $v(x)$, then it is easy to see that the partial numerical character $(d_1, n - \lambda_1, n - \tau_1, d_2)$ will drop. If $v(x)$ or $v(y)$ is the only minimal value among the four variables. The discussions will be routine and omitted. Furthermore, in the case that $v(T), v(z) > \text{minimal} = v(y), v(x)$ we may treat the variable z as a dummy variable, the materials in section 4 can be easily modified to yield a reduction. Henceforth, we will assume one of the following two cases: Case 1, $v(T) > v(z) = v(x) = v(y) = \text{minimal}$. Case 2, $v(T)$, $v(y) > v(z) = v(x) = \text{minimal}$. Let us factor out x in the blow-up, at a resulting point p_1 we produce the following equation,

$$F = T^p + (x^{\ell'} y^{m'})^p f_0'(x, y, z), \tag{6.51}$$
$$f_0' = f_{0pd_2}'(x, y, z) + f_{opd_2+1}'(x, y, z) + \cdots . \tag{6.52}$$

where $m' = 0$ in the Case 1. Just as in the previous section, we have the leading form no jumping situation and the leading form jumping situation.

6.1.1   Leading form no jumping

Since the original leading form $f_{0pd_2}(x, y, z)$ will produce some terms for the new leading form f_{0pd_2}', it is easy to see that we must have $f_{0pd_2}'(0, y, z) \neq 0$. Furthermore, if $f_{0pd_2}'(0, y, z)$ does involve y, then we will have $\lambda_2' \geq 2$ and it will be covered by our previous sections. Thus we shall only deal with the that $f_{0pd_2}'(x, y, z) = cz^{pd_2} + \cdots$, where where $c \neq 0$ and $p | pd_2$. The above equations (6.51) and (6.52) may be rewritten in the following form,

$$F = T^p + (x^{\ell'} y^{m'})^p f_0'(x, y, z), \tag{6.53}$$
$$f_0' = cz^{pd_2} + \sum (x^\alpha y^\beta)^{pd_2 - j_2} f_{0j_2}'(x, y) z^{j_2} + f'^r. \tag{6.54}$$

For any blow-up series, if it stays in the situation of leading form no-jumping, then the above form is stable. The arguments for an reduction in this situation is very similar to the materials in Section 4. The main differences are that we have to consider a double clean, i.e., after every blow-up, eq. (6.54) can be cleaned by changing z and changing T. We will not get into the details of the arguments here and simply state that the partial numerical character $(d_1, n - \lambda_1, n - \tau_1, pd_2)$ will eventually drop.

6.1.2   Leading form jumping, δ-type and γ-type

Let us consider the case of leading form jumping. Then it follows from the jumping lemma that a free variable will be created for f_{pd_2+1}' and moreover f_{pd_2+1}' is not in the pth power of the free varible. Let us consider the following examples,

Example 6-1. Let the ground field be of characteristic $p = 2$. Given an equation $F = 0$ as follows

$$F = T^2 + xz(z^2 + xz + x^2 + \cdots).$$

Let us blow up (x, y, z, T) and consider a substitution π of variables as follows,

$$\pi: \begin{cases} x & \mapsto & x, \\ y & \mapsto & x(y+1), \\ z & \mapsto & x(z+1), \\ T & \mapsto & xT. \end{cases}$$

Then the proper transform $\overline{F}$ of F is as follows

$$\overline{F} = T^2 + x^2(z+1)[z^2 + z + 1 + \cdots].$$

After using cleaning process to remove p-th power terms, we have the following maximal clean equation

$$F' = T^2 + x^2[z^3 + \cdots].$$

Originally the numerical d_2 for F is 1 and the new numerical d_2' for F' is $3/2$. Apparently there is an increase (against our wishes) of d_2. However, there is an improvement of the form of the equation due to the term z^3 consist of free variables only. Note that $\lambda_2^* = 2$. We will denote the singularity as δ-type (see below).

Example 6-2. Let the ground field be of characteristic $p = 2$. Given the following equation $F = 0$,

$$F = T^2 + xyz(z + \cdots),$$

let us blow up (x, y, z, T) and consider a substitution π of variables as follows

$$\pi: \begin{cases} x & \mapsto & x, \\ y & \mapsto & x(y+1), \\ z & \mapsto & xz, \\ T & \mapsto & xT. \end{cases}$$

Then the proper transform $\overline{F}$ of F is as follows

$$\overline{F} = T^2 + x^4(y+1)z(z + \cdots).$$

With a substitution $T \mapsto T - x^2 z$, the above equation becomes

$$F' = T^2 + x^4 z(yz + \cdots).$$

Although there is an increment of d_2, i.e., $d_2 = 1/2 < d_2' = 1$, while the singularity is not necessarily of δ-type (see below). Note that $\lambda_2^* = 1$ and it is of γ-type (see below).

The above two examples are typical. In the situation of leading form jumping, if the resulting proper transform F' after maximal clean is of the following form

1. $F' = T^p + x^{\ell' p} y^{m' p} f_0'(x, y, z)$,

2. $f_0' = f_{0pd_2}'(x, y, z) + f_{0pd_2+1}'(x, y, z) + \ldots$,

3. $\lambda_2^* = 2$,

4. $f_{0pd_2+1}'(0, 0, z) \notin k[z^p]$, i.e., z is free, and $p < pd_2 + 1$,

then we define the singularity to be of δ-type. Furthermore, if $f_{0pd_2}' = 0$, then the singularity is said of special δ-type. Although just after a leading form jumping, we have $\ell' \in \mathbb{Z}$, however, for the inductive process which will be deleveoped later this condition might be violated by further blow-ups. Thus we will not require $\ell' \in \mathbb{Z}$ in general.

In the situation of leading form jumping, if the resulting proper transform F' after maximal clean is of the following form

1. $F' = T^p + (x^{\ell'} z^n)^p f_0'(x, y, z)$,

2. $f_0' = f_{0pd_2}'(x, y, z) + f_{0pd_2+1}'(x, y, z) + xh'(x, y, z)$,

3. $\lambda_2^* = 1$,

4. $z^{np} f_{0pd_2+1}'(0, y, z) \in k[y, z^p] \backslash k[y^p, z^p]$,

then the singularity is said to be of γ-type. Furthermore, if $f_{0pd_2}' = 0$, then the singularity is said of special δ-type. Although just after a leading form jumping, we have $\ell' \in \mathbb{Z}$, however, for the inductive process which will be developed later, this condition might be violated by further blow-ups. Thus we will not require $\ell' \in \mathbb{Z}$ in general.

We have the following proposition with the proof omitted.

Proposition 6-3. *Let an equation $F = 0$ with $\lambda_2 = 1$ be given as follows*

$$
\begin{aligned}
F &= T^p + x^{\ell p} y^{mp} z^{np} f_0 \\
&= T^p + x^{\ell p} y^{mp} z^{np} \big(f_{opd_2}(x, y, z) + \cdots \big).
\end{aligned}
$$

Then there is a sequence of blow-ups such that at all resulting points, one of the following holds:

1. *There is no leading form jumping.*

2. *The resulting singularity is of β-type.*

3. *The resulting singularity is of δ-type.*

4. *The resulting singularity is in γ-type.*

We have the following stability propositions for the special δ-type and the special γ-type.

Proposition 6-4. *Suppose that an equation $F = 0$ is of special δ-type. Let us blow up (x, y, z, T). Then the resulting proper transform $F' = 0$ satisfies one of the following*

1. *The partial numerical character* $(d_1, n - \lambda_1, n - \tau_1, d_2, n - \lambda_2)$ *drops.*

2. *The above partial numerical character stays and F' is of δ-type with the non-decreasing λ_2^*.*

PROOF. Due to the condition $pd_2 + 1 > p$, we have $pd_2 > 1$. The proof is routine.   □

Proposition 6-5. *Suppose that an equation $F = 0$ is of special γ-type. Let us blow up (x, y, z, T). Then the resulting proper transform $F' = 0$ satisfies one of the following*

1. *The partial numerical character $(d_1, n - \lambda_1, n - \tau_1, d_2, n - \lambda_2)$ drops.*

2. *The above partial numerical character stays and F' is of γ-type with the non-decreasing λ_2^*.*

PROOF. It is routine.   □

6.2   Uniformization for δ-type singularities

Recall that we have $\lambda_2^* = 2$. The equation $F = 0$ must be of the following form

$$F = T^p + x^{\ell p} y^{mp}(f_{0pd_2}(x, y, z) + az^{pd_2+1} + \cdots).$$

where $a \neq 0$ and $p \nmid pd_2 + 1$. If $f_{0pd_2}(0, 0, z) \neq 0$, then we have $\lambda_2 \geq 2$ and it has been solved in Section 5. We may thus assume that

$$f_{0pd_2}(0, 0, z) = 0.$$

Note that we still have two subcases: it is of special δ-type, i.e., $f_{0pd_2} = 0$ or it is of non-special δ-type, i.e., $f_{0pd_2} \neq 0$. Let us consider the special subcase first. Then the system of equations is as follows,

$$F = T^p + x^{\ell p} y^{mp}(cz^{pd_2+1} + \cdots),$$

where $c \neq 0$ and $p \nmid pd_2+1$. There will be no jumping for blow-ups. The materials in Section 5 may be applied to a finite sequence of blow-ups to reduce to the situation $f_{0pd_2}(x, y, z) \neq 0$, i.e., the non-special subcase.

Let us consider the non-special subcase, i.e., $f_{0pd_2} \neq 0$. Here we must further classify all those singularities into two types,

1. $\tau_2 = 3$, i.e., $f_{0pd_2}(x, y, z) \notin k[x]$ and $\notin k[y]$.

2. $\tau_2 = 2$, i.e., $f_{0pd_2}(x, y, z) \in k[x]$ or $k[y]$.

6.2.1 $\tau_2 = 3$

The uniformization for $\tau_2 = 3$ type singularities involves the uniformization of the following singularities.

Lemma 6-6. *Let the equation $F = 0$ be given as follows with $\lambda_2 = 1$,*

$$
\begin{aligned}
F &= T^p + x^{\ell p} y^{mp} z^{np} f_0(x, y, z) \\
&= T^p + x^{\ell p} y^{mp} z^{np} \big(f_{0pd_2}(x, y, z) + \cdots \big),
\end{aligned}
$$

where $f_{0pd_2}(x, y, z) \neq 0$ and $f_{0pd_2}(0, y, z) = 0$. If $f_0 \in (x, y)^{pd_2}$ (resp. $(x, z)^{pd_2}$), let us blow up (x, y, T) (resp. (x, z, T)) or (x, y, z, T) otherwise. If $v(x) = $ minimal, then there is a drop of the partial numerical character $(d_1, n - \lambda_1, n - \tau_1, pd_2, n - \lambda_2)$.

PROOF. When we consider the leading form, the variable x may be factored out. Thus the order will drop by 1 firstly, and then it may increase by 1 due to the jumping phenomenon. □

Lemma 6-7. *Let an equation $F = 0$ be given as follows with $\lambda_2 = 1$*

$$
\begin{aligned}
F &= T^p + x^{\ell p} y^{mp} z^{np} f_0 \\
&= T^p + x^{\ell p} y^{mp} z^{np} \big(b x^{pd_2} + a y^\alpha z^\beta + h(x, y, z) \big)
\end{aligned}
$$

where $a \neq 0, b \neq 0, h(0, y, z) = 0, 0 < \alpha < pd_2, \beta \leq pd_2$. Then after finitely many blow ups with suitable centers we have either a drop of the partial numerical character $(d_1, n - \lambda_1, n - \tau - 1, d_2, n - \lambda_2)$ or the first three numbers of the preceding sequence stay, ord f_0 jump by 1 and $\lambda_2^* \geq 3$.

PROOF. Omitted. □

The following proposition will be stated without proof.

Proposition 6-8. *Given a $\tau_2 = 3$ type singularity as above, then after finitely many blow ups with suitable centers we have either a drop of the partial numerical character $(d_1, n - \lambda_1, n - \tau - 1, d_2, n - \lambda_2)$ or the first three numbers of the preceding sequence stay, ord f_0 jump by 1 and $\lambda_2^* \geq 3$.*

6.2.2 $\tau_2 = 2$

Let us consider a singularity with $\tau_2 = 2$ as above. We have the following two possible forms of the equation $F' = 0$, where ord $q' > pd_2 + 1$,

1. $F' = T^p + x^{\ell p} y^{mp} \big(c x^{pd_2} + b z^{pd_2+1} + q'(y, z) + h'(x, y, z) \big),\ 0 < \ell p < p$.

2. $F' = T^p + x^{\ell p} y^{mp} \big(c y^{pd_2} + b z^{pd_2+1} + q'(y, z) + h'(x, y, z) \big),\ 0 < mp < p$.

There are some differences due to the asymmetry of the variables x and y. The above form 1 is slightly difficult. The uniformization for singularities with $\tau_2 = 2$ as above depends on the concrete analysis of some singularities. In fact we have the following,

Lemma 6-9. *Given the following equation*

$$
\begin{aligned}
F &= T^p + x^{\ell p} y^{mp} z^{np} f_0 \\
&= T^p + x^{\ell p} y^{mp} z^{np} \big(c x^{p d_2} + b y^{p d_2} + a z y^{p d_2} + h(x,y,z) \big),
\end{aligned}
$$

where a, b, $c \neq 0$, $0 < \ell p < p$, $0 < mp < p$, $h(0,y,z) = 0$ *and* $h(x,0,z) = 0$. *Let the the center for blow-up be given as follows, (1) if* $f_0 \in (x,y)^{p d_2}$, *then the center* $= (x,y,T)$, *(2) otherwise the center* $= (x,y,z,T)$. *If we blow up the center, then we have one of the following conclusions:*

1. *There is a drop of the partial numerical character* $(d_1, n - \lambda_1, n - \tau_1, d_2, n - \lambda_2)$

2. *The above partial numerical character stays, and* F' *is of* β-type (i.e., $\lambda_2'^* \geq 3$).

3. *The resulting equation* F' *has the same form.*

   Moreover, if we repeat the above procedure finitely many times, then we must have one of the above conclusions 1 or 2.

PROOF. Omitted. □

Lemma 6-10. *Given the following equation* $F = 0$

$$
\begin{aligned}
F &= T^p + x^{\ell p} y^{mp} f_0 \\
&= T^p + x^{\ell p} y^{mp} \big(c x^{p d_2} + b y z^{p d_2} + h(x,y,z) \big),
\end{aligned}
$$

with $c, b \neq 0$, $\operatorname{ord} h(x,y,z) \geq p d_2$, $h(0,y,z) = 0$, $p d_2 > 1$ *and* $0 < \ell p < p$. *Then after blowing up with center* (x,z,T) *if* $f_0 \in (x,z)^{p d_2}$ *and with center* (x,y,z,T) *if* $f_0 \notin (x,z)^{p d_2}$, *one of the following is true:*

1. *There is a drop of the partial numerical character* $(d_1, n - \lambda_1, n - \tau_1, d_2, n - \lambda_2)$

2. *The above partial numerical character stays, and* F' *is of the same form.*

   Moreover, if we repeat the above procedure finitely many times, then we must have the conclusions 1.

PROOF. Omitted. □

Lemma 6-11. *Let an equation* $F = 0$ *be given as follows with* $\lambda_2 = 1$, $p \nmid p d_2 + 1 > p$,

$$
\begin{aligned}
F &= T^p + x^{\ell p} y^{mp} f_0 \\
&= T^p + x^{\ell p} y^{mp} \big(b x^{p d_2} + a y^\alpha z^{p d_2 + 1} + h(x,y,z) \big),
\end{aligned}
$$

where $0 < \ell p < p$, $a \neq 0$, $b \neq 0$, $0 < \alpha < p d_2$, $h(0,y,z) = 0$. *Let us blow up* (x,y,z,T). *Then we have one of the following*

1. $v(x) = minimal \Rightarrow \operatorname{ord} F' < \operatorname{ord} F$ or $\operatorname{ord} f_0' < \operatorname{ord} f$ or $\operatorname{ord} f_0' = \operatorname{ord} f$, $\lambda_2' > 1$.

2. $v(x) > v(y) = minimal \Rightarrow$ (a) if $v(z) > v(y)$, then the resulting equation $F' = 0$ will maintain the same form with α replaced by $\alpha + 1 \leq pd_2$. (b) if $v(z) = v(y)$ and $\alpha < pd_2 - 1$, then $\operatorname{ord} f_0' < \operatorname{ord} f_0$ (c) if $v(z) = v(y)$ and $\alpha = pd_2 - 1$ then the conditions of Lemma 6-9 will be satisfied.

3. $v(z) < v(x), v(y). \Rightarrow$ the conditions of Lemma 6-7 will be satisfied by the resulting equation $F' = 0$.

PROOF. Conclusion 1 follows from Lemma 6-6. Conclusion 2 is evident due to the fact $0 < \ell p < p$ and $p | p d_2 + 1$. Conclusion 3 is trivial. $\qquad \square$

Remark 6-12. *According to our preceeding Lemma 6-11, only conclusion 2(a) is unsettled. Inductively we may assume that y is factored out without translation until $\alpha = pd_2$.*

Now we may discuss the uniformization for singularities with $\tau_2 = 2$. The equation $F' = 0$ is of one of the following two forms,

$$
\begin{aligned}
F' &= T^p + x^{\ell' p} y^{m' p} f_0' \\
   &= T^p + x^{\ell' p} y^{m' p} \left(b x^{p d_2} + c z^{p d_2 + 1} + q'(y, z) + h'(x, y, z) \right),
\end{aligned}
\tag{6.55}
$$

where $0 < \ell' p < p$, $h'(0, y, z) = 0$ and $q'(y, z)$ homogeneous of degree $p d_2 + 1$, with $q'(0, z) = 0$ or

$$
\begin{aligned}
F' &= T^p + x^{\ell' p} y^{m' p} f_0' \\
   &= T^p + x^{\ell' p} y^{m' p} \left(b y^{p d_2} + c z^{p d_2 + 1} + q'(x, z) + h'(x, y, z) \right),
\end{aligned}
\tag{6.56}
$$

where $0 < m' p < p$, $h'(x, 0, z) = 0$ and $q'(x, z)$ a homogeneous polynomial of degree $p d_2 + 1$ with $q'(0, z) = 0$.

We have the following proposition.

Proposition 6-13. *With the assumptions as above, let the equation $F' = 0$ be of form (6.55) (resp. (6.56)). If $f_0' \in (x, z)^{p d_2 - 1}$ (resp. $(y, z)^{p d_2 - 1}$), then we blow up (x, z, T) (resp. (y, z, T)). Otherwise we blow up (x, y, z, T). Then one of the following holds for the resulting equation, with notations abused, for $F = 0$.*

1. A drop of the partial numerical character $(d_1, n - \lambda_1, n - \tau_1, d_2, n - \lambda_2)$.

2. The equation $F = 0$ satisfies the assumptions of Lemma 6-10.

3. The equation $F = 0$ satisfies the assumptions of Lemma 6-11.

In any case after finitely many blow-ups, there will be a reduction of ord F'.

6.3 Uniformization for γ-type singularities

6.3.1 Introduction

Let us consider an equation $F = 0$ of γ-type as follows,

$$
\begin{aligned}
F &= T^p + (x^\ell z^n)^p f_0(x, y, z), \\
f_0 &= f_{0pd_2}(x, y, z) + f_{0pd_2+1}(x, y, z) + h(x, y, z),
\end{aligned}
$$

where

$$
n \neq 0 \quad \text{and} \quad z^{np} f_{0pd_2+1}(0, y, z) \in k[y, z^p] \backslash k[y^p, z^p].
$$

Recall the definition of σ in 5.1. We will extend it naturally for γ-type singularities as follows,

$$
\sigma = \ell\big(f_{0pd_2+1}(0, y, z)\big) - 1.
$$

Then we always have $\sigma \geq 0$. It is not hard to establish a decreasing property of σ under blowing-ups from positive to zero. Let us study the case $\sigma = 0$ in detail.

6.3.2 Preparations for the unifomization for $\sigma = 0$

From now on we will discuss the case of $\sigma = 0$, i.e., an equation of the following form

$$
F = T^p + x^{\ell p} z^{np} \big(f_{0pd_2} + (y + \cdots)z^{pd_2} + h(x, y, z)\big),
$$

where

$$
\begin{aligned}
np + pd_2 &\equiv O \pmod{p}, \\
\operatorname{ord} h(x, y, z) &\geq pd_2 + 1.
\end{aligned}
$$

We want to show that in the blow-up process, we have either (1) an obvious reduction of the numerical character or the resulting surface is of α-type, β-type or δ-type, or (2) the situation is similar to that in subsections 5.4 and 5.5. Certainly by our previous materials, case (1) is done. Thus we will clarify and determine the situations which will not produce the simplifications as case (1) stated above.

For the special γ-type or the non-special γ-type singularities, we can establish the only cases for the above 1 not happening are the following,

1. Factor out y (without translation), i.e., $v(y) < v(T), v(z), v(x)$.

2. Factor out x without translation involving z, i.e.,

$$
\pi : \begin{cases}
x &\mapsto & x, \\
z &\mapsto & xz, \\
y &\mapsto & x(y + \mu_2), \\
T &\mapsto & xT.
\end{cases}
$$

However, the singularities of γ-type with $\sigma = 0$, $0 \neq f_{0pd_2}(x,y,z) \in (x)$ are peculiar and we have to dispose them first. After we establish that case by some special methods, we may assume that in the blow-up process, the singularities will stay γ-type with $\sigma = 0$ and either $f_{0pd_2}(x,y,z) = 0$ or $f_{0pd_2}(x,y,z) \in k[x,z]$, $f_{0pd_2}(0,y,z) = cz^{pd_2}$ (otherwise, $\lambda_2 \geq 2$).

It follows from our knowledge about the substitutions (see above 1 and 2) that z, T will serve as dummy variable in the further blow-ups, i.e., either

$$
\begin{array}{ccc}
z & \to & xz \\
T & \to & xT
\end{array}
\quad \text{or} \quad
\begin{array}{ccc}
z & \to & yz \\
T & \to & yT.
\end{array}
$$

Thus any changing variables involving T, z will be unnecessary! We may rename the following γ-type as $\gamma^{(1)}$-type,

$$
\begin{aligned}
F &= T^p + x^{\ell p} z^{np} f_0 \qquad\qquad\qquad\qquad\qquad\qquad (6.57)\\
&= T^p + x^{\ell p} z^{np}\left(g_{pd_2}(x,y)z^{pd_2} + \sum (x^\alpha)^{pd_2-i} g_i(x,y)z^i + g^r(x,y,z) \right)
\end{aligned}
$$

as a $\beta^{(1)}$-type equation as follows

$$
\begin{aligned}
F &= T^p + x^{\ell p}(z^{np} f_0) \qquad\qquad\qquad\qquad\qquad\qquad (6.58)\\
&= T^p + x^{\ell p}\bigg(h_{pd_2+np}(x,y)z^{pd_2+np} \\
&\qquad + \sum (x^\alpha)^{pd_2-i} h_{i+np}(x,z), z^{i+np} + z^{np} g^r(x,y,z) \bigg).
\end{aligned}
$$

Note that since z serves as a dummy variable we may freely think of the same equation as either γ-type or β-type. Thus any intermediate equation $F' = 0$ must be of the following form

$$
\begin{aligned}
F' &= T^p + x^{\ell p} y^{mp} z^{np} \widetilde{f_0'} \qquad\qquad\qquad\qquad\qquad (6.59)\\
&= T^p + x^{\ell p} y^{mp} z^{np}\left(f_{0pd_2}'(x,y,z) + \cdots \right).
\end{aligned}
$$

Let us discuss the form of $f_{0pd_2}'(x,y,z)$. Note that once we factor out y then we have

$$
f_{0pd_2}'(0,0,z) = cz^{pd_2} \neq 0. \qquad\qquad\qquad\qquad\qquad (6.60)
$$

On the other hand if we keep factoring out x, we will reach the above condition (6.60) again. Once the equations (6.59) and (6.60) are reached, there are two cases depending on the value of mp,

Case 1: $mp = 0$, we name it as $\epsilon^{(1)}$-type, and express $F'' = 0$ in the following presentation

$$
\begin{aligned}
F'' &= T^p + x^{\ell p} z^{np} f_0''. \qquad\qquad\qquad\qquad\qquad\qquad (6.61)\\
f_0'' &= z^{pd_2} + \sum x^{\alpha''(pd_2-i)} f_{0_i}''(x,y)z^i + f_0'''^r(x,y,z).
\end{aligned}
$$

Case 2: $mp \neq 0$, we name it as $\epsilon^{(2)}$-type, and express $F = 0$ in the following presentation

$$F = T^p + x^{\ell p} y^{mp} z^{np} f_0. \tag{6.62}$$
$$f_0 = z^{pd_2} + \sum (x^\alpha y^\beta)^{pd_2 - i} f_{0i}(x,y) z^i + f_0^\tau(x,y,z).$$

Comparing the above with section 5, clearly our three sets of singularities: $\gamma(1)$-type, i.e., eq. (6.57) (=eq. (6.58)), $\epsilon^{(1)}$-type, i.e., eq. (6.61), and $\gamma^{(2)}$-type, i.e., eq. (6.62), are nothing but the $\beta(1)$-type, the $\alpha^{(1)}$ type, the $\alpha^{(2)}$ type of 5.4. Precisely, we have the following correspondences

$$\gamma^{(1)}\text{-type} \quad \text{eq. (6.58)} \rightleftarrows \beta^{(1)} \text{ type eq. (5.45) (5.46)}$$
$$\epsilon^{(1)}\text{-type} \quad \text{eq. (6.61)} \rightleftarrows \alpha^{(1)} \text{ type eq. (5.47) (5.48)}$$
$$\epsilon^{(2)}\text{-type} \quad \text{eq. (6.62)} \rightleftarrows \alpha^{(2)} \text{ type eq. (5.43) (5.44)}.$$

It is then routine to use 5.4, 5.5 to prove the following proposition and theorem.

Proposition 6-14. *Given a γ-type singularity. Then there is a number n such that along any valuation, the mulitiplicity p of the hypersurface will be smaller in n steps of blow-ups.*

PROOF. See above discussions.   □

Theorem 6-15. *Given any singularity defined by the following equation,*

$$F = T^p - f(x,y,z).$$

Then there is a number n such that along any valuation, the multiplicity will be smaller in n steps of blow-ups.

PROOF. See all previous propositions.   □

References

[1] S. Abhyankar. Local uniformization on algebraic surfaces over ground fields of characteristic $p \neq 0$. Ann. of Math. Vol 63. Pages 491–526. 1956.

[2] S. Abhyankar. An algorithm on polynomials in one indeterminate with coefficients in a two dimensional regular local domain. Ann. Mat. Pura Appl. vol 71. Pages 25–60. 1966.

[3] S. Abhyankar. *Resolution of Singularities of Embedding Algebraic Surfaces*. Academic Press. New York and London. 1966.

[4] G. Albanese paper Transformazione birazionale di una superficie algebrica qualunque in un'altra priva di punti multipli.. Rend. Circolo Matem. Palermo. vol 48. pages 321–332. 1924

[5] B. Bennett. On the characteristic functions of a local ring. Ann. Math. vol. 91. pages 25–87 1970.

[6] E. Bierstone and P. Milman. Uniformization of Analytic Spaces. J. AMS. Oct 1989.

[7] V. Cossart Thesis. Orsay. 1987.

[8] V. Cossart Forme normale d'une fonction sur un k-schema de dimension 3 et caraxteristique positive. Geometrie algebrique et applications, vol 1 . pages 1– 22. 1988.

[9] V. Cossart. Desingularization of Embedded Excellent Surfaces. Tohoku Math. vol 33. pages 25–33. 1981.

[10] J. Giraud. Etude locale des singularités. Cours de 3ème Cycle. Univ. de Paris XI, U.E.R. Math. 91-Orsay. vol 26 . 1971–72.

[11] J. Giraud. Contact maximal en caracteristique positive. Ann. Ens, 4-eme serie, t.8, fasc. 2. 1975.

[12] H. Hironaka Resolution of singularities of an algebraic variety over a field of characteristic zero. Ann. Math. vol 79 pages 109–326. 1964

[13] H. Hironaka. Characteristic polyhedra of singularities. J. Math. Kyoto Univ. vol 7. pages 251–293. 1968 .

[14] H. Hironaka. Certain numerical characters of singularities. J. Math. Kyoto Univ. vol 10. pages 151–187. 1970.

[15] H. Hironaka. Additive groups associated with points of a projective space Ann. Math. vol 92. pages 327–334. 1970.

[16] M. Herrmann and U. Orbanz. Remark on a paper by B. Singh on certain numerical characters of singularities. J. Pure Appl. Algebra. vol 24. pages 151–156. 1982.

[17] J. Lipman. Desingularization of two dimensional schemes. Ann. of Math. vol 107. pages 151–207. 1978.

[18] T. T. Moh. On a stability theorem for local uniformization in characteristic p. Journal of RIM. vol 23 no.6 pages 965–973. 1987.

[19] T. T. Moh. Canonical uniformization of hypersurface singularities of characteristic zero. Manuscript.

[20] T. Oda. Hironaka's additive group scheme. Number Theory, Algebraic Geometry and Commutative Algebra in honor of Y. Akizuki. Kinokuniya. Tokyo. 1973. (Y. Kusunoki et al., eds.)

[21] T. Oda. Hironaka's additive group scheme, II. Publ. Res. Inst. Math. Sci. Kyoto Univ. vol 19, pages 1163–1179. 1983.

[22] B. Singh. Effect of a permissible blowing-up on the local Hilbert functions. Invent. Math. vol 26 pages 201–212. 1974.

[23] M. Spivakovsky. A counterexample to Hironaka's "hard" polyhedra game. Publ. Rest. Inst. Math. Sci. Kyoto Univ. vol 18 pages 1009–1012. 1982.

[24] M. Spivakovsky A solution to Hironaka's polyhedra game. Arithmetic and Geometry, Vol. II. Geometry, Progress in Math. Birkhäuser. Boston. 1983, pp. 419–432 vol 36.Papers dedicated to I. R. Shafarevich on the occasion of his sixtieth birth (M. Artin and J. Tate, eds.).

[25] O.E. Villamayor. Construtiveness of Hironaka's resolution. Ann. Serent. Ec. Nonn.Sup. vol 4^e serie, t.22 pages 1–32. 1989.

[26] R. J. Walker. Reduction of singularities of an algebraic surface. Ann. of Math. vol 36 pages 336–365. 1935.

[27] Boris Youssin. Newton Polyhedra without coordinates. Dissertation, Harvard. 1988

[28] O. Zariski. The reduction of the singularities of an algebraic surface. Annals of Math. vol 40 pages 639–689. 1939.

[29] O. Zariski. Local uniformization on algebraic varieties. Annals of Math. vol 41 pages 859–896. 1940.

[30] O. Zariski. A simplified proof for the resolution of singularities of an algebraic surface. Annals of Math. 1942 vol 43 pages 583–593.

[31] O. Zariski. The compactness of the Riemann manifold of an abstract field of algebraic functions. Bull. Amer. Math. Soc. vol 45 pages 683–691 1944.

[32] O. Zariski. Reduction of the singularities of algebraic three-dimensional varieties. Annals of Math. vol 45 pages 472–542. 1944.

Address of author:

Dept. of Mathematics
Purdue University
West Lafayette
IN 47907 U.S.A.
Email: ttm@math.purdue.edu

GEOMETRY OF PLANE CURVES VIA
TOROIDAL RESOLUTION

Mutsuo Oka[1]

1 Introduction

Let $C = \{f(x,y) = 0\}$ be a germ of a reduced plane curve. As examples of
the basic invariants of a plane curve, we have the Milnor number, the number
of irreducible components, the resolution complexity, the Puiseux pairs of the
irreducible components and their intersection multiplicities. In fact, the Puiseux
pairs of the irreducible components and their intersection multiplicities are
enough to describe the embedding topological type of C (see [17], [9]).

It is well known that C can be resolved by a composition of (ordinary)
blowing-ups. However this process is usually too long and even though we know
the information of the composition of these blowing-ups, it is not so easy to read
from these informations how the Puiseux pairs and the intersection multiplicity
behave.

Instead of ordinary blowing-ups, we study the toroidal resolution of C,
which is a resolution consisting of a finite composition of admissible toric
blowing-ups. Though a toric blowing-up is a finite composition of ordinary
blowing-ups, as a package, it contains more information. In fact, it turns out
that the toroidal resolution contains information which is perfectly suitable for
the determination of the above invariants. In [16], [10], we have proved the
existence of the toroidal resolution and a basic theorem about the complexity
of the resolution.

The purpose of this paper is to show how we can read Puiseux pairs and
the intersection multiplicities among the irreducible components using the data
of the toroidal resolution. See Theorem 7-1 and Theorem 7-6 in section 7. As an
application, we consider a plane curve C which is a plane curve obtained as an
n-times iterated generic hyperplane section of a non-degenerate hypersurface.
We will show that the resolution complexity of such a curve is at most $n + 1$
and each irreducible component has at most one Puiseux pair (Theorem 8-1).

2 Toric blowing-up and a tower of toric blowing-ups

First we recall the definition of a toric blowing-up (or a toric modification). Let

$$\sigma = \begin{pmatrix} \alpha & \beta \\ \gamma & \delta \end{pmatrix}$$

---
[1]This work is done when the author is visiting at Mathematisches Institut, University of
Basel in 1991. He thanks to the Institut for their support.

be a unimodular integral 2×2 matrix. We associate to σ a birational morphism

$$\pi_\sigma \colon (\mathbb{C}^*)^2 \to (\mathbb{C}^*)^2,$$

by $\pi_\sigma(x,y) = (x^\alpha y^\beta, x^\gamma y^\delta)$. If $\alpha, \gamma \geq 0$ (respectively $\beta, \delta \geq 0$), this map can be extended to $x = 0$ (resp. $y = 0$). Note that if τ is another unimodular 2×2 matrix,

$$\pi_\sigma \circ \pi_\tau = \pi_{\sigma\tau} \quad \text{and} \quad (\pi_\sigma)^{-1} = \pi_{\sigma^{-1}}.$$

Let N be a real vector space of dimension two with a fixed basis $\{E_1, E_2\}$. Through this basis, we identify N with $\mathbb{R}^2$. We denote a vector in N by a column vector. So E_1 and E_2 correspond to ${}^t(1,0)$ and ${}^t(0,1)$ respectively. Let N^+ be the space of positive vectors of N. Let $\{P_1, \ldots, P_m\}$ be given positive primitive integral vectors in N^+. Let $P_i = {}^t(a_i, b_i)$ and assume that $\det(P_i, P_{i+1}) = a_i b_{i+1} - a_{i+1} b_i > 0$ for each $i = 0, \ldots, m$. Here $P_0 = E_1$, $P_{m+1} = E_2$. We associate to $\{P_1, \ldots, P_m\}$ a simplicial cone subdivision Σ^* which has $m+1$ cones $\mathrm{Cone}(P_i, P_{i+1})$ of dimension two where

$$\mathrm{Cone}(P_i, P_{i+1}) := \{tP_i + sP_{i+1}; t, s \geq 0\}.$$

We call $\{P_0, \ldots, P_{m+1}\}$ the vertices of Σ^*. We say that Σ^* is a regular simplicial

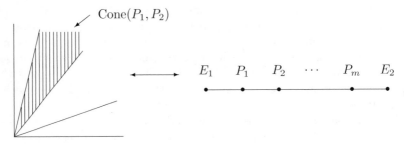

Figure 2.1: A cone subdivision.

cone subdivision of N^+ if $\det(P_i, P_{i+1}) = 1$ for each $i = 0, \ldots, m$.

Assume that Σ^* is a given regular simplicial cone subdivision with vertices $\{P_0, P_1, \ldots, P_m, P_{m+1}\}$, $(P_0 = E_1, P_{m+1} = E_2)$ and let $P_i = {}^t(a_i, b_i)$. For each cone, $\mathrm{Cone}(P_i, P_{i+1})$, we associate the unimodular matrix

$$\sigma_i := \begin{pmatrix} a_i & a_{i+1} \\ b_i & b_{i+1} \end{pmatrix}$$

We often identify $\mathrm{Cone}(P_i, P_{i+1})$ and the unimodular matrix σ_i. Let (x, y) be a system of coordinates at the origin O of $\mathbb{C}^2$. Then we consider, for each σ_i, an affine space $\mathbb{C}^2_{\sigma_i}$ of dimension two with coordinates $(x_{\sigma_i}, y_{\sigma_i})$ and the birational map $\pi_{\sigma_i} \colon \mathbb{C}^2_{\sigma_i} \to \mathbb{C}^2$. Now we construct a non-singular algebraic variety X. First we consider the disjoint union $\bigcup_i \mathbb{C}^2_{\sigma_i}$ and we define the variety X as the quotient of this union by the following identification: Two points $(x_{\sigma_i}, y_{\sigma_i}) \in \mathbb{C}^2_{\sigma_i}$ and $(x_{\sigma_j}, y_{\sigma_j}) \in \mathbb{C}^2_{\sigma_j}$ are identified if and only if the birational map $\pi_{\sigma_j}^{-1}{}_{\sigma_i}$ is defined at the point $(x_{\sigma_i}, y_{\sigma_i})$ and $\pi_{\sigma_j}^{-1}{}_{\sigma_i}(x_{\sigma_i}, y_{\sigma_i}) = (x_{\sigma_j}, y_{\sigma_j})$.

As can be easily checked, X is non-singular and the maps $\{\pi_{\sigma_i} \colon \mathbb{C}^2_{\sigma_i} \to \mathbb{C}^2 : 0 \leq i \leq m\}$ glue into a proper analytic map $\pi \colon X \to \mathbb{C}^2$.

Definition 2-1. *The map* $\pi \colon X \to \mathbb{C}^2$ *(or its restriction on an open neighborhood of the origin) is called the* toric blowing-up *(or* toric modification*) associated with* $\{\Sigma^*, (x, y), O\}$ *where* Σ^* *is a regular simplicial cone subdivision of* N^+ *and* (x, y) *is a coordinate system of* $\mathbb{C}^2$ *centered at the origin* O.

This modification enjoys the following properties.

1. $\{\mathbb{C}^2_{\sigma_i}, (x_{\sigma_i}, y_{\sigma_i})\}$, $(0 \leq i \leq m)$ give coordinate charts of X and we call them the toric coordinate charts of X.

2. Two affine divisors $\{y_{\sigma_{i-1}} = 0\} \subset \mathbb{C}^2_{\sigma_{i-1}}$ and $\{x_{\sigma_i} = 0\} \subset \mathbb{C}^2_{\sigma_i}$ glue together to make a compact divisor isomorphic to $\mathbb{P}^1$ for $1 \leq i \leq m$. We denote this divisor by $\widehat{E}(P_i)$. We call $\{\mathbb{C}_{\sigma_i}; (x_{\sigma_i}, y_{\sigma_i})\}$ (or, respectively $\{\mathbb{C}_{\sigma_{i-1}}; (x_{\sigma_{i-1}}, y_{\sigma_{i-1}})\}$) the right (resp. the left) toric coordinate chart of $\widehat{E}(P_i)$ hereafter.

3. $\pi^{-1}(O) = \bigcup_{i=1}^m \widehat{E}(P_i)$ and $\pi \colon X - \pi^{-1}(O) \to \mathbb{C}^2 - \{O\}$ is an isomorphism. The non-compact divisor $x_{\sigma_0} = 0$ (respectively $y_{\sigma_m} = 0$) is mapped isomorphically onto the divisor $x = 0$ (resp. $y = 0$). We denote the divisor $x_{\sigma_0} = 0$ by $\widehat{E}(P_0)$ and the divisor $y_{\sigma_m} = 0$ by $\widehat{E}(P_{m+1})$ respectively.

4. $\widehat{E}(P_i) \cap \widehat{E}(P_j) \neq \emptyset$ if and only if $i - j = \pm 1$. If $i - j = \pm 1$, they intersect transversely at a point. Similarly $\widehat{E}(P_i) \cap \mathbb{C}_{\sigma_j} \neq \emptyset$ if and only if $j = i \pm 1$. Thus the configuration graph of the exceptional divisors of $\pi \colon X \to \mathbb{C}^2$ is a line graph with m vertices.

5. The self-intersection numbers $\widehat{E}(P_i)^2$ can be easily computed using the well-known property $(\pi^* x) \cdot \widehat{E}(P_i) = 0$ (see [8], Theorem (2.6)): $\widehat{E}(P_i)^2 = -(a_{i-1} + a_{i+1})/a_i$. Here $(\pi^* x)$ is the divisor associated to the function $\pi^* x = x \circ \pi$.

In the case of $\Sigma^* = \{\mathrm{Cone}(E_1, E_2)\}$ $(m = 0)$, $\pi \colon X \to \mathbb{C}^2$ is nothing but the identity map and we say π is trivial.

Now we consider the composition of non-trivial toric blowing-ups

$$\mathcal{T} \colon X_k \xrightarrow{\ p_k\ } X_{k-1} \xrightarrow{\ p_{k-1}\ } \cdots \longrightarrow X_1 \xrightarrow{\ p_1\ } X_0 \tag{2.1}$$

where each $X_{i+1} \xrightarrow{p_{i+1}} X_i$ is the toric blow-up associated with $\{\Sigma_i^*, (u_i, v_i), \xi_i\}$ where $\xi_i \in X_i$ is the center of the blowing-up, (u_i, v_i) are local coordinates of X_i

in a neighborhood U_i of ξ_i, so that ξ_i corresponds to $(0,0)$ in these coordinates, and Σ_i^* is a regular simplicial cone subdivision of N^+. Let $E_{i,1}, \ldots, E_{i,s_i}$ be the exceptional divisors of $p_i\colon X_i \to X_{i-1}$. By abuse of the notation, we denote by the same $E_{i,j}$, the strict transforms of $E_{i,j}$ to X_ℓ for any $\ell \geq i$. Thus the exceptional divisors of the composition $p_1 \circ \cdots \circ p_k\colon X_k \to X_0$ is given by the union of $\{E_{i,j}\}$ for $1 \leq i \leq k$, $1 \leq j \leq s_i$.

We call $\mathcal{T}$ a tree of admissible toric blowing-ups if the following conditions are satisfied.

(i) X_0 is an open neighborhood of the origin of $\mathbb{C}^2$ and $\xi_0 = O$.

(ii) For $i > 0$, $p_{i+1}\colon X_{i+1} \to X_i$ is non-trivial and there exists α_i, β_i with $1 \leq \alpha_i \leq i$, $1 \leq \beta_i \leq s_{\alpha_i}$ so that $\xi_i \in E_{\alpha_i,\beta_i}$ and $\xi_i \notin E_{\alpha,\beta}$ for any (α,β) with $(\alpha,\beta) \neq (\alpha_i,\beta_i)$, $1 \leq \alpha \leq i$ and $1 \leq \beta \leq s_\alpha$.

(iii) Let $\{\mathbb{C}^2_{\sigma_i}, (x_{\sigma_i}, y_{\sigma_i})\}$ be the right toric coordinate chart of E_{α_i,β_i} and let $\xi_i = (0, \gamma_i)$ in this chart. Recall that $E_{\alpha_i,\beta_i} = \{x_{\sigma_i} = 0\}$ by the definition of the right toric coordinates. Then the coordinate function u_i on U_i is the restriction of x_{σ_i} to U_i and $v_i = y_{\sigma_i} - \gamma_i + h_i(x_{\sigma_i})$ for some analytic function h_i with $h_i(0) = 0$.

A tree of admissible toric blowing-ups

$$\mathcal{T}\colon X_k \xrightarrow{p_k} X_{k-1} \xrightarrow{p_{k-1}} \cdots \longrightarrow X_1 \xrightarrow{p_1} X_0$$

is called a tower of admissible toric blowing-ups if in addition of the above conditions, we have

(iv) $p_i(\xi_i) = \xi_{i-1}$. In other words, $\alpha_i = i$.

Let $\mathcal{T}$ be a tree of admissible toric blowing-ups as in (2.1). We define an oriented graph which we call the hierarchy graph, $\mathcal{HG}(\mathcal{T})$, as follows. The vertices of $\mathcal{HG}(\mathcal{T})$ are $\{v_i : 0 \leq i \leq k-1\}$ which corresponds bijectively to the centers xi_i, $i = 0, \ldots, k-1$ of the blowing-ups $p_{i+1}\colon X_{i+1} \to X_i$, $i = 0, \ldots, k-1$. For each v_i with $i > 1$, we join v_i with v_{α_i-1}. We orient this edge so that the positive direction is from v_{α_i-1} to v_i. The orientation of the edges generates also an partial order over the vertices so that v_0 is the minimal element. Note that $\mathcal{HG}(\mathcal{T})$ is a line graph if and only if $\mathcal{T}$ is a tower.

Take a vertex v_a. Then there is a unique subgraph $\mathcal{HG}(v_a)$ which is a totally ordered line graph, and whose end vertices are v_0 and v_a. Let $\{v_{i_0}, v_{i_1}, \ldots, v_{i_{\ell-1}}\}$ be the vertices of $\mathcal{HG}(v_a)$ ($i_0 = 0$, $i_{\ell-1} = a$). Then

$$i_\nu = \alpha_{i_{\nu+1}} - 1, \qquad \nu = 0, \ldots, \ell - 2. \tag{2.2}$$

$\mathcal{HG}(v_a)$ is called the hierarchy graph of the vertex v_a. ℓ is called the hierarchy length of the vertex v_a. For any exceptional divisor $E_{a+1,b}$, we also define the hierarchy graph $\mathcal{HG}(E_{a+1,b})$ of $E_{a+1,b}$ by $\mathcal{HG}(E_{a+1,b}) := \mathcal{HG}(v_a)$. The following diagram indicates the situation. Put $\beta_{i_\nu} = j_{\nu-1}$.

$$
\begin{array}{ccccccc}
\cdots \longrightarrow & X_{i_\nu} & \longrightarrow & X_{i_{\nu-1}+1} & \xrightarrow{p_{i_{\nu-1}+1}} & X_{i_{\nu-1}} & \longrightarrow \cdots \\
& \uparrow & & \uparrow & & \uparrow & \\
\cdots \longmapsto & \xi_{i_\nu} \in E_{i_{\nu-1}+1,\beta_{i_\nu}} \xmapsto{\simeq} & & \xi_{i_\nu} \in E_{i_{\nu-1}+1,\beta_{i_\nu}} \longmapsto & & \xi_{i_{\nu-1}} \in E_{i_{\nu-2}+1,\beta_{i_{\nu-1}}} \longmapsto \cdots
\end{array}
$$

There exists a tower of admissible toric blowing-ups which corresponds to the graph $\mathcal{HG}(v_a)$ or $\mathcal{HG}(E_{a+1,b})$:

$$
\mathcal{HT}(E_{a+1,b}): Y_\ell \xrightarrow{q_\ell} Y_{\ell-1} \xrightarrow{q_{\ell-1}} \cdots \longrightarrow Y_1 \xrightarrow{q_1} Y_0, \tag{2.3}
$$

and surjective birational morphisms $\varphi_t \colon X_{i_t} \to Y_t$, $0 \le i \le \ell-1$, $\varphi_\ell \colon X_{i_{\ell-1}+1} \to Y_\ell$ which satisfy the following conditions:

(i) $Y_0 = X_0$, $Y_1 = X_1$, $\varphi_0 = \mathrm{Id}_{X_0}$ and $\varphi_1 = \mathrm{Id}_{X_1}$.

(ii) The following diagram is commutative (where $p^{(t)}$ is the composition $p_{i_{t-1}+1} \circ \cdots \circ p_{i_t}$).

$$
\begin{array}{ccc}
X_{i_t} & \xrightarrow{p^{(t)}} & X_{i_{t-1}} \\
\downarrow{\varphi_t} & & \downarrow{\varphi_{t-1}} \\
Y_t & \xrightarrow{q_t} & Y_{t-1}
\end{array}
$$

(iii) The map φ_s is a birational map and the exceptional divisors $\{E_{i_t+1,\beta}\}$, $1 \le t \le s-1$, $1 \le \beta < s_{i_t+1}$ are mapped isomorphically onto the exceptional divisors of the toric modification $q_t \colon Y_{t+1} \to Y_t$ for $1 \le t \le s-1$, and the other divisors are collapsed to points.

In fact, one can construct this tower inductively starting from $Y_0 = X_0$, taking the associated toric blowing-up corresponding to $\{\Sigma_{i_t}^*; (u_{i_t}, v_{i_t}), \xi_{i_t}\}$ for $t = 0, \ldots, \ell-1$. We omit the details for this construction. We call the tower $\mathcal{HT}(E_{a+1,b})$ the hierarchy tower of the divisor $E_{a+1,b}$. This is a tower of admissible blowing-ups. For any point $\xi_{a+1,b} \in E'_{a+1,b}$, where

$$
E'_{a,b} = E_{a+1,b} - \bigcup_{(i,j) \ne (a+1,b)} E_{i,j},
$$

the local behavior of the composition $p_1 \circ \cdots \circ p_k \colon X_k \to X_0$ in a neighborhood of $\xi_{a+1,b}$ is equivalent to that of the composition $q_1 \circ \cdots \circ q_\ell \colon Y_\ell \to Y_0$. This fact will be used to compute the Puiseux pairs of an irreducible component C_j of a given germ of a plane curve C whose proper transform $\widetilde{C}_j$ is smooth and transversal to $E'_{a+1,b}$ in sections 6 and 7.

3 Dual Newton diagram and an admissible toric blowing-up

Let f be a given complex analytic function of two variables defined on an open neighborhood U of the origin O of $\mathbb{C}^2$ and suppose that $f(O) = 0$. Let $f(x,y) = \sum a_{\alpha,\beta} x^{\alpha} y^{\beta}$ be the Taylor expansion of f at the origin $O = (0,0)$ with respect to a local coordinate system (x,y) centered at the origin O. We assume that $f(x,y)$ is reduced as a germ.

The Newton polygon $\Gamma_+(f;x,y)$ of f at the point O relatively to the coordinate system (x,y) is the convex hull of the set $\bigcup_{a_{\alpha,\beta} \neq 0} \{(\alpha,\beta) + \mathbb{R}_+^2\}$ and the Newton boundary $\Gamma(f;x,y)$ of f at O is the union of compact faces of the boundary of the Newton polygon of f at O. For each compact face Δ of $\Gamma(f;x,y)$ we denote by $f_{\Delta}(x,y)$ the polynomial $\sum_{(\alpha,\beta)\in\Delta} a_{\alpha,\beta} x^{\alpha} y^{\beta}$. This is a weighted homogeneous polynomial.

Let $M := \mathbb{R}^2$ be the euclidean space where the Newton polygon $\Gamma_+(f;x,y)$ is contained and we fix (u,v) as coordinates. Let N be the dual vector space of M and let N^+ be the space of the non-negative dual vectors. For any non-negative dual vector $P = {}^t(a,b)$, we define $d(P;f)$ to be the smallest value of the restriction of the linear function P to the Newton polygon $\Gamma_+(f;x,y)$ and $\Delta(P;f)$ be the face where this smallest value is taken. Namely,

$$
\begin{cases}
d(P;f) = \min\{P(u,v) = au + bv; (u,v) \in \Gamma_+(f;x,y)\}, \\
\Delta(P;f) = \{(u,v) \in \Gamma_+(f;x,y); P(u,v) = d(P;f)\}.
\end{cases}
$$

Let $f_P(x,y) = f_{\Delta(P;f)}(x,y)$. We also call $f_P(x,y)$ the face function of $f(x,y)$ with respect to the dual vector P. By the definition, $f_P(x,y)$ is a weighted homogeneous polynomial of weight (a,b) with degree $d(P;f)$. For each face Δ of dimension 1 there is a unique primitive integral dual vector $P = {}^t(a,b)$ such that $\Delta(P;f) = \Delta$. The Newton boundary $\Gamma(f;x,y)$ contains only a finite number of faces of dimension one. Let $\Delta_1, \ldots, \Delta_m$ be the one-dimensional faces.

Let $P_i = {}^t(a_i,b_i)$ be the corresponding primitive integral dual vector, i.e., $\Delta(P_i;f) = \Delta_i$. Then we can write

$$
f_{P_i}(x,y) = c_i x^{r_i} y^{s_i} \prod_{j=1}^{k_i} (y^{a_i} - \gamma_{i,j} x^{b_i})^{\nu_{i,j}}, \tag{3.4}
$$

with distinct non-zero complex numbers $\gamma_{i,1}, \ldots, \gamma_{i,k_i}$. We order the compact faces Δ_i ($1 \leq i \leq m$) so that $\det(P_i, P_{i+1}) = a_i b_{i+1} - a_{i+1} b_i > 0$ for $i = 1, \ldots, m-1$.

Definition 3-1. We say that f is non-degenerate on Δ_i if the restriction $f_{\Delta_i}: \mathbb{C}^{*2} \to \mathbb{C}$ has no critical points (see [6]). This is equivalent to $\nu_{i,j} = 1$ for each $j = 1, \ldots, k_i$. f is non-degenerate if f is non-degenerate on any face of $\Gamma(f;x,y)$.

In the space N^+ of positive dual vectors, we introduce an equivalence relation $\sim$ defined by $P \sim Q$ if and only if $\Delta(P; f) = \Delta(Q; f)$. The equivalence classes of this equivalence relation define a conical subdivision of N^+. This gives a simplicial cone subdivision of N^+ with $m + 2$ vertices $\{P_0, \ldots, P_{m+1}\}$. We denote this subdivision by $\Gamma^*(f; x, y)$ and we call it the dual Newton diagram of f with respect to the system of coordinates (x, y). $\Gamma^*(f; x, y)$ has $m + 1$ two-dimensional cones $\mathrm{Cone}(P_i, P_{i+1})$, $i = 0, \ldots, m$ $(P_0 = E_1, P_{m+1} = E_2)$.

Note that $\Gamma^*(f; x, y)$ need not to be regular.

Definition 3-2. *A regular simplicial cone subdivision Σ^* is admissible for $f(x, y)$ with respect to the system of coordinate (x, y) if Σ^* is a subdivision of $\Gamma^*(f; x, y)$. The corresponding toric blowing-up $\pi \colon X \to \mathbb{C}^2$ is called an admissible toric blowing-up for $f(x, y)$ with respect to the coordinate system (x, y). In Lemma (3.6) of [13], we have shown that there exists a unique canonical regular simplicial cone subdivision. We denote this subdivision by $\Sigma^*(f; x, y)$ and we call the corresponding toric blowing-up the canonical toric blowing-up with respect to the function $f(x, y)$ and the coordinate system (x, y).*

Let $\{P_0, P_1, \ldots, P_m, P_{m+1}\}$ $(P_0 = E_1, P_{m+1} = E_2)$ be the vertices of the dual Newton diagram $\Gamma^*(f; x, y)$ and let Σ^* be an admissible regular simplicial cone subdivision and let $T_{i,1}, \ldots, T_{i,\ell_i}$ be the new vertices which are added in $\mathrm{Cone}(P_i, P_{i+1})$ in this order. We consider the associated toric bowing-up $\pi \colon X \to \mathbb{C}^2$. Basic properties of this toric blowing-up are:

1. The exceptional divisor $\widehat{E}(T_{i,j})$ does not intersect with the proper transform $\widetilde{C}$ for $0 \le i \le m$, $1 \le j \le \ell_i$. If Σ^* is equal to the canonical subdivision $\Sigma^*(f; x, y)$, the self-intersection number of the divisor $\widehat{E}(T_{i,j})$ is strictly less than -1 for $1 \le j \le \ell_i$ and $0 \le i \le m$.

2. The exceptional divisor $\widehat{E}(P_i)$ intersects with $\widetilde{C}$ at k_i points. In the right toric coordinates $\{\mathbb{C}_{\sigma_i}, (x_{\sigma_i}, y_{\sigma_i})\}$ of $\widehat{E}(P_i)$, where $\sigma_i = \mathrm{Cone}(P_i, T_{i,1})$, the intersection points are $\{(0, \gamma_{i,1}), \ldots, (0, \gamma_{i,k_i})\}$. Here $T_{i,\ell_i+1} = P_{i+1}$ in the case of $\ell_i = 0$.

3. The divisor of the function $\pi^* f$ is given by

$$(\pi^* f) = \sum_{i=1}^{m} \sum_{\ell=1}^{k_i} \widetilde{C}_{i,\ell} + \sum_{i=0}^{m+1} \sum_{j=0}^{\ell_i} d(T_{i,j}; f) \widehat{E}(T_{i,j}), \qquad (3.5)$$

$(\ell_{m+1} = 0)$, where $\widetilde{C}_{i,\ell}$ is the union of components of $\widetilde{C}$ which pass through $(0, \gamma_{i,\ell})$.

4. The curve $\widetilde{C}_{i,j}$ is reduced, smooth and intersects transversely with $\widehat{E}(P_i)$ if and only if $\nu_{i,j} = 1$. Thus if $f(x, y)$ is non-degenerate, $\pi \colon X \to \mathbb{C}^2$ is a good resolution of the function $f(x, y)$ as a germ at the origin.

As a corollary of property 2, we have the following well-known proposition.

Proposition 3-3. *Assume that $f(x,y)$ is irreducible as a germ of a function at the origin. We assume that the singularity of f at the origin is not normal crossing. Then $\Gamma(f;x,y)$ has only one face of dimension one touching both axis and the corresponding face function has only one factor. In the notation of (3.4), this implies $m = 1$ and $r_1 = s_1 = 0$ and $k_1 = 1$.*

Definition 3-4. *Now we consider a tree of admissible toric blowing-ups as in section 2:*

$$\mathcal{T}: X_k \xrightarrow{p_k} X_{k-1} \xrightarrow{p_{k-1}} \cdots \longrightarrow X_1 \xrightarrow{p_1} X_0 \ .$$

Assume that $p_{i+1}: X_{i+1} \to X_i$ is the toric blowing-up associated with $\{\Sigma_i^*, (u_i, v_i), \xi_i\}$ as in section 2. We say that $\mathcal{T}$ is is admissible for $f(x,y)$ if each toric blowing-up $p_{i+1}: X_{i+1} \to X_i$ is an admissible for the function $(p_1 \circ \cdots \circ p_i)^* f$ with respect to the system of coordinates (u_i, v_i) for $i = 0, \ldots, k-1$.

4   Resolution complexity

Let $\mathcal{F}$ be a graph which is a tree. Let $V(\mathcal{F})$ be the set of vertices of $\mathcal{F}$. For any vertex $v \in V(\mathcal{F})$, let $\delta(v)$ be the number of edges meeting at v. The complexity of the graph $\varrho(\mathcal{F})$ is defined ([10]) as

$$\varrho(\mathcal{F}) := 1 + \sum_{v \in V(\mathcal{F})} \max\big(\delta(v) - 2, 0\big).$$

Let $f(x,y)$ be a reduced analytic function defined on a neighborhood U of O in $\mathbb{C}^2$ and let $C = \big\{f(x,y) = 0\big\}$. Recall that a map $p: Y \to U$ is a (good) resolution of the function f or C as a germ at the origin if there exists an open neighborhood V of O with $V \subset U$ and

(i) Y is non-singular;

(ii) p is a proper surjective analytic mapping and the restriction $p: p^{-1}(V) - p^{-1}(O) \to V - \{0\}$ is biholomorphic;

(iii) the divisor $(p^* f)$ defined by the pull-back $p^* f := f \circ p$ has only normal crossings singularities and irreducible components are non-singular (not necessarily reduced) in $p^{-1}(V)$.

It is known that each component of the divisor $p^{-1}(O)$ is isomorphic to $\mathbb{P}^1$. To a given resolution $p: Y \to V$ of f, we associate a graph, which is usually called the dual resolution graph $\mathcal{G}(p)$ of p, in the following way. Let E_i for $1 \leq i \leq s$, be the irreducible components of the exceptional divisor $p^{-1}(O)$. To each E_i we associate a vertex v_i of $\mathcal{G}(p)$ and we give an edge joining v_i and v_ℓ if $E_i \cap E_\ell \neq \emptyset$. The dual resolution graph is a tree. The complexity $\varrho(p; f)$ of the resolution $p: Y \to V$ is defined by $\varrho(p; f) := \varrho\big(\mathcal{G}(p)\big)$.

Let $p\colon Y \to V$ be a resolution of f. An exceptional divisor E_i is called collapsible with respect to f if the self-intersection number E_i^2 is -1 and E_i intersects at most two components of the divisor (p^*f). A resolution $p\colon Y \to V$ is called minimal if and only if there is no collapsible exceptional divisor. It is known that there exists a unique minimal resolution up to isomorphism. The resolution complexity $\varrho(f)$ of f is defined by $\varrho(p; f)$ for a minimal resolution $p\colon Y \to V$. We also write $\varrho(C)$ instead of $\varrho(f)$. If f has a normal crossing singularity at O, $\varrho(f) = 0$ by definition. When we have already a resolution $p\colon Y \to V$, a minimal resolution is given by blowing down collapsible exceptional divisors one by one as long as possible. To blow down an exceptional divisor E_i corresponds to omit the vertex v_i in the dual graph. Thus by an easy induction, we have

Proposition 4-1. Let $p\colon Y \to V$ be a resolution of f. Then $\varrho(p; f) \geq \varrho(f)$.

Definition 4-2. A modification $p\colon Z \to V$ is called a toroidal resolution of the function $f(x, y)$ if there exists a tree of admissible toric blowing-ups for f

$$T\colon X_k \xrightarrow{\;p_k\;} X_{k-1} \xrightarrow{\;p_{k-1}\;} \cdots \longrightarrow X_1 \xrightarrow{\;p_1\;} X_0 \,,$$

such that $Z = X_k$ and $V = X_0$ and

(i) the map p is the composition of the morphisms $p_1 \circ \cdots \circ p_k$ and

(ii) the map $p\colon Z \to V$ is a resolution of f in a neighborhood of O.

5 Characteristic power and Puiseux Pairs

5.1 Characteristic powers

Let $\mathcal{O}_1$ be the ring of germs of holomorphic functions in the variable t. Let $\mathcal{O}_1^* = \{h(t) \in \mathcal{O}_1 : h(0) \neq 0\}$ be the set of the unit functions and let $\mathcal{M}_1$ be the maximal ideal.

Let $h(t) = \sum_{i=0}^{\infty} a_i t^i$ be the Taylor expansion of $h(t)$, $h(t) \in \mathcal{O}_1$. Let n be a given positive integer with $n \geq 2$. Assume first that there exists a $j > 0$ such that $j \not\equiv 0 \bmod n$ and $a_j \neq 0$. We consider the following integer:

$$P_1\big(h(t); n\big) = \min\{j > 0 : a_j \neq 0, \; j \not\equiv 0 \bmod n\}.$$

We call $P_1\big(h(t); n\big)$ the first characteristic power with respect to n. If there does not exists any $j > 0$ such that $j \not\equiv 0 \bmod n$ and $a_j \neq 0$, we define $P_1\big(h(t); n\big) = \infty$. In this case, $h(t)$ can be written as $H(t^n)$ for some $H \in \mathcal{O}_1$.

Note that the first characteristic power depends on the choice of the variable t.

Let $\nu_1 = P_1\big(h(t); n\big)$. We define $D_1\big(h(t); n\big) = \gcd(n, \nu_1)$ and we call it the first characteristic common divisor of $h(t)$ with respect to n. Let $n^{(1)} =$

$D_1(h(t); n)$. In the case of $\nu_1 = \infty$, $n^{(1)} = n$ by definition. Thus $1 \leq n^{(1)} \leq n$. If $1 < n^{(1)} < n$, we define the second characteristic power ν_2 as $\nu_2 = P_1(h(t); n^{(1)})$ and let $n^{(2)} = \gcd(n^{(1)}, \nu_2)$. We continue this operation until either $n^{(k)} = n^{(k-1)}$ or $n^{(k)} = 1$. The first case occurs if and only if $\nu(k) = \infty$ i.e.,

$$h(t) = H\left(t^{n^{(k-1)}}\right) \qquad \text{for some } H \in \mathcal{O}_1.$$

Thus there exist non-negative integers k, $\nu_1 < \cdots < \nu_k \leq \infty$ and $n > n^{(1)} > \ldots > n^{(k-1)} \geq n^{(k)} \geq 1$ so that

$$\nu_i = P_1(h(t); n^{(i-1)}), \quad n^{(i)} = \gcd(n^{(i-1)}, \nu_i), \qquad 1 \leq i \leq k, \quad n^{(0)} = n. \tag{5.6}$$

Let

$$P(h(t); n) = \{\nu_1, \ldots, \nu_k\}, \qquad D(h(t); n) = \{n^{(1)}, \ldots, n^{(k)}\}.$$

We call $P(h(t); n)$ and $D(h(t); n)$ the characteristic powers and the characteristic common divisors of $h(t)$ respectively with respect to the integer n. By definition, they are related by (5.6).

Lemma 5-1.

(i) Let $h(t) = \sum_{i=0}^{\infty} a_i t^i \in \mathcal{O}_1^*$. For any $r \in \mathbb{Q}$, $r \neq 0$, we have $P_1(h^r(t); n) = P_1(h(t); n)$. In particular, this implies $P(h^r(t); n) = P(h(t); n)$.

(ii) Let $h_1(t)$, $h_2(t) \in \mathcal{O}_1^*$. Then

$$P_1(h_1(t)h_2(t); n) \geq \min(P_1(h_1(t); n), P_1(h_2(t); n)),$$

and equality holds if $P_1(h_1(t); n) \neq P_1(h_2(t); n)$.

(iii) Let $P(h(t); n) = \{\nu_1, \ldots, \nu_k\}$. Let m be an integer which is divisible by n. Then $P(t^m h(t); n) = P(h(t); n) + m$ and $D(t^m h(t); n) = D(h(t); n)$ where $P(h(t); n) + m$ is by definition $\{\nu_1 + m, \ldots, \nu_k + m\}$.

PROOF.

Let $\nu_1 = P_1(h(t); n)$ and assume first that r is a positive integer. Then it is easy to see that the coefficient of t^j in $h^r(t)$ is zero for any j such that $j \not\equiv 0 \bmod n$ and $j < \nu_1$. On the other hand, the coefficient of t^{ν_1} in $h^r(t)$ is $ra_0 a_{\nu_1} \neq 0$. Thus the assertion (i) follows immediately in this case. Let $r = p/q$ with p, $q \in \mathbb{N}$. Then the assertion follows from the above case as

$$P_1(h(t); n) = P_1(h(t)^{1/q}; n) = P_1(h(t)^{p/q}; n).$$

Consider the equality $h(t)h(t)^{-1} = 1$. This implies that $P_1(h(t)^{-1}; n) = P_1(h(t); n)$. Thus for any negative rational number $r \in \mathbb{Q}$, the assertion follows from the above argument and the equality $P_1(h^{-1}; n) = P_1(h; n)$. The second assertions (ii) and (iii) can be proved by an easy calculation. $\qquad \square$

Lemma 5-2. *Consider the change of the parameter:* $\tau = \tau(t) = t\tau_0(t)$ *where* $\tau_0(t) \in \mathcal{O}_1^*$. *We can also write* $t = t(\tau) = \tau t_0(\tau)$ *with* $t_0(\tau) \in \mathcal{O}_1^*$. *Then*

(i) $P_1\big(\tau_0(t); n\big) = P_1\big(t_0(\tau); n\big)$. *In particular* $\mathcal{P}\big(\tau_0(t); n\big) = \mathcal{P}\big(t_0(\tau); n\big)$.

(ii) *Let* $h(t) \in \mathcal{O}_1$ *and assume that*

$$P_1\big(\tau_0(t); n\big) \geq P_1\big(h(t); n\big).$$

Then $P_1\big(h(t(\tau)); n\big) = P_1\big(h(t); n\big)$.

PROOF. Let $\tau_0(t) = \sum_{i=0}^{\infty} \alpha_i t^i$ and $t_0(\tau) = \sum_{i=0}^{\infty} \beta_i \tau^i$ and let $\nu = P_1\big(\tau_0(t); n\big)$ and $\ell = P_1\big(t_0(\tau); n\big)$. We show a contradiction by assuming that $\ell < \nu$. From the equality $t = t\big(\tau(t)\big)$, we obtain

$$1 = \tau_0(t) \cdot \left(\sum_{j=0}^{\infty} \beta_j t^j \tau_0(t)^j \right) \qquad \text{or} \qquad \tau_0(t)u(t) = 1, \qquad (5.7)$$

where $u(t) = \sum_{j=0}^{\infty} \beta_j t^j \tau_0(t)^j$. By (5.7) and Lemma 5-1, we have $\mathcal{P}\big(u(t); n\big) = \mathcal{P}\big(\tau_0(t); n\big)$. For any $0 < j < \ell$ with $\beta_j \neq 0$, we have $j \equiv 0 \bmod n$ by the assumption. Thus $P_1\big(\beta_j t^j \tau_0(t)^j; n\big) = j + \nu$ for any such j's. On the other hand, $P_1\big(\beta_\ell t^\ell \tau_0(t)^\ell; n\big) = \ell$. Therefore $P_1\big(u(t); n\big) = \ell$ and the corresponding coefficient is $\beta_\ell \alpha_0^\ell$. This contradicts the equality $P_1\big(u(t); n\big) = P_1\big(\tau_0(t); n\big)$ which results from (5.7). The case $\nu < \ell$ is also impossible. (Use the equality $\tau = \tau\big(t(\tau)\big)$.) This proves the assertion (i) of Lemma 5-2.

Now we consider the assertion (ii). Let $h(t) = \sum_{i=0}^{\infty} \gamma_i t^i$ and let $\xi = P_1\big(h(t); n\big)$. Assume that $\ell \geq \xi$. Then $h\big(t(\tau)\big) = \sum_{i=0}^{\infty} \gamma_i \tau^i t_0(\tau)^i$. Take any $i < \xi$ with $\gamma_i \neq 0$. Then $i \equiv 0 \bmod n$. Therefore for any such i, $P_1\big(\gamma_i \tau^i t_0(\tau)^i; n\big) = i + \ell$. On the other hand, $P_1\big(\gamma_\xi \tau^\xi t_0(\tau)^\xi; n\big) = \xi$. Thus we get the equality $P_1\big(h(t(\tau)); n\big) = \xi$. $\qquad\square$

5.2 Puiseux Pairs

Let $f(x, y)$ be a germ of an irreducible function such that $f(0, 0) = 0$. Then the curve $C := \{f(x, y) = 0\}$ can be parametrized as:

$$x(t) = t^n, \qquad y(t) = \sum_{i=m}^{\infty} a_i t^i, \qquad (5.8)$$

where m and n are positive integers. Hereafter we assume that $x = 0$ is not the tangential direction of $f(x, y) = 0$ at the origin, so $m \geq n$. C is non-singular if and only if $n = 1$.

Assume that $n > 1$ and let $\mathcal{P}(y(t); n) = \{\nu_1, \ldots, \nu_k\}$ and $\mathcal{D}(y(t); n) = \{n^{(1)}, \ldots, n^{(k)}\}$ be respectively the characteristic powers and the characteristic common divisors of $y(t)$ with respect to n. They are defined inductively by

$$\nu_i = P_1\big(y(t); n^{(i-1)}\big), \quad n^{(i)} = \gcd\big(n^{(i-1)}, \nu_i\big), \quad 1 \le i \le k, n^{(0)} = n. \quad (5.9)$$

As (5.8) is a parametrization of C, we must have $n^{(k)} = 1$. We define inductively positive integers $m_i, n_i \ge 2$ $(1 \le i \le k)$ by the property:

$$n_i = n^{(i-1)}/n^{(i)}, \quad m_i = \nu_i/n^{(i)}, \quad i = 1, \ldots, k. \quad (5.10)$$

By the definition, we have also

(i) $n^{(i-1)} = n_i \ldots n_k$, $i = 1, \ldots, k$

(ii) $\gcd(m_i, n_i) = 1$ and $m_i > m_{i-1} n_i$, $i = 1, \ldots, k$ $(m_0 = 1)$.

The pairs $\{(m_1, n_1), \ldots, (m_k, n_k)\}$ are called the Puiseux pairs of f. We can eliminate the parameter t using the equality $t = x^{1/n}$ so that we can write

$$y = x^{m_1/n_1} h_1(x^{1/n_1}) + x^{m_2/n_1 n_2} h_2(x^{1/n_1 n_2}) + \cdots + x^{m_k/n_1 \cdots n_k} h_k(x^{1/n_1 \cdots n_k}),$$

where $h_i(s)$ is a suitable function in $\mathcal{O}_1^*$ for $i = 1, \ldots, k$.

Remark 5-3. *The Puiseux pairs do not depend on the choice of coordinates (x, y) which can be easily proved using Lemma 5-1 and Lemma 5-2.*

Consider a plane curve $C = \{f(x, y) = 0\}$ *which is not necessarily irreducible at the origin as a germ of a curve. Let C_i $(i = 1, \ldots, r)$ be the irreducible components. It is known that the Puiseux pairs of the irreducible components C_i, $i = 1, \ldots, r$ and the intersection multiplicities $\{I(C_i, C_s; O) : i \ne s\}$ determine the topology of the embedded curve (U, C) where U is a sufficiently small neighborhood of the origin. See [1, 18, 19, 9].*

6   The Puiseux pairs of normal slice curves

In this section, we consider the Puiseux pairs of the normal slice curves of a given tree of admissible toric blowing-ups.

6.1   A tower of toric blowing-ups

First we consider a tower of admissible toric blowing-ups:

$$T: X_k \xrightarrow{p_k} X_{k-1} \xrightarrow{p_{k-1}} \cdots \longrightarrow X_1 \xrightarrow{p_1} X_0 .$$

Let $E_{i,1}, \ldots, E_{i,s_i}$ be the exceptional divisors of $p_i: X_i \to X_{i-1}$. As in section 2, we denote by the same $E_{i,j}$, the strict transforms of $E_{i,j}$ to X_ℓ for any $\ell \geq i$. Thus the exceptional divisors of $p_1 \circ \cdots \circ p_k \colon X_k \to X_0$ is the union $\{E_{i,j} : 1 \leq i \leq k, 1 \leq j \leq s_i\}$.

We assume that $p_{i+1} \colon X_{i+1} \to X_i$ is the toric blowing-up associated with $\{\Sigma_i^*, (u_i, v_i), \xi_i\}$ where $\xi_i \in X_i$ is the center of the blowing-up, (u_i, v_i) is a local coordinate of X_i in a neighborhood U_i of ξ_i such that $\xi_i = (0,0)$ in these coordinates. Σ_i^* is a regular simplicial cone subdivision of N^+.

We assume that the following are satisfied as in section 2:

(i) X_0 is an open neighborhood of the origin of $\mathbb{C}^2$ and $\xi_0 = O$.

(ii) For $i > 0$, $p_i \colon X_i \to X_{i-1}$ is non-trivial and there exists β_i with $1 \leq \beta_i \leq s_i$ so that $\xi_i \in E_{i,\beta_i}$ and $\xi_i \notin E_{\alpha,\beta}$ for any (α, β), with $1 \leq \alpha \leq i$, $1 \leq \beta \leq s_\alpha$ and $(\alpha, \beta) \neq (i, \beta_i)$.

(iii) Let $\{\mathbb{C}^2_{\sigma_i}, (x_{\sigma_i}, y_{\sigma_i})\}$ be the right toric coordinate chart of the divisor E_{i,β_i} in X_i such that $E_{i,\beta_i} = \{x_{\sigma_i} = 0\}$ and $\xi_i = (0, \gamma_i)$ in this chart. Then the coordinate function u_i on U_i is the restriction of x_{σ_i} to U_i and v_i is written as

$$v_i = y_{\sigma_i} - \gamma_i + h_i(x_{\sigma_i}), \qquad (6.11)$$

for some analytic function h_i with $h_i(0) = 0$.

Let

$$\sigma_i = \begin{pmatrix} a_i & a_i' \\ b_i & b_i' \end{pmatrix}, \qquad a_i b_i' - a_i' b_i = 1. \qquad (6.12)$$

Let $E'_{k,\beta_k} = E_{k,\beta_k} - \bigcup_{(\alpha,\beta) \neq (k,\beta_k)} E_{\alpha,\beta}$. We choose an exceptional divisor E_{k,β_k} and a point $\xi_k \in E'_{k,\beta_k}$. We consider the following sequence:

$$\xi_k \in E_{k,\beta_k} \overset{p_k}{\longmapsto} \xi_{k-1} \in E_{k-1,\beta_{k-1}} \longmapsto \cdots \longmapsto \xi_1 \in E_{1,\beta_1} \overset{p_1}{\longmapsto} \xi_0 = O.$$

Let $\{\mathbb{C}^2_{\sigma_k}, (x_k, y_k)\}$ be the right toric coordinate of the divisor E_{k,β_k} and let

$$\sigma_k = \begin{pmatrix} a_k & a_k' \\ b_k & b_k' \end{pmatrix}, \qquad a_k b_k' - a_k' b_k = 1.$$

Definition 6-1. We call $^t(a_i, b_i)$ the *dual vectors* of the divisor E_{i,β_i}. The collection of dual vectors

$$\left\{ \begin{matrix} a_1 \\ b_1 \end{matrix} , \ldots, \begin{matrix} a_k \\ b_k \end{matrix} \right\},$$

are called the dual vectors of the hierarchy of the divisor E_{k,β_k}.

Let us consider a germ of a smooth curve $\widetilde{C}$ which intersects with E_{k,β_k} at ξ_k transversely and let $C = (p_1 \circ \cdots \circ p_k)(\widetilde{C})$ which is a germ of a curve at the origin $O \in X_0$. We call $\widetilde{C}$ a generic normal slice curve of E_{k,β_k} at ξ_k. Let $\mathcal{I} = \{1 \leq i \leq k : a_i > 1\}$ and let $\mathcal{I} = \{r_1, r_2, \ldots, r_\ell\}$ where $r_i < r_{i+1}$. Then we have:

Theorem 6-2.

Let C be the image of a generic normal slice curve of E_{k,β_k} as above. Then C has the following parametrization:

$$\begin{cases} x = \psi(t), & \psi(t) = t^{a_1 \cdots a_k}, \\ y = \varphi(t), & \varphi(t) = t^{b_1 a_2 \cdots a_k} \rho(t), \quad \rho(0) = \gamma_1^{1/a_1}, \end{cases} \tag{6.13}$$

such that the characteristic powers and the characteristic common divisors of $\varphi(t)$ with respect to $a_1 \cdots a_k$ are given by

$$\mathcal{P}\big(\varphi(t); a_1 \cdots a_k\big) = \{\nu_{r_1}, \ldots, \nu_{r_\ell}\}, \qquad \mathcal{D}\big(\varphi(t); a_1 \cdots a_k\big) = \{n^{(r_1)}, \ldots, n^{(r_\ell)}\},$$

where

$$n^{(i)} = a_{i+1} \cdots a_k, \qquad \nu_i = (b_i + b_{i-1} a_i + \cdots + b_1 a_2 \cdots a_i) a_{i+1} \cdots a_k. \tag{6.14}$$

PROOF. We prove the Theorem by induction on k.

Step 1. Assume that $k = 1$. As $\tilde{C}$ is a smooth curve which is transversal to E_{1,β_1} at ξ_1, $\tilde{C}$ is parametrized as

$$\begin{cases} x_1 = \psi_1(t), & \psi_1(t) = t, \\ y_1 = \varphi_1(t), & \varphi_1(0) = \gamma_1, \end{cases}$$

where (x_1, y_1) is the right toric coordinate chart of E_{1,β_1}. Thus composing with $p_1 \colon X_1 \to X_0$, C is parametrized as

$$\begin{cases} x = \psi_0(t), & \psi_0(t) = \psi_1(t)^{a_1} \varphi_1(t)^{a_1'}, \\ y = \varphi_0(t), & \varphi_0(t) = \psi_1(t)^{b_1} \varphi_1(t)^{b_1'}. \end{cases}$$

As $\psi_0(t) = t^{a_1} \varphi_1(t)^{a_1'}$ and $\varphi_1(t)^{a_1'} \in \mathcal{O}_1^*$, we can take the new parameter

$$s = \big(\psi_1(t)^{a_1} \varphi_1(t)^{a_1'}\big)^{1/a_1} = t \varphi_1(t)^{a_1'/a_1}.$$

Thus $x = s^{a_1}$ in this parameter. Note that

$$\varphi_0(t) = \psi_0(t)^{b_1/a_1} \varphi_1(t)^{1/a_1}. \tag{6.15}$$

by the equality $a_1 b_1' - a_1' b_1 = 1$. Thus it is easy to see that $y = s^{b_1} \rho(s)$ where $\rho(s) = \varphi_1(t)^{1/a_1} \in \mathcal{O}_1^*$ and $\rho(0) = \gamma_1^{1/a_1}$. Let $\psi(s) = s^{a_1}$ and $\varphi(s) = s^{b_1} \rho(s)$. As $\mathcal{P}_1\big(\varphi(s); a_1\big) = b_1$ and $\gcd(a_1, b_1) = 1$, $\mathcal{P}\big(\varphi(s); a_1\big) = \{b_1\}$ if $a_1 > 1$ and $= \emptyset$ if $a_1 = 1$. Thus

$$x = \psi(s) = s^{a_1}, \qquad y = \varphi(s) = s^{b_1} \rho(t),$$

which is the desired parametrization.

Step 2. Assume that $k > 1$. Let C_1 be the image of $\widetilde{C}$ by $p_2 \circ \cdots \circ p_k \colon X_k \to X_1$. By the induction's assumption, there is a parametrization of C_1 of the following type:

$$\begin{cases} u_1 = \psi_1(t), & \psi_1(t) = t^{a_2 \cdots a_k}, \\ v_1 = \varphi_1(t), & \varphi_1(t) = t^{b_2 a_3 \cdots a_k} \rho_1(t), \quad \rho_1(t) \in \mathcal{O}_1^*, \end{cases}$$

and

$$\begin{aligned} \mathcal{P}\big(\varphi_1(t); a_2 \cdots a_k\big) &= \{\bar{\nu}_i : 2 \le i \le k, a_i > 1\}, \\ \mathcal{D}\big(\varphi_1(t); a_2 \cdots a_k\big) &= \{n^{(i)} : 2 \le i \le k, a_i > 1\}, \end{aligned}$$

where

$$n^{(i)} = a_{i+1} \cdots a_k, \qquad \bar{\nu}_i = (b_i + b_{i-1} a_i + \cdots + b_2 a_3 \cdots a_i) a_{i+1} \cdots a_k. \quad (6.16)$$

Note that

$$\nu_i = \bar{\nu}_i + b_1 a_2 \cdots a_k.$$

To compose with p_1, we first rewrite this parametrization with respect to the right toric coordinate (x_1, y_1). Recall that $x_1 = u_1$ and $y_1 = v_1 + \gamma_1 - h_1(x_1)$ for some $h_1 \in \mathcal{M}_1$. Therefore

$$\begin{cases} x_1 = \psi_1(t), & \psi_1(t) = t^{a_2 \cdots a_k}, \\ y_1 = \rho_1(t), & \rho_1(t) = \gamma_1 + \varphi_1(t) - h_1(t^{a_2 \cdots a_k}). \end{cases}$$

As $h_1(t^{a_2 \cdots a_k})$ is divisible by $t^{a_2 \cdots a_k}$, we have

$$\mathcal{P}\big(\rho_1(t); a_2 \cdots a_k\big) = \mathcal{P}\big(\varphi_1(t); a_2 \cdots a_k\big). \quad (6.17)$$

Composing with the projection p_1, we get the parametrization

$$\begin{cases} x = \psi_0(t), & \psi_0(t) = \psi_1(t)^{a_1} \rho_1(t)^{a_1'}, \\ y = \varphi_0(t), & \varphi_0(t) = \psi_1(t)^{b_1} \rho_1(t)^{b_1'}. \end{cases}$$

As $\psi_0(t) = t^{a_1 \cdots a_k} \rho_1(t)^{a_1'}$ and $\rho_1(t) \in \mathcal{O}_1^*$, we can take the new parameter

$$s = \psi_0(t)^{1/a_1 \cdots a_k}. \quad (6.18)$$

Let us consider the change of parameter:

$$t = s t_0(s), \quad s = t s_0(t), \qquad t_0(s), s_0(t) \in \mathcal{O}_1^*, \quad (6.19)$$

as $y = \psi_0(t)^{b_1/a_1} \rho_1(t)^{1/a_1}$ by (6.15), we can write

$$y = \varphi(s) = s^{b_1 a_2 \cdots a_k} \rho(s), \qquad \rho(s) = \rho_1\big(s t_0(s)\big)^{1/a_1}. \quad (6.20)$$

It is easy to see that $\rho(0) = \rho_1(0)^{1/a_1} = \gamma_1^{1/a_1}$. We will show that this parametrization

$$\begin{cases} x = \psi(s), & \psi(s) = s^{a_1\cdots a_k}, \\ y = \varphi(s), & \varphi(s) = s^{b_1 a_2\cdots a_k}\rho(s), \end{cases} \tag{6.21}$$

satisfies the assertion of Theorem 6-2.

Assertion 6-3. $\mathcal{P}\big(\rho(s); a_2\cdots a_k\big) = \mathcal{P}\big(\varphi_1(t); a_2\cdots a_k\big)$.

Assume this assertion for a while.

First we consider the case that $a_1 = 1$. Then as $\mathcal{P}\big(\varphi(s); a_2\cdots a_k\big) = b_1 a_2\cdots a_k + \mathcal{P}\big(\rho(s); a_2\cdots a_k\big)$, by (iii) of Lemma 5-1, (6.21) satisfies the assertion.

We consider now the case that $a_1 > 1$. Note that $r_1 = 1$ in this case. As $\rho(s) \in \mathcal{O}_1^*$, we have from (6.20)

$$P_1\big(\varphi(s); a_1\cdots a_k\big) = b_1 a_2\cdots a_k = \nu_1, \quad D_1\big(\varphi(s); a_1\cdots a_k\big) = a_2\cdots a_k = n^{(1)}.$$

Thus using the equality $\nu_i = \bar{\nu}_i + b_1 a_2\cdots a_k, i \geq 2$, we have that

$$\begin{aligned} \mathcal{P}\big(\varphi(s); a_1\cdots a_k\big) &= \{b_1 a_2\cdots a_k, b_1 a_2\cdots a_k + \mathcal{P}\big(\rho(s); a_2\cdots a_k\big)\}, \\ &= \{\nu_1, \nu_{r_2}, \ldots, \nu_{r_\ell}\}, \\ \mathcal{D}\big(\varphi(s); a_1\cdots a_k\big) &= \{n^{(1)}, \mathcal{D}\big(\rho(s); a_2\cdots a_k\big)\}, \\ &= \{n^{(1)}, n^{(r_2)}, \ldots, n^{(r_k)}\}, \end{aligned}$$

where $\nu_1, \ldots, \nu_k$ and $n^{(1)}, \ldots, n^{(k)}$ are as in the assertion of Theorem 6-2. This completes the proof of Theorem 6-2. □

PROOF. (OF ASSERTION 6-3) Assume first that $a_1' = 0$. Then $b_1' = a_1 = 1$ and $s = t$ and $\rho(s) = \rho_1(s)$. Thus the assertion is immediate from (6.17). So we assume now that $a_1' > 1$. By (6.17), Lemma 5-1 and Lemma 5-2, we obtain

$$\begin{aligned} P_1\big(t_0(s); a_2\cdots a_k\big) &= P_1\big(s_0(t); a_2\cdots a_k\big) & \text{by 5-2,} \\ &= P_1\big(s_0(t)^{a_1\cdots a_k}; a_2\cdots a_k\big) & \text{by 5-1,} \\ &= P_1\big(s(t)^{a_1\cdots a_k}; a_2\cdots a_k\big) - a_1\cdots a_k & \text{by 5-1,} \\ &= P_1\big(\psi_0(t); a_2\cdots a_k\big) - a_1\cdots a_k & \\ &= P_1\big(\rho_1(t); a_2\cdots a_k\big) & \text{by 5-1,} \\ &= P_1\big(\varphi_1(t); a_2\cdots a_k\big) = \bar{\nu}_{r_2} & \text{by (6.14).} \end{aligned}$$

Thus by (6.17), we get $P_1\big(t_0(s); a_2\cdots a_k\big) = P_1\big(\rho_1(t); a_2\cdots a_k\big)$. Applying (ii) of Lemma 5-2, we get

$$\begin{aligned} P_1\big(\rho(s); a_2\cdots a_k\big) &= P_1\big(\rho_1(t(s)); a_2\cdots a_k\big), \\ &= P_1\big(\rho_1(t); a_2\cdots a_k\big), \\ &= P_1\big(\varphi_1(t); a_2\cdots a_k\big) = \bar{\nu}_{r_2}, \\ &= (b_{r_2} + b_{r_2-1}a_{r_2} + \cdots + b_2 a_3\cdots a_{r_2})a_{r_2+1}\cdots a_k. \end{aligned}$$

Therefore $P_1\big(\rho(s); a_2 \cdots a_k\big) = \bar{\nu}_{r_2}$ and $D_1\big(\rho(s); a_2 \cdots a_k\big) = a_{r_2+1} \cdots a_k$. Using a similar argument and by induction, we conclude that

$$\mathcal{P}\big(\rho(s); a_2 \cdots a_k\big) = \{\bar{\nu}_{r_2}, \ldots, \bar{\nu}_{r_\ell}\}$$

and

$$\mathcal{D}\big(\rho(s); a_2 \cdots a_k\big) = \{n^{(r_2)}, \ldots, n^{(r_\ell)}\}.$$

This completes the proof of 6-3. $\square$

Let $\tilde{a}_1 = \min(a_1, b_1)$ and $\tilde{b}_1 = \max(a_1, b_1)$ and $\tilde{a}_i = a_i$ for $2 \leq i \leq k$. Define the integers n_i, m_i, $i = 1, \ldots, k$ by

$$(n_i, m_i) = \begin{cases} (\tilde{a}_1, \tilde{b}_1) & i = 1 \\ (a_i, b_i + b_{i-1}a_i + \cdots + b_2 a_3 \cdots a_i + \tilde{b}_1 a_2 \cdots a_i) & i \geq 2 \end{cases}$$

Corollary 6-4. Let C be as in Theorem 6-2. The Puiseux pairs of C are given by $\{(n_i, m_i) : \tilde{a}_i > 1\}$.

PROOF. By Theorem 6-2, we have the parametrization for C:

$$\begin{cases} x = \psi(t), & \psi(t) = t^{a_1 \cdots a_k}, \\ y = \varphi(t), & \varphi(t) = t^{b_1 a_2 \cdots a_k} \rho(t), \quad \rho(t) \in \mathcal{O}_1^*, \end{cases} \tag{6.22}$$

such that the characteristic powers and the characteristic common divisors of $\varphi(t)$ with respect to $a_1 \cdots a_k$ are given by $\mathcal{P}\big(\varphi(t); a_1 \cdots a_k\big) = \{\nu_{r_1}, \ldots, \nu_{r_\ell}\}$ and $\mathcal{D}\big(\varphi(t); a_1 \cdots a_k\big) = \{n^{(r_1)}, \ldots, n^{(r_\ell)}\}$ where

$$n^{(i)} = a_{i+1} \cdots a_k, \qquad \nu_i = (b_i + b_{i-1}a_i + \cdots + b_2 a_3 \cdots a_i) a_{i+1} \cdots a_k. \tag{6.23}$$

In the case of $a_1 \leq b_1$, the assertion is immediate from the definition of Puiseux pairs.

Assume that $b_1 < a_1$. We take the new coordinate

$$s = \varphi(t)^{1/b_1 a_2 \cdots a_k} = t s_0(t)$$

where $s_0(t) = \rho(t)^{1/b_1 a_2 \cdots a_k}$. Then $s_0(t) \in \mathcal{O}_1^*$. Write $t = s t_0(s)$, $t_0(s) \in \mathcal{O}_1^*$. In this parameter, we have

$$y = s^{b_1 a_2 \cdots a_k}, \qquad x = s^{a_1 \cdots a_k} \eta(s),$$

where $\eta(s) = t_0(s)^{a_1 \cdots a_k}$. By Lemma 5-1 and Lemma 5-2, we have

$$\begin{aligned} \mathcal{P}\big(\eta(s); a_2 \cdots a_k\big) &= \mathcal{P}\big(t_0(s); a_2 \cdots a_k\big), \\ &= \mathcal{P}\big(s_0(t); a_2 \cdots a_k\big), \\ &= \mathcal{P}\big(s_0(t)^{b_1 a_2 \cdots a_k}; a_2 \cdots a_k\big), \\ &= \mathcal{P}\big(\rho(t); a_2 \cdots a_k\big). \end{aligned}$$

Thus the proof is completely parallel to that of Theorem 6-2. $\square$

Remark 6-5. *Note that in Theorem 6-2, the role of a_i and b_i is unchangeable for $i \geq 2$ though a_1 and b_1 can be changed by a permutation of the coordinates. The reason is that a_i corresponds to the weight of the toric variable u_{i-1} which is the defining equation of the divisor $E_{i-1,\beta_{i-1}}$.*

6.2   General Tree

We consider now a tree of admissible toric blowing-up of the origin:

$$T : X_k \xrightarrow{p_k} X_{k-1} \xrightarrow{p_{k-1}} \cdots \longrightarrow X_1 \xrightarrow{p_1} X_0 .$$

Let $E_{i,1}, \ldots, E_{i,s_i}$ be the exceptional divisors of $p_i : X_i \to X_{i-1}$ which is the toric blowing-up associated with $\{\Sigma_i^*, (u_i, v_i), \xi_i\}$, where $\xi_i \in X_i$ is the center of the blowing-up. For $i > 0$, there exists α_i, β_i with $1 \leq \alpha_i \leq i$, $1 \leq \beta_i \leq s_{\alpha_i}$ so that $\xi_i \in E_{\alpha_i,\beta_i}$ and $\xi_i \notin E_{\alpha,\beta}$ for any (α, β) with $(\alpha, \beta) \neq (\alpha_i, \beta_i)$, $1 \leq \alpha \leq i$ and $1 \leq \beta \leq s_\alpha$. Let us consider an exceptional divisor $E_{a,b}$ and take a point $\xi_{a,b} \in E'_{a,b}$ where $E'_{a,b} = E_{a,b} - \bigcup_{(\alpha,\beta) \neq (a,b)} E_{\alpha,\beta}$.

We consider a germ of a smooth curve $\widetilde{C}_{a,b}$ which intersects with $E_{a,b}$ at $\xi_{a,b}$ transversely and we consider the image $C_{a,b} = (p_1 \circ \cdots \circ p_k)(\widetilde{C}_{a,b})$, which is a germ of a curve at the origin O. We call $\widetilde{C}_{a,b}$ a generic normal slice curve of $E_{a,b}$ at $\xi_{a,b}$ as in the case of the tower. Let $\{v_{i_0}, \ldots, v_{i_{\ell-1}}\}$ be the vertices of the hierarchy subgraph $\mathcal{HG}(E_{a,b})$ where $\alpha_{i_t} = i_{t-1} + 1$ and $i_0 = 0$, $i_{\ell-1} = a - 1$. Recall that this corresponds to

$$\xi_{a,b} \in E_{a,b} \xmapsto{p'} \xi_{i_{\ell-1}} \in E_{i_{\ell-2}+1,\beta_{\ell-1}} \longmapsto \cdots \longmapsto \xi_{i_1} \in E_{i_0+1,\beta_{i_1}} \xmapsto{p_1} \xi_0$$

Then $C_{a,b}$ can be understood as the image of the normal slice curve $\widetilde{C}_{a,b}$ by the composition $q_1 \circ \cdots \circ q_k$ of the hierarchy tower

$$\mathcal{HT}(E_{a,b}) : Y_\ell \xrightarrow{q_k} Y_{\ell-1} \longrightarrow \cdots \longrightarrow Y_1 \xrightarrow{q_1} Y_0 .$$

Let $\{{}^t(a_i, b_i) : i = 1, \ldots, k_a\}$ be the dual vectors of the hierarchy of the divisor $E_{a,b}$. In the words usen in the original tower, $P_s = {}^t(a_s, b_s)$ is the dual vector of $E_{i_{s-1}+1,\beta_{i_s}}$ in the right toric coordinate chart. Let $\xi_{i_1} = (0, \gamma_1)$ in the right toric coordinates of $E_{1,\beta_{i_1}}$. Applying Theorem 6-2, we obtain:

Theorem 6-6. *Let $C_{a,b}$ be the image of a generic normal slice curve of $E_{a,b}$ as above and let $\{P_i = {}^t(a_i, b_i) : i = 1, \ldots, k_a\}$ be the dual vectors of the hierarchy of $E_{a,b}$. Then $C_{a,b}$ has the following parametrization:*

$$\begin{cases} x = \psi(t), & \psi(t) = t^{a_1 \cdots a_{k_a}}, \\ y = \varphi(t), & \varphi(t) = t^{b_1 a_2 \cdots a_{k_a}} \rho(t), \quad \rho(0) = \gamma_1^{1/a_1}; \end{cases} \qquad (6.24)$$

such that the characteristic powers and the characteristic common divisors of $\varphi(t)$ with respect to $a_1 \cdots a_{k_a}$ are given by

$$\mathcal{P}\big(\varphi(t); a_1 \cdots a_{k_a}\big) = \{\nu_i; a_i > 1\}, \qquad \mathcal{D}\big(\varphi(t); a_1 \cdots a_{k_a}\big) = \{n^{(i)}; a_i > 1\}$$

where

$$n^{(i)} = a_{i+1} \cdots a_{k_a}, \qquad \nu_i = (b_i + b_{i-1}a_i + \cdots + b_1 a_2 \cdots a_i) a_{i+1} \cdots a_{k_a}, \quad (6.25)$$

Let $\widetilde{a}_1 = \min(a_1, b_1)$ and $\widetilde{b}_1 = \max(a_1, b_1)$ and $\widetilde{a}_i = a_i$ for $2 \le i \le k_a$. Let n_i, m_i be as before:

$$(n_i, m_i) = \begin{cases} (\widetilde{a}_1, \widetilde{b}_1), & i = 1 \\ (a_i, b_i + b_{i-1}a_i + \cdots + b_2 a_3 \cdots a_i + \widetilde{b}_1 a_2 \cdots a_i) & i > 1. \end{cases}$$

Corollary 6-7. Let $C_{a,b}$ be as in Theorem 6-4. Then the Puiseux pairs of $C_{a,b}$ are given by $\{(n_i, m_i); \widetilde{a}_i > 1\}$.

7 Geometry of plane curves via a toroidal resolution

Let $C = \{f(x,y) = 0\}$ be a given reduced germ of plane curve at the origin and let

$$\mathcal{T} : X_k \xrightarrow{p_k} X_{k-1} \xrightarrow{p_{k-1}} \cdots \longrightarrow X_1 \xrightarrow{p_1} X_0 , \qquad p = p_1 \circ \cdots \circ p_k,$$

be a tree of toric blowing-ups which gives an admissible toroidal resolution $p : X_k \to X_0$ of C as is defined in Sections 2 and 4. $p_{i+1} : X_{i+1} \to X_i$ is the toric blowing-up with respect to $\{\Sigma_i^*, (u_i, v_i), \xi_i\}$ where ξ_i is the center. Let $E_{i+1,1}, \ldots, E_{i+1,s_{i+1}}$ be its exceptional divisors. We assume that $\xi_i \in E_{\alpha_i, \beta_i}$ as before. Let $p^{(i)} = p_1 \circ \cdots \circ p_i$.

We assume also that the singularity of $p^{(i)*}f$ at ξ_i is not normal crossing for each $i = 0, \ldots, k-1$. Let $C_1, \ldots, C_r$ be the irreducible components of C and let $C_j = \{f_j(x,y) = 0\}$, where $f_j(x,y)$ is a germ of irreducible function.

Let $\widetilde{C}_1, \ldots, \widetilde{C}_r$ be the proper transforms of $C_1, \ldots, C_r$ to X_k. For each $\widetilde{C}_j$, there is a unique exceptional divisor, say E_{d_j, e_j}, such that $\widetilde{C}_j$ meets transversely with E_{d_j, e_j}. Let $\{\eta_j\} = \widetilde{C}_j \cap E_{d_j, e_j}$. As $p : X_k \to X_0$ is a resolution of $f(x,y)$, $\widetilde{C}_j$ is a generic slice curve of the exceptional divisor E_{d_j, e_j} at η_j and therefore we can apply the results of Section 6.

7.1 Representation of C_j and its Puiseux pairs

Let $\{v_{\nu(j,0)}, \ldots, v_{\nu(j,k_j-1)}\}$ be the vertices of the hierarchy graph $\mathcal{HG}(E_{d_j, e_j})$ of the exceptional divisor E_{d_j, e_j} where $\nu(j,0) = 0, \nu(j, k_j - 1) = d_j - 1$ and let

$$\mathcal{HT}(E_{d_j, e_j}) : Y_{j,k_j} \xrightarrow{q_{j,k_j}} Y_{j,k_j-1} \longrightarrow \cdots \longrightarrow Y_{j,1} \xrightarrow{q_{j,1}} Y_{j,0}$$

be the hierarchy tower of E_{d_j, e_j}.

By definition, $q_{j,\ell+1}:Y_{j,\ell+1} \to Y_{j,\ell}$ is equivalent to the toric blowing-up associated with $\{\Sigma^*_{\nu(j,\ell)}, (u_{\nu(j,\ell)}, v_{\nu(j,\ell)}), \xi_{\nu(j,\ell)}\}$. We denote by $C_j^{(i)}$ the strict transform of C_j to $Y_{j,i}$. Let

$$\left\{P_{j,i} = {}^t(a_{j,i}, b_{j,i}); i = 1, \ldots, k_j\right\}, \qquad (7.26)$$

be the dual vectors of the hierarchy of the divisor E_{d_j, e_j}. Here k_j is the hierarchy length of the divisor E_{d_j, e_j}. Let $E_{1, \beta_{\nu(j,1)}}$ be the support divisor of the center $\xi_{\nu(j,1)}$ of the second blowing-up $q_{j,2}:Y_{j,2} \to Y_{j,1}$ of $\mathcal{HT}(E_{d_j, e_j})$.

By the definition of the hierarchy tower, this implies that $\xi_{\nu(j,1)} := C_j^{(1)} \cap E_{1, \beta_{\nu(j,1)}}$. Let $(0, \gamma_j)$ be the right toric coordinate of $\xi_{\nu(j,1)}$. As we use the product $a_{j,\ell} \cdots a_{j,k_j}$ so often in this section, we use the following notations to shorten our formula.

$$A_{j,\ell} := a_{j,\ell} a_{j,\ell+1} \cdots a_{j,k_j}, \qquad 1 \le \ell \le k_j, 1 \le j \le r. \qquad (7.27)$$

Let

$$n^{(j,i)} = A_{j,i+1}, \quad \nu_{j,i} = b_{j,i} A_{j,i+1} + b_{j,i-1} A_{j,i} + \ldots + b_{j,1} A_{j,2}, \quad i = 1, \ldots, k_j.$$

Theorem 7-1. C_j has the following parametrization:

$$\begin{cases} x = \psi_j(t), & \psi_j(t) = t^{A_{j,1}}, \\ y = \varphi_j(t), & \varphi_j(t) = t^{b_{j,1} A_{j,2}} \rho_j(t), \quad \rho_j(0) = \gamma_j^{1/a_{j,1}}, \end{cases} \qquad (7.28)$$

such that the characteristic powers and the characteristic common divisors of $\varphi_j(t)$ with respect to $A_{j,1}$ are given by

$$\mathcal{P}(\varphi(t); A_{j,1}) = \{\nu_{j,t}; a_{j,t} > 1\}, \qquad \mathcal{D}(\varphi(t); A_{j,1}) = \{n^{(j,t)}; a_{j,t} > 1\}.$$

Recall that the multiplicity of the curve C_j at the origin is defined by $d(P; f_j)$ where $P = {}^t(1, 1)$. We denote the multiplicity of C_j at the origin by $m(C_j; O)$. Let $\tilde{a}_{j,1} = \min(a_{j,1}, b_{j,1}), \tilde{b}_{j,1} = \max(a_{j,1}, b_{j,1})$ and $\tilde{a}_{j,i} = a_{j,i}$ for $2 \le i \le k_j$. It is well-known that the defining function f_j of C_j (up to multiplication by units) is given by

$$\begin{aligned} f_j(x, y) &= \prod_{h=1}^{A_{j,1}} \left(y - \varphi_j(t\zeta_j^h)\right) \\ &= (y^{a_{j,1}} - \gamma_j x^{b_{j,1}})^{A_{j,2}} + f_j'(x, y), \end{aligned}$$

where $\zeta_j = \exp(2\pi i / A_{j,1})$.

The product in the right hand side in the first equality is an analytic function in the variables y and $t^{A_{j,1}}$ and thus it defines a single valued analytic function of x, y replacing $t^{A_{j,1}}$ by x. We assert that the remainder in the second equality, $f_j'(x, y)$ satisfies $d(P_{j,1}; f_j') > b_{j,1} A_{j,1}$.

In fact, let $P'_{j,1} = {}^t(1/A_{j,2}, b_{j,1})$ be a weight vector for t and y and let $F_j(t,y) = \prod_{h=1}^{A_{j,1}} \left(y - \varphi_j(t\zeta_j^h)\right)$. It is easy to see that

$$d\left(P_{j,1}; f_j(x,y)\right) = d\left(P'_{j,1}; F_j(t,y)\right) = b_{j,1} A_{j,1}.$$

By (7.28), the face function of F_j with respect to $P'_{j,1}$ is

$$
\begin{aligned}
(F_j)_{P'_{j,1}}(t,y) &= \prod_{h=1}^{A_{j,1}} \left(y - \gamma_j^{1/a_{j,1}} (t\zeta_j^h)^{b_{j,1} A_{j,2}}\right) \\
&= \left(\prod_{h=1}^{a_{j,1}} \left(y - \gamma_j^{1/a_{j,1}} t^{b_{j,1} A_{j,2}} \exp(2\pi i h b_{j,1}/a_{j,1})\right)\right)^{A_{j,2}} \\
&= (y^{a_{j,1}} - \gamma_j x^{b_{j,1}})^{A_{j,2}}.
\end{aligned}
$$

Thus

$$b_{j,1} A_{j,1} < d\left(P'_{j,1}; F_j(t,y) - (F_j)_{P'_{j,1}}(t,y)\right) = d\left(P_{j,1}; f'_{j,1}(x,y)\right).$$

This proves the assertion.

Let

$$
(n_i, m_i) = \begin{cases}
(\tilde{a}_{j,1}, \tilde{b}_{j,1}) & i = 1, \\
\begin{aligned}(a_{j,i}, b_{j,i} + b_{j,i-1}a_{j,i} + \cdots \\ + b_{j,2}a_{j,3}\cdots a_{j,i} + \tilde{b}_{j,1}a_{j,2}\cdots a_{j,i})\end{aligned} & i > 1,
\end{cases}
$$

as before. Then we have:

Corollary 7-2. The multiplicity of C_j at O is given by

$$m(C_j; O) = \tilde{a}_{j,1} A_{j,2} = \min(a_{j,1}, b_{j,1}) A_{j,2}. \tag{7.29}$$

The Puiseux pairs of C_j are given by $\{(n_i, m_i); \tilde{a}_{j,i} > 1\}$.

Remark 7-3. *Assume that C is an irreducible germ of a curve. Assume that*

$$\mathcal{T}: X_k \xrightarrow{p_k} X_{k-1} \xrightarrow{p_{k-1}} \cdots \longrightarrow X_1 \xrightarrow{p_1} X_0 \,, \quad p = p_1 \circ \cdots \circ p_k,$$

is an admissible tower which gives a toroidal resolution $p: X_k \to X_0$. Let

$$\left\{P_i = {}^t(a_i, b_i); i = 1, \ldots, k\right\}$$

be the dual vectors of the hierarchy. We may assume that $a_1 < b_1$ and

$$a_i > 1, \qquad i = 1, \ldots, k.$$

Then the Puiseux pairs are given by $\left\{(n_i, m_i); i = 1, \ldots, k\right\}$ where

$$(n_i, m_i) = (a_i, b_i + b_{i-1}a_i + \cdots + b_1 a_2 \cdots a_i), \qquad 1 \le i \le k.$$

In [10], we have shown the existence of such a resolution. The number k is equal to the complexity $\varrho(f)$. See Theorem (3.12) of [10] or [16].

7.2   Intersection multiplicity

Let X be a complex surface. We use the following notation for the intersection number. Let D_1, D_2 be two divisors. Let $\xi \in X$. The local intersection number of D_1 and D_2 at ξ is denoted by $I(D_1, D_2; \xi)$. In the case that one of D_1 or D_2 is a compact divisor, the global intersection number is also defined. We denote the global intersection number of D_1 and D_2 by $D_1 \cdot D_2$. Let h be a rational function on X and assume that D is a compact divisor. Then we have the following criterion (see for example [8], Theorem (2.6)).

$$(h) \cdot D = 0 \tag{7.30}$$

This is useful to compute the intersection numbers.

We now reformulate the well-known formula for the intersection multiplicity for the ordinary blowing-ups in our toric blowing-up situation. See Chapter I of [17] for the classical case. The situation is same as in Theorem 7-1. By the formula (3.5), we have

$$(q^*_{j,1} f_j) = C_j^{(1)} + D_{j,1}, \tag{7.31}$$

where $D_{j,1}$ is a linear combination of $E_{1,s}, 1 \le s \le s_1$. The the coefficient of the support divisor $E_{1,\beta_{\nu(j,1)}}$ is of particular importance. By (3.5) and 7-2, we can write:

$$D_{j,1} = b_{j,1} A_{j,1} E_{1,\beta_{\nu(j,1)}} + D'_{j,1}, \tag{7.32}$$

where $D'_{j,1}$ does not contain the divisor $E_{1,\beta_{\nu(j,1)}}$. Let $L = \{x = 0\}$. Then

$$I(L, C_j; O) = A_{j,1}. \tag{7.33}$$

This follows immediately from $I(L, C_j; O) = \mathrm{val}_y\big(f_j(0, y)\big)$ and (7.29). Applying (7.33) for $E_{1,\beta_{\nu(j,1)}}$ and $C_j^{(1)}$, we get

$$E_{1,\beta_{\nu(j,1)}} \cdot C_j^{(1)} = A_{j,2}, \qquad D_{j,1} \cdot C_j^{(1)} = b_{j,1} A_{j,1} A_{j,2}. \tag{7.34}$$

Definition 7-4. *We say that C_i and C_j have the same toric tangential direction at O with respect to $\mathcal{T}$ if $\nu(i,1) = \nu(j,1)$. This implies that $C_i^{(1)}$ and $C_j^{(1)}$ still intersect. More generally, we say that C_i and C_j has the same toric tangential direction of depth μ with respect to $\mathcal{T}$ if their hierarchy graph is same up to μ-th vertices. That is,*

$$\nu(j,t) = \nu(i,t), \quad 0 \le t \le \mu \qquad \text{and} \qquad \nu(j, \mu+1) \ne \nu(i, \mu+1). \tag{7.35}$$

In the case of $\mu = k_i - 1$ or $k_j - 1$, the last condition is to be understood as $\eta_i \notin C_j^{(\mu+1)}$ or $\eta_j \notin C_i^{(\mu+1)}$ respectively. In particular (7.35) implies that in the corresponding hierarchy towers, $Y_{j,t} = Y_{i,t}$ and $q_{j,t} = q_{i,t}$ for $t \le \mu + 1$ up to isomorphism and

$$P_{j,t} = P_{i,t}, \quad 1 \le t \le \mu \qquad \text{and} \qquad C_i^{(\ell)} \cap C_j^{(\ell)} \begin{cases} \ne \emptyset & \ell \le \mu, \\ = \emptyset & \ell = \mu + 1. \end{cases}$$

Here $C_j^{(\ell)}$ is the proper transform of C_j to $Y_{j,\ell}$.

Lemma 7-5.

(i) *Assume that C_i and C_j has the same toric tangential direction. Then*

$$I(C_i, C_j; O) = I\big(C_i^{(1)}, C_j^{(1)}; \xi_{\nu(j,1)}\big) + b_{j,1} A_{j,1} A_{i,2}. \tag{7.36}$$

(ii) *Assume that C_i and C_j have different toric tangential directions. Then*

$$I(C_i, C_j; O) = \min(a_{j,1} b_{i,1}, a_{i,1} b_{j,1}) \cdot A_{j,2} A_{i,2}. \tag{7.37}$$

PROOF. Assume first that C_i and C_j has the same toric tangential direction. By (7.31) and (7.30), we have

$$
\begin{aligned}
I(C_i, C_j; O) &= I\big(C_i^{(1)}, C_j^{(1)}; \xi_{\nu(j,1)}\big) + D_{i,1} \cdot C_j^{(1)} \\
&= I\big(C_i^{(1)}, C_j^{(1)}; \xi_{\nu(j,1)}\big) + D_{j,1} \cdot C_i^{(1)}.
\end{aligned}
$$

Combining with (7.32) and (7.34), we obtain that

$$D_{i,1} \cdot C_j^{(1)} = a_{j,1} b_{j,1} A_{j,2} A_{i,2}. \tag{7.38}$$

Note that $a_{j,1} = a_{i,1}, b_{j,1} = b_{i,1}$ in this case. This proves the assertion (i).

Now we assume that C_i and C_j has different toric tangential directions. By Theorem 7-1, C_i and C_j have the following representations:

$$C_j : \begin{cases} x = \psi_j(s), & \psi_j(s) = s^{A_{j,1}}, \\ y = \varphi_j(s), & \varphi_j(s) = s^{b_{j,1} A_{j,2}} \rho_j(s), \quad \rho_j(0) = \gamma_j^{1/a_{j,1}}, \end{cases}$$

and

$$C_i : \begin{cases} x = \psi_i(t), & \psi_i(t) = t^{A_{i,1}}, \\ y = \varphi_i(t), & \varphi_i(t) = t^{b_{i,1} A_{i,2}} \rho_i(t), \quad \rho_i(0) = \gamma_i^{1/a_{i,1}}. \end{cases}$$

Thus by the definition of intersection number, we have

$$
\begin{aligned}
I(C_I, C_j; O) &= \mathrm{val}_t\big(f_i(\psi_j(t), \varphi_j(t))\big) \\
&= \mathrm{val}_t\big(f_i(\psi_j(\tau^{A_{i,1}}), \varphi_j(\tau^{A_{i,1}}))\big) \big/ A_{i,1} \\
&= \mathrm{val}_\tau\bigg(\prod_{h=1}^{A_{i,1}} \big(\varphi_j(\tau^{A_{i,1}}) - \varphi_i(\zeta_i^h \tau^{A_{j,1}})\big)\bigg) \Big/ A_{i,1} \quad \text{by (7.29).}
\end{aligned}
$$

Assume that $P_{j,1} \neq P_{i,1}$ and assume $a_{j,1} b_{i,1} < a_{i,1} b_{j,1}$ for instance. Then the lowest term of the above product is

$$\prod_{h=1}^{A_{i,1}} \gamma_i^{1/a_{i,1}} \zeta_i^{h b_{i,1} A_{i,2}} \tau^{b_{i,1} A_{i,2} A_{j,1}} = \gamma_i^{A_{i,2}} \times \tau^{a_{j,1} b_{i,1} a_{i,1} A_{i,2}^2 A_{j,2}}.$$

Thus the assertion follows immediately. Assume that $P_{i,1} = P_{j,1}$ and $\gamma_j \neq \gamma_i$. Then the above product starts from

$$(\gamma_j - \gamma_i)^{A_{i,2}} \times \tau^{a_{j,1} b_{i,1} a_{i,1} A_{i,2}^2 A_{j,2}}.$$

Thus the assertion is also clear. $\qquad \square$

Applying Lemma 7-5 inductively, we obtain

Theorem 7-6. *Assume that C_i and C_j has the same toric tangential direction of depth μ in $\mathcal{T}$. Then*

$$I(C_i, C_j; O) = \sum_{\ell=1}^{\mu} b_{j,\ell} A_{j,\ell} A_{i,\ell+1} + \min(a_{j,\mu+1} b_{i,\mu+1}, a_{i,\mu+1} b_{j,\mu+1}) A_{j,\mu+2} A_{i,\mu+2}.$$

8   Iterated generic hyperplane section curves

A germ of plane curve $C = \{f(x, y) = 0\}$ at the origin is called a non-degenerate n-times iterated generic hyperplane section plane curve if there exists a non-degenerate complete intersection variety $V = \{\mathbf{z} \in U; F(\mathbf{z}) = \ell_1(\mathbf{z}) = \cdots = \ell_n(\mathbf{z}) = 0\}$ at the origin where U is a open neighborhood of the origin of $\mathbb{C}^{n+2}$ and $\mathbf{z} = (z_1, \ldots, z_{n+2})$ such that the function $\ell_i(\mathbf{z})$, $i = 1, \ldots, n$, are generic linear forms and $f(z_1, z_2) \equiv F(\mathbf{z})$ modulo the ideal $(\ell_1, \ldots, \ell_n)$. See [5, 14, 15] for the definition of non-degenerate complete intersection varieties. Then we assert

Theorem 8-1. *Assume that C is a non-degenerate n-times iterated generic hyperplane section plane curve. Then the resolution complexity $\varrho(f)$ is at most $n + 1$. Each irreducible component of C has at most one Puiseux pair. No Puiseux pair implies smoothness.*

PROOF. For the proof, we need the theory of toric resolution of a germ of non-degenerate complete intersection variety. Let $S = \{\mathbf{z} \in U; \ell_1(\mathbf{z}) = \cdots = \ell_n(\mathbf{z}) = 0\}$. S is a smooth linear subspace of dimension two. By the genericity of $\ell_i(\mathbf{z})$, the projection $q_{i,j} : S \to \mathbb{C}^2$, which is defined by $q_{i,j}(\mathbf{z}) = (z_i, z_j)$, is an isomorphism for each $i, j, 1 \leq i < j \leq n + 2$. In other words, (z_i, z_j) is also a system of coordinates of the plane S.

Let N^+ be the space of weight vectors (or dual vectors) of dimension $n+2$. N^+ is a cone over $n + 2$ vertices $E_1 = {}^t(1, 0, \ldots, 0), \ldots, E_{n+2} = {}^t(0, \ldots, 0, 1)$. Let Σ^* be a regular simplicial cone subdivision of N^+. As we have done for $n = 0$ in Section 2, we can construct a complex manifold X of dimension $n + 2$ and a birational morphism $\widehat{\pi} : X \to \mathbb{C}^{n+2}$ as follows. Let $\sigma = \mathrm{Cone}(P_1, \ldots, P_{n+k})$ is an $(n+2)$-dimensional simplicial cone of Σ^* where each $P_i = {}^t(p_{1,i}, \ldots, p_{n+2,i})$ is a primitive integral column vector. By the regularity of Σ^*, the matrix $(P_1, \ldots, P_{n+2})$ is an unimodular matrix. $P_1, \ldots, P_{n+2}$ are called vertices of the simplex σ.

To each σ, we associate $(n + 2)$-dimensional affine space $\mathbb{C}_\sigma^{n+2}$ with coordinates $\mathbf{y}_\sigma := (y_{\sigma,1}, \ldots, y_{\sigma,n+2})$ and a birational morphism $\pi_\sigma : \mathbb{C}_\sigma^{n+2} \to \mathbb{C}^{n+2}$ which is defined by $\pi_\sigma(\mathbf{y}_\sigma) = \mathbf{z}$ where $z_i = y_{\sigma,1}^{p_{i,1}} \cdots y_{\sigma,n+2}^{p_{i,n+2}}$, $i = 1, \ldots, n + 2$. X is a quotient space of the disjoint union of $\{\mathbb{C}_\sigma^{n+2}; \sigma \in \Sigma^*\}$, obtained by identifying two points $\mathbf{u}_\sigma \in \mathbb{C}_\sigma^{n+2}$ and $\mathbf{u}_\tau \in \mathbb{C}_\tau^{n+2}$ if and only if the map $(\pi_\sigma)^{-1} \circ \pi_\tau = \pi_{\sigma^{-1}\tau} : \mathbb{C}_\tau^{n+2} \to \mathbb{C}_\sigma^{n+2}$ is defined and $\pi_{\sigma^{-1}\tau}(\mathbf{u}_\tau) = \mathbf{u}_\sigma$. A dual vector $P = {}^t(p_1,\ldots,p_{n+2})$ is called strictly positive if $p_i > 0$ for each $i = 1,\ldots,n+2$. For each strictly positive vertex P of Σ^*, there exists a compact exceptional divisor $\widehat{E}(P)$ of $\widehat{\pi}\colon X \to \mathbb{C}^{n+2}$. For further generality about the toric variety X, we refer to [2, 4, 7, 12].

Now assume that Σ^* is a regular (or unimodular) simplicial cone subdivision of the dual Newton diagram $\Gamma^*(F,\ell_1,\ldots,\ell_n)$ and let $\widetilde{S}$ and $\widetilde{V}$ be the proper transform of S and V. Then we have the following diagram.

$$
\begin{array}{ccccc}
\widetilde{S} & \xrightarrow{\pi'} & S & \xrightarrow{\cong} & \mathbb{C}^2 \\
\uparrow & & \uparrow & & \uparrow \\
\widetilde{V} & \xrightarrow{\pi''} & V & \xrightarrow{\cong} & C
\end{array}
\qquad
\pi' = \widehat{\pi}|_{\widetilde{S}}, \pi'' = \widehat{\pi}|_{\widetilde{V}}.
$$

Then by the non-degeneracy of S and V and by the admissibility of Σ^* with the dual Newton diagram $\Gamma^*(F,\ell_1,\ldots,\ell_n)$, $\pi'\colon \widetilde{S} \to S$ is a resolution of our curve $V \cong C$. We use hereafter the same notation as in Section 3 or [15]. For the details, see [15].

Let Δ be the $(n+1)$-dimensional simplex with vertex $A_1 = (1,0,\ldots,0)$, $\ldots$, $A_{n+2} = (0,\ldots,0,1)$. Δ is the unique $(n+1)$-dimensional face of $\Gamma(\ell_i)$ (for any i). Let $P = {}^t(p_1,\ldots,p_{n+2}) \in \mathrm{Vertex}(\Sigma^*)$ and let $E(P) = \widehat{E}(P) \cap \widetilde{S}$ and $D(P) = \widehat{E}(P) \cap \widetilde{V}$. Let $p_{min} = \min\{p_i; 1 \le i \le n+2\}$ and let $I(P) = \{i : p_i = p_{min}\}$. Then $\Delta(P;\ell_i)$ is a simplex which is generated by $\{A_i : i \in I(P)\}$. In [14, 15], we have shown the following.

1. For a (strictly positive) vertex P of Σ^*, $E(P)$ is a non-empty compact divisor of $\widetilde{S}$ if and only if $\dim \Delta(P;\ell_i) \ge n$, or equivalently $|I(P)| \ge n+1$. Let $E_0 = {}^t(1,\ldots,1)$. Thus this is the case if and only if $P \in \mathrm{Cone}(E_0 E_i)$ for some $i = 1,\ldots,n+2$. If $E(P)$ is non-empty, $E(P)$ is a rational sphere. The restriction $\pi' : \widetilde{S} - \cup E(P) \to S - \{O\}$ is biholomorphic where the union is taken for all strictly positive vertex P on $\bigcup_{i=1}^{n+2} \mathrm{Cone}(E_0 E_i)$. $D(P)$ is non-empty if and only if $|I(P)| \ge n+1$, $\dim(\Delta(P;F)) \ge 1$ and $\dim(\Delta(P;\ell_i) + \Delta(P;F)) = n+1$. In this case, $D(P)$ consists of finite simple points.

2. $E(P) \cap E(Q) \ne \emptyset$ if and only if $P,Q \in \mathrm{Cone}(E_0 E_i)$ for some $i = 1,\ldots n+2$. In particular, the configuration graph $\mathcal{G}(\pi')$ of the exceptional divisor of $\pi' : \widetilde{S} \to S$ is a star graph with center E_0 and at most $n+2$ branches on $\overline{E_0 E_i}$, $i = 1,\ldots,n+2$. Thus by proposition 4-1, we conclude that $\varrho(C) \le \varrho(\pi') \le n+1$. This proves the first assertion.

Now we show the second assertion of Theorem 8-1. Let $C_1,\ldots,C_r$ be irreducible components of C and let $V_1,\ldots,V_r$ be the corresponding components of V. We fix j and let $\widetilde{V}_j$ be the proper transform of V_j. Then there exists a unique strictly positive vertex $P \in \mathrm{Vertex}(\Sigma^*)$ such that $E(P) \cap \widetilde{V}_j \ne \emptyset$. As $\pi'\colon \widetilde{S} \to S$ is a resolution of V, $\widetilde{V}_j$ intersects transversely with the divisor $E(P)$ in $\widetilde{S}$. Assume that $P \in \mathrm{Cone}(E_0\widehat{E}_\iota)$ for some $\iota, 1 \leq \iota \leq n+2$. Let $P = {}^t(p_1, \ldots, p_{n+2})$. Then there exists positive integers $a \leq b$ such that $p_\iota = b$ and $p_i = a$ for $i \neq \iota$. As P is a primitive integral vector, $\gcd(a, b) = 1$. Let $\sigma = \mathrm{Cone}(P_1, \ldots, P_{n+2})$ be a simplicial cone in Σ^* such that $P = P_1$. Let $\{\xi_j\} = \widehat{E}(P) \cap \widetilde{V}_j$. Then in these toric coordinates, we can write $\xi_j = (0, \gamma_2, \ldots, \gamma_{n+2})$ for some non-zero complex numbers $\gamma_2, \ldots, \gamma_{n+2}$. Let $L_i = \{\ell_i(\mathbf{z}) = 0\}$ and let $\widetilde{L}_i$ be the strict transform of L_i. Then $\widetilde{L}_i$ is a non-singular divisor in $\mathbb{C}_\sigma^{n+2}$. Let $\widetilde{\ell}_{\sigma,i}(y_{\sigma,1}, \ldots, y_{\sigma,n+2}) = 0$ be the defining function of $\widetilde{L}_i$ in $\mathbb{C}_\sigma^{n+2}$. The divisor $\widehat{E}(P)$ is defined by $y_{\sigma,1} = 0$. By the admissibility of Σ^*, $\widehat{E}(P), \widetilde{L}_1, \ldots, \widetilde{L}_n$ intersect transversely at ξ_j.

Thus in a neighborhood of ξ_j, we can take $y_{\sigma,1}$ and $y_{\sigma,\alpha}$ as a system of coordinates of the surface $\widetilde{S}$ for some $\alpha, 2 \leq \alpha \leq n+2$. That is, we can solve the equations $\widetilde{\ell}_1(\mathbf{y}_\sigma) = \cdots = \widetilde{\ell}_n(\mathbf{y}_\sigma) = 0$ as $y_{\sigma,i} = \psi_i(y_{\sigma,1}, y_{\sigma,\alpha})$, $i \neq 1, \alpha$ where $\psi(0, \gamma_\alpha) = \gamma_i$. We start with a parametrization of $\widetilde{V}_j$ with respect to the coordinates $(y_{\sigma,1}, y_{\sigma,\alpha})$:

$$\widetilde{V}_j \quad y_{\sigma,1} = t, \quad y_{\sigma,\alpha} = \rho_\alpha(t), \quad \rho_\alpha(0) = \gamma_\alpha.$$

Substituting this in ψ_i, $\widetilde{V}_j$ is parametrized as a curve in $\mathbb{C}_\sigma^{n+2}$:

$$\widetilde{V}_j: \quad y_{\sigma,1} = t, \quad y_{\sigma,i} = \rho_i(t), \quad \rho_i(0) = \gamma_i, \quad 2 \leq i \leq n+2.$$

Now we compose this parametrization and the projection $\pi_\sigma \colon \mathbb{C}_\sigma^{n+2} \to \mathbb{C}^{n+2}$. We choose $(z_{\iota-1}, z_\iota)$ as a system of coordinates of S ($z_0 = z_{n+2}$ in the case of $\iota = 1$). Let $(p_{i,k})$ be the corresponding unimodular matrix of σ. By the definition of π_σ, we get

$$V_j: \begin{cases} z_{\iota-1} = t^a \rho_2(t)^{p_{\iota-1,2}} \cdots \rho_{n+2}(t)^{p_{\iota-1,n+2}} := t^a \eta_1(t), & \eta_1 \in \mathcal{O}_1^*, \\ z_\iota = t^b \rho_2(t)^{p_{\iota,2}} \cdots \rho_{n+2}(t)^{p_{\iota,n+2}} := t^a \eta_2(t), & \eta_2 \in \mathcal{O}_1^*. \end{cases}$$

By the same argument as in section 6, we conclude that V_j can be reparametrized as

$$V_j: \begin{cases} z_{\iota-1} = s^a, \\ z_\iota = s^b \rho(s), \quad \rho(0) \neq 0. \end{cases}$$

Therefore in the case of $b > a > 1$, (a, b) is the unique Puiseux pair of $V_j \cong C_j$. In the case of $a = 1$, V_j is smooth. This completes the proof. $\qquad\square$

References

[1] K. Brauner Klassifikation der Singularitäten algebroider Kurven. Abh. Math. Semin. Hamburg Univ. vol 6, 1928.

[2] V.I. Danilov The geometry of toric varieties Russian Math. Surveys. Pages 97–154 vol 33:2, 1978.

[3] P. Griffiths and J. Harris *Principles of Algebraic Geometry.* A Wiley-Interscience Publication. New York-Chichester-Brisbane-Toronto. 1978.

[4] A.G. Khovanskii Newton polyhedra and toral varieties. Funkts. Anal. Prilozhen. vol 11, No.4, pages 56–67 1977.

[5] A.G. Khovanskii Newton polyhedra and the genus of complete intersections Funkts. Anal. Prilozhen. vol 12, No.1, pages 51–61, 1977.

[6] A.G. Kouchnirenko Polyèdres de Newton et Nombres de Milnor Inventiones Math. vol 32, pages 1–32, 1976.

[7] G. Kempf, F. Knudsen, D. Mumford and B. Saint-Donat *Toroidal Embeddings, Lecture Notes in Math.* 339. Springer-Verlag. Berlin-Heidelberg-New York 1973.

[8] H.B. Laufer *Normal Two-Dimensional Singularities.* Annals of Math. Studies, 71. Princeton Univ. Press. Princeton 19971.

[9] D.T. Lê Sur un critére d'equisingularité. C.R.Acad.Sci.Paris, Ser. A-B. vol 272. pages 138–140 1971

[10] D.T. Lê and M. Oka On the Resolution Complexity of Plane Curves Titech-Mathfg preprints series 10–93. 1993.

[11] J. Milnor *Singular Points of Complex Hypersurface.* Annals Math. Studies vol 61 Princeton Univ. Press. Princeton 1968.

[12] T. Oda *Convex Bodies and Algebraic Geometry.* Springer-Verlag. Berlin-Heidelberg-New York. 1987.

[13] M. Oka On the Resolution of Hypersurface Singularities. Advanced Study in Pure Mathematics. vol 8, pages 405–436, 1986

[14] M. Oka Principal zeta-function of non-degenerate complete intersection singularity J. Fac. Sci., Univ. of Tokyo vol 37, No. 1. pages 11–32. 1990.

[15] M. Oka On the topology of full non-degenerate complete intersection variety Nagoya Math. J. vol 121. pages 137–148, 1991.

[16] M. Oka Note on the resolution complexity of plane curve. Proceeding of the Workshop on Resolution of Singularities vol 1.82 (742) pages 125–148. 1993

[17] O. Zariski *Algebraic surfaces.* Springer-Verlag. Berlin Heidelberg New York. 1934

[18] O. Zariski Studies in equisingularity, I. Amer. J. Math. vol 31 pages 507–537. 1965.

[19] O. Zariski *Le Probleme des modules pour les branches planes.* Cours donné au Centre de Mathématiques de l'Ecole Polytechnique. 1973.

Address of author:

Department of Mathematics,
Tokyo Institute of Technology,
Oh-Okayama, Meguro-ku, Tokyo
Email: oka@math.titech.ac.jp

INTRODUCTION TO THE ALGORITHM OF RESOLUTION

Orlando Villamayor U.[1]

Dedicated to Prof. Heisuke Hironaka
on his sixtieth birthday.

1 Introduction

Suppose that one is confronted with the problem of resolving singularities of a particular and perhaps very simple polynomial equation over a field of characteristic zero. There is a theorem that states that such a resolution does exist ([6]), but if we want to know how to resolve the singularities the theorem falls short for providing an algorithm.

What makes this point even more striking is the fact that in essence the idea behind inductive resolution of singularities is (very simple and) quite algorithmic: "express the equation in a Weierstraß polynomial form with respect to a privileged variable and transfer the original problem of reduction to a problem involving only the coefficients of this polynomial".

By a constructive theorem of resolution we mean a theorem that provides an algorithm for resolving singularities. This is the achievement of [12], where the algorithm was introduced.

The purpose of this work is precisely to focus on the algorithm itself; to outline the main ideas and to illustrate how it works.

But beyond the fun of explicit computation and that of computing examples, we also insist here on other adventages of this "algorithmic" proof of desingularization, which might be of use for further development and for applications.

In fact the constructive resolution of singularities has some nice properties such as:

(a) If a group is acting on a scheme then the action lifts an action on the desingularization of the scheme defined by the algorithm (equivariance).

(b) If two singular points are formally isomorphic, then there are convenient isomorphic "neighborhoods" of the points which undergo the same procedure of resolution.

(c) It simplifies the understanding of patching local procedures as part of a global one as opposed to the argument used in Hironaka's theorem.

---

[1]Partially supported by DGICYT PB 91-0370-CO2-02

The expression "resolving singularites explicitly" or the word "algorithm" used before, require some clarification and preciseness, which by the way brings us back to Zariski's notion of local uniformization.

This procedure consists on defining at a singular point x a value, $\Psi(x)$, on a fixed and totally ordered set I. This function Ψ from the singular points to I, takes only finitely many values in I, is upper semi-continuous and defines a stratification; each stratum defined as

$$\{z \mid z \text{ is a singular point and } \Psi(z) = \alpha\},$$

for $\alpha \in \operatorname{Img} \Psi \subset I$. Furthermore, the stratum through x is the center of a permissible monoidal transformation. Now, for any exceptional and singular point y mapping to x, value $\Psi(y)$ is strictly smaller than $\Psi(x)$ (a step by step improvement).

So I already claim that the function Ψ is well defined after this monoidal transformation and furthermore with the same properties as before. In particular, if y is singular, Ψ defines a stratum through y and we are back with the original set up.

The point is that:

(i) After applying this process finitely many times, the exceptional point y will be smooth and in this case the procedure comes to an end (local uniformization).

(ii) The function Ψ is well defined globally so that any process as in (i) arises from a unique and global procedure of resolution of singularities (all local uniformizations patch).

In this way Ψ defines an "algorithm" of resolution which is also canonical. Section 6 is devoted to simple but explicit examples.

2   Stating the problem of resolution of singularities

Let us fix some notations and conventions before we state the theorem of embedded resolution of singularities.

Notation 2-1. If $\pi\colon X \to Y$ is a birational morphism of schemes, set $E(\pi) \subset X$ to be the exceptional locus of π.

Example 2-2. Set $Y = \operatorname{Spec}(\mathbb{C}[x_1 \ldots, x_n])$ and $\pi\colon X \to Y$ the quadratic transformation with center at the origin $O \equiv x_1 = \cdots = x_n = 0$, then $E(\pi) = \pi^{-1}(O)$ is in this case a smooth irreducible hypersurface isomorphic to $\mathbb{P}^{n-1}_{\mathbb{C}}$.

Notation 2-3. (Normal crossings) Let W be a regular scheme so that $\mathcal{O}_{W,x}$ is a local regular ring for any $x \in W$. Two regular subschemes Y_1, Y_2 of W have normal crossing at a point x if $x \notin Y_1 \cap Y_2$ or if $x \in Y_1 \cap Y_2$ and there exists a regular system of parameters, say $\{x_1, \ldots, x_n\}$ at the regular local ring $\mathcal{O}_{W,x}$ such that $I(Y_i)$, the proper ideal defining Y_i locally at x, can be expressed as the ideal spanned by a subset of $\{x_1, \ldots, x_n\}$, both for $i = 1$ and $i = 2$ simultaneously.

Example 2-4.

(a) $x = 0 \in \mathbb{C}^3$, $Y_1 = \langle x_1 \rangle$, $Y_2 = \langle x_2, x_3 \rangle$.

(b) $x = 0 \in \mathbb{C}^3$, $Y_1 = \langle x_1 \rangle$, $Y_2 = \langle x_1, x_2 \rangle$.

An analogous definition holds for a point x which belongs to the intersection of more then two regular subschemes. Finally we say that the regular subschemes $Y_1, \ldots, Y_r$ for W have normal crossing if they have normal crossings at any point.

Example 2-5.

(c) The coordinate planes of $\mathbb{R}^3$ have normal crossing.

(d) Consider at $\mathbb{C}^2$ the subschemes $Y_1 = \{x = 0\}$, $Y_2 = \{y = 0\}$, $Y_3 = \{x + y = 0\}$. Now Y_1, Y_2 and Y_3 do not have normal crossing at the origin.

Within the class of regular schemes we will deal with those which are smooth schemes of finite type over a field k.

Examples of smooth schemes over k are the spectrum of a polynomial ring over k, non-singular projective schemes, and open subsets of non-singular projective schemes (over k).

This frame should not discourage those who would prefer to work on an analytic set up, the theorem of resolution is an application of analytic methods in algebraic geometry.

Theorem 2-6. (of embedded resolution of singularities) *Let X be a reduced and closed subscheme of a smooth noetherian scheme Z. If Z is smooth and of finite type over a field k of characteristic zero then there exists a sequence:*

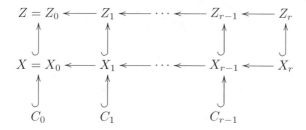

where each C_i is a closed and smooth subscheme, $\pi_i \colon Z_i \to Z_{i-1}$ is the monoidal transformation with center C_i, and $C_i \subset \operatorname{Sing} X_i$ where X_i is the strict transform of X_{i-1}, such that:

(a) X_r is a smooth subscheme of Z_r.

(b) $E(\pi_r \circ \cdots \circ \pi_1) = \{H_1, \ldots, H_r\}$ are smooth hypersurfaces with normal crossings at Z_r.

(c) X_r and $\{H_1, \ldots, H_r\}$ have normal crossings.

Remark 2-7. *If Z is a smooth scheme of finite type over a field k, and $C \subset Z$ is a closed and smooth subscheme then the monoidal transform with center C_i, say $\pi\colon Z' \to Z$, is such that:*

(a) *Z' is also smooth and of finite type over k.*

(b) *$E(\pi) = H'$ is a smooth hypersurface of Z'.*

The hypersurfaces $H_i \in \{H_1, \ldots, H_r\}$ of part (b) of our theorem, are the ultimate strict transforms of $H_i' = E(\pi_i) \subset Z_i$ to the last smooth scheme Z_r.

Remark 2-8. *The condition on X being reduced means that there is an open and dense subset of points, which are regular points, called $\mathrm{Reg}(X) = X - \mathrm{Sing}(X)$. So $\overline{\mathrm{Reg}(X)} = X$.*

If we now restrict our attention to the X_i's, then $X_{i+1} \to X_i$ is the blow up at the closed center $C_i \subset \mathrm{Sing}\, X_i$. The concatenation of all the proper maps $\overline{\pi}_i\colon X_{i+1} \to X_i$ induces a proper map $\overline{\pi}\colon X_r \to X = X_0$, and moreover from the condition $C_i \subset \mathrm{Sing}\, X_i$ we conclude that $\overline{\pi}$ restricts an isomorphism $\overline{\pi}\colon \overline{\pi}^{-1}\big(\mathrm{Reg}(X)\big) \widetilde{\to} \mathrm{Reg}(X)$.

It is clear from the setup of the theorem that the exceptional locus of $\overline{\pi}\colon X_r \to X$ is given by the intersection of the exceptional locus of $\pi = \pi_r \circ \cdots \circ \pi_1\colon Z_r \to Z$ with $X_r\colon E(\overline{\pi}) = E(\pi) \cap X_r$.

In the case that X is reduced, condition (c) of the theorem states that $H_i \cup X_r$ (at Z_r) is either empty or a smooth hypersurface of X_r, and moreover all hypersurfaces of X_r arising in this way are smooth and have normal crossings at X_r.

So the proper map $\overline{\pi}\colon X_r \to X$ is such that:

(i) X_r is smooth.

(ii) $\overline{\pi}$ is the identity map over an open dense set both of X_r and X which is

$$\mathrm{Reg}(X) = X - \mathrm{Sing}(X).$$

(iii) The exceptional locus $E(\overline{\pi}) = \overline{\pi}^{-1}\big(\mathrm{Sing}(X)\big)$ is a set of smooth hypersurfaces with normal crossings in X_r.

A recurrent idea in resolution.

There is a recurrent idea in the frame of resolution of singularities which we will work out conveniently to obtain an explicit algorithm defining a resolution for any $X \subset Z$ as in the theorem. This is what we call a constructive resolution of singularities, which is simpler then the original "existencial" theorem stated before.

This idea should be stated, at least roughly, so that the reader is aware of it, and can enjoy and follow our line starting with an apparently very ample frame and ending with a concrete and canonical algorithm of resolution.

The point is to define a convenient upper semi-continuous function along the singular points.

To begin with we fix a totally ordered set $(I, \geq)$ and define a function from any pair $X \subset Z$ (as in the theorem) to the ordered set, say

$$\varphi_X \colon X \longrightarrow (I, \geq),$$

such that:

(i) φ is upper semi-continuous.

(ii) The function φ takes only finitely different values. In other words, Img φ is a finite subset of I.

Since Img φ is finite on a totally ordered set then:

(a) There is an element $f_0 = \mathrm{Max}(\varphi_X)$ which is the biggest among those elements of Img φ_X.

Moreover since φ is uper semicontinuous,

(b) If f_0 is as in (a), the subset

$$S_{f_0} = \left\{ f \in X \mid \varphi_X(y) = f_0 \right\}$$

is a non-empty and closed subset of X.

But we will consider not only one fix datum, say $X \subset Z$, we shall also define a notion of transformations of data:

$$
\begin{array}{ccc}
Z & \longleftarrow & Z' \\
\uparrow & & \uparrow \\
X & \longleftarrow & X'
\end{array}
$$

After all, the statement of our theorem already involves some form of transformations of data.

c) If

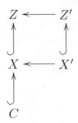

is a transformation of our datum $X \subset Z$ then

$$\operatorname{Max} \varphi_{X'} \leq \operatorname{Max} \varphi_X = f_0.$$

In particular if $\operatorname{Max} \varphi_{X'}$ is again equal to f_0 then one can define a non-empty closed subset

$$S_{f_0}(X') = \{y \in X' \mid \varphi_{X'}(y) = f_0\},$$

and the whole problem of resolution reduces to "forcing f_0 to drop", in other words, in defining a transformation as in (c) (or a concatenation of them) such that

$$\operatorname{Max} \varphi_{X'} < \operatorname{Max} \varphi_X = f_0.$$

A basic object on $X \subset Z$ will be something attached to $f_0 = \operatorname{Max} \varphi_X$, and it will naturally describe the closed subset $S_{f_0} \subset X$.

Moreover a notion of transformation of basic objects is defined so that if

is as in c), and suppose in addition that $\operatorname{Max} \varphi_{X'} = f_0 = \operatorname{Max} \varphi_X$, then the transform of the basic object describes the subset $S_{f_0}(X') \subset X'$. Therefore the "lowering of f_0" is obtained by what we call a resolution of the basic object.

3   Auxiliary result: Idealistic pairs

We attach to a coherent sheaf of ideals J in a smooth scheme W, $(J \subset \mathcal{O}_W)$ and to an integer $b \in \mathbb{Z} > 0$, a pair (J, b) called a couple.

We also define

$$\operatorname{Sing}(J, b) = \{x \in W \mid \gamma_x(J_x) \geq b\},$$

where $\gamma_x(J_x)$ denotes the order of the ideal J_x at the local regular ring $\mathcal{O}_{W,x}$.

We will always assume that $J_x \neq 0$ for all $x \in W$, so $\mathrm{Sing}(J,b)$ is a locally proper subset of W, and moreover it is proper on any connected component of the smooth scheme W.

Here W is a scheme of finite type over a field so $\mathrm{Sing}(J,b)$ is a proper and closed subset.

Example 3-1.

1. $J = \langle z^2 + x^3 y^3 \rangle$, $b = 2$, $W = \mathrm{Spec}(\mathbb{R}[z,x,y])$. Then

$$\mathrm{Sing}(J,2) = \{z = 0 \wedge x = 0\} \cup \{z = 0 \wedge y = 0\}.$$

2. Idem as before but with $b = 1$, then

$$\mathrm{Sing}(J,1) = \{z^2 + x^3\, y^3 = 0\} \subset W$$

Definition 3-2. *A basic object will be a datum of the form* $\big(W, E, (J,b)\big)$ *where: W is a smooth scheme of finite type over a field k of characteristic zero, $E = \{H_i\}$ is a finite set of smooth hypersurfaces having only normal crossings; $J \subset \mathcal{O}_W$ is a coherent sheaf of ideals such that $J_x \neq 0$ for all $x \in W$ and b is a natural number.*

Recall that (J,b) is called a couple and

$$\mathrm{Sing}(J,b) = \big\{x \in W \mid \nu_x(J_x) \geq b\big\}$$

is a closed and strictly proper subset of W.

(3-3) Now we want to define "transformations of basic objects". So fix a basic object $\big(W, E, (J,b)\big)$ and let C denote a closed and smooth subscheme of W such that:

(a) C has normal crossings with the hypersurfaces H_i's of $E = \{H_i\}$.

(b) $C \subset \mathrm{Sing}(J,b)$.

Set

$$\pi \colon W_1 \longrightarrow W$$

to be the blow-up at C.

Since C is closed and smooth W_1 is also smooth of finite type over k. Now we define a finite set E_1 of smooth hypersurfaces at W_1 as follows:

$$E_1 = \big\{ \text{ strict transform of } H_i, \text{ for all } H_i \in E \big\} \cup \big\{H = \pi^{-1}(C)\big\}.$$

Condition (a) (on C) will guarantee that E_1 consists again of smooth hypersurfaces having only normal crossings at W_1.

Assume now for simplicity that C is irreducible and let b' denote the order of J at a generic point of C. Since $C \subset \mathrm{Sing}(J,b)$ (condition (b)), it is clear that $b' \geq b$. One can check that at $\mathcal{O}_{W_1}$ there is a unique expression of the form:

$$J\mathcal{O}_{W_1} = I(H)^{b'} \underline{J},$$

where $I(H)$ denotes sheaf of ideals defining H, and $\underline{J}$ is a coherent sheaf of ideals, and moreover b' is the highest power of the (invertible) sheaf of ideals $I(H)$ that one can factor out of $J\mathcal{O}_{W_1}$.

Now set

$$J_1 = I(H)^{b'-b} \underline{J}, \tag{3.1}$$

which is a coherent sheaf of ideals in $\mathcal{O}_{W_1}$, and define (J_1, b) to be the transform of (J, b). The data

$$\big(W, E, (J, b)\big) \longleftarrow \big(W_1, E_1, (J_1, b)\big)$$

is called a permissible transformation of $\big(W, E, (J, b)\big)$.

Definition 3-4.

1. *A permissible tree over $\big(W, E, (J, B)\big)$ is a concatenation of permissible transformations say:*

$$\big(W, E, (J, b)\big) \longleftarrow \big(W_1, E_1, (J_1, b)\big) \longleftarrow \cdots \longleftarrow \big(W_s, E_s, (J_s b)\big) \tag{3.2}$$

An extension of formula (3.1) states that:

$$J_k = I(H_1)^{\alpha(1)} \ldots I(H_k)^{\alpha(k)} \underline{J}_k,$$

where $\underline{J}_k$ is "not divisible" by any $I(H_i)$, $i = 1, \ldots, k$. We denote the product

$$I(H_1)^{\alpha(1)} \ldots I(H_k)^{\alpha(k)}$$

by $\mathcal{M}_k$ which we call a monomial sheaf of ideals, so:

$$J_k = \mathcal{M}_k \cdot \underline{J}_k. \tag{3.3}$$

2. *A resolution of $\big(W, E, (J, b)\big)$ is a permissible tree (as (3.2)) such that $\mathrm{Sing}(J_s, b) = \emptyset$.*

Let us now state very roughly a theorem that will be clarified in the next sections.

Theorem 3-5. (of local idealistic presentation) *Given $X \subset W$, a subscheme of a smooth scheme over a field, the theorem of embedded resolution of singularities of X in W reduces "locally" to resolutions of basic objects.*

We shall clarify the meaning of the theorem in two stages, first we discuss it in the case when X is hypersurface and then in the general case.

(3-6) (the hypersurface case) Consider in $W = \mathrm{Spec}(\mathbb{C}[X, Y, Z]) = \text{``}\mathbb{C}^3\text{''}$ the subscheme $X = \{f = z^2 + x^2 y^2 = 0\}$. Of course the singular locus of this subscheme $\mathrm{Sing}(X)$ is the zero set of the sheaf of ideals

$$
\begin{aligned}
F_2(f) &= \left\langle f, \frac{\partial f}{\partial z}, \frac{\partial f}{\partial x}, \frac{\partial f}{\partial y} \right\rangle, \\
&= \langle z, xy^2, x^2 y \rangle,
\end{aligned}
$$

i.e., $\mathrm{Sing}(X) = \{z = 0 \wedge x = 0\} \cup \{z = 0 \wedge y = 0\}$.

So first we will attach to the problem of resolution of singularities of X, the problem of resolution of singularities of the basic object (3-4, 2) $\big(W, E, (J, b)\big)$ where $W = \mathbb{C}^3$, $E = \emptyset$, $J = J(X) = \langle f \rangle$ and $b = 2$.

At least one point is clear: the singular locus of (J, b) is in this case the singular locus of X, but this is a particular feature of the example, while the main property (which holds in our example) is what we shall call property B.P. (basic property).

Definition 3-7. *A basic object $\big(W, E, (J, b)\big)$ satisfies the basic property B.P. if*

$$
\nu_x(J) = b, \qquad \forall\, x \in \mathrm{Sing}(J, b).
$$

Theorem 3-8. *If a basic object $\big(W, E, (J, b)\big)$ satisfies the basic property and $\big(W, E, (J, b)\big) \xleftarrow{\pi} \big(W_1, E_1, (J_1, b)\big)$ is a permissible transformation of basic objects (3-3), then either*

(i) $\mathrm{Sing}(J_1, b) = \emptyset$, *in which case π is a resolution of the basic object (3-4, 2), or*

(ii) *the basic object $\big(W_1, E_1, (J_1, b)\big)$ satisfies B.P.*

Going back to our example where $W = \mathbb{C}^3$, $J = \langle f \rangle$, $b = 2$, $E = \emptyset$ we fix $\big(W, E, (J, 2)\big) \leftarrow \big(W_1, E_1, (J_1, 2)\big)$ a permissible transformation of basic objects and make the following observations:

(a) $J_1 \subset \mathcal{O}_{W_1}$ is the sheaf of ideals defining the strict transform of X, say X_1 at W_1.

(b) The theorem asserts that either (i) $\mathrm{Sing}(J_1, 2) = \emptyset$, in which case the hypersurface X_1 has no point of order ≥ 2, or (ii) $\mathrm{Sing}(J_1, 2) \neq \emptyset$ in which case B.P. holds for $\big(W_1, E_1, (J_1, 2)\big)$. This means that the biggest possible order of the hypersurface X_1 at any points is again $b(= 2)$.

As far as our example is concerned either X_1 is non-singular or $\big(W_1, (J_1, 2), E_1\big)$ also satisfies B.P.

Assertion 3-9. Let

$$
\big(W, E, (J, 2)\big) \longleftarrow \cdots \longleftarrow \big(W_r, E_r, (J_r, 2)\big) \tag{3.4}
$$

be a resolution (3-4, 2). Then for each $i \in \{0, 1, \cdots, r\}$, J_i denotes the sheaf of ideals defining the strict transform X_i of $X \subset W$ at W_i. Then X_r has no point of order ≥ 2.

Caution. Although X_r is a non-singular hypersurface, embedded resolution is not yet achieved. In fact the smooth hypersurface X_r might not have normal crossings with the exceptional hypersurfaces introduced by the r monoidal transformations between W_r with W.

We shall define later in a natural way a new basic object $\big(W_r, (\ell_r, b), E_r\big)$, (at W_r) so that the concatenation of (3.4) with a resolution of $\big(W_r, E_r, (\ell_r, b)\big)$ say

$$\big(W, E, (J, b)\big) \longleftarrow \cdots \longleftarrow \big(W_r, E_r, (J_r, b)\big)$$

$$\big(W_N, E_N, (\ell_N, b)\big) \longrightarrow \cdots \longrightarrow \big(W_r, E_r, (\ell_r, b)\big)$$

is such that the final strict transform X_N of X at W_N, is a smooth subscheme having normal crossings with the hypersurfaces of E_N.

Remark 3-10. *Consider now* $f = z^3 + x^2 y^2$, $X = \{f = 0\} \subset W = \mathbb{C}^3$. *In this case the highest possible order of the hypersurface X at a point is 3.*

Consider the basic object $\big(\mathbb{C}^3, \emptyset, (f, 3)\big)$ *and a resolution of the basic object:*

$$\big(\mathbb{C}^3, E = \emptyset, (J = \langle f \rangle, 3)\big) \longleftarrow \cdots \longleftarrow \big(W_s, E_s, (J_s, 3)\big).$$

Now for each $i \in \{0, 1, \ldots, s\}$ *the basic object* $\big(W_i, E_i, (J_i, b)\big)$ *is such that:*

(i) J_i *is the sheaf of ideals defining X_i: the strict transform of X at W_i.*

(ii) $X_s \subset W_s$ *is a hypersurface with no points of order bigger than 2, in fact we have that* $\mathrm{Sing}(J_s, 3) = \emptyset$.

So now at the stage s we start off with the resolution of a new basic object, $\big(W_s, E_s, (J_s, 2)\big)$ *and argue as in our first example.*

(3-11) On the theorem of local idealistic presentation: the general case.

In the case when X is a hypersurface of a smooth scheme W, we associated to each point $x \in X$ the multiplicity of X at x. This map has two properties (i) it is an upper semicontinuous function and (ii) it takes only finitely many different values.

We look for the maximal possible value, say b_0, of this upper semi-continuous function. Assume that W is noetherian so this maximum is reached, and then we defined a basic object $\big(W, E, (J, b)\big)$, where $J = I(X)$ the sheaf of ideals of X and $b = b_0$: the maximal possible multiplicity at points of X.

Clearly: $\operatorname{Sing}(I(X), b_0) = \{x \in X \mid \gamma_x(I(X)) = b_0\}$. Moreover if

$$\left(W, E, (J, b_0)\right) \longleftarrow \cdots \longleftarrow \left(W_i, E_i, (J_i, b_0)\right) \longleftarrow \cdots \longleftarrow \left(W_i, E_i, (J_i, b_0)\right)$$

is a resolution of our basic object, and if $X_i \subset W_i$ denotes the strict transform of X to W_i, then:

(a) X_i is the subscheme of W_i defined by the sheaf of ideals J_i. And

(b) $\operatorname{Sing}(J_i, b_0) = \{x \in X_i \mid \gamma_x(I(X_i)) = b_0\}$.

In particular

(c) $\{x \in X_r \mid \gamma_x(I(X_r)) = b_0\} = \emptyset$.

(3-12) Now suppose that Y is a subscheme of W which is not a hypersurface. So the first point is to find a good analogue of the notion of multiplicty in the hypersurface case. In this way we introduce a function

$$\mathrm{HS} \colon Y \to \mathbb{Z}^{\mathbb{Z}} = \{\text{functions } f \mid f \colon \mathbb{Z} \to \mathbb{Z}\}$$

so that if x is a closed point of Y, $\mathrm{HS}(x) = f$ where $f(k) = l(A/\mathcal{M}^{k+1})$; $A = \mathcal{O}_{Y,x}$; $\mathcal{M}$ the maximal ideal of the local ring A; and $l(A/\mathcal{M}^{k+1})$ stands for the length of $A/\mathcal{M}^{k+1}$ if $k \geq 0$ and $l(A/\mathcal{M}^{k+1}) = 0$ if $k < 0$.

Definition 3-13. *Given $x \in Y$ as before, the function $f = \mathrm{HS}(x)$ is called the Hilbert-Samuel function of Y at the point x.*

Result 3-14. There is a natural extension of HS to any point of Y such that $\mathrm{HS} \colon Y \to \mathbb{Z}^{\mathbb{Z}}$ is locally finite (takes only finitely different values if Y is noetherian) and upper semi-continuous.

Now we want to attach to the upper semi-continuous map $\mathrm{HS} \colon Y \to \mathbb{Z}^{\mathbb{Z}}$, an idealistic object. Before stating this result properly we need to define an appropriate notion of "restrictions".

Let $\varphi \to W_1 \to W$ be a map of smooth schemes of finite type over a field k of characteristic zero.

Definition 3-15. *φ is etale at $x \in W_1$ if:*

(i) *the extension of residue fields $k(\varphi(x)) \to k(x)$ induced by the morphism of local rings*

$$\varphi^* \colon \mathcal{O}_{W,\varphi(x)} \to \mathcal{O}_{W_1,x}$$

is a finite algebraic extension.

(ii) *if $\{y_1, \cdots, y_n\}$ is a regular system of parameters at the regular ring $\mathcal{O}_{W,\varphi(x)}$, then*

$$\{\varphi^*(y_1), \ldots, \varphi^*(y_n)\}$$

is a regular system of parameters at the local regular ring $\mathcal{O}_{W_1,x}$.

The map φ^* of local rings defines a map of the corresponding graded rings, say:

$$\operatorname{gr}\varphi^*\colon \operatorname{gr}\mathcal{O}_{W,\varphi(x)} \to \operatorname{gr}\mathcal{O}_{W_1,x},$$

but graded rings of local regular rings are polynomial rings and we think of

(a)

$$\operatorname{gr}\varphi^*\colon \operatorname{gr}\mathcal{O}_{W,\varphi(x)} = k\big(\varphi(x)\big)\big[\operatorname{In}(y_1),\cdots,\operatorname{In}(y_n)\big] \longrightarrow$$
$$k(x)\big[\operatorname{In}\big(\varphi^*(y_1)\big),\cdots,\operatorname{In}\big(\varphi^*(y_n)\big)\big] = \operatorname{gr}\mathcal{O}_{W_1,x}$$

simply as an extension of the coefficient field $k\big(\varphi(x)\big) \to k(x)$ of the first polynomial ring.

If $f \in \mathcal{O}_{W,\varphi(x)}$, then one can check that $\operatorname{In}\big(\varphi^*(f)\big) = \operatorname{gr}\varphi^*\big(\operatorname{In}(f)\big)$. In particular $\gamma_{\varphi(x)(f)}$ (the order of f at $\mathcal{O}_{W,\varphi(x)}$) is the same as $\gamma_x\big(\varphi^*(f)\big)$ (order at $\mathcal{O}_{W_1,x}$).

Furthermore, if $\varphi(x) \in Y$ a subscheme of W, defined locally (at $\mathcal{O}_{W,\varphi(x)}$) by the proper ideal J then:

(b) $\operatorname{gr}(\mathcal{O}_{Y,\varphi(x)}) = \operatorname{gr}(\mathcal{O}_{W,\varphi(x)})/\operatorname{In}(J)$, ($\operatorname{In}(J)$ the homogeneous ideal spanned by all $\operatorname{In}(f)f \in J$).

$\varphi^{-1}(Y) \subset W_1$ is defined locally at x (at $\mathcal{O}_{W_1,x}$) by $\varphi^*(J)$ so

$$\operatorname{gr}\big(\mathcal{O}_{\varphi^{-1}(Y),x}\big) = \operatorname{gr}(\mathcal{O}_{W_1,x})/\operatorname{In}\varphi^*(J).$$

(c) $\operatorname{In}\varphi^*(J) \subset \operatorname{gr}(\mathcal{O}_{W_1,x})$ is the ideal generated by $\operatorname{gr}\varphi^*\big(\operatorname{In}(J)\big)$.

Now from (a) (b) and (c) we conclude that

(i) the map

$$\operatorname{gr}\overline{\varphi}^*\colon \operatorname{gr}\mathcal{O}_{Y,\varphi(x)} \to \operatorname{gr}\mathcal{O}_{\varphi^{-1}(Y),x}$$

is simply the coefficient field extension $k\big(\varphi(x)\big) \to k(x)$ of $\operatorname{gr}\mathcal{O}_{Y,\varphi(x)}$. In particular:

(ii) for any k, the lengths ($= l$)

$$l\big(\mathcal{O}_{Y,\varphi(x)}/\mathcal{M}_{\varphi(x)}^{k+1}\big) = \sum_{l=0}^{k} \dim_{k(\varphi(x))}\big(\operatorname{gr}\mathcal{O}_{Y,\varphi(x)}\big)_l$$

and

$$l\big(\mathcal{O}_{\varphi^{-1}(Y),x}/\mathcal{M}_x^{k+1}\big) \sum_{l=0}^{k} \dim_{k(x)}\big(\operatorname{gr}\mathcal{O}_{\varphi^{-1}(Y)x}\big)_l$$

are the same.

Definition 3-16. *A map* $\varphi\colon W_1 \to W$ *is etale, if it is etale at every* $x \in W_1$.

Remark 3-17. *Any open inmersion* $\varphi\colon W_1 \to W$ *is of course an etale map*

If φ is etale, it is etale at any closed point $x \in W_1$ and the Hilbert-Samuel function of Y at $\varphi(x)$ called $\mathrm{HS}(\varphi(x)) \in \mathbb{Z}^{\mathbb{Z}})$ is the same as that of $\varphi^{-1}(Y)$ at x (called $\mathrm{HS}(x)$).

Since HS is an upper semi-continuous function we may assume after restricting φ to an open neighborhood of y (which is again etale) that $f_0 = \mathrm{HS}(y)$ is the maximal value taken by the function HS So the set

$$S = \{z \in Y \mid \mathrm{HS}(z) = f_0 = \mathrm{HS}(y)\}$$

(the Samuel stratum of the function f_0) is locally closed at y, and

$$S' f_0 = \{z \in \varphi^{-1}(Y) \mid \mathrm{HS}(z) = f_0 = \mathrm{HS}(y)\}$$

is closed at W_1.

Theorem 3-18. (of local idealistic presentation.) *Fix a closed subscheme* Y *of a smooth scheme* W *and a closed point* $y \in Y$, *and set* $f_0 = HS(y)$. *Then an etale map* $\varphi\colon W_1 \to W$ *can be chosen together with a point* $x \in W_1$ *such that* $\varphi(x) = y$ *and:*

(a) *there is a basic object at* W_1, *say* $(W_1, E = \emptyset, (J_1, b))$ *so that*

$$\begin{aligned}\mathrm{Sing}(J_1, b) &= \{z \in \varphi^{-1}(Y) \mid \mathrm{HS}(z) = f_0\}, \\ &= S_{f_0},\end{aligned}$$

where S_{f_0} *is the the Samuel statum of* f_0.

(b) *if*

$$(W_1, E, (J_1, b)) \longleftarrow \cdots \longleftarrow (W_r, E_r, (J_r, b)) \qquad (3.5)$$

is any concatenation of permissible transformations of basic objects, and if $Y_i \subset W_i$ *denotes the strict transform of* $Y_1 = \varphi^{-1}(Y) \subset W_1$ *then:*

(i) $\mathrm{Sing}(J_i, b) = \{z \in Y_i \mid \mathrm{HS}(z) = f_0\}$.

(ii) f_0 *is the biggest possible value of* HS *along the points of* Y_i.

(iii) *If (3.5) is a resolution of* $(W_1, E_1, (J_1, b))$, *then* $\mathrm{HS}(z) < f_0$ *for all* $z \in Y_r$.

(c) *the basic object* $(W_1, (J_1, b), E_1)$ *defined at (b) is such that* (J_1, b) *satisfies the basic property B.P. (3-7).*

Before any further discussion about the effectiveness of this theorem for resolution, we state the following result of Hironaka.

Theorem 3-19. *Let X be a closed subscheme of a smooth scheme Z of finite type over a field k. Consider an infinite sequence*

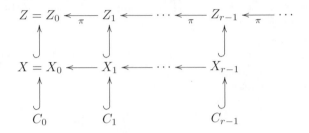

where:

(i) *C_i is a closed and smooth subscheme of Z_i.*

(ii) *$\pi_i \colon Z_i \to Z_{i-1}$ is the blow up at C_i.*

(iii) *X_i is the strict transfrom of X to Z_i.*

Assume furthermore that

$$C_i \subset \underline{\mathrm{Max}}\,\mathrm{HS}_{x_i} \subset X_i,$$

where $\underline{\mathrm{Max}}\,\mathrm{HS}_{X_i} = \{x \in X_i \mid \mathrm{HS}_{x_i}(x) = f_i\}$ and f_i denotes the maximal value achieved by $\mathrm{HS}_{X_i} \colon X_i \to \mathbb{Z}^{\mathbb{Z}}$. And that all centers C_i are strictly included in X_i (so that no X is empty). Then for some index N big enough:

$$f_N = f_{N+1} = \cdots = f_{N+k} = \cdots, \qquad k \geq 0.$$

Remark 3-20. *Theorem 3-19 is obvious if X is a hypersurface since the Hilbert-Samuel function will rely entirely on the multiplicity which is non-increasing by permissible transformations.*

In the general case, assume that $\underline{\mathrm{Max}}\,\mathrm{HS}_{x_s}$ is a proper subset of X_s and let f_s denote the Hilbert-Samuel function along its closed points. Suppose furthermore that we could give an algorithm for $X_s \subset Z_s$ and f_s so as to force f_s to drop after a finite sequence of monoidal transformations.

If we repeat this procedure and take the concatenation of all these sequences of monoidal transformations, then we are forcing the maximums of the Hilbert-Samuel functions (the f_i's) to drop again and again.

Theorem 3-19 states that this procedure can be done only be finitely many times, which means that at some stage s, $\underline{\mathrm{Max}}\,\mathrm{HS}_{X_s}$ can not be a proper subset. Since X_s is irreducible and reduced this implies that X_s is smooth.

In other words, the theorem states that if you can force the maximum Hilbert-Samuel function to drop by some algorithm, then you can solve singularities.

Recapitulation. We want to resolve singularities of a given closed sub-scheme Y embedded in a smooth scheme W. Since

$$\text{HS}: Y \to \mathbb{Z}^{\mathbb{Z}}$$

is an upper semi-continuous function, we first look at the biggest value of the form $HS(x)$ for $x \in Y$, call it f_0.

Set $S_{f_0}(Y) = \{x \in Y \mid \text{HS}(x) = f_0\}$, S_{f_0} is a non-empty and closed subset of Y.

According to the last theorem, if $y \in S_{f_0}(Y)$ is a closed point, there is an etale map

$$\varphi: (W_1, x) \to (W, y) \qquad (\varphi(x) = y),$$

and a basic object at W_1 $\big(W_1, E_1, (J_1, b)\big)$, so that

$$S_{f_0}\big(\varphi^{-1}(Y)\big) = \varphi^{-1}(S_{f_0}) = \big\{z \in \varphi^{-1}(Y) \mid \text{HS}(z) = f_0\big\}$$

is the closed subset of W_1 defined also as $\text{Sing}(J_1, b)$.

Moreover the theorem of local idealistic presentation can be stated as follows: "the lowering of f_0 as the maximal possible H-S function at points of $\varphi^{-1}(Y)$ is achieved by a resolution of the basic object $\big(W_1, E_1, (J_1, b)\big)$".

Now at this stage we can formulate explicitly an important questions which still must be answered. Indeed if the maps $\varphi: W_1 \to W$ of theorem 3-18 would be open immersions (which is the case in the analytic context) then we would be willing to accept the theorem as a local "simplification" or at least as a reformulation of the problem of resolution. But still, we would be facing the problem of defining resolutions for the different local "basic objects" in such a way that all these local resolutions somehow patch to a sequence of monoidal transformations at closed and smooth centers over $Y \subset W$.

Recall that here $W_1 \to W$ is an etale map, so first of all we must give a very settle notion of patching of all these "local" definitions of basic object and furthermore if we can define resolutions of these basic objects we must be careful so that these resolutions also patch.

(3-21) What do we mean by patching? Suppose we are given a sequence of monoidal transformations starting with $Y \subset W$

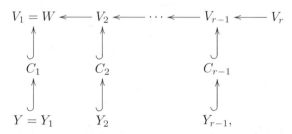

each $\pi_i: V_{i+1} \to V_i$ being a monoidal transformation at a smooth and closed center $C_i \subset Y_i \subset V_i$, where Y_i denotes the strict transform of $Y = Y_1$.

Now one can check that an etale map $\varphi\colon W_1 \to W = V$ induces a sequence of monoidal transformations at smooth centers, say:

$$
\begin{array}{ccccccc}
W_1 & \longleftarrow & U_2 & \longleftarrow & \cdots & \longleftarrow & U_{r-1} & \longleftarrow & U_r \\
\downarrow{\scriptstyle \varphi_1} & & \downarrow{\scriptstyle \varphi_2} & & & & \downarrow{\scriptstyle \varphi_{r-1}} & & \downarrow{\scriptstyle \varphi_r} \\
W & \longleftarrow & V_2 & \longleftarrow & \cdots & \longleftarrow & V_{r-1} & \longleftarrow & V_r
\end{array}
$$

where each $\varphi_i\colon U_i \to V_i$ is etale and $U_i \leftarrow U_{i+1}$ is the blow-up at $\varphi_i^{-1}(C_i)$.

Remark 3-22. *If $\varphi\colon W_1 \to W = V$, is an open inmersion, the sequence*

$$
W \longleftarrow V_2 \longleftarrow \cdots \longleftarrow V_{r-1} \longleftarrow V_r
$$

is the restriction to the open set W_1 and its pull backs.

Definition 3-23. We shall say that the resolutions of the different idealistic objects of our last theorem will "patch", if, conversely, there exists a fixed sequence of monoidal transformations over W such that each resolution of basic objects arises from that one, as we have stated above.

In the context of the theorem of local idealistic presentation, we start with a singular and closed subscheme Y of a smooth scheme W.

Let now f_0 denote the biggest Hilbert-Samuel function at points of Y, so

$$
S_{f_0} = \big\{ z \in Y \mid \mathrm{HS}(z) = f_0 \big\}
$$

(the Samuel stratum) is a closed subset of W.

To each point closed $y \in S_{f_0} (\subset Y)$ we attach in some way and etale map

$$
\varphi\colon W_1 \to W
$$

together with a point $y \in W_1$ such that $\varphi(x) = y$, and a basic object at W_1, $\big(W_1, (J_1, b), E\big)$.

This can be done for any $y \in S_{f_0}$. Now the different etale maps $\varphi\colon W_1 \to W$ define a notion of a "topology" over W called the etale topology of W.

We will attach to each such $\big(W_1, (J_1, b), E\big)$ a sheaf of sets over W_1, say $\mathcal{F}_{W_1}$, but considering on W_1 its etale topology.

The point is that these different sheaves of sets patch naturally to a sheaf of sets $\mathcal{F}_W$ over W with etale topology.

The next step is then to define a very concrete algorithm to obtain a resolution of each basic object $\big(W_1, (J_1, b), E\big)$ in such a way that the procedure to be done locally at $x \in \mathrm{Sing}(J_1, \varphi) = \varphi^{-1}(S_{f_0})$ depends entirely on the stalk of the sheaf of sets $\mathcal{F}_W$ at the point $y = \varphi(x)$ i.e., at $\mathcal{F}_{W,y}$. But then (since these sheaves patch well) the resolutions of the different basic objects arising from theorem 3-18 will patch to sequence of monoidal transformations over $Y \subset W$ in the sense of 3-23.

4 Constructive resolutions

Let $(I, \geq)$ be a set with a total order and $f: S \to I$ a map which takes only finitely many different values. Then set:

$$\text{Max} f = \text{the maximal value achieved by} f$$
$$\underline{\text{Max}} f = \{x \in S \mid f(x) = \text{Max} f\}$$

Definition 4-1. *Given a permisible tree*

$$\big(W, E, (J, b)\big) \leftarrow \cdots \leftarrow (W_k, E_k, (J_k, b)),$$

each $J_i = \mathcal{M}_i J_i$ *as in (3.3), define:* $w - \text{ord}^i(): \text{Sing}(J_i, b) \to (\mathbb{Q}, \geq)$; $w - \text{ord}^i(x) = \big(\nu_x(J_i)\big) / b$

Lemma 4-2. *Suppose inductively that* $W_i \leftarrow W_{i+1}$, *the blow-up at* C_i, *is such that* $C_i \subset \underline{\text{Max}} w - \text{ord}^i$. *Then*

$$\text{Max} w - \text{ord}^0 \geq \text{Max} w - \text{ord}^1 \geq \cdots \geq \text{Max} w - \text{ord}^k.$$

Remark 4-3. (The monomial case.) $\text{Max} w - \text{ord}^k = 0$ *if and only if* $J_k = \mathcal{M}_k$ *(see (3.3)) or equivalently, if and only if* $\underline{J_k} = \mathcal{O}_{W_k}$. *In this case we shall see later that it is simple to extend the tree to a resolution of* $\big(W, E, (J, b)\big)$.

Definition 4-4.

1. *A permissible tree over* $\big(W, E, (J, b)\big)$ *which fulfils the conditions of 4-2 will be called a* w-tree *over* $\big(W, E, (J, b)\big)$.

2. *In the conditions of 1, let the* birth *of* $\big(W_k, E_k, (J_k, b)\big)$ *be the smallest index* j *such that* $\text{Max} w - \text{ord}^j = \text{Max} w - \text{ord}^k$. *That means that for some index* k_0,

   $$\text{Max} w - \text{ord}^0 \geq \text{Max} w - \text{ord}^1 \geq \cdots > \text{Max} w - \text{ord}^{k_0} = \cdots = \text{Max} w - \text{ord}^k$$

   and k_0 *is called the* birth *of* $\big(W_k, E_k, (J_k, b)\big)$.

3. *Set* $E_k = E_k^+ \cup E_k^-$ *(disjoint union) such that* E_k^- *consist exactly of those hypersurfaces of* E_k *which are strict transforms of hypersurfaces of* E_{k_0} *(k_0 as in 2).*

4. *Define*
   $$n_k: \text{Sing}(J_k, b) \to (\mathbb{Z}, \geq)$$
   $$n_k(x) = \sharp\{H_i \in E_k^- \mid x \in H_i\}$$
   if $w - \text{ord}^k(x) = \text{Max} w - \text{ord}^k$, *and*

   $$n_k(x) = \sharp\{H_i \in E_k \mid x \in H_i\},$$

   where $\sharp\{\}$ *denotes the number of elements of the set.*

5. *And finally set:*

$$t_{d^k} : \mathrm{Sing}(J_k, b) \to (\mathbb{Q} \times \mathbb{Z}, \geq_{lex})$$

$$1t_{d^k}(x) \quad = \bigl(w - \mathrm{ord}^k(x), n_k(x)\bigr),$$

where $d = \dim W_k$.

(4-5) We formulate here the main idea for constructive resolutions; a double induction on $\mathrm{Max}\, t_d$ and on codim $\underline{\mathrm{Max}}\, t_d$.

We fix a w-tree of k terms say $\bigl(W, E, (J, b)\bigr) \leftarrow \cdots \leftarrow \bigl(W_k, E_k, (J_k, b)\bigr)$. For any index $i \in \{1, \ldots, k\}$, we obtain a new w-tree by looking only at the first i terms of the tree.

Theorem 4-6. *For a fixed w-tree as before:*

1. *each $t_{d^i} : \mathrm{Sing}(J_i, b) \to \mathbb{Q} \times \mathbb{Z}$ is an upper semi-continuous function which takes only finitely many different values.*

2. $\mathrm{Max}\, t_{d^0} \geq \mathrm{Max}\, t_{d^1} \geq \cdots \geq \mathrm{Max}\, t_{d^k}$

3. *Suppose that $\mathrm{Max}\, w - \mathrm{ord}^k > 0$, then set:*

$$\underline{\mathrm{Max}}\, t_{d^k} = R(1) \cup F$$

where $R(1) = \bigl\{ x \in \underline{\mathrm{Max}}\, t_{d^k} \mid \mathrm{codim}_x(\underline{\mathrm{Max}}\, t_{d^k} = 1) \bigr\}$

Then:

3a. *$R(1)$ is (if not empty) a smooth hypersurface which is the center of a permissible transformation*

$$\bigl(W_k, E_k, (J_k, b)\bigr) \leftarrow \bigl(W_{k+1}, E_{k+1}, (J_{k+1}, b)\bigr)$$

and after blowing up $R(1)$ we may assume that $R(1) = \emptyset$ (the blow up at a hypersurface is the identity map, however the couple (J_k, b) does undergo a non-trivial transformation to a couple (J_{k+1}, b)).

3b. *(At W_{k+1} so that $R(1) = \emptyset$) locally at any $x \in \underline{\mathrm{Max}}\, t_{d^{k+1}} = F$ there is an open neighborhood W'_{k+1}, a smooth hypersurface V_1, a closed immersion $i_1 : V_1 \hookrightarrow W'_{k+1}$ and a basic object $\bigl(V_1, E_1, (\mathcal{A}_1, b')\bigr)$ such that*

   i) *$i_1\bigl(\mathrm{Sing}(\mathcal{A}_1, b')\bigr) = \underline{\mathrm{Max}}\, t_{d^{k+1}} / W'_{k+1}$. ($\underline{Max}\, t_{d^{k+1}}$ restricted to W'_{k+1}.)*
   Moreover any permissible tree:

$$l\bigl(V_1, E_1, (\mathcal{A}_1, b')\bigr) \leftarrow \cdots \leftarrow \bigl(V_s, E_s, (\mathcal{A}_s, b')\bigr) \qquad (4.6)$$

   induces a sequence of blowing ups at closed centers and immersions

$$
\begin{array}{ccc}
W'_{k+1} & \longleftarrow \cdots \longleftarrow & W'_{k+s} \\[2pt]
{\scriptstyle i_1} \Big\uparrow & & {\scriptstyle i_s} \Big\uparrow \\[2pt]
V_1 & \longleftarrow \cdots \longleftarrow & V_s
\end{array}
$$

ii) *Any tree (4.6) induces a permissible tree*

$$\big(W'_{k+1}, E'_{k+1}, (J_{k+1}, b)\big) \leftarrow \cdots \leftarrow \big(W'_{k+s}, E'_{k+s}, (J_{k+s}, b)\big),$$

where $\mathrm{Max}\, t_{d^{k+s}} = \mathrm{Max}\, t_{d^{k+1}}$ *if and only if* $\mathrm{Sing}(\mathcal{A}_s, b') \neq \emptyset$ *in which case*

$$\underline{\mathrm{Max}}\, t^{k+s} = i_s\big(\mathrm{Sing}(\mathcal{A}_s, b')\big).$$

On the proof of 4-6: The main point here is that for the fixed value $\mathrm{Max}\, t_d$ we attach to $\big(W_k, E_k, (J_k, b)\big)$ a new basic objet $\big(W_k, E'_k, (J'_k, b')\big)$ such that:

1. $\mathrm{Sing}(J'_k, b') = \underline{\mathrm{Max}}\, t_d$,

2. property 1 is stable by any extensions of the tree as long as it is a w-tree and $\mathrm{Max}\, t_d$ does not improve,

3. $\big(W_k, E'_k, (J'_k, b')\big)$ satisfies

$$R(1) \cap F = \emptyset.$$

The idea of the proof of 3b (if $x \in F$) is a "reduction to coefficients" of the problem (Tschirnhausen) and I suggest to look at our last example in section 6.

(4-7) Three morals out of 4-6

1. The lowering of $\mathrm{Max}\, t_{d^{k+1}}$ is "locally" equivalent to a resolution of a basic object $\big(V_1, E_1, (\mathcal{A}_1, b')\big)$ where now $\dim V_1 = \dim W_{k+1} - 1 = d - 1$.

2. Suppose we could delete the word "locally" in 1, then an inductive argument would allow us to enlarge the tree of length $k+1$ to a tree of lengh $k+s$, so that we can force $\mathrm{Max}\, t_{d^{k+1}} > \mathrm{Max}\, t_{d^{k+s}}$ (by taking (4.6) to be a resolution).

   So in that case an inductive argument of resolutions would tell us that we can force $\mathrm{Max}\, t_d$ to drop.

   Now if we force $\mathrm{Max}\, t_d$ to drop again and again, then the first coordinate ($\mathrm{Max}\, w - \mathrm{ord}$) must drop in the long run. But then at some point we achieve the nice condition $\mathrm{Max}\, w - \mathrm{ord} = 0$ and we shall see that an extension of such a permissible tree to a resolution will be simple.

3. The smooth subscheme $R(1)$ is, *par excellence*, the center to be chosen. This is a key point for the uniqueness of our constructive resolution.

Definition 4-8. *A constructive resolution of a basic object* $\big(W, E, (J, b)\big)$ *with values at* $(I, \geq)$ *will mean:*

1. *An upper semi-continuous function*

$$\Psi_0 \colon \mathrm{Sing}(J,b) \to I$$

(see beginning of section 4) such that $\underline{\mathrm{Max}}\,\Psi_0$ *is the center of a permissible transformation (3-3):*

$$\bigl(W, E, (J,b)\bigr) \leftarrow \bigl(W_1, E_1, (J_1,b)\bigr),$$

and either $\mathrm{Sing}(J_1,b) = \emptyset$ *or an upper semi-continuous function*

$$\Psi_1 \colon \mathrm{Sing}(J_1,b) \to I$$

is defined.

2. *Given inductively the permissible tree*

$$\bigl(W, E, (J,b)\bigr) \leftarrow \cdots \leftarrow \bigl(W_{k-1}, E_{k-1}, (J_{k-1},b)\bigr),$$

together with upper semi-continuous functions $\Psi_i \colon \mathrm{Sing}(J_i,b) \to I$, $(0 \le i < k-1)$ *so that* $\underline{\mathrm{Max}}\,\Psi_i$ *is the center of* $W_{i+1} \to W_i$, *then either* $\mathrm{Sing}(J_{k-1},b) = \emptyset$ *or*

$$\Psi_{k-1} \colon \mathrm{Sing}(J_{k-1},b) \to (I, \ge)$$

is defined as an upper semi-continuous function so that $\underline{\mathrm{Max}}\,\Psi_{k-1}$ *is a permissible center for*

$$\bigl(W_{k-1}, E_{k-1}, (J_{k-1},b)\bigr) \leftarrow \bigl(W_k, E_k, (J_k,b)\bigr).$$

3. *If* $W_i \leftarrow W_{i+1} \colon \pi_i$ *is the blow-up at* $\underline{\mathrm{Max}}\,\Psi_i$, *and* $x \in \mathrm{Sing}(J_{i+1},b)$ *then:*

   a) $\pi_i(x) \in \mathrm{Sing}(J_i,b)$

   b) $\Psi_{i+1}(x) \le \Psi_i(\pi_i(x))$ *with equality if and only if* $\pi_i(x) \notin \underline{\mathrm{Max}}\,\Psi_i$

4. *For some index* N, *the set* $\mathrm{Sing}(J_N,b) = \emptyset$ *so that:*

$$\bigl(W, E, (J,b)\bigr) \leftarrow \cdots \leftarrow \bigl(W_N, E_N, (J_N,b)\bigr)$$

is a resolution of $\bigl(W, E, (J,b)\bigr)$ *(see 3-4,2).*

(4-9) The construction of the function will grow essentially from the development of 4-8, where $t_{d^i} \colon \mathrm{Sing}(J_i,b) \to \mathbb{Q} \times \mathbb{Z}$ was defined together with an expression

$$\underline{\mathrm{Max}}\,t_{d^i} = R(1) \cup F$$

Suppose now that $d = \dim W$ is 1. In this case any point of $\underline{\mathrm{Max}}\,t_1$ is of codimension 1 i.e., $\underline{\mathrm{Max}}\,t_1 = R(1)$. One can check that $\Psi_i = t_{1^i}$ with values at $I = \mathbb{Q} \times \mathbb{Z}$ satisfies conditions 1, 2, and 3 of 4-8 as long as $\mathrm{Max}\,w - \mathrm{ord}^1 > 0$. Before going into the other cases we first fix two conventions:

1. If $(I, \geq)$ is a set with a total order, we assume the existence of a "formal element" say ∞, such that $\infty \geq x$ for all $x \in I$.

2. Given $(I_1, \geq_1), (I_2, \geq_2)$, then $I_1 \times I_2$ is ordered lexicographically.

For each dimension $d = \dim W$ see associate first a set with a total order

$$I(d) = \mathbb{Q} \times \mathbb{Z} \times \wedge$$

(for some ordered set $\wedge$), and we also extend the function $t_{d^i} : \operatorname{Sing}(J_i, b) \to \mathbb{Q} \times \mathbb{Z}$ to a function $f_{d^i} : \operatorname{Sing}(J_i, b) \to I(d)$

$$f_{d^i}(x) = \big(w - \operatorname{ord}^i(x), n_i(x), H_i(x)\big).$$

The behaviour of the functions H_i (Hironaka's function) will be clarified in 4-10, but for the time being set $H_i(x) = \infty$ if $w - \operatorname{ord}^i(x) > 0$.

Case of $d = \dim W > 1$ and $\operatorname{Max} w - \operatorname{ord}^k > 0$.

Recall the expression $\underline{\operatorname{Max}} t_d = R(1) \cup F$. Locally at any point $x \in F$ there is a $d - 1$ dimensional basic object $\big(V_1, E_1, (\mathcal{A}_1, b')\big)$ and a closed immersion $i_1 : V_1 \hookrightarrow W'_k$ so that $i_1\big(\operatorname{Sing}(\mathcal{A}_1, b')\big) = \underline{\operatorname{Max}} t_{d^k} / W'_k$.

We assume the existence of constructive resolutions of basic objects of dimension $d - 1$ by functions $\Psi^i_{d-1} : \operatorname{Sing}(\mathcal{A}_i, b') \to I_{d-1}$

Now set:

$$I = I_d = I(d) \times I_{d-1},$$

and define $\Psi : \underline{\operatorname{Max}} t_d \to I$ as follows:

If $x \in F$ set:

$$
\begin{aligned}
f_d(x) &= \big(w - \operatorname{ord}(x), n(x), \infty\big) \in I(d), \\
\Psi_d(x) &= \big(f_d(x), \Psi^i_{d-1}(x)\big) \in I(d) \times I_{d-1}.
\end{aligned}
$$

If $x \in R(1)$ set $f_d(x)$ as before and $\Psi_d(x) = (f_d(x), \infty) \in I(d) \times I_{d-1}$.

Since Ψ_{d-1} will be defined inductively in terms of t_{d-1} and H, it turns out that $\underline{\operatorname{Max}} \Psi = R(1)$ if $R(1)$ is not empty, so we are following the rules of 4-6: blow-up first of all $R(1)$ if this set is not empty.

After doing so $R(1)$ will be empty, and clearly $\underline{\operatorname{Max}} \Psi \subset F \subset \underline{\operatorname{Max}} w - \operatorname{ord}$, so we are producing a w-tree (see Def. 4-4,1)).

Now 4-7 says that repeating this again and again we come to the

Case of $\underline{\operatorname{Max}} w - \operatorname{ord}^k = 0$:

In this case (J_k, b) is such that $J_k = \mathcal{M}_k$ (see 4-3) and Hironaka defines now a function

$$H_k : \operatorname{Sing}(J_k, b) \to \wedge$$

which fulfils conditions 1–4 of 4-8 over the basic object $\big(W_k, E_k, (J_k, b)\big)$.

In this case we disregard the function $n(x)$ and set:

$$
\begin{aligned}
f_{d^k}(x) &= \big(w - \operatorname{ord}^k(x) = 0, n_k(x) = \infty, H_k(x)\big) \in I(d), \\
\Psi_{d^k}(x) &= \big(f_{d^k}(x), \infty\big) \in I(d) \times I_{d-1} = I.
\end{aligned}
$$

Since in this monomial case we are not interested in any form of induction on the dimension the value of Ψ relies entirely on the function H.

(4-10) The function H.

(I) We start now with a basic object $(W, E, (J, b))$ and the additional assumption that J is monomial:

$$J = \mathcal{M} = I(H_1)^{a(1)} \ldots I(H_n)^{a(n)}, \qquad (4.7)$$

$E = \{H_1, \ldots H_n\}$ and $a(i) \in \mathbb{N} \cup \{0\}$. It is clear that in this case

$$\mathrm{Sing}(J, b) = \cup (H_{i_1} \cap \ldots \cap H_{i_s}),$$

where $i_1 < \cdots < i_s$ are elements of $\{1, \ldots, n\}$ subject to the conditions:

(i) $a(i_1) + \ldots + a(i_s) \geq b$.

(ii) $(H_{i_1} \cap \ldots \cap H_{i_s}) \neq \emptyset$.

Hironaka's proposal of resolution of these couples is the following.

(a) Let k_0 denote the smallest possible s so that there is $\{i_1, \ldots, i_s\} \subset \{1, \ldots, n\}$ satisfying i) and ii). Once k_0 is defined let $\mathcal{F} \subset P(\{1, \ldots, n\})$ (the set of subsets of $\{1, \ldots, n\}$) consist of:

$$\mathcal{F} = \{S \subset \{1, \ldots, n\} \mid \sharp S = k_0$$
$$\text{and } S = \{i_1, \ldots, i_n\} \text{ fulfills both i) and ii)}\}$$

(b) For each $S \in \mathcal{F}$ define the weight of S to be: $a(i_1) + \cdots + a(i_{k_0})$ if $S = \{i_1, \ldots, i_{k_0}\}$ and set b':

$$b' = \max_{S \in \mathcal{F}} \{\text{weight of } S\};$$

and finally define G as

$$G = \{S \in \mathcal{F} \mid \text{weight of } S \text{ is } b'\}.$$

Example 4-11. $W = \mathbb{C}^4$; $E = \{H_1, H_2, H_3, H_4\}$, $H_i = \{x_i = 0\}$ (the coordinate planes), $J = I(H_1)^2 \cdot I(H_2)^3 \cdot I(H_3) \cdot I(H_4)$; $b = 2$. Clearly $\mathrm{Sing}(J, b) = H_1 \cup H_2 \cup (H_3 \cap H_4)$, so k_0 here is one and:

$$\mathcal{F} = \{\{1\}, \{2\}\}$$

corresponding to H_2 and to H_1.

Now the weight of $\{1\}$ is 2 (the exponent of $I(H_1)$) and the weight of $\{2\}$ is 3. So $b' = 3$ and

$$G = \{\{2\}\}.$$

(c) Set $f = \sharp G$.

Proposition 4-12. *Given* $I = \{i_1, \ldots, i_{k_0}\} \in G$ *the smooth scheme* $F = H_{i_1} \cap \cdots \cap H_{i_{k_0}}$ *has normal crossing with* E *and is permissible for* $(W, E, (J, b))$.

Let $(W, E, (J, b)) \leftarrow (W_1, E_1, (J_1, b))$ *denote such permissible transformation. Then:*

1. J_1 *is monomial.*

2. (*suppose* $\mathrm{Sing}(J_1, b) \neq \emptyset$). *If* $\bar{k}_0$, $\bar{b}'$, $\bar{f}$ *are defined as in a), b) and c) but now for* (J_1, b), *then:* $\bar{k}_0 > k_0$ *or* $\bar{b}' < b'$ *or* $\bar{f} < f$; *i.e.,* $(-\bar{k}_0, \bar{b}', \bar{f}) < (-k_0, b', f)$ *in* $\mathbb{Z}^3$ *with lexicographic ordering.*

Example 4-13. In the case of our last example the only choice for F is $F = H_2$. In this case $\bar{k}_0 = k_0 = 1$ but $\bar{b}' = 2 < b' = 3$. In fact

$$J_1 = I(H_1)^2 \cdot I(H_5) \cdot I(H_3) \cdot I(H_4),$$

where our convention is to denote the strict transform of H_i, if not empty, as H_i again. Since $H_2 \subset W$ was the center, then the strict transform via the identity map is empty and the exceptional locus $(\mathrm{id}^{-1}(H_2))$ is what we denote here by H_5.

Now $H_1 \subset \mathrm{Sing}(J_1, b)$ and

$$\bar{k}_0 = 1, \quad \mathcal{F} = G = \{\{1\}\} \qquad \text{and} \qquad b' = 2.$$

(II) How to pursue Hironaka's procedure constructively.

With the notation and assumption as in part I) of this note 4-10, for each index i (for each $H_i \in E$) define: $\alpha(i): \mathrm{Sing}(J, b) \to \mathbb{Q}$,

$$\alpha(i)(x) = \begin{cases} (a(i))/b & \text{if } x \in H_i \\ 0 & \text{if } x \notin H_i \end{cases}$$

Assume that each index i belongs to $L = \mathbb{Z}$ and for each $x \in \mathrm{Sing}(J, b)$, set:

$$L_x = \{i \in L \mid x \in H_i\}$$

of course each L_x is a finite and totally ordered set of $\mathbb{Z}$.

Now define $a(x) = -b(x) \in \mathbb{Z}$ where

$$b(x) = \min\left\{k \;\Big|\; \exists i_1 < \cdots < i_k \;\Big|\; \sum \alpha(i_j)(x) \geq 1\right\} \in \mathbb{N}$$

and for $b = b(x)$ define

$$c(x) = \max\left\{\alpha(i_j)(x) + \cdots + \alpha(i_b)(x) \mid i_1 < \cdots < i_b\right\} \in \mathbb{Q}.$$

Consider $(L_x)^b \subset (\mathbb{Z})^b$ $(b = b(x))$ "included" in $\mathbb{Z}^\mathbb{N}$ via

$$(i_1, \ldots, i_b) \to (i_1, \ldots, i_b, \infty, \infty, \ldots, \infty, \ldots),$$

and define

$$\begin{aligned}
\beta(x) &= (\bar{\beta}_1, \ldots, \bar{\beta}_b) \\
&= \max\{(\beta_1, \ldots, \beta_b) \mid \beta_1 > \cdots > \beta_b \\
&\qquad \text{and } \alpha(\beta_1)(x) + \cdots + \alpha(\beta_b)(x) = c(x)\}
\end{aligned}$$

where the order involved in the definition of β is the lexicographic order.
Finally set

$$\wedge = \mathbb{Z} \times \mathbb{Q} \times \mathbb{Z}^\mathbb{N} \quad \text{and} \quad H(x) = \big(a(x), c(x), \beta(x)\big).$$

So even if the invariant f is bigger then one (see c) of part (I)), Max H will select a unique center.

One can check that the function H satisfies all the conditions 1–4 of 4-8 if (4.7) holds.

(4-14) We are left at this stage with three questions.

i) is Ψ_{d^k} well defined along $\underline{\text{Max}}\, t_d$? As it stands this is not clear since this value depends on the closed immersion

$$i_1 : V_1 \hookrightarrow W'_{k+1}$$

where W'_{k+1} is a convenient neighborhood of a given point, and also on other choises.

So first we must show that our constructions patch where the different W'_{k+1} overlap. However this will be clarified at the very end 5-2, so we accept this point for the time being.

ii) how to define Ψ_{d^k} along all $\text{Sing}(J_k, b)$? In fact we have defined Ψ_{d^k} only at points of $\underline{\text{Max}}\, t_d$ (if Max $w - \text{ord}^k > 0$).

Now (4-7, 2) says that we will achieve the case Max $w - \text{ord}^N = 0$ for some $N \gg 0$ by blowing up conveniently at centers included in $\underline{\text{Max}}\, t_d = \underline{\text{Max}}\, f_d$.

Moreover, Hironaka's procedure for the monomial case Max $w - \text{ord}^N = 0$, says that by blowing up at $\underline{\text{Max}}\, H = \underline{\text{Max}}\, f_d$ again and again, then a resolution is achieved and in a unique way.

For any index k and any $x \in \text{Sing}(J_k, b)$, if $x \notin \underline{\text{Max}}\, f_{d^k}$ we can identify x with a point say x_{k+1} of $\text{Sing}(J_{k+1}, b)$.

Now since the permissible tree we have constructed is a resolution (3-4, 2), for some index the center should contain x (otherwise a resolution is never achieved). So there is a well defined index $k_0 \geq k$ such that $x`` = ``x_{k_0} \in \underline{\mathrm{Max}}\, f_{d^{k_0}}$.

Now set $\Psi_{d^k}(x) = \Psi_{d^{k_0}}(x_{k_0})$.

iii) why is the function $\Psi_{d^k} \colon \mathrm{Sing}(J_k, b) \to I$ an upper semi-continuous function? We do know that $t_{d^k} \colon \mathrm{Sing}(J_k, b) \to \mathbb{Q} \times \mathbb{Z}$ is an upper semi-continuous function which takes only finitely many values (Therorem 4-6). To show that both these conditions also hold for

$$f_{d^k} \colon \mathrm{Sing}(J_k, b) \to I(d) = \mathbb{Q} \times \mathbb{Z} \times \wedge,$$

we would undergo the same argument that we give below, so we accept this and prove that

$$\Psi_{d^k} \colon \mathrm{Sing}(J_k, b) \to I = I(d) \times I_{d-1}$$

is an upper semi-continuous function which takes only finitely many values.

Recall from 4-9 that

$$\begin{aligned}\Psi_{d^k}(x) &= (f_{d^k}(x), \infty) \in I(d) \times I_{d-1}, \qquad \text{or:} \\ \Psi_{d^k}(x) &= (f_{d^k}(x), \Psi_{d-1^i}(x)) \in I(d) \times I_{d-1}\end{aligned}$$

Fix $(a, b) \in I(d) \times I_{d-1}$, we want to show that

$$C(a, b) = \{x \in \mathrm{Sing}(J_k, b) \mid \Psi(x) \geq (a, b)\}$$

is a closed subset of $\mathrm{Sing}(J_k, b)$.

Since f_{d^k} is an upper semi-continuous function which takes only finitely different values not only sets of the form $\{x \in \mathrm{Sing}(J_k, b) \mid f_{d^k}(x) \geq a\}$ are closed but also those with strict inequality, say:

$$B(a) = \{x \in \mathrm{Sing}(J_k, b) \mid f_{d^k}(x) > a\}.$$

If $\mathrm{Max}\, f_{d^k} < a$, then clearly $C(a, b) = \emptyset$. If $\mathrm{Max}\, f_{d^k} \geq a$ let k_0 be the smallest index $\geq k$, such that $\mathrm{Max}\, f_d^{k_0} = a$ (we can always reduce to the existence of such k_0 by eventually changing a).

According to our construction, at k_0, the set

$$D(b) = \{x \in \underline{\mathrm{Max}}\, f_d^{k_0} \subset \mathrm{Sing}(J_{k_0}, b) \mid \Psi_{d-1}(x) \geq b\}$$

is a closed subset of $\underline{\mathrm{Max}}\, f_d^{k_0}$ (of W_{k_0}).

Now let $\pi \colon W_{k_0} \to W_k$ be the composition of all intermediate maps (if $k_0 > k$) or $\pi = id$ (if $k_0 = k$). Now from 4-8, 3 we know that

$$C(a, b) = B(a) \cup \pi(D(b))$$

which is a union of closed sets since π is a proper mar map.

5   The language of groves and the problem of patching ([13])

Recall again that at 3b) of 4-6 we assigned to $x \in F \subset \underline{\mathrm{Max}}\, t_{d^{k+1}}$ an open neighborhood W'_{k+1} (of $x \in W_{k+1}$),a closed immersion $i_1 \colon V_1 \hookrightarrow W'_{k+1}$ and a basic object $\big(V_1, E_1, (\mathcal{A}_1, b')\big)$. Moreover the values of the function $\Psi_{d-1}(x)$ were defined along F in terms of this datum. In this section we go back to i) of 4-14 which questions the good definition of $\Psi_{d-1}(x)$ there where the data overlap.

It turns out that the notion of groves is suitable for this purpose. A grove will be a sheaf of sets which we relate to our problem in the following sense:

i)   to the basic object $\big(V_1, E_1, (\mathcal{A}_1, b')\big)$ we attach the set where the elements are all the permissible trees over the basic object (3-4,1)).

ii) via the closed immersion of V_1 in W'_{k+1} we can define a set over W'_{k+1} induced by that of i).

iii) to an etale map $W_1 \to W'_{k+1}$, the notion of restriction introduced in 3-21 induces a restriction of elements, and therefore a notion of sheaf of sets on W'_{k+1} with etale topology,

iv) if two such open neighborhoods W'_{k+1} happen to overlap, then the corresponding sheaves of sets will patch (as sheaves) to define a global sheaf of sets on W_{k+1} say $\mathcal{F}_F$.

Theorem 5-1. *For any $x \in F$ the value $\Psi_{d-1}(x)$ is expressable in terms of the stalk of $\mathcal{F}_F$ at the point x i.e., in terms of $\mathcal{F}_{F,x}$.*

Remark 5-2. *Now i) of 4-14 has a positive answer and the development of Section 4 asserts the existence of constructive resolutions of basic objects in the sense of 4-8.*

(5-3) We come back now to Theorem 3-5 where the word "locally" means that there are etale maps $W' \to W$ and basic objects $\big(W', E', (J', b')\big)$, for a covering of W in the sence of the etale topology. And if one can achieve resolutions of all these basic objects (3-4,2)), in such a way that all these procedures patch (now in the sence of etale topology), then one ultimately achieves the conditions of embedded resolution of singularities.

Now that there is a notion of constructive resolution for each such basic object $\big(W', E', (J', b')\big)$ (5-2) we are confronted again with our original question: why and in what context do the procedures of constructive resolutions of the different $\big(W', E', (J', b')\big)$ patch?

To answer to this question we argue as before, roughly:

1. Each basic object $\big(W', E', (J', b')\big)$ induces a sheaf of sets (as pointed out before), moreover these sheaves patch (glue) to a global sheaf $\mathcal{F}$ on W, but now considering on W the etale topology.

2. Both

    i) the formulas: $x \in \mathrm{Sing}(J', b')$; $\mathrm{codim}_x(\mathrm{Sing}(J', b')) = 1$ and

    ii) the value $\Psi_d(x)$

are expressable in terms of the sheaf $\mathcal{F}$, more precisely in terms of the stalk of $\mathcal{F}$ at the point x (i.e., in terms of $\mathcal{F}_x$).

Furthermore

3. If $\Theta\colon X \to X$ is an isomorphism, it induces for each $x \in X$ a natural bijection

$$\Theta_x\colon \mathcal{F}_x \to \mathcal{F}_{\Theta(x)}$$

which preserves expressions of the form i) and ii) of 2 (for x replaced by $\Theta(x)$).

Indeed, the outcome of 1 and 2 is that the different procedures of constructive resolutions of the basic objects will patch since the functions Ψ_d (defined for each of them) will patch.

An outcome of 3 will be that the sets $\underline{\text{Max}}\,\Psi_d$ are Θ-invariants which will justify both points (a) and (b) of the introduction.

6   Examples

Our first example is intended to elucidate the relation between the resolution of basic objects with the theorem of embedded resolutions of singularities.

Example 6-1.   Take on $W = \mathbb{C}^2$ the smooth curve $C = \{y - x^2 = 0\}$ and $E = \{H_1\}$ with $H_1 = \{y = 0\}$.

Here H_1 is tangent to C at $(0,0)$. The curve C is smooth but we will need (as will become clear with our next examples) to apply a number of permissible transformations so that the final strict transform of the curve C has normal crossings both with the strict transform of H_1 and with all the exceptional hypersurfaces introduced in this procedure.

So we set $\big(W, E, (J, b)\big)$ with $W = \mathbb{C}^2$; $E = \{H_1\}$; $J = I(C)$ (the sheaf of ideals defining C); and $b = 1$.

Clearly $\text{Sing}\big(I(C), 1\big) = C$ and for any transform

$$\big(W, E, (I(C), 1)\big) \leftarrow \big(W_1, E_1, (I(C)_1, 1)\big),$$

the ideal $I(C)_1$ is $I(C_1)$, $I(C_1)$ being the sheaf of ideals defining the strict transform say C_1 of C.

The function $w - \text{ord}$ is constantly equal to one along $\text{Sing}\big(I(C), 1\big) = C$, so $\text{Max}\,w - \text{ord} = 1$ and $\underline{\text{Max}}\,w - \text{ord} = \text{Sing}\big(I(C), 1\big) = C$.

However the function $n(x)$ will distinguish the origen i.e., $\text{Max}\,t_2 = (1, 1)$ and $\underline{\text{Max}}\,t_2$ the point $(0, 0)$ of $\mathbb{C}^2$. So the center assigned in this case by the constructive resolution is clearly the origen. After such quadratic transformation we are left with:

a) The strict transform of C

b) The strict transfrom of H_1

c) H_2 : the exceptional locus of the transformation

d) A point q which is the intersection of the three curves mentioned in a) b) and c).

The transform of the basic object is

$$\big(W, E, (I(C), 1)\big) \leftarrow \big(W_1, E_1, (I(C), 1)\big),$$

where $E_1 = \{H_1, H_2\}$ and we denote again the strict transform of C by C.

Now $\operatorname{Max} t_2 = (1, 1)$ and $\underline{\operatorname{Max}} t_2$ is the point $q = H_1 \cap H_2 \cap C$.

Applying once again a quadratic transformation now at q, consider

e) H_3: the exceptional locus of the transformation

The strict transforms of the curves of a) b) and c) (called again C, H_1 and H_2) intersect H_3 at three different points.

Finally $\operatorname{Max} t_2$ has dropped, in fact now $\operatorname{Max} t_2 = (1, 0)$ and $\underline{\operatorname{Max}} t_2 = \operatorname{Sing}\big(I(C), 1\big)$. So $n(x)$ is constantly zero along the smooth scheme C and one can check now that C has normal crossing with $E_3 = \{H_1, H_2, H_3\}$. In fact, the condition $n(x) = 0$ for a smooth scheme in this setup is a sufficient condition for the smooth scheme to have normal crossing with the exceptional hypersurfaces (with $E = \{H_1, \ldots, H_s\}$).This is why we call $n(x)$ the "obstruction function".

Remark 6-2. *According to 4-7, 1), the lowering of $\operatorname{Max} t_2$ (since $R(1) = \emptyset$ in our case), is a one dimensional problem. We have not bothered with induction since $\underline{\operatorname{Max}} t_2$ were themselves points. However at the very beginning we could have chosen $V \hookrightarrow W$ with $V = H_1$. And the lowering of $\operatorname{Max} t_2 = (1, 1)$ could also have been done as in 4-6, 3) by taking the constructive resolution of: $\big(V, \mathbb{E}_1 = \emptyset, (\mathcal{A}, d)\big); \mathcal{A} = \langle x^2 \rangle; d = 1$. Here $V = \mathbb{C}$ and the constructive resolution consists of two quadratic transformations (identity maps) centered at the origen, say:*

$$\big(V, \mathbb{E}_1 = \emptyset, (\langle x^2 \rangle, 1)\big) \leftarrow \big(V, \mathbb{E}_2, (\langle x \rangle, 1)\big) \leftarrow \big(V, \mathbb{E}_3, (\mathcal{O}_V, 1)\big) \qquad (6.8)$$

In fact, $\operatorname{Sing}(\mathcal{O}_V, 1) = \emptyset$. One can check that the two quadratic transformations of Example 6-1 are induced by (6.8) in the sense of 4-6,3b).

Example 6-3. Here we want to illustrate the fact that in order to obtain an embedded resolution of singularities we might need the resolution of several basic objects, one after the other, all this without affecting the uniqueness of the whole procedure!!

In other words the conditions for embedded resolutions of singularities will require (in general) a concatenation of different resolutions of basic objects. Set $W = \mathbb{C}^2$; $C = \{x^3 + y^8 = 0\}$.

The highest multiplicity is 3 so set $J = \langle x^3 + y^8 \rangle$; $b = 3$. Now $\operatorname{Sing}(J, 3)$ is the set of points of maximal multiplicity, and for any permissible transformation

$$\big(W, E = \emptyset, (J, 3)\big) \leftarrow \big(W_1, E_1, (J_1, 3)\big),$$

J_1 will be the sheaf of ideals defining C_1 (the strict transform of C), and again $\operatorname{Sing}(J_1, 3)$ will be the set of points of multiplicity 3, which (if $\operatorname{Sing}(J_1, 3) \neq \emptyset$) is the highest possible multiplicity at points of C_1.

In our example a resolution of the original basic object is obtained by two quadratic transformations.

Step 0:
$$W_1 = \mathbb{C}^2$$
$$C = \{x^3 + y^8 = 0\}$$
$$\mathrm{Sing}\left(\langle x^3 + y^8 \rangle, 3\right) = (0,0) \in \mathbb{C}^2$$

so we blow-up at $(0,0)$.
Step I:
$$C_1 = \{x_1^3 + y_1^5 = 0\}$$
$$\mathrm{Sing}\left(\langle x_1^3 + y_1^5 \rangle, 3\right) = H_1 \cap C_1$$

where $x_1 = x$, $x_1 y_1 = y$. So we blow-up at $H_1 \cap C_1$.
Step II:
$$C_2 = \{x_2^3 + y_2^2 = 0\};$$
$$\mathrm{Sing}(< x_2^3 + y_2^2 >, 3) = \emptyset$$

where $x_2 = x_1$; $x_2 y_2 = y_1$.
Now the heighest multiplicity at points of C_2 is two so we start again with:

$$\big(W_3, E_3 = \{H_1, H_2\}, (I(C_2), 2)\big)$$

and we produce a resolution of this last basic object. The concatenation of
both resolutions will be a resolution of singularities but still not an embedded
resolution of singularities. In fact at that stage we are exactly in the conditions
of example 6-1, so that other basic objects are to be solved before the final
strict transform of C has normal crossing with all the exceptional hypersurfaces
introduced in the whole procedure.

Example 6-4. We want to solve the singularities of $f = z^2 + (x^2 + y^3)^2 = 0 \subset \mathbb{C}^3$.
So take the couple $(\langle f \rangle, 2)$ where $\mathrm{Max}\, t_3 = (1,0)$ (see below). Here $\underline{\mathrm{Max}\, t_3}$ is a
curve (therefore of codimension two in three space) so $R(1) = \emptyset$, and now the
lowering of $\mathrm{Max}\, t_3$ is a two dimensional problem (3b of 4-6).

   This two dimensional problem will be given at $\{z = 0\}$ " $= \mathbb{C}^2$" by a new
basic object.

   From the beginning (stage 0) we will consider a 3-dimensional problem
(called stage 0_a)) and a two dimensional problem (called stage 0_b) together
with a natural immersion:

$$[\text{Stage } 0a] \quad \supset \quad [\text{Stage } 0b]$$

given by the immersion

$$\mathbb{C}^3 \supset \mathbb{C}^2 \quad \text{via} \quad \mathbb{C}^2 = \{Z = 0\}$$

   When we say, as above, that the lowering of $\mathrm{Max}\, t_3$ is a two dimensional
problem (see 4-6), we mean that $\mathrm{Max}\, t_3$ (defined for the three dimensional
problem), will drop (or improve) by a resolution of a basic object on a the
smooth scheme $\mathbb{C}^2 (\simeq \{Z = 0\})$. So the first observation is that a sequence of

monoidal transformations over $\{Z = 0\}(\subset \mathbb{C}^3)$ induces naturally a sequence of monoidal tranformations over $\mathbb{C}^3$ together with a closed immersions:

$\bar{W}$ being the strict transform of $\{Z = 0\}$ at W. So after k monoidal transformations we will have an immersion:

$$[\text{Stage } k\text{a}] \quad \supset \quad [\text{Stage } k\text{b}]$$

Along part b) of each stage we apply our algorithm in dimension 2, and passing from stage kb to stage $(k+1)$b we introduce a new exceptional hypersurface (at the smooth 2-dimensional scheme) called: H_{k+1}.

$$X = \{Z^2 + (X^2 + Y^3)^2 = 0\} \subset \mathbb{C}^3$$

Stage 0a (dimension 3)
$W = \mathbb{C}^3$, $\text{Max}\,\omega - \text{ord}(J, 2) = 1$
$E = \emptyset$, $\text{Max}\,t_3(J, 2) = (1, 0)$
$J = \langle f \rangle$
$\text{Sing}(J, 2) = \underline{\text{Max}}\,t_3(J, 2) =$
$(J, 2), = \{z = 0\} \wedge \{x^2 + y^3 = 0\}$

Stage 0b (dimension 2)
$W' = \mathbb{C}^2$
$E' = \emptyset$
$J' = I(C)^2 = (x^2 + y^3)^2$
$(J', 2)$
$\text{Max}\,t_2 = (2, 0)$
$\text{Max}\,\omega - \text{ord} = 2$
$\underline{\text{Max}}\,t_2 = \emptyset, (0, 0) \in C^2$
is the center of the first
monoidal transformation. . .

Stage 1a (dimension 3)
$\text{Max}\,t_3(J_1, 2) = (1, 0)$
$\underline{\text{Max}}\,t_3(J_1, 2) = \text{Sing}(J, 2)$

Stage 1b (dimension 2)
$J_1' = \big(I(H_1)^2\big)I(C)^2$, $(J_1', 2)$
$E_1' = (E_1')^- = \{H_1\}$
$\text{Max}\,\omega - \text{ord} = 1$
$\text{Max}\,t_2 = (1, 1)$
$\underline{\text{Max}}\,t_2 = C \cap H_1$ is the center
of the next monomial
transformation. . .

Stage 2a (dimension 3)
$\text{Max}\,t_3 = (1, 0)$, $(J_2, 2)$
$\underline{\text{Max}}(t_2) = \text{Sing}(J_2, 2)$

Stage 2b (dimension 2)
$J_2' = I(H_1)^2\big\{I(H_2)^2\big\}I(C)^2$
$E_2' = \{H_1, H_2\}$
$(E_2')^- = \{H_1\}$
$\text{Max}\,t_2 = (1, 1)$
$\underline{\text{Max}}\,t_2 = H_1 \cap H_2$

Stage 3a (dimension 3)
Max $t_3 = (1,0)$, $(J_3, 2)$
$\underline{\text{Max}}\, t_3 = \text{Sing}(J_3, 2)$

Stage 3b (dimension 2)
$J'_3 =$
$\left(I(H_1)^2\, I(H_2)^2\, I(H_3)^2\right) I(C)^2$
$E'_3 = \{H_1, H_2, H_3\}$
$(E'_3)^- = \{H_1\}$
Max $t_2 = (1,0)$
$\underline{\text{Max}}\, t_2 = C$

Stage 4a (dimension 3)
Max $t_3 = (1,0)$, $(J_4, 2)$
$\underline{\text{Max}}\, t_3 = \text{Sing}(J_4, 2)$

Stage 4b (dimension 2)
$J'_4 = I(H_1)^2\, I(H_2)^2\, I(H_3)^4$
$E'_4 = \{H_1, H_2, H_3, H_4(= C)\}$
Max $H = (-1, 2, 3)$
$\underline{\text{Max}}\, H = H_3$

Stage 5a (dimension 3)
Max $t_3 = (1,0)$
$\underline{\text{Max}}\, t_3 = \text{Sing}(J_5, 2)$

Stage 5b (dimension 2)
$J'_5 = I(H_1)^2\, I(H_2)^2\, I(H_3)^2$
$E'_5 = \{H_1, H_2, H_4, H_5\}$
Max $H = (-1, 1, 5)$ $\underline{\text{Max}}\, H = H_5$

Stage 6 a (dimension 3)
Max $t_3 = (1,0)$
$\underline{\text{Max}}\, t_3 = \text{Sing}(J_6, 1)$

Stage 6b (dimension 2)
$J'_6 = I(H_1)^2\, I(H_2)^2$
$E'_6 = \{H_1, H_2, H_4, H_6\}$
Max $H = (-1, 1, 2)$
$\underline{\text{Max}}\, H = H_2$

Stage 7a (dimension 3)
Max $t_3 = (1,0)$
$\underline{\text{Max}}\, t_3 = \text{Sing}(J_7, 2)$

Stage 7b (dimension 2)
$J'_7 = I(H_1)^2$
$E'_7 = \{H_1, H_4, H_6, H_7\}$
Max $H = (-1, 1, 1)$
$\underline{\text{Max}}\, H = H_1$

Stage 8a (dimension 3)
Resolution

Stage 8b (dimension 2)
$J'_8 = \mathcal{O}_{\overline{w}_8}$
$\text{Sing}(J'_8, 2) = \emptyset$

At part (b) of each stage we are applying our algorithm in dimension two. At 3b $\underline{\text{Max}}\, t_2 = C = R(1)$.

After blowing up $R(1)$ (stage 4b) we are left with the monomial case.

At the very end, not only have we lowered Max t_3 of our original 3-dimensional problem, but we have also solved (in this case) the singularities of $f = 0$.

References

[1] S.S. Abhyankar, *Good points of a hypersurface*. Advances in Math. Vol 68 No. 2 (1988).

[2] S.S. Abhyankar and T.T. Moh, Newton-Puiseux expansion and generalized Tschirnhausen transformation, Crelle journal, 260 (1973), 47–83 and 261 (1973), 29–54.

[3] V. Cossart, J. Giraud, U. Orbanz, *Lecture Notes in Mathematics*. Springer Verlag no. 1101 (1984).

[4] J. Giraud, Sur la theorie du contact maximal, Math. Zeit. 137 (1974), 285–310.

[5] J. Giraud, *Analysis Situs*, Sem. Bourbaki 1962/63, No. 256.

[6] H. Hironaka, Resolution of singularities of an algebraic variety over a field of characteristic zero I–II Ann. Math. 79 (1964) 109–326.

[7] H. Hironaka, *Idealistic exponent of a singularity*. Algebraic Geometry. The Johns Hopkins centennial lectures, p. 52–125. Baltimore: Johns Hopkins University Press 1977.

[8] J. Lipman, Introduction to resolution of singularities. Proc. Symp. in Pure Math., 29 (1975), 187–230.

[9] M. Lejeune, B. Teissier, *Quelques calculs utiles pour la resolution des singularites*. Centre de Mathematique de l'Ecole Polytechnique (1972).

[10] T.T. Moh, Canonical uniformization of hypersurface singularities of characteristic zero. (to appear) **Journal of Pure and Applied Algebra**.

[11] T. Oda, *Infinitely Very Near-Singular Points*. Complex Analytic Singularities, p. 363–404. Advanced Studies in Pure Mathematics 8, (1986).

[12] O.E. Villamayor, Constructiveness of Hironaka's resolution. Ann. Scient. Ec. Norm. Sup., 4^e serie, t. 22, p. 1–32 (1989).

[13] O.E. Villamayor, Patching local uniformizations. (to appear) Ann. Scient. Ec. Norm. Sup. Vol. 25 (1992).

[14] O. Zariski, Local uniformization on algebric varieties Ann. Math., 41 (1941), 852–896.

Address of author:

Dep. de Matemáticas, Fac. de Ciencias
Universidad Autónoma de Madrid
28049 Cantoblanco - Madrid
Spain
Email: villa@roble.sdi.uam.es

Part II

Complex Singularities and Differential Systems

POLARITY WITH RESPECT TO A FOLIATION

J. García[1] and A. J. Reguera[2]

1 Introduction

Let $\mathcal{F}$ be a reduced foliation defined on $\mathbb{P}_2$—the complex projective plane—given in homogeneous coordinates $(X:Y:Z)$ by a 1-form of degree q,

$$\Omega = A\,dX + B\,dY + C\,dZ,$$

i.e., such that A, B, C are homogeneous polynomials of degree q, with no common factors and satisfying Euler's equation $XA + YB + ZC \equiv 0$. In this way we have a rational map

$$\begin{array}{rccc} \Phi & : & \mathbb{P}_2 & \longrightarrow & \mathbb{P}_2^\vee \\ & & Q & \longmapsto & \Phi(Q) \end{array}$$

defined on $\mathbb{P}_2 \backslash \mathrm{Sing}(\mathcal{F})$ which associates to each Q the point in $\mathbb{P}_2^\vee$ corresponding to the line defined by $\mathcal{F}$ at Q.

Thinking geometrically, the map Φ will be called the polarity map of $\mathcal{F}$. Moreover, one has a 2-dimensional linear system of curves $\{\lambda A + \nu B + \rho C\}$, defining Φ, which will be called the net of polars of $\mathcal{F}$. The singular set of $\mathcal{F}$ consists exactly of the points of indetermination of Φ, i.e., the base points of the net of polars.

The purpose of this paper is to study the relationship between the properties and behaviour of the foliation and the geometry of the net of polars, providing in this way a framework to think geometrically when one deals with questions relative to foliations.

Our motivation comes from problems which have been treated recently in the literature, like the existence of global solutions and the classification of foliations relating global and local invariants (see [2] and [10]) and, in particular, relating the degree of a global integral curve with the degree of the foliation ([1], [3] and [7]). We have observed the geometrical ideas coming from polarity to be useful in order to understand questions as above.

Polarity has been classically a powerful tool, very useful for classification problems in geometry, appearing in the work by Enriques, Severi, Tood or Chern. In the 80's, Lê and Teissier introduced the corresponding local tools in connection with equisingularity problems ([9], [11] and [12] for a historical review) and polarity is also used to deal with the singularities of maps (see [4]).

In this paper, we study the polarity relative to a foliation in $\mathbb{P}_2$ in a classical geometric way. We obtain a result that allows us to explain the behaviour

[1]Supported by Iberdrola
[2]Supported by D.G.I.C.Y.T PB91-0210-C02-01

of the integral lines to the foliation. To complete the discussion, we need to eliminate the indeterminancy locus of the polarity map and also consider the behaviour of the exceptional lines which are integral for the foliation. In sections 3 and 4 we show how the polarity map from the projective plane to its dual suggests to generalize Plücker's Formula for a foliation and a curve as a way to relate its global and local invariants. Finally, in sections 5 and 6, by means of the net of polars, we get explicit formulas to compute those invariants. We look for the extension of these results to foliations in arbitrary projective spaces in a forthcoming work.

2   Preliminaries on linear systems

Let $\Gamma = \{\lambda_0 f_0 + \cdots + \lambda_n f_n\}$ be a linear system of two variable power series, free of common factors, with complex coefficients. We define the multiplicity of Γ as $m = \min\{\mathrm{ord}(f_i)\}$.

(2-1) Let $f_i = f_i^{(m)} + f_i^{(m+1)} + \cdots$ be the homogeneous decomposition of f_i (where $f_i^{(m)} \neq 0$) and let D be the greatest common divisor of the $f_i^{(m)}$. We say that Γ has a base point if $m \neq 0$. If, on the contrary, $m = 0$ then some f_i must be a unit and Γ is resolved. Let r denote the diference $m - \mathrm{gr}(D)$, then we say that Γ is dicritical if r is not null and Γ is non-diciritical if $r = 0$.

Let $X_0 = \mathrm{Spec}\,\mathbb{C}[\![x,y]\!]$, where $\mathbb{C}[\![x,y]\!]$ is the ring of power series in two variables and let $\pi\colon X_1 \to X_0$ be the blowing-up at the origin of X_0. The transform Γ'_P of Γ at a point P of the exceptional divisor $\pi^{-1}(O)$ is the linear system consisting on the virtual transforms of the elements of Γ with virtual multiplicity m. The strict transform of the generic element of Γ is the generic element of Γ'_P.

Theorem 2-2. ([6], 1.1.3) *There exist a finite sequence of blowing-ups*

$$X_p \longrightarrow X_{p-1} \longrightarrow \cdots \longrightarrow X_0,$$

such that the transform of Γ is resolved in X_p.

The resolution process of Γ is described by means of points infinitely near O. We call points of the first neighbourhood of $O \in X_0$, denoted V_1, the points of $\pi^{-1}(O)$. Inductively, we call points of the $i-th$ neighbourhood, denoted V_i, the points of the $(i-1)-th$ neighbourhood of some point of V_1. The points in some neighbourhood of O, denoted by V, are called points infinitely near O. The set V is partially ordered in a natural way: $P < P'$ if and only if P' is infinitely near of P.

We say that a subset $S \subset V$ is a simple cluster with origin at O if and only if:

1. $O \in S$.
2. The restriction of the order of V to S is a total order.
3. Let $P, P' \in V$ and $P < P'$. If $P' \in S$ then $P \in S$.

We define a cluster with root O to be a finite union of simple clusters with origen at O. Let Γ be a not resolved linear system, then the resolution cluster of Γ is the set of pairs (P, r_P) such that

1. $P \in V$.
2. P is a base point of a transform of Γ.
3. r_P is the value of r at P (see 2-1).

We denote $C(\Gamma)$ this cluster.

(2-3) Let $I(\Gamma)$ be the ideal generated in $\mathbb{C}[\![x, y]\!]$ by the elements of Γ and let $\overline{I(\Gamma)}$ be its integral closure. Then, using the theory of complete ideals introduced by Zariski ([14], App.4-7) one can prove the following result.

Theorem 2-4. ([6], 1.5.4) *Let Γ_1, Γ_2 be two linear systems. Then*

$$C(\Gamma_1) = C(\Gamma_2) \quad \text{if and only if} \quad \overline{I(\Gamma_1)} = \overline{I(\Gamma_2)}.$$

3 The polarity map

Let $P \in \text{Sing}(\mathcal{F})$ be a base point of the net of polars. Blowing up this point we get the transform linear system. On the points of the exceptional divisor where the transform linear system resolved, we have a natural extension Φ_1 of the map Φ. In other cases, the indetermination of the extended map persists. Blowing-up successively, the indetermination points of Φ are removed after finitely many steps as in theorem 2-2.

In this way we obtain a surface $\widetilde{\mathbb{P}}_2$ and an extension $\widetilde{\Phi}$ of Φ

In the exceptional divisor $\pi^{-1}(\text{Sing}(\mathcal{F}))$ we have a finite number of components each one isomorphic to $\mathbb{P}_1$. Their images via $\widetilde{\Phi}$ are either irreducible rational curves or else points in $\mathbb{P}_2^\vee$. Let E be an irreducible component of the exceptional divisor over a singular point P. The points in $\widetilde{\Phi}(E)$ correspond to lines in $\mathbb{P}_2$ passing through P, and so $\widetilde{\Phi}(E)$ is either the dual line of P or else a point in this line.

Remark 3-1. *Let a, b, c be the germs of A, B, C at the origin $(0:0:1)$ and let (x, y) be local coordinates. The cluster of the net of germs $\{\lambda a + \mu b + \delta(-xa - yb)\}$ depends only on the ideal $(a, b, -xa - yb)$ (in fact on the complete ideal defined in 2-3), and so they are the same points as those given by the pencil $\{\lambda a + \mu b\}$. In general, given a germ of a foliation $\omega = a\,dx + b\,dy$ where a and b have no common factors, we call linear system of local polars the linear system of germs $\{\lambda a + \mu b\}$ (see [5], 2.10). Although the linear system of local*

160 J. García and A. J. Reguera

polars depends on the local analytic coordinates, its cluster of infinitely near base points only depends on the germ of foliation because the ideal (a, b) is invariant by analytic transformations.

Let $\Gamma = \{\lambda a(x, y) + \mu b(x, y)\}$ be a pencil of germs of analytic functions in the origin. Let s be its multiplicity and a_s, b_s the homogeneous forms of degree s of a, b. Then we say Γ is dicritical (resp. non-dicritical) if a_s, b_s are linearly independent (resp. linearly dependent). Each point in the cluster gives rise to an irreducible component of the exceptional divisor. We say that this component is dicritical (resp. non-dicritical) if the point is (resp. is not) dicritical.

Theorem 3-2. *The irreducible curves in $\widetilde{\mathbb{P}}_2$ whose image via $\widetilde{\Phi}$ is contracted to a point are precisely the non-dicritical divisor and the strict transforms of the integral lines of the foliation $\mathcal{F}$ in $\mathbb{P}_2$.*

PROOF. Previously, we will see how Φ acts in coordinates. Let (x, y) be the affine coordinates of $(X : Y : Z)$ and let $a(x, y) = A(x : y : 1)$, $b(x, y) = B(x : y : 1)$. We write $[U : V : W]$ for the dual coordinates in $\mathbb{P}_2^{\vee}$. For $P = (x : y : 1)$ we have

$$\Phi(P) = [a(x, y) : b(x, y) : -xa(x, y) - yb(x, y)].$$

Let s be the multiplicity of the pencil $\{\lambda a + \mu b\}$, and let $a_s(x, y)$, $b_s(x, y)$, be the homogeneous forms of degree s of $a(x, y)$, $b(x, y)$ respectively. Let $\pi_1 : \mathbb{P}_2^{(1)} \to \mathbb{P}_2$ be the blowing-up of $O = (0 : 0 : 1)$ and $\Phi_1 = \Phi \circ \pi_1 : \mathbb{P}_2^{(1)} \to \mathbb{P}_2^{\vee}$. The restriction map of Φ_1 to $\pi^{-1}(O)$ is a map valued on the line $W = 0$ of $\mathbb{P}_2^{\vee}$. Let $x' = x$, $y' = y/x$ be coordinates at a point of $\pi^{-1}(O)$. We have

$$\Phi_1(Q) = [a_s(1, y') : b_s(1, y') : 0], \qquad Q = (0, y').$$

If $\pi^{-1}(O)$ is dicritical, then the linear forms a_s, b_s are linearly independent and Φ_1 is onto. Reciprocally, if $\pi^{-1}(O)$ is non-dicritical, then a_s, b_s are linearly dependent, i.e., $\alpha a_s + \beta b_s \equiv 0$ for $\alpha, \beta \in \mathbb{C}$ and the image of Φ_1 is only the point $[\beta : -\alpha : 0]$. The integral lines are contracted to its dual points by the definition of Φ.

To finish the proof, let us observe that if $H \subset \widetilde{\mathbb{P}}_2$ is an irreducible curve contracted to a point via $\widetilde{\Phi}$, then either $\pi(H)$ is a point or the curve $\pi(H)$ contracts to a point via Φ. In the first case H is a component of E and in the second case $\pi(H)$ is an integral line. $\square$

Example 3-3. Let $\Omega = Y dX - X dY$. In this case the pencil of polars $\{\lambda X + \mu Y\}$ consists of the lines of the pencil passing through $P = (0 : 0 : 1)$. All these lines are integral lines and so they will concentrate in a point. The only singular point of the foliation is P and this is the only base point of the cluster of the local polar pencil $\{\lambda x + \mu y\}$ at P. The point P gives rise to a dicritical divisor E_P whose image via $\widetilde{\Phi}$ is the dual line of P. Therefore, Im $\widetilde{\Phi}$ consists on a line, the dual line of P, and there are infinitely many lines whose image is a point.

We will call radial foliation a foliation as the one before, but related, in general, to a pencil of lines passing through an arbitrary point in the plane.

Proposition 3-4. *Given the foliation $\mathcal{F}$, the following conditions are equivalent:*
(a) $\dim(\operatorname{Im}\widetilde{\Phi}) = 1$.
(b) *There are infinitely many integral lines.*
(c) $\mathcal{F}$ *is a radial foliation.*

PROOF. $(a) \Rightarrow (b)$ is obvious because the fibers of $\widetilde{\Phi}$ have dimension one and so there are curves whose irreducible components are contracted via $\widetilde{\Phi}$. As there are a finite number of non-dicritical divisors, there must be an infinite number of integral lines.

$(b) \Rightarrow (c)$. Given two different integral lines, its intersection point must be singular for $\mathcal{F}$. As the number of singular points is finite, and each pair of integral lines have one of these points in common, there must be infinitely many lines passing through a certain singular point P. Now, any other integral line must pass through P because if not, intersecting the preceding ones, it will produce infinitely many singular points. Let us suppose $P = (0\!:\!0\!:\!1)$, and let $\omega = a\,dx + b\,dy$ be the expression of the foliation at P. If there are infinitely many integral lines $y = \lambda x$, then

$$0 = a(x,\lambda x)\,dx + \lambda b(x,\lambda x)\,dx = \big(a(x,\lambda x) + \lambda b(x,\lambda x)\big)\,dx$$

for infinitely many values of λ. Writing

$$a(x,y) = a_0(x) + a_1(x)y + a_2(x)y^2 + \cdots,$$
$$b(x,y) = b_0(x) + b_1(x)y + b_2(x)y^2 + \cdots,$$

we have

$$a_0(x) + \sum_{i\geq 1}\big(a_i(x)x^i + b_{i-1}(x)x^{i-1}\big)\lambda^i = 0.$$

Interpreting this as a polynomial in λ with coefficients in $k(x)$ we conclude

$$a(x,y) = yp(x,y), \qquad b(x,y) = -xp(x,y),$$

$p(x,y)$ being a polynomial in $k[x,y]$. As the foliation is reduced, then $p(x,y)$ is a unit and we can suppose $\omega = y\,dx - x\,dy$ that is, a radial foliation.
$(c) \Rightarrow (a)$ is obvious. $\qquad\square$

We will say that a curve H and the foliation $\mathcal{F}$ are tangent in a point $P \in \mathbb{P}_2$ if they have the same tangent line at P.

Proposition 3-5. *Let H be a non-integral curve of degree m. The number of points where H and $\mathcal{F}$ are tangent, counting multiplicity, is $m(q + m - 2)$.*

PROOF. We take coordinates in $\mathbb{P}_2$ in such a way that $(Z = 0) \cap H$ are m different points such that $\mathcal{F}$ and H are non-singular and transversal. Let $F = 0$ be a reduced equation of H, then

$$\Omega \wedge dF = J_1 dX \wedge dY + J_2 dX \wedge dZ + J_3 dY \wedge dZ,$$

where $J_1 = AF_Y - BF_X$, $J_2 = AF_Z - CF_X$, $J_3 = BF_Z - CF_Y$. The points of the affine part of H where $\Omega \wedge dF$ is zero are $(J_1 = 0) \cap H \cap (Z \neq 0)$ counted with multiplicities, and $(J_1 = 0) \cap H \cap (Z = 0)$ are m points where $\mathcal{F}$ and H are non-singular and transversal. Therefore the number of points in H where $\Omega \cap dF$ is null is

$$m(q + m - 1) - m = m(q + m - 2). \qquad \square$$

Corollary 3-6. If $\mathcal{F}$ is a non-radial foliation, then

$$\deg(\Phi) = q - 1.$$

Indeed, $\deg(\Phi)$ is the number of points where a generic line is tangent to $\mathcal{F}$.

4   Plücker's formula

Let $H \in \mathbb{P}_2$ be an irreducible curve and $\widetilde{H}$ its normalization. Although in H there may be points of indetermination of Φ, the restriction map $\Phi|_H$ can be extended to a well defined morphism ϕ on $\widetilde{H}$. So $H^* = \phi(H)$ is an irreducible curve in $\mathbb{P}_2^\vee$ (or else a point if H is an integral line). Let $\widetilde{H^*}$ be the normalization of H^*, and $\widetilde{\phi}: \widetilde{H} \to \widetilde{H^*}$ the induced map, then the following diagram is commutative.

$$
\begin{array}{ccc}
\widetilde{H} & \xrightarrow{\phi} & \mathbb{P}_2^\vee \\
\downarrow{\widetilde{\phi}} & & \uparrow{i^*} \\
\widetilde{H^*} & = & \widetilde{H^*}
\end{array}
$$

We will define three invariants for the branches of $\widetilde{H}$ and $\widetilde{H^*}$.

Ramification of ϕ in $\gamma \in \widetilde{H}$.

Let $\gamma \in \widetilde{H}$ be a branch of H and let us take affine coordinates $\{v, w\}$ at the point $\phi(\gamma) \in \mathbb{P}_2^\vee$. Let $[v(t), w(t)]$ be the local expresion of ϕ respect to a local parameter t of the branch γ, then we define

$$\beta(\gamma) = \min\{\mathrm{ord}_t\, v'(t), \mathrm{ord}_t\, w'(t)\}.$$

The number $\beta(\gamma)$ is independent of the choice of t and $\{v, w\}$, and we call it ramification index of ϕ in γ.

Ramification of i^* in $\gamma^* \in \widetilde{H^*}$.

Let $\gamma^* \in \widetilde{H^*}$ be a branch of H^* and let us take affine local coordinates $\{v_1, w_1\}$ at the point $i^*(\gamma^*) \in \mathbb{P}_2^\vee$. Let $[v_1(s), w_1(s)]$ be the local expression of i^* respect to a local parameter s of the branch γ^*, then we define

$$\beta^*(\gamma^*) = \min\{\mathrm{ord}_s\, v_1'(s), \mathrm{ord}_s\, w_1'(s)\}.$$

Again $\beta^*(\gamma^*)$ is independent of the choice of s and $\{v_1, w_1\}$ and we call it the ramification index of i^* in γ^*.

Ramification of $\widetilde{\phi}$ in $\gamma \in \widetilde{H}$.

Let $\gamma \in \widetilde{H}$ and let t and s be local parameters of the branches γ and $\widetilde{\phi}(\gamma)$ respectively. The morphism $\widetilde{\phi}$ can be locally expressed as $s = s(t)$. We define

$$\nu(\gamma) = \mathrm{ord}_t\, s(t),$$

which is independent of t and s and we call it ramification index of $\widetilde{\phi}$ in γ.

Finally, let us define two global invariants associated to the curve H and to the map Φ.

Class of H respect to Φ.

Given a generic point $R \in \mathbb{P}_2$, the class of H respect to Φ , denoted by $\mathrm{cl}_\Phi(H)$, is the number of points $\gamma \in \widetilde{H}$ such that R belongs to $\widetilde{\phi}(\gamma)$, or equivalently, the number of points $G \in H$ such that the tangent line to the foliation passes through R. This number can be expressed as

$$\mathrm{cl}_\Phi(H) = \delta(\widetilde{\phi}) \cdot \deg(H^*),$$

where $\delta(\widetilde{\phi})$ is the degree of the field extension $\left[K(\widetilde{H}):K(\widetilde{H^*})\right]$ induced by $\widetilde{\phi}$.

Second class of H respect to Φ.

We consider the morphism j obtained by composing the dualization of H^* with the inclusion of $(H^*)^\vee$ in $\mathbb{P}_2$

$$j: H^* \longrightarrow (H^*)^\vee \subset \mathbb{P}_2.$$

Now we call the second class of H respect to Φ, denoted by $\mathrm{cl}^*_\Phi(H)$, the number of points $\gamma \in \widetilde{H}$ such that $j\big(\phi(\gamma)\big)$ belongs to a generic line $L \subset \mathbb{P}_2$, or equivalently, the number of points $Q \in H$ such that the dual point of the tangent line to the foliation at Q belongs to a generic line in $\mathbb{P}_2$. This number can be expressed as

$$\mathrm{cl}^*_\Phi(H) = \delta(\widetilde{\phi}) \cdot \deg\big((H^*)^\vee\big).$$

Now, Plücker's formula says:

Theorem 4-1. *We have*

$$2g_H - 2 = -2\,\mathrm{cl}_\Phi(H) + \mathrm{cl}^*_\Phi(H) + \sum_{\gamma \in \widetilde{H}} \beta(\gamma), \qquad (4.1)$$

where g_H is the genus of H and the other invariants are just defined.

PROOF. Hurwitz's theorem applied to $\widetilde{\phi}$ gives

$$2g_H - 2 = \delta(\widetilde{\phi}) \cdot (2g_{H^*}) + \sum_{\gamma \in \widetilde{H}} (\nu(\gamma) - 1), \qquad (4.2)$$

and Plücker's theorem applied to $\widetilde{i^*}$ says

$$2g_{H^*} - 2 = -2\deg(H^*) + \deg\big((H^*)^\vee\big) + \sum_{\gamma^* \in \widetilde{H^*}} \beta^*(\gamma^*). \qquad (4.3)$$

Substituting (4.3) in (4.2) we obtain

$$2g_H - 2 = -2\operatorname{cl}_\Phi(H) + \operatorname{cl}_\Phi^*(H) + \left[\delta(\widetilde{\phi}) \cdot \sum_{\gamma^* \in \widetilde{H^*}} \beta^*(\gamma^*) + \sum_{\gamma \in \widetilde{H}} (\nu(\gamma) - 1)\right]. \quad (4.4)$$

and so, we only need to prove that the bracketed expression in (4.4) is the global ramification of ϕ, that is, $\sum_{\gamma \in \widetilde{H}} \beta(\gamma)$.

Let us fix $\gamma \in \widetilde{H}$. For a regular parameter t of γ we can take local coordinates $[v(t), w(t)]$, $s = s(t)$ and $[v_1(s), w_1(s)]$ of ϕ, $\widetilde{\phi}$ and i^* respectively such that

$$\big[v_1\big(s(t)\big), w_1\big(s(t)\big)\big] = [v(t), w(t)].$$

Then, we have

$$\begin{aligned} \beta(\gamma) &= \min\{\operatorname{ord}_t v'(t), \operatorname{ord}_t w'(t)\}, \\ &= \operatorname{ord}_t s(t)\{\beta^*(\widetilde{\phi}(\gamma)) + 1\} - 1, \\ &= \nu(\gamma) \cdot \beta^*(\widetilde{\phi}(\gamma)) + (\nu(\gamma) - 1), \end{aligned}$$

and so

$$\sum_{\gamma \in \widetilde{H}} \beta(\gamma) = \sum_{\gamma^* \in \widetilde{H^*}} \left(\sum_{\gamma \in \widetilde{\phi}^{-1}(\gamma^*)} \nu(\gamma)\right) \beta^*(\gamma^*) + \sum_{\gamma \in \widetilde{H}} (\nu(\gamma) - 1).$$

But for any $\gamma^* \in \widetilde{H^*}$,

$$\delta(\widetilde{\phi}) = \sum_{\gamma \in \widetilde{\phi}^{-1}(\gamma^*)} \nu(\gamma),$$

and so (4.1) is proved. $\qquad\square$

PROOF. (SECOND PROOF) The general Plücker's formula for the morphism $\phi\colon \widetilde{H} \to \mathbb{P}_2^{\vee}$ says

$$2g_H - 2 = -2\deg(\phi) + \deg(\phi_2) + \sum_{\gamma \in \widetilde{H}} \beta(\gamma),$$

where ϕ_2 is defined as follows: If ϕ is given locally by the vector function $v(z) = [v_0(z)\colon v_1(z)\colon v_2(z)]$ then ϕ_2 is locally given by $v(z) \wedge v'(z)$. Therefore $\phi_2(z)$ is an element of $\mathbb{P}(\wedge^2\mathbb{C}^3)^{\vee}$, but this space is canonically isomorphic to $\mathbb{P}_2$, and hence ϕ_2 is the composition of ϕ and j

$$\phi_2\colon \widetilde{H} \xrightarrow{\phi} H^* \xrightarrow{j} (H^*)^{\vee}.$$

We denote by $\deg(\phi)$ and $\deg(\phi_2)$ the degree of these morphisms between Riemann surfaces and therefore

$$\deg(\phi) = \mathrm{cl}_{\Phi}(H), \qquad \deg(\phi_2) = \mathrm{cl}_{\Phi}^*(H)$$

and the theorem is proved.   □

Remark 4-2.

  1. In the proof we do not use Euler's relation for the polynomials A, B and C. Therefore (4.1) is true for any rational morphism Φ between two projective planes and any irreducible curve H.

  2. If Φ is the polarity map of a foliation and H is an integral curve, then $H^* = H^{\vee}$ and $\widetilde{\phi}$ is the dual map. Therefore, $\widetilde{\phi}$ is birational and $\delta(\widetilde{\phi}) = 1$. Equation (4.1) is in this case the classic Plücker's formula

$$2g_H - 2 = -2\,\mathrm{cl}(H) + \deg(H) + \sum_{\gamma \in \widetilde{H}} \beta(\gamma).$$

  3. Let D_1, D_2 be two generic elements of the polar system $\{x_0 A + y_0 B + z_0 C\}$. The intersection cycle contains $\mathrm{Sing}(\mathcal{F})$ as fixed points and some free points. As $\phi(D_1)$ and $\phi(D_2)$ are generic lines in $\mathbb{P}_2^{\vee}$, their intersection is only one point. So, if $Q^* \in \phi(D_1) \cap \phi(D_2)$ then $\phi^{-1}(Q^*)$ are the free points and therefore

$$q^2 = D_1 \cdot D_2 = \sum_{P \in \mathrm{Sing}(\mathcal{F})} I_P(D_1, D_2) + \deg(\phi).$$

  4. Let $H \in \mathbb{P}_2$ be an irreducible curve which is not an integral line. Then $\widetilde{\phi_H}\colon \widetilde{H} \to \widetilde{H}^*$ is a finite morphism of degree $\delta(\phi_H)$ satisfying

$$1 \leq \delta(\phi_H) \leq q - 1.$$

For a generic point in H^*, the number of points in its inverse image via Φ is $\delta(\Phi) = q - 1$. In fact, $\delta(\phi_H)$ can reach arbitrary values between 1 and $q - 1$.

For example, we consider an element of the polar net $\{x_0 A + y_0 B + z_0 C\}$ and let $H_1, \ldots, H_r$ be the irreducible components of this element. Then

$$\delta(\phi_{H_r}) + \cdots + \delta(\phi_{H_r}) = q - 1,$$

because for a generic point $P \in \phi(H_1 \cap \cdots \cap H_r)$ the fiber $\Phi^{-1}(P)$ is contained in $H_1 \cap \cdots \cap H_r$ and we can choose the point P so that no point of $\Phi^{-1}(P)$ is in the intersection any two H_i's (If $r = 1$ then $\delta(\phi_H) = q - 1$).

5　The net of polars

Let $P = (x_0 : y_0 : z_0)$ be a point in $\mathbb{P}_2$. We call polar curve of P with respect to $\mathcal{F}$ the curve $\mathcal{P}(P)$ defined by

$$\mathcal{P}(P) \ : \ x_0 A + y_0 B + z_0 C = 0,$$

that is, the closure of the set of points $Q \in \mathbb{P}_2 \setminus \operatorname{Sing}(\mathcal{F})$ such that P belongs to $\Phi(Q)$. Let us remark that for any $Q \in \mathbb{P}_2 - \operatorname{Sing}(\mathcal{F})$ the line $\Phi(Q)$ is given by

$$\Phi(Q) = \{P \in \mathbb{P}_2 \mid Q \in \mathcal{P}(P)\},$$

that is, the line of equation

$$X A(Q) + Y B(Q) + Z C(Q) = 0.$$

Let H be an irreducible curve, Q a point in H and $\gamma \in \tilde{H}$ a branch of H at Q. Then we define

$$
\begin{aligned}
m_1(\gamma) &= \inf\{I_Q(\mathcal{P}(P), \gamma) \mid P \in \mathbb{P}_2\} \\
&= \min\{I_Q(A = 0, \gamma), I_Q(B = 0, \gamma), I_Q(C = 0, \gamma)\}. \quad (5.5)
\end{aligned}
$$

Let us note that if $Q \notin \operatorname{Sing}(\mathcal{F})$ then $m_1(\gamma) = 0$.

The set of points $P \in \mathbb{P}_2$ in wich $I_Q(\mathcal{P}(P), \gamma) > m_1(\gamma)$ is a line in $\mathbb{P}_2$ (to prove it, we take a parametrization of γ, and after substituing in $\mathcal{P}(P)$, we impose the first coefficient to be zero). If Q is a non-singular point for $\mathcal{F}$, this line is the set of points whose polars go through Q, that is, $\Phi(Q)$. It does not depends on the branch γ but on the point Q.

If Q is a singular point for $\mathcal{F}$, then $m_1(\gamma) > 0$ and for a local parametriza-tion $(x(t), y(t))$ for γ, the equation of the line is

$$\lim_{t \to 0} \frac{1}{t^{m_1}} \left[a(t) X + b(t) Y + c(t) Z \right] = 0,$$

where $a(t) = A(x(t), y(t), 1)$, $b(t) = B(x(t), y(t), 1)$, $c(t) = C(x(t), y(t), 1)$. This is exactly the way $\tilde{\phi}$ acts on the branches of H centered at a singular point of Φ. Therefore

$$\tilde{\phi}(\gamma) = \{P \in \mathbb{P}_2 \mid I_Q(\mathcal{P}(P), \gamma) > m_1(\gamma)\}.$$

In this way the foliation $\mathcal{F}$ and the morphism $\widetilde{\phi}$ can be obtained from the system of polars.

If H is an irreducible curve which is not an integral line for the foliation, and $\gamma \in \widetilde{H}$ is a branch of H centered at $Q \in H$, then we define

$$m_2(\gamma) = \inf\{I_Q(\mathcal{P}(P), \gamma) \mid P \in \phi(\gamma)\}. \tag{5.6}$$

Lemma 5-1. *There exist only one point $P_\gamma \in \phi(\gamma)$ for which the contact at Q of its polar curve with γ is bigger than $m_2(\gamma)$.*

PROOF. To prove it, we can take coordinates so that $\phi(\gamma)$ is the line $X = 0$ and $Q = (0{:}0{:}1)$. Then, we have

$$\begin{aligned}
m_1(\gamma) &= \operatorname{ord}_t a(t), \\
m_2(\gamma) &= \inf_{(y:z)} \{\operatorname{ord}_t (b(t)y + c(t)z)\}.
\end{aligned}$$

Taking

$$\sigma = \lim_{t\to 0} \frac{b(t)}{t^{m_2}}, \quad \text{and} \quad \tau = \lim_{t\to 0} \frac{c(t)}{t^{m_2}},$$

it is clear that $P_\gamma = (0{:}\tau{:}{-}\sigma)$. $\qquad\square$

We call $m_3(\gamma)$ this greatest intersection multiplicity

$$m_3(\gamma) = I_Q(\mathcal{P}(P_\gamma), \gamma). \tag{5.7}$$

Let $(H^*)^\vee \subset \mathbb{P}_2$ be the dual curve of $H^* \subset \mathbb{P}_2^\vee$. We associate to the point $\phi(\gamma) \in H^*$ the point $\phi_2(\gamma) = j(\phi(\gamma)) \in \mathbb{P}_2$ representing the tangent line of H^* at $\phi(\gamma)$. In these conditions we have

Proposition 5-2. *The tangent line of H^* at $\widetilde{\phi}(\gamma)$ is $P_\gamma = \phi_2(\gamma)$.*

PROOF. As before, we suppose $Q = (0{:}0{:}1)$ and $\widetilde{\phi}(\gamma)$ is given by $X = 0$. We take $[U{:}V{:}W]$ coordinates of $\mathbb{P}_2^\vee$ dual of $(X{:}Y{:}Z)$. Now, let

$$v(t) = \frac{b(t)}{a(t)}, \qquad w(t) = \frac{c(t)}{a(t)};$$

then $[1{:}v(t){:}w(t)]$ is a parametrization of $\widetilde{\phi}(\gamma)$. We proceed to calculate $\phi_2(\gamma)$ and P_γ. One of the following three situations must be happen:

1. $\operatorname{ord}_t b(t) < \operatorname{ord}_t c(t)$, then $P_\gamma = (0{:}0{:}1)$. In this case, we have $0 < \operatorname{ord}_t v(t) < \operatorname{ord}_t w(t)$ and so the tangent line of H^* at $\widetilde{\phi}(\gamma)$ is $W = 0$, therefore $\phi_2(\gamma) = (0{:}0{:}1) = P_\gamma$.
2. $\operatorname{ord}_t b(t) > \operatorname{ord}_t c(t)$, the reasoning is the same as in the preceding case, and $\phi_2(\gamma) = (0{:}1{:}0) = P_\gamma$.
3. $\operatorname{ord}_t b(t) = \operatorname{ord}_t c(t)$. The tangent line of H^* at $\widetilde{\phi}(\gamma)$ is $\tau V - \sigma W = 0$ and so $\phi_2(\gamma) = (0{:}\tau{:}{-}\sigma) = P_\gamma$. $\qquad\square$

Therefore, to any branch $\gamma \in \tilde{H}$ we associate a line $\tilde{\phi}(\gamma)$ passing through Q and a point P_γ in this line.

Remark 5-3. If H is an integral curve of $\mathcal{F}$—not a line—, then $H^* = H^\vee$ and $(H^*)^\vee = H$ so that $Q = P_\gamma = \phi_2(\gamma)$ for any branch $\gamma \in \tilde{H}$ centered at $Q \in H$.

6 Some calculus

The net of polars of $\mathcal{F}$ and the local invariants defined in (5.5), (5.6) and (5.7), help us to compute some of the global invariants appearing in Plücker's formula.

Theorem 6-1. We have

$$\mathrm{cl}_\Phi(H) \;=\; m \cdot q - \sum_{\gamma \in \tilde{H}} m_1(\gamma), \tag{6.8}$$

$$\beta(\gamma) \;=\; m_2(\gamma) - m_1(\gamma) - 1. \tag{6.9}$$

PROOF. (6.8). Let $R \in \mathbb{P}_2$ be a generic point and let $\mathcal{P}(R)$ be its polar. Applying Bezout's theorem, we obtain

$$\begin{aligned} m \cdot q \;&=\; H \cdot \mathcal{P}(R) \\ &=\; \sum_{Q \in H - \mathrm{Sing}(\Phi)} I_Q\big(H, \mathcal{P}(R)\big) + \sum_{Q \in \mathrm{Sing}(\Phi)} I_Q\big(H, \mathcal{P}(R)\big). \end{aligned} \tag{6.10}$$

The first sum in the right member is $\mathrm{cl}_\Phi(H)$. When Q is a base point of the linear system

$$I_Q\big(H, \mathcal{P}(R)\big) = \sum_{\gamma \to Q} I_Q\big(\gamma, \mathcal{P}(R)\big),$$

where $\gamma \to Q$ means that γ is a branch of H centered at Q. As $\mathcal{P}(R)$ is generic, we have $I_Q\big(\gamma, \mathcal{P}(R)\big) = m_1(\gamma)$ and so

$$I_Q\big(H, \mathcal{P}(R)\big) = \sum_{\gamma \to Q} m_1(\gamma),$$

for all base points $Q \in H$. But we know that $m_1(\gamma) = 0$ when γ is centered in a point in $H - \mathrm{Sing}(\Phi)$, therefore the second sum in (6.10) is

$$\sum_{\gamma \in \tilde{H}} m_1(\gamma),$$

and equality (6.8) is proved.

(6.9). Let us suppose $Q = (0:0:1)$ and $\tilde{\phi}(\gamma)$ is the line $X = 0$. Then we have

$$\begin{aligned} m_1(\gamma) \;&=\; \mathrm{ord}_t\, a(t), \\ m_2(\gamma) \;&=\; \min\big\{\mathrm{ord}_t\, b(t), \mathrm{ord}_t\, c(t)\big\} - \end{aligned}$$

Let also

$$v(t) = \frac{b(t)}{a(t)} \quad \text{and} \quad w(t) = \frac{c(t)}{a(t)}.$$

Then, $\big[1 : v(t) : w(t)\big]$ is a local parametrization of $\widetilde{\phi}(\gamma)$ and

$$\begin{aligned}
\operatorname{ord}_t v'(t) &= \operatorname{ord}_t b(t) - m_1(\gamma) - 1, \\
\operatorname{ord}_t w'(t) &= \operatorname{ord}_t c(t) - m_1(\gamma) - 1.
\end{aligned}$$

Therefore

$$\beta(\gamma) \;=\; \min\big\{\operatorname{ord}_t v'(t), \operatorname{ord}_t w'(t)\big\} \;=\; m_2(\gamma) - m_1(\gamma) - 1. \qquad \square$$

Remark 6-2. *Let H be an integral curve. Then, $H^* = H^\vee$ and the morphism $\widetilde{\phi}$ is the dualization of H. Given a branch $\gamma \in \widetilde{H}$ parametrized by $\big(x(t), y(t)\big)$, the ramification index of the natural morphism $i\colon \widetilde{H} \to \mathbb{P}_2$ at γ is*

$$\beta_0 = \min\big\{\operatorname{ord}_t x'(t), \operatorname{ord}_t y'(t)\big\},$$

and in this case we have $\operatorname{cl}_\Phi(H) = \operatorname{cl}(\Phi)$ and $\operatorname{cl}_\Phi^*(H) = m$. If we define

$$e(\gamma) = m_1(\gamma) - \beta(\gamma),$$

then, using Plücker's formula for the morphism i and equality (6.9), we obtain

$$\sum_{\gamma \in \widetilde{H}} e(\gamma) = m(q - 2) - \kappa(H), \qquad (6.11)$$

where $\kappa(H)$ is the Euler-Poincaré characteristic of the curve H. The number $e(\gamma)$ is called the Poincaré-Hopf index of γ. In certain cases, equality (6.11) allows us to give bounds for the degree of an integral curve in function of the degree of the differential equation. This was stated by Poincaré (see [13]) and recently treated in [1, 3, 7].

References

[1] CARNICER, M. The Poincaré problem in the non-dicritical case. Preprint, Univ. Valladolid, (1992).

[2] CERVEAU, D. Equations différentielles algébriques: Remarques et problèmes. J. Fac. Sci. Univ.T okyo, Sec. IA, Vol. 36, No. 3, 665–680, (1989).

[3] CERVEAU, D., LINS NETO, A. Holomorphic foliations in $\mathbb{CP}_2$ having an invariant algebraic curve. Ann. Institut Fourier 41, Fas. 4, 883–903, (1991).

[4] GAFFNEY, T. Integral closure of modules and Withney equisingularity. Inv. Math., Vol. 107, Fasc. 2, 301–322,(1992).

[5] GARCÍA, J. Geometría de los sistemas lineales de series de potencias en dos variables. Tesis, Univ. Valladolid, (1989).

[6] GARCÍA, J. Géométrie des systèmes linéaires planes locaux. Pinceaux et différentielles meromorphes. Preprint, Univ. Valladolid,(1990).

[7] GARCÍA, J. Divisor of a foliation on a separatrix. The degree of the separatrix. Preprint, Univ. Valladolid,(1990).

[8] GÓMEZ MONT, X. On foliations in surfaces tangent to an algebraic curve. Preprint, UNAM, Mexico, (1990).

[9] Lê, D. T., TEISSIER, B. Variétés polaires locales et classes de Chern des variétés singulières. Ann. Math. 114, 457–491, (1981).

[10] LINS NETO, A. Algebraic solutions of polynomial differential equations and foliations in dimension two. Holomorphic Dynamics, Springer LNM 1345, (1988).

[11] TEISSIER,B. Varietés polaires, I. Inv. Math. 40, 267–292 (1977)

[12] TEISSIER, B. Quelques points de l'histoire des variétés polaires, de Poncelet à nos jours. Sémin. Analyse 1897-1988. Exp. No. 4, 12 pp., Univ. Clermont-Ferrand II,(1990).

[13] POINCARÉ, H. Èquations différentielles du premier ordre et du premier degré. Red. Circ. Mat. Palermo, t5(1891) y t11(1897).

[14] ZARISKI, O., SAMUEL, P. Conmutative Algebra II. Springer-Verlag, (1960).

Addresses of authors:

Dep. de Algebra, Geometría y Topología
Fac. de Ciencias, Univ. de Valladolid
Prado de la Magdalena
47005 Valladolid, Spain
Email: areguera@cpd.uva.es

Email: jgarcia@cpd.uva.es

On Moduli Spaces of Semiquasihomogeneous Singularities

Gert-Martin Greuel and Gerhard Pfister

1 Introduction

Let $A = \mathbb{C}[\![x_1, \ldots, x_n]\!]/(f)$ be the complete local ring of a hypersurface singularity. A is called semiquasihomogeneous with weights $w_1, \ldots, w_n$ if $f = f_0 + f_1$, f_0 a quasihomogeneous polynomial defining an isolated singularity, and $\deg f_0 < \deg f_1$. We assume that $w_1, \ldots, w_n$ are positive integers and let deg always denote the weighted degree, i.e., $\deg X^\alpha = w_1 \alpha_1 + \cdots + w_n \alpha_n$ for a monomial $X^\alpha = X_1^{\alpha_1} \ldots X_n^{\alpha_n}$. For an arbitrary power series f, $\deg f$ denotes the smallest weighted degree of a monomial occurring in f. By definition, all monomials of a quasihomogeneous polynomial have the same degree. The singularity with local ring $A_0 = \mathbb{C}[\![x_1, \ldots, x_n]\!]/(f_0)$ is called the principal part of A. If the moduli stratum of A_0 has dimension 0, i.e., the τ-constant stratum in the semiuniversal deformation of A_0 is a reduced point, then A_0 is uniquely determined by the weights. Let $H^i = H^i(\mathbb{C}[\![x_1, \ldots, x_n]\!])$ be the ideal generated by all quasihomogeneous polynomials of degree $\geq iw$, $w := \min\{w_1, \ldots w_n\}$. This (weighted) degree-filtration defines a Hilbert-function $\underline{\tau}$ on the Tjurina algebra of A by

$$\tau_i(A) := \dim_{\mathbb{C}} \mathbb{C}[\![x_1, \ldots, x_n]\!] \Big/ \left(f, \frac{\partial f}{\partial x_1}, \ldots, \frac{\partial f}{\partial x_n}, H^i \right).$$

We call f or A a semi Brieskorn singularity if the principal part is of Brieskorn-Pham type, i.e., $f_0 = x_1^{m_1} + \cdots + x_n^{m_n}$, $\gcd(m_i, m_j) = 1$, for $i \neq j$. Then f_0 is quasihomogeneous with weight $\underline{w} = (w_1, \ldots, w_n)$, where $w_i = m_1 \ldots \hat{m}_i \ldots m_n$, and degree $d = m_1 \ldots m_n$, the moduli stratum is of dimension zero, and hence f_0 is uniquely determined by its weights (cf. [5]). We are mainly interested in the classification of such singularities with respect to contact equivalence, i.e., in isomorphism classes of the local algebra A. With respect to this equivalence relation we shall prove:

Theorem 1-1. *There exists a coarse moduli space $\mathcal{M}_{\underline{w}, \underline{\tau}}$ for all semiquasihomogeneous singularities with fixed principal part A_0, weight $\underline{w}$ and Hilbert function $\underline{\tau}$. $\mathcal{M}_{\underline{w}, \underline{\tau}}$ is an algebraic variety, locally closed in a weighted projective space.*

We follow the general method to construct such moduli spaces (cf. [5], [3]):

1. We prove that the versal μ-constant deformation $\tilde{X}_\mu \to \underline{H}_\mu$ of A_0 contains already all isomorphism classes of semiquasihomogeneous singularities with principal part A_0. (If we take the quotient of $\underline{H}_\mu$ by a natural action of the group of d-th roots of unity we obtain already a coarse moduli space with respect to right equivalence.)

Progress in Mathematics, Vol. 134
© 1996 Birkhäuser Verlag Basel/Switzerland

2. This family contains analytically trivial subfamilies. They are the integral manifolds of a Lie-algebra V_μ, the kernel of the Kodaira-Spencer map of the family. We prove that two singularities are isomorphic if and only if they are in one integral manifold of V_μ.

3. The integral manifolds of the (infinite dimensional) Lie-algebra V_μ can be identified with the orbits of a solvable algebraic group G. Now the results of [4] can be applied. We prove that the stratification $\{\underline{H}_{\mu,\underline{\tau}}\}$ of $\underline{H}_\mu$ by fixing the Hilbert function has the properties required in [4], i.e., $\underline{H}_{\mu,\underline{\tau}} \to \underline{H}_{\mu,\underline{\tau}}/G$ is a geometric quotient and a coarse moduli space of all semiquasihomogeneous singularities with weight $\underline{w}$, Hilbert function $\underline{\tau}$ and principal part A_0.

2   Versal μ-constant deformations and kernel of Kodaira-Spencer map

In this part we recall some known facts about the versal μ-constant deformation and the kernel of the Kodaira-Spencer map.

Let $f_0 = x_1^{m_1} + \cdots + x_n^{m_n}$, $n \geq 2$, $m_i \geq 2$ and $\gcd(m_i, m_j) = 1$ if $i \neq j$.

Let $w_i = m_1 \ldots \hat{m}_i \ldots m_n$, $i = 1, \ldots, n$ and $d = m_1 \ldots m_n$ then f_0 is a quasihomogeneous polynomial with weight $\underline{w} = (w_1, \ldots, w_n)$ of degree d. Let $A_0 = \mathbb{C}[\![x]\!]/(f_0)$, $x = (x_1, \ldots, x_n)$ and consider the deformation functor $\mathrm{Def}_{A_0 \to \mathbb{C}}$ which consists of isomorphism classes of deformations of the residue morphism $A_0 \to \mathbb{C}$. Geometrically, an element of $\mathrm{Def}_{A_0 \to \mathbb{C}}$ is represented by a "deformation with section" of the singularity defined by f_0 (cf. [2]). It is not difficult to see (cf. [5]) that $\mathrm{Def}_{A_0 \to \mathbb{C}}(\mathbb{C}[\epsilon]) = (x)/(f_0 + (x)(\partial f_0/\partial x_1, \ldots, \partial f_0/\partial x_n))$, where (x) is the ideal generated by $x_1, \ldots, x_n$. This vector space has a unique monomial base $\{x^\alpha | \alpha \in B\}$, $\alpha = (\alpha_1, \ldots, \alpha_n)$, $x^\alpha = x^{\alpha_1} \ldots x^{\alpha_n}$ where $B = \{\alpha \in \mathbf{N}^n \backslash \{0\} \mid \alpha_i \leq m_i - 2\} \cup \{(0, \ldots, m_i - 1, 0, \ldots) \mid i = 1, \ldots, n\}$ (see figure 2.1).

$\mathrm{Def}_{A_0 \to \mathbb{C}}$ has a hull, the semiuniversal deformation, given on the ring level by $H \to H[\![x]\!]/F$ with

$$F = F(T) = f_0 + \sum_{\alpha \in B} T_{d-|\alpha|} x^\alpha,$$

$$H = \mathbb{C}[T],$$

where $T = (T_{d-|\alpha|})_{\alpha \in B}$ and $| \alpha | = \sum_{i=1}^n w_i \alpha_i$ which is by definition the degree of x^α.

Notice that F is quasihomogeneous if we define $\deg T_i = i$. We put $\underline{H} :=$ Spec $H \cong \mathbb{C}^N$, $N = \#B = \prod_{i=1}^n (m_i - 1) + n - 1$, the base space of the semiuniversal deformation.

The moduli stratum, i.e., the τ-constant stratum, is the zero point in $\underline{H}$.

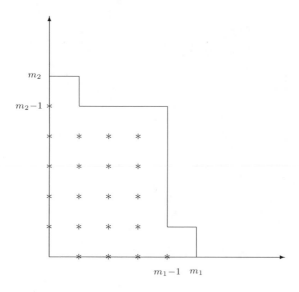

Figure 2.1: B $(n = 2)$

Let $\mathrm{Def}_{A_0 \to \mathbb{C}, \mathbb{C}^*}$ denote the functor of $\mathbb{C}^*$-equivariant deformations of $A_0 \to \mathbb{C}$ (cf. [8]) and let $\mathrm{Def}^{\mu}_{A_0 \to \mathbb{C}} = \Im(\mathrm{Def}_{A_0 \to \mathbb{C}, \mathbb{C}^*} \to \mathrm{Def}_{A_0 \to \mathbb{C}})$. $\mathrm{Def}^{\mu}_{A_0 \to \mathbb{C}}$ gives the μ-constant deformations over a reduced base space. The functor $\mathrm{Def}_{A_0 \to \mathbb{C}, \mathbb{C}^*}$ has a hull, the semiuniversal μ-constant deformation, given by $H_\mu \to H_\mu[\![x]\!]/(F_\mu)$ with

$$F_\mu = F_\mu(T) \quad = \quad f_0 + \sum_{\alpha \in B_-} T_{d-|\alpha|} x^\alpha,$$

$$H_\mu \quad = \quad \mathbb{C}[\{T_{d-|\alpha|}\}_{\alpha \in B_-}],$$

where $B_- = \{\alpha \in B, \ d - |\alpha| < 0\}$ (see figure 2.2).

Remark 2-1.

1. The assumption $\gcd(m_i, m_j) = 1$ implies that except on the axes, there are no extra integral points on the hyperplane $|\alpha| - d = 0$, i.e., f_0 has no moduli. Moreover, it follows that on each hyperplane $|\alpha| = d'$, $\alpha \in B$, there is at most one monomial x^α, hence the elements of B can be numbered by degree which turns out to be very convenient.

2. For any $t \in \underline{H}_\mu := \mathrm{Spec}\, H_\mu$ we have that $F_\mu(t) = f_0 + f_1 \in \mathbb{C}[\![x]\!]$ is semiquasihomogeneous, with principal part f_0. The natural $\mathbb{C}^*$-actions,

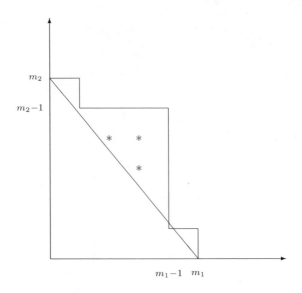

Figure 2.2: B_- $(n = 2)$

$c \circ x = (\ldots, c^{w_i} x_i, \ldots)$ and $c \circ t = (\ldots, c^j t_j, \ldots)$, $c = \mathbb{C}^*$, have the property $F_\mu(c \circ t)(c \circ x) = c^d F_\mu(t)(x)$, in particular, $F_\mu(c \circ t)$ and $F_\mu(t)$ are right equivalent if $c^d = 1$.

3. The action of μ_d on $\underline{H}_\mu - \{0\}$ is faithful since μ_d acts with degree 0 and the T_i have different degrees. This implies: if $\mathcal{X} \to S$ is any μ-constant deformation of $A = \mathbb{C}[\![x]\!]/(f_0 + f_1)$, then there is an open covering $\{\mathcal{U}_i\}$ of S such that $\mathcal{X}|_{\mathcal{U}_i}$ is obtained via some base change $\varphi_i : \mathcal{U}_i \to \underline{H}_\mu$. By the following proposition $\varphi_i \circ \varphi_j^{-1}$ is equal to the $\mathbb{C}^*$-action given by some d-th root of unity c_{ij}. Since μ_d acts faithfully $\{c_{ij}\}$ defines a 1-Čech cocycle of μ_d on S. Hence, if $H^1(S, \mathbf{Z}/d\mathbf{Z}) = 0$, the φ_i can be glued such that $\mathcal{X} \to S$ is globally obtained by some base change $S \to \underline{H}_\mu$.

Proposition 2-2. We have

1. For any semiquasihomogeneous polynomial $f = f_0 + f_1$ with principal part f_0 there is an automorphism $\varphi \in \operatorname{Aut} \mathbb{C}[\![x]\!]$ and $t \in \underline{H}_\mu$ such that $\varphi(f) = F_\mu(t)$.

2. If $F_\mu(t)$ and $F_\mu(t')$ are right equivalent for t, $t' \in \underline{H}_\mu$ then there is a d-th root of unity c, such that $c \circ t = t'$.

Corollary 2-3. Let μ_d denote the group of d-th roots of unity acting on $\underline{H}_\mu$ as above, then $\underline{H}_\mu/\mu_d$ is a coarse moduli space for semiquasihomogeneous polynomials f with principal part f_0 and with respect to right equivalence.

For the notion of (coarse) moduli spaces see [6] and [7]. The fact that 2-3 is a corollary of 2-2 follows from general principals (cf. [7]; the assumption made there that all spaces are reduced is not necessary). See also remark 4-5.

PROOF. (OF 2-2) 1 is proved in [1], 12.6, theorem (p. 209).

For 2, first notice that roots of unity cannot be avoided: take $f = x^5 + y^{11} + xy^9$, $c_1^5 = c_2^{11} = 1$ and $c = c_1c_2$. The automorphism $x \mapsto c^{11}x$, $y \mapsto c^5y$ maps f to $x^5 + y^{11} + c^{56}xy^9$. The statement of 2 will follow from the following two lemmas: □

Lemma 2-4. *Let f, g be semiquasihomogeneous with principal part f_0 as above and $\varphi \in \mathrm{Aut}\,\mathbb{C}[\![x]\!]$ such that $\varphi(f) = g$. Then there is a d-th root of unity c such that*

$$\varphi(x_i) = c^{w_i}x_i + h_i, \ \deg h_i > w_i.$$

PROOF. Let $w_1 < \ldots < w_n$. By proposition 4-1 we have $\deg \varphi \geq 0$, hence

$$\varphi(x_i) = \sum_{j \geq i} c_{ij}x_j + \text{ higher order terms.}$$

Since φ is an automorphism, $\prod_i c_{ii} \neq 0$, and $\varphi(x_i) = c_{ii}x_i + h_i, \deg h_i > w_i$. From $x_1^{m_1} + \ldots + x_n^{m_n} = c_{11}^{m_1}x_1^{m_1} + \cdots + c_{nn}^{m_n}x_n^{m_n}$ we deduce $c_{ii}^{m_i} = 1$ and putting $c = \prod_i c_{ii}$ we obtain the result. □

Lemma 2-5. *Let $\varphi \in \mathrm{Aut}\,\mathbb{C}[\![x]\!]$, $\deg \varphi > 0$, and t, $t' \in \underline{H}_\mu$ such that $\varphi\big(F_\mu(t)\big) = F_\mu(t')$. Then $t = t'$.*

PROOF. By lemma 2-4, $\varphi(x_i) = x_i + h_i$. Hence $\varphi_s(x_i) := x_i + sh_i$ is a family of automorphisms of positive degree which connects φ with the identity. Then $\varphi_s\big(F_\mu(t)\big)$ is a $\mathbb{C}^*$-equivariant family of isolated singularities, joining $F_\mu(t)$ and $F_\mu(t')$. This family may not be contained in $\underline{H}_\mu$ but it can be induced from $\underline{H}_\mu$ by a suitable base change (remark 2-1). But since $\underline{H}_\mu$ is everywhere miniversal and does, therefore, not contain trivial subfamilies with respect to right equivalence, $t = t'$ as desired. □

The Kodaira-Spencer map (cf. [5]) of the functor $\mathrm{Def}_{A_0 \to \mathbb{C}, \mathbb{C}^*}$ and of the family $H_\mu \to H_\mu[\![x]\!]/F_\mu$,

$$\rho : \mathrm{Der}_\mathbb{C}\,H_\mu \longrightarrow (x)H_\mu[\![x]\!] \Big/ \left(F_\mu + (x)\left(\frac{\partial F_\mu}{\partial x_1}, \ldots, \frac{\partial F_\mu}{\partial x_n}\right) \right),$$

is defined by $\rho(\delta) = \mathrm{class}(\delta F_\mu) = \mathrm{class}(\sum_{\alpha \in B_-} \delta(T_{d-|\alpha|})x^\alpha)$.

Let $\mathbf{V}_\mu$ be the kernel of ρ. $\mathbf{V}_\mu$ is a Lie-algebra and along the integral manifolds of $\mathbf{V}_\mu$ the family is analytically trivial (cf. [5]).

In our situation it is possible to give generators of $\mathbf{V}_\mu$ as H_μ-module: Let $I_\mu = (x)H_\mu[\![x]\!]/(x)(\partial F_\mu/\partial x_1,\ldots,\partial F_\mu/\partial x_n)$, then I_μ is a free H_μ-module and $\{x^\alpha\}_{\alpha\in B}$ is a free basis. Multiplication by F_μ defines an endomorphism of I_μ and $F_\mu I_\mu \subseteq \oplus_{\alpha\in B_-} x^\alpha H_\mu$.

Especially, for $\alpha \in B$, define $h_{i,j}$ by

$$x^\alpha F_\mu = \sum_{\beta\in B_-} h_{|\alpha|,d-|\beta|}x^\beta \in I_\mu.$$

Then h_{ij} is homogeneous of degree $i + j$. This implies $h_{ij} = 0$ if $i + j \geq 0$, in particular $h_{ij} = 0$ if $i \geq (n-1)d-2\sum w_i$. For $\alpha \in B$ and $|\alpha| < (n-1)d-2\sum w_i$ let $\delta_{|\alpha|} := \sum_{\beta\in B_-} h_{|\alpha|,d-|\beta|}(\partial/\partial T_{d-|\beta|})$.

Proposition 2-6. (cf. [5], proposition 4.5):

1. $\delta_{|\alpha|}$ is homogeneous of degree $|\alpha|$.

2. $\mathbf{V}_\mu = \sum_\alpha H_\mu \delta_{|\alpha|}$.

Now there is a non-degenerate pairing on I_μ (the residue pairing) which is defined in our situation by $\langle h, k\rangle = \mathrm{hess}(h \cdot k)$. Here for $h = \sum_{\alpha\in B} h_\alpha x^\alpha \in I_\mu$, $\mathrm{hess}(h) = h_{(m_1-2,\ldots,m_n-2)}$ which is the coefficient belonging to the Hessian of f.

Let the numbering of the elements of $B_- = \{\alpha_1,\cdots,\alpha_k\}$, be such that $|\alpha_1| < \ldots < |\alpha_k|$ and denote by $\beta_i = \alpha_{k-i+1}^\vee$, $i = 1,\ldots,k$, the dual exponents induced by the pairing, i.e., if $\gamma = (\gamma_1,\ldots,\gamma_n)$ then $\gamma^\vee = (m_1-2-\gamma_1,\ldots,m_n-2-\gamma_n)$.

Using the pairing one can prove the following

Proposition 2-7. There are homogeneous elements $m_1,\ldots,m_k \in H_\mu[\![x]\!]$ with the following properties:

1. $\deg m_i = |\beta_i|$.

2. If $m_i F_\mu = \sum_{j=1}^k \tilde{h}_{ij}x^{\alpha_j}$ in I_μ then $\tilde{h}_{ij} = \tilde{h}_{k-j+1,k-i+1}$.

3. If $\tilde{\delta}_{|\beta_i|} := \sum_{j=1}^k \tilde{h}_{ij}(\partial/\partial T_{d-|\alpha_j|})$ then $\tilde{\delta}_{|\beta_i|}$ is homogeneous of degree $|\beta_i|$ and $\mathbf{V}_\mu = \sum_{i=1}^k H_\mu \tilde{\delta}_{|\beta_i|}$.

In [5] (proposition 5.6) this proposition is proved for $n = 2$. The proof can easily be extended to arbitrary n. The important fact is the symmetry, expressed in 2.

Let L be the Lie-algebra generated (as Lie-algebra) by $\{\tilde{\delta}_{|\beta_1|},\ldots,\tilde{\delta}_{|\beta_k|}\}$. Then L is finite dimensional and solvable, $L_0 := [L, L]$ is nilpotent and $L/L_0 = \mathbb{C}\tilde{\delta}_{|\beta_1|}$, where $\tilde{\delta}'_{|\beta_1|} = \sum_{i=1}^k (|\alpha_i|-d)T_{d-|\alpha_i|}(\partial/\partial T_{d-|\alpha_i|})$ is the Euler vector field (cf. [5]).

Corollary 2-8. The integral manifolds of $\mathbf{V}_\mu$ coincide with the orbits of the algebraic group $\exp(L)$.

Now consider the matrix

$$M(T) := \big(\tilde{\delta}_{|\beta_i|}(T_{d-|\alpha_j|})\big)_{i,j=1,\ldots,k} = (\tilde{h}_{ij})_{i,j=1,\ldots,k}.$$

Evaluating this matrix at $t \in \underline{H}_\mu$ we have

$$
\begin{aligned}
\operatorname{rank} M(t) \;=\; & \text{dimension of a maximal integral manifold of} \\
& \mathbf{V}_\mu \text{ (resp. of the orbit of } \exp(\mathrm{L})\text{) at } t, \\
=\; & \mu - \tau(t),
\end{aligned}
$$

where $\tau(t)$ denotes the Tjurina number of the singularity defined by t i.e., of $F(x,t)$.

3 Existence of a geometric quotient for fixed Hilbert function of the Tjurina algebra

We want to apply theorem 4.7 from [4] to the action of L_0 on $\underline{H}_\mu$.

Theorem 3-1. ([4]) *Let A be a noetherian $\mathbb{C}$-algebra and $L_0 \subseteq \operatorname{Der}_{\mathbb{C}}^{\mathrm{nil}} A$ a finite dimensional nilpotent Lie algebra. Suppose A has a filtration*

$$F^\bullet : 0 = F^{-1}(A) \subset F^0(A) \subset F^1(A) \subset \ldots$$

by subvector spaces $F^i(A)$ such that

$$\delta F^i(A) \subseteq F^{i-1}(A) \qquad \text{for all } i \in \mathbf{Z}, \delta \in L_0. \tag{3.1}$$

Suppose moreover, L_0 has a filtration

$$Z_\bullet : L_0 = Z_0(L_0) \supseteq Z_1(L_0) \supseteq \ldots \supseteq Z_e(L_0) \supseteq Z_{e+1}(L_0) = 0,$$

by sub Lie algebras $Z_j(L_0)$ such that

$$[L_0, Z_j(L_0)] \subseteq Z_{j+1}(L_0) \qquad \text{for all } j \in \mathbf{Z}. \tag{3.2}$$

Let $d : A \to \operatorname{Hom}_{\mathbb{C}}(L_0, A)$ be the differential defined by $d(a)(\delta) = \delta(a)$ and let $\operatorname{Spec} A = \cup U_\alpha$ be the flattening stratification of the modules

$$\operatorname{Hom}_{\mathbb{C}}(L_0, A)/A\, d\big(F^i(A)\big) \qquad i = 1, 2, \ldots,$$

and

$$\operatorname{Hom}_{\mathbb{C}}(Z_j(L_0), A)/\pi_j\big(A(dA)\big) \qquad j = 1, \ldots, e,$$

where π_j denotes the projection $\operatorname{Hom}_{\mathbb{C}}(L_0, A) \to \operatorname{Hom}_{\mathbb{C}}(Z_j(L_0), A)$. Then U_α is invariant under the action of L_0 and $U_\alpha \to U_\alpha/L_0$ is a geometric quotient which is a principal fibre bundle with fibre $\exp(L_0)$.

To apply the theorem we have to construct these filtrations and interpret the corresponding stratification in terms of the Hilbert function of the Tjurina algebra.

There are natural filtrations $H^\bullet(\mathbb{C}[\![x]\!])$ (resp. $F^\bullet(H_\mu)$) on $\mathbb{C}[\![x]\!]$ (resp. H_μ) defined as follows: Let $F^i(H_\mu) \subseteq H_\mu$ be the $\mathbb{C}$-vector space generated by all quasihomogeneous polynomials of degree $> -(i+1)w$ and $H^i(\mathbb{C}[\![x]\!])$ be the ideal generated by all quasihomogeneous polynomials of degree $\geq iw$, where

$$w := \min\{w_1, \ldots, w_n\}.$$

For $t \in \underline{H}_\mu$ the Hilbert function of the Tjurina algebra

$$\mathbb{C}[\![x]\!] \Big/ \left(F_\mu(t), \frac{\partial F_\mu(t)}{\partial x_1}, \ldots, \frac{\partial F_\mu(t)}{\partial x_n}\right),$$

corresponding to the singularity defined by t with respect to $H^\bullet$ is by definition the function,

$$n \mapsto \tau_n(t) := \dim_\mathbb{C} \mathbb{C}[\![x]\!] \Big/ \left(F_\mu(t), \frac{\partial F_\mu(t)}{\partial x_1}, \ldots, \frac{\partial F_\mu(t)}{\partial x_n}, H^n\right).$$

Notice that $\tau_n(t) = \tau(t)$ if n is large and $\tau_n(t)$ does not depend on t for small n. On the other hand,

$$\mu_n := \mu_n(t) := \dim_\mathbb{C} \mathbb{C}[\![x]\!]/(\partial F_\mu(t)/\partial x_1, \ldots, \partial F_\mu(t)/\partial x_n, H^n)$$

does not depend on $t \in H_\mu$ and

$$\mu_n - \tau_n(t) = \mathrm{rank}\left(\tilde{\delta}_{|\beta_i|}(T_{d-|\alpha_j|})(t)\right)_{|\alpha_j| < nw}.$$

This is an immediate consequence of the following fact: Let

$$T^n := H_\mu[\![x]\!] \Big/ \left(F_\mu, \frac{\partial F_\mu}{\partial x_1}, \ldots, \frac{\partial F_\mu}{\partial x_n}, H^n\right),$$

then the following sequence is exact and splits:

$$0 \longrightarrow \bigoplus_{\substack{\alpha \in B \\ |\alpha| \leq d}} H_\mu x^\alpha \longrightarrow T^{\frac{d}{w}+i} \longrightarrow \mathrm{Der}_\mathbb{C} H_\mu \Big/ \left(v_\mu + \sum_{|\beta| \geq d+iw} H_\mu \frac{\partial}{\partial T_{d-|\beta|}}\right) \longrightarrow 0$$

$$x^\alpha \longmapsto \mathrm{class}(x^\alpha)$$

$$\mathrm{class}(x^\beta) \longmapsto \mathrm{class}\left(\frac{\partial}{\partial T_{d-|\beta|}}\right),$$

and with the identification

$$\sum_{|\beta|<d+iw} H_\mu \frac{\partial}{\partial T_{d-|\beta|}} \simeq H_\mu^{N_i},$$

we get

$$\mathrm{Der}_{\mathbb{C}} H_\mu \Big/ \left(\mathbf{V}_\mu + \sum_{|\beta|\geq d+iw} H_\mu \frac{\partial}{\partial T_{d-|\beta|}} \right) \simeq H_\mu^{N_i}/M_i,$$

where M_i is the H_μ-submodule generated by the rows of the matrix

$$\left(\tilde\delta_{|\beta_\ell|}(T_{d-|\alpha_j|}) \right)_{|\alpha_j|<d+iw}.$$

The filtration $F^\bullet(H_\mu)$ has the property (3.1) because every homogeneous vector field of L_0 is of degree $\geq w$ (since L/L_0 is the Euler vector field, cf. section 2) and

$$H_\mu dH_\mu = H_\mu dF^s H_\mu, \quad s = \left[\frac{(n-1)d - 2\sum w_i}{w} \right]$$

(since $nd - 2\sum w_i$ is the degree of the Hessian of f and $T_{d-(nd-2\Sigma w_i)}$ is the variable of smallest degree).

To define $Z_i(L_0)$ we use the duality defined in section 2,

$$\alpha \mapsto \alpha^\vee = (m_1 - 2 - \alpha_1, \ldots, m_n - 2 - \alpha_n),$$

and set $Z_i(L_0) :=$ the Lie algebra generated by

$$\left\{ \tilde\delta_{|\alpha|} \in L_0 \mid T_{d-|\alpha^\vee|} \in F^{s-i} \right\}.$$

$Z_\bullet(L_0)$ has the property (3.2) (for the definition of $\tilde\delta_{|\alpha|}$, see proposition 2-3). We have $F \in H^n$, hence $\mu_n = \tau_n$, if $n \leq d/w$ and $H^n \subset (\partial F_\mu/\partial x_1, \ldots, \partial F_\mu/\partial x_n)$ hence $\mu_n - \tau_n(t)$ is independent of n if $n \geq d/w + s + 1$, $s = [((n-1)d - 2\Sigma w_i)/w]$ and equal to $\mu - \tau(t)$. Therefore, we have $s+1$ relevant values for τ_i, and we denote

$$\underline{\tau}(t) := \left(\tau_{\frac{d}{w}+1}(t), \ldots, \tau_{\frac{d}{w}+s+1}(t) \right),$$
$$\underline{\mu} := (\mu_{\frac{d}{w}+1}, \ldots, \mu_{\frac{d}{w}+s+1}).$$

Moreover, let $S = \{ \underline{r} := (r_1, \ldots, r_{s+1}) | \exists t \in \underline{H}_\mu$ such tthat $\underline{\mu} - \underline{\tau}(t) = \underline{r} \}$ and $\underline{H}_\mu = \cup_{\underline{r} \in S} U_{\underline{r}}$ be the flattening stratification of the modules $T^{\frac{d}{w}+1}$, $\ldots, T^{w+1}$, i.e., $\{U_{\underline{r}}\}$ is the stratification of $\underline{H}_\mu$ defined by fixing the Hilbert function $\underline{\tau} = \underline{\mu} - \underline{r}$ with the scheme structure defined by the flattening property.

We obtain:

Lemma 3-2.

1. $(0,\ldots,0,1)$ and $(0,\ldots,0) \in S$. $U_{(0,\ldots,0)} = \{0\}$ is a smooth point and $U_{(0,\ldots,1)}$ is defined by $T_{d-|\beta|} = 0$ for $|\beta| < nd - 2\sum w_i$ and $T_{2\Sigma w_i - (u-1)d} \neq 0$ (and hence is smooth).

2. Let $\bar{S} = S \backslash \{(0,\ldots,0)\}$ and for $\underline{r} \in \bar{S}$ put

$$\bar{U}_{\underline{r}} = \begin{cases} U_{\underline{r}} & \text{if } \underline{r} \neq (0,\ldots,0,1), \\ U_{(0,\ldots 0,1)} \cup U_{(0,\ldots,0)} & \text{if } \underline{r} = (0,\ldots,0,1). \end{cases}$$

Then $\{\bar{U}_{\underline{r}}\}_{\underline{r} \in \bar{S}}$ is the flattening stratification of the modules

$$\{\mathrm{Hom}_{\mathbb{C}}(L_0, H_\mu)/H_\mu dF^i H_\mu\} \quad \text{and} \quad \{\mathrm{Hom}_{\mathbb{C}}(Z_i(L_0), H_\mu)/\pi_i(H_\mu dF H_\mu)\}.$$

As a corollary we obtain the following theorem (recall that $\mathbf{V}_\mu$ denotes the kernel of the Kodaira Spencer map, cf. section 2):

Theorem 3-3. For $\underline{r} \in S$, $\bar{U}_{\underline{r}}$ is invariant under the action of $\mathbf{V}_\mu$ and $\bar{U}_{\underline{r}} \to \bar{U}_{\underline{r}}/\mathbf{V}_\mu$ is a geometric quotient. $\bar{U}_{\underline{r}}/\mathbf{V}_\mu$ is locally closed in a weighted projective space.

PROOF. Using the lemma and theorem 3-1 we obtain that $\bar{U}_{\underline{r}}$ is invariant under the action of L_0 and $\bar{U}_{\underline{r}} \to \bar{U}_{\underline{r}}/L_0$ is a geometric quotient. $L/L_0 = \mathbb{C}\delta_0$ acts on $\bar{U}_{\underline{r}}/L_0$. By corollary 2-4, $\bar{U}_{\underline{r}}/\bar{V}_\mu = \bar{U}_{\underline{r}}/L$. If $\underline{r} \neq (0,\ldots,1)$, then $\bar{U}_{\underline{r}}/L_0 \to \bar{U}_{\underline{r}}/L$ is a geometric quotient embedded in the corresponding weighted projective space. If $\underline{r} = (0,\ldots,1)$ then $\bar{U}_{\underline{r}}/L_0 = \bar{U}_{\underline{r}}$ and the geometric quotients $U_{(0,\ldots,1)} \to U_{(0,\ldots,1)}/\mathbf{V}_\mu$, $U_{(0,\ldots,0)} \to U_{(0,\ldots,0)}/\mathbf{V}_\mu$ exist as smooth points. $\square$

It remains to prove the lemma.

PROOF. (OF 3-2) Because of the exact sequence above the flattening stratification of the modules $\{T^{\frac{d}{w}+i}\}$ is also the flattening stratification of

$$\left\{\mathrm{Der}_{\mathbb{C}} H_\mu \Big/ \left(\mathbf{V}_\mu + \sum_{|\beta| \geq d+iw} H_\mu \frac{\partial}{\partial T_{d-|\beta|}}\right)\right\}$$

resp. the flattening stratification of $\{H_\mu^{N_i}/M_i\}$, M_i the submodule generated by the rows of the matrix $\left(\tilde{\delta}_{|\beta_\ell|}(T_j)\right)_{-iw<j}$. Now we have

$$\tilde{\delta}_{|\beta|}(T_{d-|\alpha|}) = \tilde{\delta}_{|\alpha^\vee|}(T_{d-|\beta^\vee|}). \tag{3.3}$$

By definition of $Z_i(L_0)$ we have

$$H_\mu Z_i(L_0) = \sum_{T_{d-|\alpha^\vee|} \in F^{s-i}} H_\mu \tilde{\delta}_{|\alpha|},$$

and with the identification

$$\sum_{\beta \in B_-} H_\mu \frac{\partial}{\partial T_{d-|\beta|}} = H_\mu^N,$$

and M^i the submodule generated by the rows of the matrix

$$\left(\tilde{\delta}_{|\alpha|}(T_j)\right)_{d+(s-i+1)w>|\alpha^\vee|},$$

we obtain

$$\mathrm{Der}_{\mathbb{C}} \, H_\mu / H_\mu Z_i(L_0) \cong H_\mu^N / M^i.$$

Equation (3.3) implies that the flattening stratification of the modules $\{T^{\frac{d}{w}+1}, \ldots, T^s\}$, which is $\underline{H}_\mu = \cup_{r \in \bar{s}} \bar{U}_r$, is the flattening stratification of the modules $\{\mathrm{Der}_{\mathbb{C}} \, H_\mu / H_\mu Z_i(L_0)\}_{i=1,\ldots,s}$.

On the other hand, the flattening stratification of the modules $\{H_\mu^N / M^i\}$ for $i = 1, \ldots, s$ is the flattening stratification of the modules

$$\{\mathrm{Hom}_{\mathbb{C}}\left(Z_i(L_0), H_\mu\right) / \pi_i(H_\mu dH_\mu)\},$$

because

$$H_\mu Z_i(L_0) = \sum_{T_{d-|\alpha^\vee|} \in F^{s-i}} H \tilde{\delta}_{|\alpha|}.$$

Furthermore the modules

$$\{\mathrm{Hom}_{\mathbb{C}}(L_0, H_\mu) / H_\mu dF^i H_\mu\} \quad \text{and} \quad \left\{\mathrm{Der}_{\mathbb{C}} \, H_\mu / H_\mu L_0 + \sum_{|\beta| \geq d+iw} H_\mu \frac{\partial}{\partial T_{d-|\beta|}}\right\},$$

have the same flattening stratification and they are flat on U_r, because

$$0 \longrightarrow H_\mu \longrightarrow \mathrm{Der}_{\mathbb{C}} \, H_\mu / H_\mu L_0 + \sum_{|\beta| \geq d+iw} H_\mu \frac{\partial}{\partial T_{d-|\beta|}} \longrightarrow$$

$$\mathrm{Der}_{\mathbb{C}} \, H_\mu / \mathbf{V}_\mu + \sum_{|\beta| \geq d+iw} H_\mu \frac{\partial}{\partial T_{d-|\beta|}} \longrightarrow 0$$

is exact and splits on $\underline{H}_\mu \backslash \{0\}$.

This proves the lemma.   □

Remark 3-4. *The main point of the lemma is that the flattening stratification of the modules* $\{\mathrm{Hom}_{\mathbb{C}}(L_0, H_\mu) / H_\mu dF^i H_\mu\}$ *is contained in the flattening stratification of the modules* $\{\mathrm{Hom}_{\mathbb{C}}(Z_j(L_0), H_\mu) / \pi_i(H_\mu dH_\mu)\}$, *hence is defined by the Hilbert function of the Tjurina algebra alone without any reference to the action of* L. *This is a consequence of the symmetry, expressed in proposition 2-3.*

4 The automorphism group of semi Brieskorn singularities

In this section we prove that the automorphism group of a semi Brieskorn singularity with principal part $f_0 = x_1^{m_1} + \ldots + x_n^{m_n}$, $\gcd(m_i, m_j) = 1$ for $i \neq j$, has no automorphisms of negative degree. A consequence of this result is that two points in $\underline{H}_\mu$ correspond to isomorphic singularities if and only if they are in one integral manifold of $\mathbf{V}_\mu$. Again, $d = m_i \ldots m_n$ denotes the degree of f_0.

Let $\mathbb{C}[\![x]\!]_m$ denote the ideal of $\mathbb{C}[\![x]\!]$ generated by power series of degree $\geq m$. An automorphism φ of $\mathbb{C}[\![x]\!]$ has degree m if

$$(\varphi - id)\mathbb{C}[\![x]\!]_i \subset \mathbb{C}[\![x]\!]_{i+m},$$

for any i. For $c \in \mathbb{C}^*$ let $\varphi_c : \mathbb{C}[\![x]\!] \to \mathbb{C}[\![x]\!]$, $\varphi_c(x_i) = c^{w_i} x_i$, $i = 1, \ldots, n$, denote the $\mathbb{C}^*$-action which is an automorphism of degree 0.

Proposition 4-1. *Let* $f = f_0 + \sum_{|\alpha| > d} a_\alpha x^\alpha$, $g = f_0 + \sum_{|\alpha| > d} b_\alpha x^\alpha$, $\varphi \in \text{Aut } \mathbb{C}[\![x]\!]$ *and* $u \in \mathbb{C}[\![x]\!]$ *a unit such that* $uf = \varphi(g)$. *Then* $\deg \varphi \geq 0$.

Remark 4-2. *Let* $\varphi \in \text{Aut } \mathbb{C}[\![x]\!]$ *be of degree* ≥ 0, f, g *as above and* $f = u\varphi(g)$ *for some unit* u. *Then* $\varphi(x_i) = c_i x_i + h_i$, $\deg h_i > w_i$, $u(0)c_i^{m_i} = 1$. *Putting* u_i *some* m_i-*th root of* $u(0)$ *and* $c = \prod_{i=1}^m u_i c_i$ *we obtain* $c^d = 1$, $c^{w_i} = u_i c_i$, *hence* $u(0)\varphi(g) = \tilde{\varphi} \circ \varphi_c(g)$ *and* $f = \tilde{u}\tilde{\varphi} \circ \varphi_c(g)$ *where* $\deg \tilde{\varphi} > 0$ *and* $\tilde{u}$ *is a unit with* $\tilde{u}(0) = 1$.

PROOF. We prove the proposition by induction on n, the case $n = 1$ being trivial. We may assume that $m_n < \cdots < m_1$. Then we can write $\varphi(x_1) = \alpha_1 x_1 + h_1$, $\alpha_1 \in \mathbb{C}$ and $\deg h_1 > w_1 = \min\{w_1, \ldots, w_n\}$.

First of all we shall see that $\alpha_1 \neq 0$. Assume $\alpha_1 = 0$ then there is an $i > 1$ such that $\varphi(x_i) = \beta x_1 + h_i$ and $\deg h_i > w_1$, $\beta \neq 0$.

Using an automorphism of non-negative degree we may assume $\varphi(x_i) = x_1$. Now

$$uf \mid_{x_1=0} = g\big(\varphi(x_1) \mid_{x_1=0}, \ldots, \varphi(x_{i-1}) \mid_{x_1=0}, 0, \varphi(x_{i+1}) \mid_{x_1=0}, \ldots\big),$$

and

$$\varphi \mid_{x_1=0} : \mathbb{C}[\![x_1, \ldots, \hat{x}_i, \ldots, x_n]\!] \quad \to \quad \mathbb{C}[\![x_2, \ldots, x_n]\!]$$

$$x_k \quad \mapsto \quad \varphi(x_k) \mid_{x_1=0}$$

is an isomorphism. Hence

$$\varphi \mid_{x_1=0} \big(g(x_1, \ldots, x_{i-1}, 0, x_{i-1}, \ldots, x_n)\big) = uf \mid_{x_1=0}.$$

But $g(x_1, \ldots, x_{i-1}, 0, x_{i+1}, \ldots, x_n)$ and $f(0, x_2, \ldots, x_n)$ define isolated singularities with different Milnor numbers (they are semiquasihomogeneous with weights $w_1, \ldots, \hat{w}_i, \ldots, w_n$ resp. $w_2, \ldots, w_n$ and degree d). This is a contradiction and implies $\alpha_1 \neq 0$. Using $\varphi_{\alpha^{-1}}$ and an automorphism of positive degree we may assume now $\varphi(x_1) = x_1$.

Let us consider again the automorphism $\varphi|_{x_1=0}$ of $\mathbb{C}[\![x_2,\ldots,x_n]\!]$. Using the induction hypothesis we may assume $deg\,\varphi|_{x_1=0} \geq 0$. Since the inverse is also of non-negative degree we may assume that $\varphi|_{x_1=0}$ is the identity, i.e.,

$$\varphi(x_1) = x_1 \quad \text{and} \quad \varphi(x_i) = x_i + x_1 h_i, \ i = 2,\ldots,n.$$

Using again an automorphism of non-negative degree we may assume now that h_i has only terms of degree $< w_i - w_1$. We have to prove that $h_i = 0$.

If h_i has only terms of degree $< w_i - w_1$ then h_i does not depend on $x_i,\ldots,x_n$. We prove now that $h_n = 0$.

We may assume that $g = x_n^{m_n}+x_n^{m_n-2}a_2+\ldots+a_{m_n}$, $a_i \in \mathbb{C}[\![x_1,\ldots,x_{n-1}]\!]$. Indeed by the Weierstrass preparation theorem $g \cdot \text{unit} = x_n^{m_n} + a_1 x_n^{m_n-1} + \cdots$. This equality implies $\deg a_1 x_n^{m_n-1} = (m_n-1)w_n + \deg a_1 > d$ and consequently the automorphism defined by $x_n \to x_n - (1/m_n)a_1$ has positive degree. We may assume $a_1 = 0$ but this changes $\varphi(x_n)$ to $\varphi(x_n) = x_n + x_1 h_n - (1/m_n)a_1$. Now $f \cdot u = \varphi(x_n^{m_n} + x_n^{m_n-2}a_2 + \ldots) = x_n^{m_n} + (m_n x_1 h_n - a_1)x_n^{m_n-1} + \cdots$ and $\deg m_n x_1 h_n x_n^{m_n-1} < d$. But this is only possible if $h_n = 0$ because this term cannot be cancelled (the other h_i do not depend on x_n). This implies $h_n = 0$.

Now $f \cdot u\,|_{x_n=0} = f(x_1, x_2 + x_1 h_2, \ldots, x_{n-1} + x_1 h_{n-1}, 0)$ because the h_i do not depend on x_n. Using again the induction hypothesis we obtain $h_i = 0$, $i = 2,\ldots,n-1$. This proves the proposition. $\qquad\square$

Corollary 4-3. If $t, t' \in \underline{H}_\mu$ define isomorphic singularities then t and t' are in the same maximal integral manifold of $\mathbf{V}_\mu$.

PROOF. Let $F_\mu(t) = u\varphi\big(F_\mu(t')\big)$, $u \in \mathbb{C}[\![x]\!]$ a unit and $\varphi \in \text{Aut}\,\mathbb{C}[\![x]\!]$. By the proposition $\deg\varphi \geq 0$. Using remark 4-2 there is a d'th root of unity c such that $F_\mu(x,t) = u\varphi\big(F_\mu(c^{-1} \circ x, t')\big) = u\varphi\big(F_\mu(x, c \circ t')\big)$ and such that $\deg\varphi > 0$ and $u(0) = 1$. Then

$$G(z) := u(z^{w_1}x_1,\ldots,z^{w_n}x_n) \cdot F_\mu\left(\frac{1}{z^{w_1}}\varphi(z^{w_1}x_1),\ldots,\frac{1}{z^{w_n}}\varphi(z^{w_n}x_n), c \circ t'\right)$$

is an unfolding of $G(0) = F_\mu(x, c \circ t')$. This unfolding can be induced by the universal unfolding by remark 2-1, i.e., there exists a family of coordinate transformations $\underline{\psi}(z,-)$ and a path v in $\underline{H}_\mu$ such that

$$G(z) = F_\mu\big(\psi_1(z,x),\ldots,\psi_n(z,x), v(z)\big),$$

and $v(0) = c \circ t'$, $\psi_i(0,x) = x$. By [1] we may assume that $\underline{\psi}(z,-)$ has positive degree.

Because $F_\mu(x,t) = F_\mu\big(\psi(1,x), v(1)\big)$ we obtain $v(1) = t$ by lemma 2-5. This implies that t and $c \circ t'$ are in an analytically trivial family, i.e., in an integral manifold of $\mathbf{V}_\mu$ which contains the $\mathbb{C}^*$-orbits (cf. section 2). Hence the result. $\qquad\square$

This finishes the second step of the approach. Together with the theorem of section 3 we obtain the theorem stated in the introduction:

Theorem 4-4. *There exists a coarse moduli space $\mathcal{M}_{\underline{w},\tau} = \bar{U}_{\mu-\underline{r}}/\mathbf{V}_\mu$ of all semi-quasihomogeneous hypersurface singularities $A = \mathbb{C}[\![\underline{x}]\!]/(f)$ with fixed principal part $A_0 = \mathbb{C}[\![\underline{x}]\!]/(f_0)$, weight $\underline{w}$ and Hilbert function $\underline{\tau}$. $\mathcal{M}_{\underline{w},\tau}$ is an algebraic variety, locally closed in a weighted projective space.*

Remark 4-5. *To be more precise, first of all $\mathcal{M}_{\underline{w},\tau}$ is a coarse moduli space for the functor which associates to any complex space germ S the set of isomorphism classes of flat families over S of quasihomogeneous hypersurface singularities with fixed principal part A_0, weight $\underline{w}$ and Hilbert function τ. The category of base spaces is that of germs since we constructed $\mathcal{M}_{\underline{w},\tau}$ from the versal family over $\underline{H}_\mu$ which has the versality property only for germs. But by remark 2-1; 3, we can actually enlarge the category of base spaces to all complex spaces S for which $H^1(S, \mathbf{Z}/d\mathbf{Z}) = 0$. The same applies to the coarse moduli space $\underline{H}_\mu/\mu_d$ for functions with respect to right equivalences (cf. corollary 2-3).*

5　Problems

We use the notations of section 2

　　In the case $n = 2$ (plane curves) the following holds (cf. [5]): let $\{S_\tau\}$ be the stratification of $\underline{H}_\mu$ by constant Tjurina number, then

1. $S_\tau \neq \emptyset$ if $\tau_{\min} \leq \tau \leq \mu$ (i.e., all possible Tjurinia numbers occur).

2. $\dim S_\tau/\mathbf{V}_\mu \geq \dim S_{\tau'}/\mathbf{V}_\mu$ if $\tau \leq \tau'$ (i.e., the number of moduli decreases when τ becomes more special).

3. $S_{\tau_{min}}/\mathbf{V}_\mu$ is a quasismooth algebraic variety.

In [5] there is an example showing that 1 and 2 are wrong in higher dimension.

Problem 5-1.　Does 3 hold in higher dimension?

Problem 5-2.　Find the dimensions of $\underline{H}_\mu/\mathbf{V}_\mu$.

　　In section 4 we proved that for semi Brieskorn singularities with principal part $f_0 = x_1^{m_1} + \cdots + x_n^{m_n}$, $\gcd(m_i, m_j) = 1$ for $i \neq j$ the automorphisms have non-negative degree.

Problem 5-3.　Is this true for all quasihomogeneous singularities with zero-dimensional moduli stratum?

　　A solution of this problem would solve the moduli problem for this class of semiquasihomogeneous singularities.

References

[1] Arnol'd, V.I.; Gusein-Zade, S.M.; Varchenko, A.N.: Singularities of Differentiable Maps, Vol. I, Boston-Basel-Stuttgart: Birkhäuser 1985.

[2] Buchweitz, A.: Thesis, Université Paris VII, (1981).

[3] Greuel, G.-M.; Pfister, G.: Moduli for singularities. Preprint 207, Kaiserslautern, 1991. To appear in the proceedings of the Lille conference on singularities, 1991.

[4] Greuel, G.-M.; Pfister, G.: Geometric quotients of unipotent group actions. To appear in Proc. Lond. Math. Soc.

[5] Laudal, O.A.; Pfister, G.: Local moduli and singularities. Lecture Notes in Math., Vol. 1310. Berlin-Heidelberg-New York: Springer 1988.

[6] Mumford, D.; Fogarty, J.: Geometric Invariant Theory. (Second, enlarged edition). Ergb. Math. Grenzgeb. Bd. 34. Berlin-Heidelberg-New York: Springer 1982.

[7] Newstead, P.E: Introduction to Moduli Problems and Orbit Spaces. Tata Inst. Fund. Res. Lecture Notes 51. Berlin-Heidelberg-New York: Springer 1978.

[8] Pinkham, H.C.: Normal surface singularities with $\mathbb{C}^*$ action. Math. Ann. **227**, 183–193 (1977).

Addresses of authors:

Prof. Gert-Martin Greuel
Universität Kaiserslautern
Fachbereich Mathematik
Erwin-Schrödinger-Straße
D - 6750 Kaiserslautern
Germany
Email: greuel@mathematik.uni-kl.de

Prof. Gerhard Pfister
Humboldt-Univerität zu Berlin
Fachbereich Mathematik
Unter den Linden 6
D - 1086 Berlin
Germany
Email: pfister@mathematik.uni-kl.de

STRATIFICATION PROPERTIES OF CONSTRUCTIBLE SETS

Zbigniew Hajto

1 Introduction

In this paper we study general properties of analytically constructible sets and their connections with constructible sets in the sense of C. Chevalley. In constructible geometry we apply the Grassmann blowing-up to rephrase the proof of the Henry-Merle Proposition [3, Proposition 1] (cf. section 4). This proposition plays an important role in the theory of polar varieties [3], [4]. After that (cf. section 5) we give a certain description of Whitney stratifications of constructible sets. Basic facts from the theory of analytically constructible sets are given in section 3 following the elementary approach of S. Łojasiewicz [6].

Acknowledgements. Part of this work was done during my stay at Regensburg University and I would like to thank Professor Manfred Knebusch and his assistants for their warm hospitality and many valuable discussions. Also I am thankful to my colleagues from Valladolid University especially to Antonio Campillo for stimulating suggestions to the final version of this paper.

2 Grassmann blowing-up

Let $\mathbb{K} = \mathbb{R}$ or $\mathbb{C}$. By $\mathbb{G}_d(\mathbb{K}^n)$ we denote the Grassmannian of d-planes in $\mathbb{K}^n$. On $\mathbb{G}_d(\mathbb{K}^n)$ we have an algebraic manifold structure introduced by a finite atlas of inverse charts:

$$\varphi_{U_i V_i} : L(U_i, V_i) \ni f \mapsto \hat{f} = \{u + f(u) : u \in U_i\} \in \Omega(V_i),$$

where the sum $U_i + V_i = \mathbb{K}^n$ is direct, $\dim U_i = d$ and $\Omega(V_i)$ denotes the set of all algebraic complements of V_i (cf. [6]).

For fixed $W \in \mathbb{G}_d(\mathbb{K}^n)$ and $k \geq d$, we define $S^k(W) = \{T \in \mathbb{G}_k(\mathbb{K}^n) : T \supset W\}$. Following [6], we call $S^k(W)$ the Schubert cycle of W. The algebraic manifold structure on $S^k(W)$ is introduced by the following atlas of inverse charts:

$$\psi_{U_i, V_i} : L(U_i, V_i) \ni f \mapsto \hat{f} + W \in \Omega_W(V_i),$$

where $W \approx \mathbb{K}^d \times 0$, the sum $U_i + V_i = 0 \times \mathbb{K}^{n-d}$ is direct, $\dim U_i = k - d$ and $\Omega_W(V_i)$ denotes the set of all algebraic complements of V_i from $S^k(W)$. Therefore $S^k(W)$ is isomorphic with $\mathbb{G}_{k-d}(\mathbb{K}^{n-d})$ and $\dim S^k(W) = (k-d)(n-k)$.

Let $\pi = \pi_{n,k}^d : E_{n,k}^d \to S^k(W)$ be the canonical k-plane bundle over the Schubert cycle $S^k(W)$. Recall that $E_{n,k}^d = \{(x, L) \in \mathbb{K}^n \times S^k(W) : x \in L\}$.

In local charts:

$$(\mathrm{id}_{\mathbb{K}^n} \times \psi_{U_i V_i})^{-1}(E_{n,k}^d) = \{(x, y, w, f) \in \mathbb{K}^n \times L(U, V) \colon y = f(x)\}, \qquad (2.1)$$

where $\mathbb{K}^n \approx U \times V \times W$. Then $E_{n,k}^d$ is an algebraic submanifold of $\mathbb{K}^n \times S^k(W)$, because in local charts it may be presented as a graph of a polynomial mapping.

Now, let us consider the map:

$$\beta = \beta_{n,k}^d \colon E_{n,k}^d \ni (x, L) \mapsto x \in \mathbb{K}^n.$$

We call β the G-blowing-up of $\mathbb{K}^n$ along W or with center W. The inverse image $S = \beta^{-1}(W) = W \times S^k(W)$ will be called the exceptional manifold or exceptional subset of the G-blowing-up.

Remark 2-1. *For $d = 0$, $\beta = \beta_{n,k}^0$ is just the G-blowing-up of $\mathbb{K}^n$ at 0 as in [5].*

If $W = \mathbb{K}^d \times 0 \subset \mathbb{K}^n$, the restriction $\beta^\Omega \colon (E_{n,k}^d)_\Omega \to \Omega$ to an open neighbourhood Ω of $0 \in \mathbb{K}^n$ will be called the local G-blowing-up.

Remark 2-2. *The local G-blowing-up can be used as a local model for defining the G-blowing-up of a manifold in a closed submanifold.*

In a standard way we define the strict transform of a closed subset $V \subset \Omega$, i.e.,

$$\tilde{V} = \text{closure of } \beta^{-1}(V) \backslash S \quad \text{in} \quad \beta^{-1}(\Omega).$$

Let us define $S_V = \tilde{V} \cap (W \times S^k(W))$ the exceptional subset and $S_{V,0} = \tilde{V} \cap (0 \times S^k(W))$ its fiber over the origin.

3 Analytically constructible sets

In this section we shall gather together some facts from the theory of analytically constructible sets which we shall use later. For more details and other notions from that theory we refer to [6], [2], [8] and [1].

Throughout this and the next sections $\mathbb{K} = \mathbb{C}$. Let M be a complex reduced analytic space of finite dimension n.

A subset $E \subset M$ is called analytically constructible if it can be described at any $c \in M$ by a finite number of holomorphic functions. More precisely, for any $c \in M$ there exists an open neighbourhood U_c such that

$$U_c \cap E = \bigcup_{i=1}^{p} \bigcap_{j=1}^{l} A_{ij},$$

where A_{ij} are of the form $\{f_{ij} = 0\}$ or $\{f_{ij} \neq 0\}$ and $f_{ij} \in \mathcal{O}(U_c)$, where $\mathcal{O}(U_c)$ denotes the ring of holomorphic functions on U_c.

The equivalent global definition is the following (cf. [6]): the family of analytically constructible subsets in M is exactly the smallest family $\mathcal{A}$ of subsets such that:

(1) Every analytic subset of M is in $\mathcal{A}$.

(2) A finite intersection of elements of $\mathcal{A}$ is in $\mathcal{A}$.

(3) The complement of an element from $\mathcal{A}$ is in $\mathcal{A}$.

Definition 3-1. *A locally finite decomposition of M into disjoint submanifolds $\{\Gamma_\nu^i\}$ is called the complex stratification of M if:*

a) $\dim \Gamma_\nu^i = i$.

b) *Every Γ_ν^i is connected.*

c) *Every boundary $\partial\Gamma_\nu^k$ is a union of certain Γ_μ^i for $i < k$.*

From the Remmert-Stein Lemma (cf. e.g. [6]) we conclude that if $U\Gamma_\nu^i = M$ is a complex stratification, then every Γ_ν^i is an analytically constructible leaf, i.e., $\bar{\Gamma}_\nu^i$ and $\partial\Gamma_\nu^i$ are analytic subsets of M.

Let $\mathcal{A}$ and $\mathcal{B}$ be two families of subsets of M we say that $\mathcal{A}$ is compatible with $\mathcal{B} \Longleftrightarrow$ for any $A \in \mathcal{A}$ and $B \in \mathcal{B}$ $A \subset B$ or $A \subset M\backslash B$.

Theorem 3-2. (see [6]) *For any locally finite family of analytically constructible sets in M there exists a complex stratification of M compatible with that family.*

Using Theorem 3-2 it is easy to prove basic topological properties of analytically constructible sets:

(4) the closure of any analytically constructible set is analytic.

(5) every connected component of an analytically constructible set is analytically constructible.

(6) the family of connected components of any analytically constructible set is locally finite.

(7) any analytically constructible set is locally connected.

(8) the interior and boundary of any analytically constructible set is analytically constructible.

For a holomorphic mapping $f\colon M \to N$ of analytic spaces we know that

(9) for any analytically constructible set $F \subset N$ the inverse image $f^{-1}(F)$ is analytically constructible in M.

(10) and according to the Chevalley-Remmert Theorem if $E \subset M$ is analytically constructible and $f_{\bar{E}}\colon \bar{E} \to N$ is proper then the image $f(E)$ is analytically constructible in N.

Lemma 3-3. (Curve Selection) *Let E be analytically constructible in M. If a point $c \in \bar{E}$ is not isolated then there exists a simple arc λ of class C^1 with one end at c, such that $\lambda \backslash c \subset E$. Moreover using a local embedding φ of M at c we can get $\varphi(c) = 0 \in \mathbb{C}^N$ and $\varphi(\lambda) = \{(t^p, h(t)) \in \mathbb{C}^N : 0 \le t \le \tau\}$, where $h \colon [0, \tau] \to \mathbb{C}^{N-1}$ is $\mathbb{C}$-analytic and $p \in \mathbb{N} \backslash 0$.*

Let Γ be a k-dimensional analytically constructible submanifold of an open subset $D \subset \mathbb{C}^N$. Further we shall need the following two facts which are essentially proved in [8, Chapitre II].

(11) Let us define the graph of the tangent map: $\mathcal{N}(\Gamma) = \{(x, T_x\Gamma) : x \in \Gamma\}$. Then $\mathcal{N}(\Gamma)$ is analytically constructible in $D \times \mathbb{G}_k(\mathbb{C}^N)$.

(12) Let us define the conormal fibration over Γ: $\mathcal{C}(\Gamma) = \{(x, H) : x \in \Gamma, T_x\Gamma \subset H \in \check{\mathbb{P}}^{N-1}\}$. Then $\mathcal{C}(\Gamma)$ is analytically constructible in $D \times \check{\mathbb{P}}^{N-1}$.

Recall that for two linear subspaces $S, T \subset \mathbb{C}^N$ we define the "distance"

$$\varepsilon(S, T) = \sup_{\substack{u \in T^\perp \\ u \in S \backslash 0}} \frac{|\langle u, v \rangle|}{|u| \cdot |v|},$$

where $\langle u, v \rangle$ is the Hermitian scalar product in $\mathbb{C}^N$ and $T^\perp$ is the orthogonal complement of T.

Theorem 3-4. (Verdier [9]) *Let X, Y be two analytically constructible submanifolds of an open subset $D \subset \mathbb{C}^N$ such that $Y \subset \bar{X} \backslash X$, then there exists a dense and Zariski open subset $W \subset Y$ such that for any $y_0 \in W$ X is (w)-regular over Y at y_0 i.e., there exists an open neighbourhood U of y_0 in $\mathbb{C}^N$ and a constant $C > 0$, such that for every $x \in X \cap U$ and $y \in Y \cap U$:*

$$\varepsilon(T_yY, T_xX) \le C|x - y|.$$

Remark 3-5. *An elementary proof of Verdier's Theorem (which does not use the resolution of singularities) can be obtained from [7] by modifying the proof in the real case.*

Remark 3-6. *In fact the deep theorem of B. Teissier [8, Théorème Principal] describes precisely the constructible set $S_w(X, Y)$ of points of Y, where (w)-regularity fails.*

Finally note that for analytically constructible sets we have the GAGA principle, i.e., any analytically constructible subset of a compact algebraic variety is algebraically constructible (cf. [6, Chapter VII]).

4   An application: the Henry-Merle Proposition

Let Z be an analytic set in an open subset $D \subset \mathbb{C}^N$ such that $0 \in Z$. We say that a linear subspace $L \subset \mathbb{C}^N$ is transverse to Z at $0 \in \mathbb{C}^N$ if for each limit of tangent spaces $\tau = \lim_{x_n \to 0} T_{x_n} Z^0$, L intersects τ transversally i.e., $L + \tau = \mathbb{C}^N$.

Proposition 4-1. *Let X be an analytically constructible submanifold in $D \subset \mathbb{C}^N$ such that $d = \dim X \geq 2, 0 \in \bar{X} \cap D$. Let Y be a line through $0 \in \mathbb{C}^N$ such that $Y \cap X = \emptyset$ and let $\beta = (\beta_{n,k+1}^1)^D$ be the local G-blowing-up of D in $Y \cap D$. Suppose that there exists $k \geq 1$ such that $S_{\bar{X},0} = 0 \times S^{k+1}(Y)$; then for every hyperplane $L \in \check{\mathbb{P}}^{N-1}$ which is transverse to Y and $\bar{X}$ at 0, there exists a dense and Zariski open subset $U \subset S^{k+1}(Y)$ with the following property:*

(P) *for every sequence $(x_n, H_n) \in \beta^{-1}(X)$ such that $(x_n, H_n) \to (0, H) \in 0 \times U$ and such that there exists $\lim_{x_n \to 0} (T_{x_n} X \cap L) = \tau \cap L$, the limit $\tau \cap L$ intersects H transversally in $\mathbb{C}^N$.*

Example 4-2. Consider the variety $V = \{z^4 - x^3 - y^2 x^2 = 0\} \subset \mathbb{C}^3$. Let Y be the y-axis and let X be the set of regular points $V^0 = V \backslash Y$. It is easily seen that $S_{\bar{X},0} = 0 \times S^2(Y)$ for $\beta = \beta_{3,2}^1$.

PROOF. (OF 4-1) Let us consider $V = \{(x, H, E): (x, H) \in \beta^{-1}(X), E = T_x X \cap L\} \subset D \times S^{k+1}(Y) \times \mathbb{G}_{d-1}(L)$. Shrinking D we can assume that L intersects $T_x X$ transversally for any $x \in X$. From properties (9) and (11) of section 3 we conclude that V is analytically constructible in $D \times S^{k+1}(Y) \times \mathbb{G}_{d-1}(L)$. Let us define $(\bar{V})_0 = \{(H, E) \in S^{k+1}(Y) \times \mathbb{G}_{d-1}(L): (0, H, E) \in \bar{V}\}$. The fiber $(\bar{V})_0$ is an algebraic subset of $S^{k+1}(Y) \times \mathbb{G}_{d-1}(L)$.

Let Z denote an algebraic subset of $S^{k+1}(Y) \times \mathbb{G}_{d-1}(L)$ defined by the property $\dim(H \cap E) > k + d - N$. The projection of $(\bar{V})_0 \cap Z$ by $p: S^{k+1}(Y) \times \mathbb{G}_{d-1}(L) \to S^{k+1}(Y)$ is an algebraic subset of $S^{k+1}(Y)$. Let us take $U = S^{k+1}(Y) \backslash p((\bar{V})_0 \cap Z)$. For such a set U condition (P) is fulfilled, and to prove the Proposition it is enough to show that U is not empty.

(4-3) To prove this let us change the coordinates in $\mathbb{C}^N$ such that $L = \mathbb{C}^{N-1} \times 0$ and $Y = 0 \times \mathbb{C}$, and such that there is no irreducible component of the germ $\bar{X}_0$ which is contained in the union of the hyperplanes $\{x_{N-k} = 0\} \cup \ldots \cup \{x_N = 0\}$. Now for each system of integers $r_1, \ldots, r_k, s \geq 1$ we consider the following mappings

$$p_{r_1 \cdots r_k s}: L(\mathbb{C}^k, \mathbb{C}^{N-k-1}) \times \mathbb{C}^k \times \mathbb{C} \to \mathbb{C}^{N-1} \times \mathbb{C},$$

where

$$p_{rs}(T, z, u) = \left(T(z_1^{r_1}, \ldots, z_k^{r_k}), z_1^{r_1}, \ldots, z_k^{r_k}, u^s \right)$$

(cf. (2.1)). Let us define $X_{r,s} = p_{rs}^{-1}(X)$. Because $S_{\bar{X},0} = 0 \times S^{k+1}(Y)$, the set $A = L(\mathbb{C}^k, \mathbb{C}^{N-k-1}) \times 0 \times 0$ is contained in the closure of $X_{r,s}$ in $p_{rs}^{-1}(D)$.

From Theorem 3-4 we know that A contains a dense Zariski open subset $A_{r,s}$ of points where (w) condition is fulfilled for A and $X_{r,s}^0$. With each $T \in A_{r,s}$ we associate a linear subspace $H_T = \{(T(z), z, u) \in \mathbb{C}^N: (z, u) \in \mathbb{C}^k \times \mathbb{C}\} \in S^{k+1}(Y)$.

From Baire's Theorem we know that $\bigcap_{r,s} A_{r,s} \neq \emptyset$.

Let us take any $T_0 \in \bigcap_{r,s} A_{r,s}$. Now we show that $H_{T_0} \in U$. Using the linear isomorphism

$$F: (x_1, \ldots, x_{N-k-1}, x_{N-k}, \ldots, x_{N-1}, y) \longmapsto$$
$$\big((x_1, \ldots, x_{N-k-1}) - T_0(x_{N-k}, \ldots, x_{N-1}), x_{N-k}, \ldots, x_{N-1}, y\big),$$

and considering $F(X)$ in the place of X, we can assume that

$$T_0 = 0 \in L(\mathbb{C}^k, \mathbb{C}^{N-k-1}) \quad \text{and} \quad H_{T_0} = \{x_1 = \ldots = x_{N-k-1} = 0\}. \quad (4.2)$$

Having made that improvement, we then write $F(X)$ again as X.

Now we fix $(0, H_{T_0}, \tau \cap L) \in 0 \times (\bar{V})_0$. By the Curve Selection Lemma there exists an arc $\lambda \subset \bar{X}$ with one end at $0 \in \bar{X}$, which is a projection of an arc $\mu \subset \bar{V}$ with one end at $(0, H_{T_0}, \tau \cap L)$, and such that $\lambda \backslash 0 \subset X$ and $\mu \backslash (0, H_{T_0}, \tau \cap L) \subset V$. Let $\gamma(t) = \big(x_1(t), \ldots, x_{N-1}(t), y(t)\big)$ be a parametrization of λ.

Taking into account property (12) of section 3, for any hyperplane $W \in \check{\mathbb{P}}^{N-1}$ which contains τ we can select an arc $\bar{\eta}(t)$ of normal vectors at $\gamma(t)$ to X such that there exists $\lim_{t \to 0} \bar{\eta}(t) = \bar{n}$ and $W = \{x \in \mathbb{C}^N : \langle x, n \rangle = 0\}$. Suppose that $r_i = \text{ord}_0 \, x_{N-k+i}(t)$ for $i = 0, \ldots, k-1$ and that $s = \text{ord}_0 \, y(t)$. From 4-3 λ can be taken in such a way that $r_i, s < \infty$.

Observe that if we identify $L(\mathbb{C}^k, \mathbb{C}^{N-k-1})$ with the space of matrices $\mathbb{C}^{k(N-k-1)}$, we can write p_{rs} in coordinates:

$$x_1 = \sum_{i=1}^{k} a_i^1 z_i^{r_i}, \ldots, x_{N-k-1} = \sum_{i=1}^{k} a_i^{N-k-1} z_i^{r_i},$$
$$x_{N-k} = z_1^{r_1}, \ldots, x_{N-1} = z_k^{r_k},$$
$$y = u^s. \quad (4.3)$$

Note that each mapping p_{rs} is a submersion if $z_i \neq 0$ for $i = 1, \ldots, k$ and $u \neq 0$. We denote by γ_{rs} a lifting of γ onto $X_{r,s}$ such that $p_{rs}(\gamma_{rs}(t)) = \gamma(t)$. Moreover note that if $\bar{\eta} = (\bar{\eta}_1, \ldots, \bar{\eta}_{N-1}, \bar{\zeta})$ is a normal vector to X at $(x_1, \ldots, x_{N-1}, y)$ then

$$\Bigg(z_1^{r_1} \eta_1, \ldots, z_k^{r_k} \eta_1, \ldots, z_1^{r_1} \eta_{N-k-1}, \ldots, z_k^{r_k} \eta_{N-k-1}, $$
$$\ldots, r_1 z_1^{r_1-1} \Big(\eta_{N-k} + \sum_{j=1}^{N-k-1} a_1^j \eta_j \Big), \ldots, r_k z_k^{r_k-1} \Big(\eta_{N-1} + \sum_{j=1}^{N-k-1} a_k^j \eta_j \Big), s u^{s-1} \zeta \Bigg)$$

is a conjugate vector to the normal one at $(a_1^1, \ldots, a_k^{N-k-1}, z_1, \ldots, z_k, u) \in X_{r,s}$.

From the (w) condition for the pair $A, X_{r,s}^0$ at $(T_0, 0, 0) \in A$ we have the inequality:

$$\varepsilon(A, T_{(a,z,u)} X_{r,s}) \leq C |(z,u)| \quad \text{for } (a,z,u) \in \Delta \cap X_{r,s}^0,$$

where Δ is an open neighbourhood of $(T_0, 0, 0) \in \mathbb{C}^{k(N-k-1)} \times \mathbb{C}^k \times \mathbb{C}$. Then in particular for $1 \leq l \leq k$ and $1 \leq j \leq N - k - 1$,

$$
\frac{|z_l^{r_l} \eta_j|}{\left| \left(r_1 z_1^{r_1 - 1} \left(\eta_{N-k} + \sum_{j=1}^{N-k-1} a_1^j \eta_j \right), \right.\right.} \leq C_1 |(z, u)|.
$$
$$
\left. \left. \ldots, r_k z_k^{r_k - 1} \left(\eta_{N-1} + \sum_{j=1}^{N-k-1} a_k^j \eta_j \right), su^{s-1} \zeta \right) \right|
$$

Therefore by taking the order relative to $t \in [0, \varepsilon]$ we conclude that

$$
\operatorname{ord}_0(z_l^{r_l} \eta_j) \geq \min_{1 \leq i \leq k} \left\{ \operatorname{ord}_0(z_i), \operatorname{ord}_0(u) \right\} +
$$
$$
\min_{1 \leq i \leq k} \left\{ \operatorname{ord}_0 \left(z_i^{r_1 - 1} \left(\eta_{N-k+1-i} + \sum_{j=1}^{N-k-1} a_i^j \eta_j \right) \right), \operatorname{ord}_0(su^{s-1}\zeta) \right\},
$$

and because $\operatorname{ord}_0(z_i) = \operatorname{ord}_0(u) = 1$ we further conclude that

$$
\operatorname{ord}_0(x_{N-k-1+l} \eta_j) \geq
$$
$$
\min_{0 \leq i \leq k-1} \left\{ \operatorname{ord}_0 x_{N-k+i} \left(\eta_{N-k+i} + \sum_{j=1}^{N-k-1} a_i^j \eta_j \right), \operatorname{ord}_0(y\zeta) \right\} \quad (4.4)
$$

Now we show that on the right hand side of (4.4) $\operatorname{ord}_0(y\zeta)$ can not be strictly minimal.

Let us abbreviate

$$
x_i' = \frac{d}{dt}(x_i \circ \gamma), \quad y' = \frac{d}{dt}(y \circ \gamma).
$$

Then

$$
\sum_{i=1}^{N-1} x_i' \eta_i + y' \zeta = 0, \quad \operatorname{ord}_0(y') = \operatorname{ord}_0(y) - 1 \quad \text{and} \quad \operatorname{ord}_0(x_i') = \operatorname{ord}_0(x_i) - 1.
$$

Therefore

$$
\operatorname{ord}_0(y\zeta) \geq 1 + \min_{1 \leq i \leq N-1} \operatorname{ord}_0(x_i' \eta_i) = \min_{1 \leq i \leq N-1} \operatorname{ord}_0(x_i \eta_i).
$$

Suppose that $(*)$ $\operatorname{ord}_0(y\zeta)$ is strictly minimal. Then

$$
\min_{1 \leq i \leq N-1} \operatorname{ord}_0(x_i \eta_i) = \min_{0 \leq i \leq k-1} \operatorname{ord}_0(x_{N-k+i} \eta_{N-k+i}),
$$

because

$$x_j = \sum_{i=1}^{k} a_i^j x_{N-k-1+i}$$

and we have (4.4) together with $(*)$.

Let

$$\alpha = \mathrm{ord}_0(x_{N-k+l_0}\eta_{N-k+l_0}) = \min_{0 \le i \le k-1} \mathrm{ord}_0(x_{N-k+i}\eta_{N-k+i})$$

and let

$$\beta = \mathrm{ord}_0(x_{N-k+l_0}(\eta_{N-k+l_0} + \sum_{j=1}^{N-k-1} a_{l_0}^j \eta_j)).$$

When we consider the three cases: $\alpha < \beta, \alpha = \beta, \alpha > \beta$ we get a contradiction with the assumption $(*)$. Therefore the minimal order is attained for a certain $i_0 \in \{0, \ldots, k-1\}$ and then

$$\mathrm{ord}_0(x_{N-k+i_0}\eta_j) \ge \mathrm{ord}_0\left(x_{N-k+i_0}\left(\eta_{N-k+i_0} + \sum_{j=1}^{N-k-1} a_{i_0}^j \eta_j\right)\right).$$

Hence

$$\mathrm{ord}_0(\eta_j) \ge \mathrm{ord}_0\left(\eta_{N-k+i_0} + \sum_{j=1}^{N-k-1} a_{i_0}^j \eta_j\right) \quad \text{for} \quad 1 \le j \le N-k-1,$$

and we conclude that

$$\mathrm{ord}_0(\eta_j) \ge \min\{\eta_{N-k+i_0}, a_{i_0}^1 \eta_1, \ldots, a_{i_0}^{N-k-1}\eta_{N-k-1}\}.$$

Finally $\mathrm{ord}_0\eta_j \ge \mathrm{ord}_0\eta_{N-k+i_0}$ for $1 \le j \le N-k-1$, and this means that for $t \to 0$, the vector

$$\frac{(\bar{\eta}_1, \ldots, \bar{\eta}_{N-1})}{|(\bar{\eta}_1, \ldots, \bar{\eta}_{N-1})|}$$

tends to the unit normal vector to $\tau \cap L$, which has a nonzero projection onto $H_{T_0} \cap L$ (cf. (4.2)). Therefore $\tau \cap L$ and $H_{T_0} \cap L$ meet transversally in L and the proof is complete.   □

One of the most spectacular applications of the Proposition was described by Henry and Merle in [4]. They considerably simplified the proof of Teissier's "Théorème Principal". Here we just recall two important corollaries from "Théorème Principal" [8], which we shall use in the next section.

Let X and Y be two disjoint analytically constructible submanifolds of an open subset $D \subset \mathbb{C}^N$. We say that X is Whitney (b)-regular over Y at $y_0 \in Y$ if the following holds:

$$\varepsilon\big(\mathbb{C}(x-y), T_x X\big) \to 0 \quad \text{for} \quad X \ni x \to y_0 \quad \text{and} \quad Y \ni y \to y_0.$$

From Teissier's Theorem we know that condition (b) and (w) are equivalent in the complex case. Moreover they are analytic invariants and therefore using local charts we can formulate them for X, Y contained in the analytic space M.

Let us denote $S(X, Y)$ the set of points $x \in Y$ where (b)-regularity fails.

Corollary 4-4. The set $S(X, Y) = S_w(X, Y)$ is analytically constructible in M, nowhere dense and closed in Y.

Corollary 4-5. If X and Y are connected and X is (b)-regular over Y at $y \in Y$, then $Y \subset \bar{X}$.

5 Canonical stratification

Let M be as in section 3, W and V let be two analytic subsets of M. We define $\tau_k(W, V) = \bigcup_i (W \cap V_i)$, where $\{V_i\}$ is the family of k-dimensional irreducible components of V, which are not included in W. Then we conclude that $\dim \tau_k(W, V) < k$. Let $\{W_j\}$ be a locally finite family of analytic subsets of M. The canonical stratification of M compatible with $\{W_j\}$ is constructed in the following way. We define inductively a descending chain of analytic sets $Z_n \supset \ldots \supset Z_{-1} = \emptyset$:

(1) $M = Z_n$.

(2) If we have $Z_n \supset \ldots \supset Z_k$ $(n \geq k \geq 0)$ with the property that $\dim Z_i \leq i$ and $Z_i \backslash Z_{i-1}$ is contained in the set of i-dimensional regular points of Z_i, then for $i = n, \ldots, k+1$ $\{\Gamma_\nu^i\}$ is the family of connected components of $Z_i \backslash Z_{i-1}$ and

$$Z_{k-1} = \bigcup_\mu Z_k^\mu \cup Z_k^* \cup \bigcup_{i>k} S(\Gamma_\nu^i, Z_k \backslash Z_k^*) \cup \bigcup_j \tau_k(W_j, Z_k),$$

where $\{Z_k^\mu\}$ is the family of irreducible components of Z_k of dimension $< k$ and Z_k^* is the set of singular points of Z_k.

Now $\dim Z_{k-1} \leq k - 1$ (see Corollary 4-4) and $Z_k \backslash Z_{k-1}$ is a subset of k-dimensional regular points of Z_k. The family $\{\Gamma_\nu^i\}$ is a complex stratification of M (see Corollary 4-5). The compatibility of $\{\Gamma_\nu^i\}$ with $\{W_j\}$ is easily seen from the following Lemma of Łojasiewicz [6].

Lemma 5-1. *Let W and V be analytic subsets of M. An open and connected subset H of k-dimensional regular points of V such that $H \cap \tau_k(W, V) = \emptyset$ is compatible with W.*

Remark 5-2. *Every E analytically constructible in M may be presented as a finite union $E = (V_0 \backslash V_1) \cup \ldots \cup (V_{2l} \backslash V_{2l+1})$, where V_i are analytic in M. Therefore we can repeat the above construction to obtain a Whitney stratification compatible with a locally finite family of analytically constructible subsets of M.*

References

[1] Grothendieck, A., Dieudonné, J.: EGA III, Publ. Math. IHES 11 (1961).

[2] Hajto, Z.: Whitney (b)-regularity for analytically constructible sets. In: Proceedings of the Conference on Algebraic Geometry, Berlin 1985, Teubner-Texte zur Mathematik 92 pp. 117–122, Teubner, Leipzig 1986.

[3] Henry, J. P., Merle, M.: Limites d'espaces tangents et transversalité de variétés polaires. In: Algebraic Geometry, Proceedings, La Rábida 1981, pp. 189–199, Lecture Notes in Mathematics 961, Springer 1982.

[4] Henry, J. P., Merle, M.: Limites de normales, conditions de Whitney et éclatement d'Hironaka. In: Proceedings of Symposia in Pure Mathematics, Vol. 40, Part 1, pp. 575–584 (1983).

[5] Kuo, T. C., Trotman, D.J.A.: On (w) and (t^s)-regular stratifications. Inventiones Math. 92, pp. 633–643 (1988).

[6] Łojasiewicz, S.: Wstep do geometrii analitycznej zespolonej, PWN Warszawa 1988.

[7] Łojasiewicz, S., Stasica, J., Wachta, K.: Stratifications sous-analytiques. Condition de Verdier, Bull. Ac. Pol.: Mathematics, Vol. 34, pp. 531–539 (1986).

[8] Teissier, B.: Variétés polaires II. In: Algebraic Geometry, Proceedings, La Rábida 1981, pp. 314–491, Lecture Notes in Mathematics 961, 1982.

[9] Verdier, J. L.: Stratifications de Whitney et théorème de Bertini-Sard. Inventiones math. 36, pp. 295–312 (1976).

Address of author:

Universidad Complutense de Madrid
Facultad de Ciencias Matemáticas
Departamento de Algebra
28040 Madrid
Spain

ON THE LINEARIZATION PROBLEM AND SOME QUESTIONS FOR WEBS IN $\mathbb{C}^2$

Alain Hénaut

This is an enlarged version of a talk given in La Rábida 91 which was entitled "Webs of maximum rank in $\mathbb{C}^2$ which are algebraic". We begin with a short survey on web geometry in $\mathbb{C}^2$. The linearization problem for webs in $\mathbb{C}^2$ is discussed and new results are given. In particular, we characterize maximum rank webs in $\mathbb{C}^2$ which are linearizable. At the end of the article, we pose some questions and we show how to use basic facts of algebraic analysis (i.e., $\mathcal{D}$-modules theory) to recover some classical results and study new problems for webs in $\mathbb{C}^2$.

1 Introduction in the form of a survey

1.1 Basic definitions

A d-web $\mathcal{W}$ in $(\mathbb{C}^2, 0)$ consists of d foliations $(d \geq 2)$ in $(\mathbb{C}^2, 0)$ by curves in general position in the sense that through every point sufficiently close to $0 \in \mathbb{C}^2$, the leaves of $\mathcal{W}$ have pairwise distinct tangents (see figure 1.1).

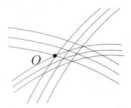

Figure 1.1: A 3-web.

Like Blaschke and his co-workers in the thirties (cf. [2] and its bibliography) and more recently Chern and Griffiths [8], we are interested in the geometry of such configurations that is, we study them up to local isomorphism $\phi : (\mathbb{C}^2, 0) \rightarrow (\mathbb{C}^2, \phi(0))$.

The leaves of $\mathcal{W}$ are the level sets of d elements $F_i \in \mathbb{C}\{x, y\}$ (i.e., the local ring of convergent power series in two variables) with $F_i(0) = 0$ and such that $dF_i \wedge dF_j(0) \neq 0$ for $1 \leq i < j \leq d$.

The most important invariant of $\mathcal{W}$ is the dimension of the $\mathbb{C}$-vector space

$$\mathcal{A} = \left\{ \left(g_i(F_i) \right)_{1 \leq i \leq d} \;\middle|\; g_i \in \mathbb{C}\{t\} \quad \text{and} \quad \sum_{i=1}^{d} g_i(F_i) dF_i = 0 \right\},$$

Progress in Mathematics, Vol. 134
© 1996 Birkhäuser Verlag Basel/Switzerland

of abelian relations of $\mathcal{W}$. It is called the rank of $\mathcal{W}$ and it is denoted by rk $\mathcal{W}$. It depends only on $\mathcal{W}$ and using, for example, elementary results on $\mathcal{D}$-modules (cf. [13] and the last paragraph below), we find the following classical bounds:

$$0 \leq \mathrm{rk}\, \mathcal{W} \leq \frac{1}{2}(d-1)(d-2). \tag{1.1}$$

A basic example is the following: let $\mathcal{H}$ be a 3-web of maximum rank 1 in $(\mathbb{C}^2, 0)$ also called an hexagonal web (this terminology is explained in [2] and [13]). We can find $\tilde{g}_i \in \mathbb{C}\{t\}$ such that

$$\tilde{g}_1(F_1) + \tilde{g}_2(F_2) + \tilde{g}_3(F_3) = 0,$$

and it follows that $\mathcal{H}$ is given, up to a local isomorphism, by $x = \text{const}$, $y = \text{const}$ and $x + y = \text{const}$ (see figure 1.2).

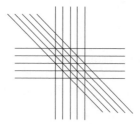

Figure 1.2: An hexagonal web.

Before we give further examples, we remark that web geometry has recently arisen in several different contexts. For instance, in papers on algebraic foliations and differential equations by Cerveau [6], and on the general theory of diagrams of mappings by Dufour [10] and Nakai [20].

1.2   Linear and algebraic webs

A d-web $\mathcal{L}$ in $(\mathbb{C}^2, 0)$ is linear if the leaves of $\mathcal{L}$ are (pieces of) straight lines in $\mathbb{C}^2$ (not necessarily parallel). A d-web $\mathcal{W}$ in $(\mathbb{C}^2, 0)$ is linearizable if, up to a local isomorphism, $\mathcal{W}$ is linear.

Every reduced algebraic curve $C = \{f(s,t) = 0\} \subset \mathbb{P}^2$ of degree d (possibly singular or reducible) defines, via duality, a linear d-web $\mathcal{L}_C$ in $\mathbb{P}^{\vee 2}$. Indeed, a general line H meets transversely C in d distinct points and by duality, we obtain $\mathcal{L}_C$ in $(\mathbb{P}^{\vee 2}, H)$ (see figure 1.3).

If C contains no lines, we note that $\mathcal{L}_C$ is given by the generic tangent lines to the dual curve $C^\vee \subset \mathbb{P}^{\vee 2}$ of C; otherwise $\mathcal{L}_C$ contains pencils of lines in $\mathbb{P}^{\vee 2}$.

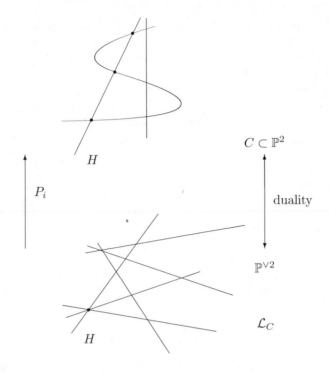

Figure 1.3: Duality of algebraic curves and linear d-webs.

Moreover, we have local branches $P_i = \big(F_i, \xi_i(F_i)\big)$ on $\mathbb{P}^{\vee 2}$ such that $H \cap C = \sum_{i=1}^{d} P_i(H)$ and if ω_C is the dualizing sheaf of C we have a $\mathbb{C}$-linear map

$$H^0(C, \omega_C) \xrightarrow{\ \theta\ } \mathcal{A} = \mathcal{A}(\mathcal{L}_C),$$

given by $\theta(\omega) = \big(g_i(F_i)\big)_i$, where $\omega = g_i(s)\,ds$ in a neighborhood of $P_i(H)$. Indeed, we have

$$\mathrm{Trace}(\omega) = \sum_{i=1}^{d} P_i^*(\omega) = \sum_{i=1}^{d} g_i(F_i)dF_i = 0,$$

by Abel's theorem. It can be proved that θ is an isomorphism. Thus, the rank of $\mathcal{L}_C$ is maximal (i.e., $\mathrm{rk}\,\mathcal{L}_C = (1/2)(d-1)(d-2)$ because $H^0(C, \omega_C)$ is generated over $\mathbb{C}$ by

$$r(s,t)\frac{ds}{\partial_t(f)}, \qquad r \in \mathbb{C}[s,t] \quad \text{and} \quad \deg r \le d-3.$$

The first fundamental result is the following:

Theorem 1-1. (Algebrization theorem, Lie, Darboux [9] ,..., Griffiths [12]) *Let $\mathcal{L}$ be a linear d-web of maximum rank in $(\mathbb{C}^2, 0)$ with $d \geq 3$, then there is a reduced algebraic curve $C \subset \mathbb{P}^2$ of degree d (possibly singular or reducible) such that $\mathcal{L} = \mathcal{L}_C$.*

A d-web $\mathcal{W}$ in $(\mathbb{C}^2, 0)$ is algebraic if, up to a local isomorphism, we have $\mathcal{W} = \mathcal{L}_C$. Thus, from the previous results, we have that a linearizable web of maximum rank in $(\mathbb{C}^2, 0)$ is algebraic.

Web geometry for webs of maximum rank in $(\mathbb{C}^2, 0)$ is, however, larger in extent than the algebraic geometry of plane curves. In fact, Bol [3] gave an example of a 5-web of maximum rank 6 in $\mathbb{C}^2$, which is non-linearizable (see figure 1.4).

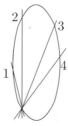

Figure 1.4: A 5-web of rank 6 $\mathcal{B}$, that is non-linarizable.

Bol's example $\mathcal{B}$ is given by four pencils of lines whose vertices are in general position in $\mathbb{C}^2$, with the fifth family consisting of the conics through the four vertices. We remark that one of the six abelian relations involves the functional relation with 5 terms given by the Euler dilogarithm $\sum_{n \geq 1}(z^n/n^2)$ (cf. [18]).

According to what we have seen before, a deep question already raised in Chern [7] is the following: determine webs in $(\mathbb{C}^2, 0)$ which are linearizable.

2   Linearization of webs in $(\mathbb{C}^2, 0)$

Let $\mathcal{W}$ be a d-web in $(\mathbb{C}^2, 0)$ not necessarily of maximum rank. By making a suitable linear change of coordinates we assume, from now on, that the leaves of $\mathcal{W}$ are integral curves of the following vector fields:

$$X_i = \partial_x + b_i \partial_y, \qquad \text{for } 1 \leq i \leq d,$$

where $b_i \in \mathbb{C}\{x, y\}$ with $b_i(0) \neq b_j(0)$ for $1 \leq i < j \leq d$. It follows that there exists an unique element

$$P_{\mathcal{W}} = \sum_{k=0}^{d-1} P_k b^k \in \mathbb{C}\{x, y\}[b],$$

called the polynomial associated to $\mathcal{W}$ such that

$$\deg P_{\mathcal{W}} \leq d - 1 \quad \text{and} \quad P_{\mathcal{W}}(x, y; b_i) = X_i(b_i), \qquad \text{for } 1 \leq i \leq d.$$

It is easy to verify that each leaf of $\mathcal{W}$ is the graph in $\mathbb{C}^2$ of element $y_i \in \mathbb{C}\{x\}$ which satisfies the following second order differential equation:

$$y'' = P_{\mathcal{W}}(x, y; y').$$

We have the following result (cf. [14]):

Theorem 2-1. For $d \geq 4$, $\mathcal{W}$ is linearizable if and only if, $\deg P_{\mathcal{W}} \leq 3$ and (P_0, P_1, P_2, P_3) satisfies the non-linear differential system:

$$\left. \begin{aligned} \partial_x^2(P_2) - 2\partial_x\partial_y(P_1) + 3\partial_y^2(P_0) \quad &+ 6P_0\partial_x(P_3) - 3P_2\partial_y(P_0) \\ - 3P_0\partial_y(P_2) + 3P_3\partial_x(P_0) + 2P_1\partial_y(P_1) - P_1\partial_x(P_2) &= 0 \\ 3\partial_x^2(P_3) - 2\partial_x\partial_y(P_2) + \partial_y^2(P_1) \quad &- 6P_3\partial_y(P_0) + 3P_1\partial_x(P_3) \\ + 3P_3\partial_x(P_1) - 3P_0\partial_y(P_3) - 2P_2\partial_x(P_2) + P_2\partial_y(P_1) &= 0. \end{aligned} \right\} \quad (2.2)$$

In fact, up to an automorphism of $\mathbb{P}^{\vee 2}$, every local isomorphism $\phi :$ $(\mathbb{C}^2, 0) \to (\mathbb{P}^{\vee 2}, \phi(0))$ giving a linearization of $\mathcal{W}$ comes from a base $[\omega_1, \omega_2, \omega_3]$ of solutions of the following second order linear differential system:

$$\left. \begin{aligned} 9\partial_x^2(\omega) &= 3P_1\partial_x(\omega) - 9P_0\partial_y(\omega) \\ &\quad + \left[9\partial_y(P_0) - 3\partial_x(P_1) - 6P_0P_2 + 2P_1^2\right]\omega \\ 9\partial_x\partial_y(\omega) &= 3P_2\partial_x(\omega) - 3P_1\partial_y(\omega) \\ &\quad + \left[3\partial_y(P_1) - 3\partial_x(P_2) - 9P_0P_3 + P_1P_2\right]\omega \\ 9\partial_y^2(\omega) &= 9P_3\partial_x(\omega) - 3P_2\partial_y(\omega) \\ &\quad + \left[3\partial_y(P_2) - 9\partial_x(P_3) - 6P_1P_3 + 2P_2^2\right]\omega \end{aligned} \right\} \quad (2.3)$$

The integrability condition of (2.3) is given by the system (2.2). In general, it would seem difficult to solve the system (2.2). Nevertheless, we shall see in the maximum rank case, how to use the abelian relations to solve the system (2.2) geometrically.

Remark 2-2.

1. It can be checked that the condition that (P_0, P_1, P_2, P_3) satisfy (2.2) is the R. Liouville-Tresse condition that ϕ transforms the second order differential equation $y'' = P_0 + P_1 \cdot y' + P_2 \cdot (y')^2 + P_3 \cdot (y')^3$ into $Y'' = 0$ (cf. [23]). Moreover, in this case all local isomorphisms ϕ are determined by (2.3). A slightly different point of view can be found in E. Cartan [5] and Arnold [1].

2. It is proved in [14] that a 3-web $\mathcal{W}$ in $(\mathbb{C}^2, 0)$ is linearizable if and only if, a certain second order, non-linear differential system has a solution. This system involves the $X_i(b_i)$ associated to $\mathcal{W}$ for $1 \leq i \leq 3$ and resembles (2.2) in part. Another point of view on the linearization problem for 3-webs in $(\mathbb{C}^2, 0)$ can be found in Goldberg [11].

Before we continue, we note an easy application of the above theorem: the previous example $\mathcal{B}$ of Bol is non-linearizable. Indeed, it can be checked that $\deg P_{\mathcal{B}} = 4$ (!).

3 Geometry of the abelian relation space and the linearization problem in the maximum rank case

From now on, let $\mathcal{W}$ be a d-web in $(\mathbb{C}^2, 0)$ with maximum rank $r = (1/2)(d-1)(d-2)$ and $d \geq 4$.

Let $\left(\gamma_i^j(F_i)\right)_{1 \leq j \leq r}^{1 \leq i \leq d}$ be a base of the abelian relation space $\mathcal{A}$ of $\mathcal{W}$. Using an idea of Poincaré [22], it is possible to define d germs of maps of rank 1

$$Z_i : (\mathbb{C}^2, 0) \rightarrow \left(\mathbb{P}^{r-1}, Z_i\,(0)\right),$$

where $Z_i(x,y) = \left[\gamma_i^1\,(F_i), \cdots, \gamma_i^r\,(F_i)\right]$. We may rewrite the abelian relations as

$$\sum_{i=1}^{d} Z_i\, dF_i = 0,$$

and the assumption of maximum rank implies that the dimension of the linear space spanned by the Z_i is $d-3$ and so we set

$$\mathbb{P}^{d-3}(x,y) = \left\{Z_i(x,y)\right\} \subset \mathbb{P}^{r-1}.$$

The d points $Z_i(x,y)$ are in general position in $\mathbb{P}^{d-3}(x,y)$, thus they lie on a unique rational normal curve $E(x,y) \subset \mathbb{P}^{d-3}(x,y)$ (up to an automorphism, $E(x,y)$ is given parametrically by $b \rightsquigarrow [1, b, b^2, \cdots, b^{d-3}]$).

The following expression suggested by Darboux and Blaschke

$$E(x,y;b) = \Pi(x,y;b) \cdot \sum_{i=1}^{d} \frac{\partial_y(F_i)}{b - b_i} Z_i,$$

where $\Pi(x,y;b) = \prod_{i=1}^{d}(b - b_i)$, gives a map

$$E : (\mathbb{C}^2, 0) \times \mathbb{P}^1 \rightarrow \mathbb{P}^{r-1},$$

which represents the family $\cup_{(x,y)\in(\mathbb{C}^2,0)} E(x,y)$ of rational normal curves in a neighborhood of $E(0)$ (essentially we use the relation $\sum_{i=1}^{d} Z_i\, dF_i = 0$ and the fact that $E(x,y;b_i) = Z_i(x,y)$).

It can be proved that

$$\dim\left\{E, \partial_b(E), \partial_y(E)\right\} = 2,$$

thus the rank of E always satisfies $\mathrm{rk}_{(x,y;b)}\, E \geq 2$.

The crucial result is the following (cf. [15]):

Lemma 3-1. Let $P_{\mathcal{W}} = \sum_{k=0}^{d-1} P_k b^k$ be the polynomial associated to $\mathcal{W}$, then the Darboux-Blaschke map E satisfies the following relation:

$$\partial_x\left(\frac{E}{\Pi}\right) + b\partial_y\left(\frac{E}{\Pi}\right) + \partial_b\left[(P_0 + P_1 b + P_2 b^2 + P_3 b^3) \cdot \frac{E}{\Pi}\right]$$

$$= P_4 \cdot \sum_{i=1}^{d} \frac{b_i^4 \partial_y(F_i)}{(b - b_i)^2} Z_i + \cdots + P_{d-1} \cdot \sum_{i=1}^{d} \frac{b_i^{d-1} \partial_y(F_i)}{(b - b_i)^2} Z_i.$$

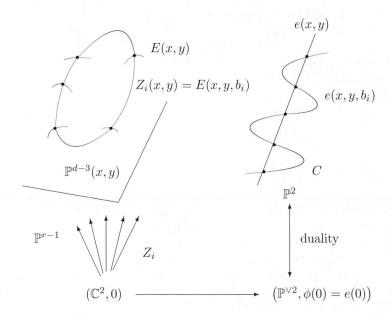

Figure 3.5: A linearization of a d-web with $\deg P_{\mathcal{W}} \leq 3$.

In particular, $\deg P_{\mathcal{W}} \leq 3$ if and only if $\mathrm{rk}_{(x,y;b)} E$ is everywhere equal to 2. Moreover, if $\deg P_{\mathcal{W}} \leq 3$, then any sufficiently close pair $E(x,y)$ and $E(x_0, y_0)$ have just one common point.

The last property comes from the second order differential equation $y'' = P_{\mathcal{W}}(x, y; y')$ which, roughly speaking, gives the fibers of E.

Theorem 3-2. Let $\mathcal{W}$ be a d-web in $(\mathbb{C}^2, 0)$ of maximum rank $r = (1/2)(d - 1)(d - 2)$ and $d \geq 4$. The following conditions are equivalent:

1. $\mathcal{W}$ is linearizable;

2. the family $\cup_{(x,y) \in (\mathbb{C}^2, 0)} E(x, y)$ lies in a Veronese surface $V \subset \mathbb{P}^{r-1}$ (up to an automorphism, V is given parametrically by

$$(b, t) \rightsquigarrow [1; b, t; b^2, bt, t^2; \cdots, t^{d-3}]);$$

3. $\deg P_{\mathcal{W}} \leq 3$.

IDEA OF THE PROOF (for details cf. [15]). $1 \Rightarrow 2$ is more or less well-known, and comes from the algebraization theorem and the description of the group $H^0(C, \omega_C)$. $1 \Rightarrow 2$ comes from the Theorem 2-1 above. $3 \Rightarrow 1$ comes from the previous lemma. Indeed, we have a map

$$e = [E_1, E_2, E_3] : (\mathbb{C}^2, 0) \times \mathbb{P}^1 \longrightarrow \mathbb{P}^2$$

with rank 2 and every algebraic curve $e(x, y)$ is a straight line in $\mathbb{P}^2$. By duality, we obtain a linearization $\phi : (\mathbb{C}^2, 0) \to (\mathbb{P}^{\vee 2}, \phi(0))$ of $\mathcal{W}$ if we set $\phi(x, y) = e(x, y)$. Up to an automorphism of $\mathbb{P}^{\vee 2}$, ϕ is unique because $d \geq 4$ (cf. [14]). Moreover (roughly speaking), we have $\phi_*(\mathcal{W}) = \mathcal{L}_C$ from Griffith's theorem [12] (see figure 3.5)

Remark 3-3. For $d = 4$, we have $\deg P_{\mathcal{W}} \leq 3$ and we recover from the above theorem the classical result due to Poincaré: every 4-web of maximal rank 3 in $(\mathbb{C}^2, 0)$ is linearizable.

4   Some questions on webs in $\mathbb{C}^2$

Let $\mathcal{W}(d)$ be a d-web in $(\mathbb{C}^2, 0)$ with leaves $\{F_i(x, y) = \text{const}\}$ in general position where $F_i \in \mathcal{O} = \mathbb{C}\{x, y\}$ and $F_i(0) = 0$ for $1 \leq i \leq d$. A fundamental problem for webs in $(\mathbb{C}^2, 0)$ is to describe non-linearizable d-webs $\mathcal{W}(d)$ t with maximum rank $(1/2)(d-1)(d-2)$. The only known example of such a web is the 5-web $\mathcal{B}$ of Bol (cf. Introduction, 1.2).

Taking into account the Bol's example $\mathcal{B}$ and certain properties of polylogarithms (cf. [18]), it would seem interesting to study abelian relations of $\mathcal{W}(d)$ of the particular form

$$\sum_{i=1}^{d} g(F_i)dF_i = 0, \qquad \text{where } g \in \mathbb{C}\{t\}.$$

Probably, such abelian relations, if there are any, will be exceptional.

From the viewpoint of the above questions and also more generally, methods for determining the rank of $\mathcal{W}(d)$ are of great interest.

We end this paper by showing how to use $\mathcal{D}$-modules to attack this problem (cf. [13] for further details and [16] for general results in the case of webs of codimension one in $\mathbb{C}^n$). Moreover, we shall see that the same methods open a new route in the study of webs in $\mathbb{C}^2$ with singularities. The leaves of $\mathcal{W}$ are the integral curves of d vector fields $X_i = A_i \partial_x + B_i \partial_y$, with coefficients in $\mathcal{O}$, which are pairwise linearly independent. We denote by $\text{Sol}\,\mathcal{R}(d)$ the $\mathbb{C}$-vector space of solutions $(f_1, \cdots, f_d) \in \mathcal{O}^d$ of the resonance differential system given by

$$\mathcal{R}(d) \equiv \begin{cases} X_i(f_i) = 0 & \text{for} \quad 1 \leq i \leq d, \\ \partial_x(f_1 + \cdots + f_d) = 0, \\ \partial_y(f_1 + \cdots + f_d) = 0. \end{cases}$$

We note that the C^∞ case corresponds to the resonance equations introduced by Joly and Rauch to study the interaction problem of non-linear oscillations. In this problem the functions f_i are real phases of oscillations and the equations $X_i(f_i) = 0$ are eikonal equations for the phases. The two other equations come from non-linear interactions between the oscillations of various modes (cf. [17]).

Using the Frobenius theorem, it can be verified that

$$\operatorname{Sol} \mathcal{R}(d) = \mathcal{A}(d) \oplus \mathbb{C}^d,$$

where $\mathcal{A}(d)$ is the $\mathbb{C}$-vector space of abelian relations of $\mathcal{W}(d)$. We recall that $\operatorname{rk} \mathcal{W}(d) = \dim_\mathbb{C} \mathcal{A}(d)$.

Let $\mathcal{D}$ be the ring of linear differential operators with coefficients in $\mathcal{O}$. We denote by $\mathcal{R}(d)$ the left $\mathcal{D}$-module associated to $R(d)$ and by $\operatorname{mult} \mathcal{R}(d)$ its multiplicity (cf. for example [19], [21]). We denote by $\mathcal{G}(d)$ the $\mathcal{O}[\xi, \eta]$-module associated to the symbol matrix defined by $R(d)$. Let $\mathcal{W}(d+1)$ be a $(d+1)$-web in $(\mathbb{C}^2, 0)$ such that $\mathcal{W}(d)$ is a d-web extracted from $\mathcal{W}(d+1)$.

With the previous notations, we have an exact sequence of left $\mathcal{D}$-modules of finite type:

$$0 \to \mathcal{D} \Big/ \left(\mathfrak{X}_{d+1}, \bigcap_{i=1}^{d} \mathfrak{X}_i \right) \to \mathcal{R}(d+1) \to \mathcal{R}(d) \to 0, \tag{4.4}$$

and an exact sequence of $\mathcal{O}[\xi, \eta]$-modules of finite type:

$$0 \to \mathcal{O}[\xi, \eta] \Big/ \left(\mathfrak{a}_{d+1}, \bigcap_{i=1}^{d} \mathfrak{a}_i \right) \to \mathcal{G}(d+1) \to \mathcal{G}(d) \to 0, \tag{4.5}$$

where $\mathfrak{X}_i$ (resp. $\mathfrak{a}_i$) is the left ideal of $\mathcal{D}$ (resp. the ideal of $\mathcal{O}[\xi, \eta]$) generated by the vector field X_i (resp. the symbol $A_i \xi + B_i \eta$ of X_i). Using $\mathcal{G}(d)$ and the general position hypothesis, one proves that the characteristic variety of $\mathcal{R}(d)$ is the zero section. By classical results on $\mathcal{D}$-modules, it follows that

$$\dim_\mathbb{C} \operatorname{Sol} \mathcal{R}(d) = \operatorname{mult} \mathcal{R}(d) = \operatorname{rk} \mathcal{W}(d) + d.$$

By induction on d, the exact sequence (4.5) gives

$$0 \leq \operatorname{mult} \mathcal{R}(d) \leq \operatorname{mult} \mathcal{G}(d) = \frac{1}{2}(d-1)(d-2) + d,$$

so, we recover the classical bounds for $\operatorname{rk} \mathcal{W}(d)$ given in (1.1).

Moreover, from the exact sequence (4.4) it follows that

$$\operatorname{rk} \mathcal{W}(d+1) + 1 = \operatorname{rk} \mathcal{W}(d) + \operatorname{mult} \mathcal{D} \Big/ \left(\mathfrak{X}_{d+1}, \bigcap_{i=1}^{d} \mathfrak{X}_i \right).$$

In particular, given the knowledge of the left ideal $(\mathfrak{X}_{d+1}, \cap_{i=1}^{d} \mathfrak{X}_i)$ of $\mathcal{D}$, the rank of a $(d+1)$-web in $(\mathbb{C}^2, 0)$ is determined by its extracted webs.

We note that the above result together with the calculational methods of Castro Jiménez [4] using Gröbner bases are well-adapted to the determination of the rank of a web in $(\mathbb{C}^2, 0)$.

If $\mathcal{W}(d+1)$ has singularities, that is if

$$\text{Sing}\,\mathcal{W}(d+1) = \left\{ \prod_{1 \le i < j \le d+1} (A_i B_j - A_j B_i)(x,y) = 0 \right\}$$

defines the germ of a curve in $(\mathbb{C}^2, 0)$, then the exact sequence (4.4) holds. An interesting problem in the singular case is to relate the characteristic cycle of $\mathcal{R}(d+1)$ to the geometry of $\mathcal{W}(d+1)$.

References

[1] V. ARNOLD, Chapitres supplémentaires de la théorie des équations différentielles ordinaires, MIR, Moscou (1980).

[2] W. BLASCHKE und G. BOL, Geometrie der Gewebe, Springer, Berlin (1938).

[3] G. BOL, Über ein bemerkenswertes Fünfgewebe in der Ebene, Abh. Hamburg **11** (1936), 387–393.

[4] F. J. CASTRO-JIMÉNEZ, Calcul de la dimension et des multiplicités d'un $\mathcal{D}$-module monogène, C.R. Acad. Sc. Paris **302** (1986), 487–490.

[5] E. CARTAN, Sur les variétés à connexion projective, Bull. Soc. Math. France **52** (1924), 205–241.

[6] D. CERVEAU, Equations différentielles algébriques: Remarques et problèmes, J. Fac. Sci. Univ. Tokyo **36** (1989), 665–680.

[7] S.S. CHERN, Web Geometry, Bull. Amer. Math. Soc. **6** (1982), 1–8.

[8] S.S. CHERN and P.A. GRIFFITHS, Abel's Theorem and Webs, Jahresberichte der deut. Math. Ver. **80** (1978), 13–110.

[9] G. DARBOUX, Leçons sur la théorie générale des surfaces, Livre I, 2-ième édition, Gauthier-Villars, Paris (1914).

[10] J.-P. DUFOUR, Familles de courbes planes différentiables, Topology **22** (1983), 449–474.

[11] V.V. GOLDBERG, On a linearizability condition for a three-web on a two-dimensional manifold, in Differential Geometry ; F.J. Carreras, O. Gil-Medrano and A.M. Naveira (Eds), Springer-Verlag Lect. Notes in Math. **1410** (1989), 223–239.

[12] P. A. GRIFFITHS, Variations on a Theorem of Abel, Invent. Math. **35** (1976), 321–390.

[13] A. HÉNAUT, $\mathcal{D}$-modules et géométrie des tissus de $\mathbb{C}^2$, Math. Scand. **66** (1990), 161–172.

[14] A. HÉNAUT, Sur la linéarisation des tissus de $\mathbb{C}^2$, Topology **32** (1993), 531–542.

[15] A. HÉNAUT, Caractérisation des tissus de $\mathbb{C}^2$ dont le rang est maximal et qui sont linéarisables, to appear in Compositio Math.

[16] A. HÉNAUT, Systèmes différentiels, nombre de Castelnuovo et rang des tissus de $\mathbb{C}^n$, Prépublication Université Bordeaux 1, Bordeaux (1993).

[17] J.-L. JOLY, G. MÉTIVIER and J. RAUCH, Resonant one dimensional non-linear geometric optics, Journal Funct. Analysis **114** (1993), 106–231.

[18] L. LEWIN, Polylogarithms and Associated Functions, North-Holland, New-York (1981).

[19] B. MALGRANGE, Séminaire Opérateurs Différentiels, Prépublication Institut Fourier, Grenoble (1975).

[20] I. NAKAI, Topology of complex webs of codimension one and geometry of projective space curves, Topology **26** (1987), 475–504.

[21] F. PHAM, Singularités des systèmes de Gauss-Manin, Progress in Math. Vol. 2, Birkhäuser, Boston (1980).

[22] H. POINCARÉ, Sur les surfaces de translation et les fonctions abéliennes, Bull. Soc. Math. France **29** (1901), 61–86.

[23] A. TRESSE, Sur les invariants différentiels des groupes continus de transformations, Acta Math. **18** (1894), 1–88.

Address of author:

Centre de Recherche en Mathématiques de Bordeaux
Université Bordeaux I et C.N.R.S.
351, cours de la Libération
33405 TALENCE Cedex
France
Email: henaut@cribx1.u-bordeaux.fr

GLOBALIZATION OF ADMISSIBLE DEFORMATIONS

Theo de Jong

1 Introduction

Consider a diagram:

$$(\Sigma, 0) \lhook\joinrel\longrightarrow (X, 0)$$

of germs of analytic spaces, with $\Sigma \hookrightarrow \mathrm{Sing}(X)$. Such a diagram we call admissible. In [2] the functor of admissible deformations $\mathrm{Def}(\Sigma, X)$ is introduced. The values on a space $(S, 0)$ consist of isomorphism classes of diagrams:

$$
\begin{array}{ccc}
\Sigma_{S,0} & \lhook\joinrel\longrightarrow & X_{S,0} \\
{\scriptstyle\text{flat}}\downarrow & & \downarrow{\scriptstyle\text{flat}} \\
(S, 0) & =\!\!=\!\!= & (S, 0)
\end{array}
$$

such that $\Sigma_{S,0} \hookrightarrow X_{S,0}$ is again admissible, with the obvious specialization property. In case of a germ of a holomorphic function:

$$(f, 0) : (\mathbb{C}^n, 0) \longrightarrow (\mathbb{C}, 0),$$

with $(\Sigma, 0) \subset (\mathrm{Sing}(f), 0)$ one similarly has a deformation functor $\mathrm{Def}(\Sigma, f)$, and a smooth forgetful morphism

$$\mathrm{Def}(\Sigma, f) \longrightarrow \mathrm{Def}(\Sigma, X),$$

where X is defined by $f = 0$. For details we refer to [2].

In this article we prove a globalization property of admissible deformations in case $(\Sigma, 0)$ is a reduced curve singularity and $(f, 0)$ (or $(X, 0)$) has generically A_∞ singularities along $(\Sigma, 0)$. By using the main result of [3] we deduce a globalization property of deformations of normal surface singularities as well. In case of a *smoothing* of a normal surface singularity the existence of a globalization has been proved by Looijenga [4]. The motivation for proving such a globalization result comes from the fact that one can prove theorems about deformations of singularities using global methods. For example, in a forthcoming paper of Steenbrink [8] the globalization is used to prove a semicontinuity property of the spectrum for germs of holomorphic functions with a one-dimensional singular locus Σ and generically A_∞ singularities along Σ.

Acknowledgment The author thanks J. H. M. Steenbrink for comments. The author is supported by a stipendium of the E. C. (SCIENCE project).

2 Compactification

The important result we need is the following Theorem of Pellikaan:

Theorem 2-1. ([6], 5.2, 7.1,7.4) *Suppose that* $(f,0) : (\mathbb{C}^n,0) \hookrightarrow (\mathbb{C},0)$ *has a one dimensional singular locus* $(\Sigma,0)$ *and generically* A_∞ *along* $(\Sigma,0)$. *Let* I *be the defining ideal of* $(\Sigma,0)$ *in* $(\mathbb{C},0)$. *Then* f *is finitely* I-*determined, i.e., for any holomorphic function* g *with* $f - g \in m^k \cap \int I$ *for* $k \gg 0$, *we have that* g *is right equivalent to* f. *Moreover* f *is right equivalent to a polynomial.*

Theorem 2-2. *Under the hypothesis of Theorem 2-1, there exists an*

$$F \in H^0\big(\mathbb{P}^n, O(l)\big), \qquad \text{for } l \gg 0,$$

such that

1. $(F,0) \sim (f,0)$ *(right equivalent).*

2. *The reduced singular locus* Σ *of* F *is a completion of* $(\Sigma,0)$.

3. F *has only smooth points,* A_∞ *or* D_∞ *singularities off 0.*

PROOF. As reduced one dimensional singularity, $(\Sigma,0)$ is algebraic. Hence we can complete $(\Sigma,0)$ to a curve Σ, smooth off 0. Let $\overline{I}$ be the homogeneous ideal of Σ. So I is the stalk at 0 of $\overline{I}$. We have that $f \in \int I \, (= I^{(2)})$. Pick a $k \gg 0$ such that f is k-determined with respect to I. This is possible after Theorem 2-1. Consider the sheaf $\mathcal{J} := m_0^k \cap \overline{I}^{(2)}$ on $\mathbb{P}^n$ (here m_0 is the maximal ideal of the point 0.) Choose $l \gg 0$ such that $\mathcal{J}(l)$ is generated by global sections. Again by Theorem 2-1, we may assume that f is a polynomial of degree $l \geq k$. Consider the homogenization $\overline{f}$ of f and the linear system

$$V := \overline{f} + H^0\big(\mathbb{P}^n, \mathcal{J}(l)\big) \subset H^0\big(\mathbb{P}^n, \overline{I}^{(2)}(l)\big).$$

By choosing l to be a multiple of the degrees of generators of $\overline{I}$ we see that the base locus of the linear system V is Σ. Hence by the Theorem of Bertini a generic element of V is smooth outside Σ. To prove that a generic element of V only has A_∞ or D_∞ singularities off 0, we cover $\mathbb{P}^n - \{0\}$ by open affines

$$\mathbb{P}^n - \{0\} = \cup_{i=1}^s U_i.$$

Because Σ is smooth off zero we may assume that locally (i.e., in each U_i) there exist coordinates $x_1, \ldots, x_n$ such that Σ is given by the ideal $(x_2, \ldots, x_n)$. Hence for every $F \in V$ we can write (locally)

$$F = \sum_{i,j=2}^n h_{ij}(x_1, \ldots, x_n) x_i x_j.$$

Because $\mathcal{J}(l)$ is generated by global sections, we can conclude that for generic $F \in V$ det $h_{ij}(x_1, 0, \ldots, 0)$ has only zeroes of multiplicity one. This means exactly that F has only A_∞ or D_∞ singularities off 0, see [7]. $\square$

Corollary 2-3. Let $(X,0) \subset (\mathbb{C}^n, 0)$ be a hypersurface singularity with reduced one dimensional singular locus $(\Sigma, 0)$ and generically A_∞ along $(\Sigma, 0)$. Then there exists a hypersurface $Y \subset \mathbb{P}^n$ such that

1. $(Y,0) \simeq (X,0)$.

2. $\mathrm{Sing}(Y)_{\mathrm{red}}$ is a completion of $(\Sigma, 0)$.

3. Y has only smooth points, A_∞ or D_∞ singularities off 0.

The proof is immediate from Theorem 2-2.

Remark 2-4. *In general one has to allow D_∞ singularities in the compactification. This happens already for the D_∞ singularity itself. A formula for the total "virtual number of D_∞ points" on a compactification is given in [1].*

3 Globalization of deformations

We consider a function $(f,0) : (\mathbb{C}^n) \to (\mathbb{C}, 0)$ with reduced one dimensional singular locus $(\Sigma, 0)$ and generically A_∞ singularities along $(\Sigma, 0)$. Let $(S,0)$ be a germ of a smooth one dimensional space and consider a deformation of $(\Sigma, 0)$

$$(\Sigma_{S,0}) \longrightarrow (S,0).$$

Let I_S be the defining ideal of $(\Sigma, 0)$ in $(\mathbb{C}^n, 0) \times (S,0)$. Let s be a parameter for $(S,0)$ and define C to be the cokernel of the injective map:

$$\cdot s : \int I_S \longrightarrow \int I_S,$$

where $\int I_S$ is the relative primitive ideal of I_S. Suppose that $f \in C$ and take a lift $f_S \in \int I_S$. By [6], 7.1, $\int I / \int I \cap J(f)$ is finite dimensional. (Here $J(f)$ is the Jacobi ideal). Because $C \subset \int I$, also $C/C \cap J(f)$ is finite dimensional. We choose functions $f_1, \dots, f_p \in C$ projecting onto a basis of this space. Choose moreover lifts $f_{1S}, \dots, f_{pS} \in \int I_S$ of $f_1, \dots, f_p$.

Proposition 3-1. *Consider the $(p+1)$-parameter admissible deformation*

$$(\mathcal{F}_S, 0) := \left(f_S + \sum s_i f_{iS}, 0 \right)$$

of f. Let $(f + sg, 0)$ be a one-parameter admissible deformation of f. Then there exists a map:

$$h : (\mathbb{C}, 0) \longrightarrow (\mathbb{C}^{p+1}, 0),$$

with $h^*(\mathcal{F}_S, 0) \simeq (f + sg, 0)$.

PROOF. We can write $(f + sg, 0) = (f_S + s\bar{g}, 0)$ and we look at the $(p + 2)$-parameter admissible deformation:

$$(\mathcal{G}, 0) = \left(f_S + \sum s_i f_{iS}, 0 \right).$$

Because $\partial\mathcal{G}/\partial t|_{t=0} \in C$, and $\partial\mathcal{G}/\partial s_i|_{s_i=0}$ generate $C/C \cap J(f)$ we deduce that there are ξ_i and η_i such that

$$\partial\mathcal{G}/\partial t = \sum \xi_i(s, s_i, t, x)\partial\mathcal{G}/\partial x_i + \sum \eta_i(s, s_i, t)\partial\mathcal{G}/\partial s_i.$$

We interprete this as having a vector field Ξ with $\Xi\mathcal{G} = 0$. By the usual argument, see e.g., [5], p. 42 we get a map:

$$\tilde{h} : (\mathbb{C}^{p+1} \times \mathbb{C}, 0) \longrightarrow (\mathbb{C}^{p+1}, 0),$$

satisfying $\tilde{h}^*(\mathcal{F}_S, 0) \simeq \mathcal{G}$. Restricting $\tilde{h}$ to $t = s$ and $s_i = 0$ we get a map $h : (\mathbb{C}, 0) \to (\mathbb{C}^{p+1}, 0)$ satisfying $h^*(\mathcal{F}_S, 0) \simeq (f + sg, 0)$. $\square$

Theorem 3-2. *Let $(f, 0) : (\mathbb{C}^n, 0) \to (\mathbb{C}, 0)$ be as in Theorem 2-1. Consider a one-parameter admissible deformation f_S of f. Then f_S can be globalized, i.e., there exist F_S homogeneous with $(f_S, 0) \sim (F_S, 0)$ as admissible deformations, such that F_S is trivial off 0.*

PROOF. By Theorem 2-2 we can globalize f to F such that F has only A_∞ or D_∞ singularities off 0. We have an induced deformation $(\Sigma_{S,0}) \to (S, 0)$ which can be globalized ([9], 4.3). Hence we have the ideal sheaf $\overline{I}_S$ of $\Sigma_S \subset \mathbb{P} \times S$. By Proposition 3-1 we only have to show that we can lift f and f_i to polynomials f_S and f_{iS}. We define $\overline{C}$ by the exact sequence of sheaves:

$$0 \longrightarrow \int \overline{I}_S \longrightarrow \int \overline{I}_S \longrightarrow \overline{C} \longrightarrow 0.$$

Because $H^1(\int \overline{I}_S(l)) = 0$ for $l \gg 0$, the existence of those lifts follows, if one takes f and f_i to be polynomials of sufficiently high degree. The triviality off 0 follows from the fact that the A_∞ or D_∞ singularities are rigid for the admissible deformation theory. $\square$

Corollary 3-3. For a germ of a hypersurface singularity $(X, 0)$ with one dimensional reduced singular locus $(\Sigma, 0)$ and generically A_∞ singularities along $(\Sigma, 0)$ we have a globalization property, i.e., for any one-parameter admissible deformation $X_{S,0} \to (S, 0)$ there exist $Y_S \subset \mathbb{P}^n \times S$ such that $Y_{S,0} \simeq X_{S,0}$ as admissible deformation and $Y_S \to S$ trivial off 0.

The proof is immediate from Theorem 2-2.

Corollary 3-4. Let $(Z, 0)$ be a normal surface singularity, $Z_{S,0} \to (S, 0)$ be a one-parameter deformation. Then there exist a compactification $\overline{Z}_S \to S$, with $\overline{Z}_{S,0} \simeq Z_{S,0}$ as deformations of $(Z, 0)$. Moreover the only singular point of the Z_0 (the fibre over $0 \in S$) is 0.

PROOF. Consider a generic projection $(Z,0) \to (X,0) \subset (\mathbb{C}^3,0)$. As the image of a general projection of a normal surface singularity to $(\mathbb{C}^3,0)$ will have a one dimensional singular locus and generically A_∞ singularities, $(X,0)$ satisfies the conditions of Corollary 3-4. By [3] 1.3 and 1.4, $\mathrm{Def}(Z \to X)$ is naturally equivalent to $\mathrm{Def}(\Sigma,X)$, and $\mathrm{Def}(Z \to X) \to \mathrm{Def}(X)$ is smooth. Hence we can take $\overline{Z}_S \to S$ to be the normalization of $Y_S \to S$ of Corollary 3-3 . The fact that Z_0 is smooth off 0 follows because A_∞ and D_∞ have smooth normalization. $\qquad\square$

References

[1] Jong, A. J. de, Jong, T. de; The Virtual Number of D_∞-Points II, Topology 29, 185–188, (1990).

[2] Jong, T. de, Straten, D. van; A Deformation Theory for Non-Isolated Singularities, Abh. Math. Sem. Hamburg 60, 177–208, (1990).

[3] Jong, T. de, Straten, D. van; Deformations of the Normalization of Hypersurfaces Math. Ann. 288, 527–547 (1990).

[4] Looijenga, E.; Riemann-Roch and Smoothings of Singularities, Topology 25, 293–302, (1986).

[5] Martinet, J.; Singularities of Smooth Functions and Maps. London Math. Soc. Lecture Note Series 58, Cambridge University Press (1983).

[6] Pellikaan, R.; Finite Determinacy of Functions with Non-Isolated Singularities, Proc. London. Math. Soc. (3) 57, 357–382 (1988).

[7] Siersma, D.; Isolated Line Singularities, In: Singularities at Arcata, Proc. Sym. Pure Math. 40 (2), 485–496 (1983).

[8] Steenbrink, J.H.M; In Preparation.

[9] Wahl, J.; Smoothings of Normal Surface Singularities, Topology 20, 219–246 (1981).

Address of author:

Afdeling Wiskunde,
Toernooiveld 1 6525 ED NIJMEGEN,
Netherlands
Email: tdejong@sci.kun.nl

CARACTÉRISATION GÉOMÉTRIQUE DE L'EXISTENCE DU POLYNÔME DE BERNSTEIN RELATIF

J. Briançon and Ph. Maisonobe

Introduction

Dans cet exposé, nous nous proposons de discuter de l'existence du polynôme de Bernstein relatif pour une famille de fonctions holomorphes, et de donner un critère géométrique de cette existence.

Soit F un germe à l'origine de $\mathbb{C}^n \times \mathbb{C}$ de fonction holomorphe nulle en 0; nous notons $\mathcal{D}_{\mathbb{C}^n \times \mathbb{C}}$ le faisceau des opérateurs différentiels linéaires sur $\mathbb{C}^n \times \mathbb{C}$, $\mathcal{D}_{\mathbb{C}^n \times \mathbb{C}/\mathbb{C}}$ le sous faisceau des opérateurs différentiels relatifs à la projection canonique $\pi_2 : \mathbb{C}^n \times \mathbb{C} \to \mathbb{C}$; on dit que F admet un polynôme de Bernstein relatif s'il existe un polynôme $B(s) \in \mathbb{C}[s]$ non nul et un germe d'opérateur différentiel relatif $P = P(x, t, \partial/\partial x, s) \in \mathcal{D}_{\mathbb{C}^n \times \mathbb{C}/\mathbb{C},0}[s]$ polynomial en s vérifiant la relation fonctionnelle:

$$B(s)F^s = PF^{s+1}.$$

Nous démontrons que l'existence de ce polynôme équivaut à la condition géométrique suivante: $\mathbb{C}^n \times 0$ n'est pas limite d'une suite d'hyperplans tangents aux surfaces de niveau de F. Dans le cas particulier où $F_{|\mathbb{C}^n \times 0}$ a un point critique isolé en zéro, les conditions précédentes sont aussi équivalentes à: $F^{-1}(0)$ est une famille d'hypersurfaces à nombre de Milnor constant le long d'une section de π_2.

Introduisons le $\mathcal{D}_{\mathbb{C}^n \times \mathbb{C}}$ module $\mathcal{O}_{\mathbb{C}^n \times \mathbb{C}}[1/F]$; nous connaissons sa variété caractéristique: si C_F est le conormal relatif à l'application F, c'est la réunion de $W_{0,F} = C_F \cap F^{-1}(0)$ et de la section nulle du fibré cotangent. La condition géométrique ci-dessus se traduit donc par: le covecteur $(0, \ldots, 0, 1)$ n'appartient pas à la variété caractéristique de $\mathcal{O}_{\mathbb{C}^n \times \mathbb{C}}[1/F]$ (au dessus de l'origine). Par ailleurs, on voit facilement que l'existence du polynôme de Bernstein relatif implique que $\mathcal{O}_{\mathbb{C}^n \times \mathbb{C}}[1/F]$ relativement cohérent (c'est à dire cohérent sur $\mathcal{D}_{\mathbb{C}^n \times \mathbb{C}/\mathbb{C}}$) au voisinage de 0. Notre équivalence n'est donc qu'un cas particulier du théorème suivant:

Théorème 0-5. *soit M un $\mathcal{D}_{\mathbb{C}^n \times \mathbb{C}}$ module holonome régulier au voisinage de l'origine; les conditions suivantes sont équivalentes:*

1. *M est relativement cohérent au voisinage de l'origine.*

2. *$\mathbb{C}^n \times 0$ est non caractéristique pour M en 0 (c'est à dire que l'intersection du conormal à $\mathbb{C}^n \times 0$ avec la variété caractéristique de M est incluse dans la section nulle).*

Progress in Mathematics, Vol. 134
© 1996 Birkhäuser Verlag Basel/Switzerland

Tous les résultats résumés ci-dessus ont été démontrés avec Y. Laurent dans [6]; nous les exposons de nouveau dans cet article avec quelques compléments et quelques exemples supplémentaires sur le polynôme de Bernstein. De plus, nous donnons des preuves directes du théorème cité, dans certains cas particuliers, preuves qui nous semblent très instructives.

1   Polynôme de Bernstein relatif

1.1   Polynôme de Bernstein associé à une singularité d'hypersurface

X désignera un voisinage de l'origine de $\mathbb{C}^n$, $x = (x_1, \ldots, x_n)$ un point de X. $\mathcal{O}_X$ désignera le faisceau des fonctions holomorphes sur X, $\mathcal{O}_{X,0}$ son germe à l'origine. $\mathcal{D}_X$ le faisceau des opérateurs différentiels sur X à coefficients dans $\mathcal{O}_X$, $\mathcal{D}_{X,0}$ son germe à l'origine. $\mathcal{D}_X[s]$ le faisceau des opérateurs différentiels sur X à coefficients dans $\mathcal{O}_X[s]$, $\mathcal{D}_{X,0}[s]$ son germe à l'origine. T^*X le fibré cotangent à X, $(x, \xi) = (x_1, \ldots, x_n, \xi_1, \ldots, \xi_n)$ un point de T^*X.

Le cas absolu

Soit $f \in \mathcal{O}_{X,0}$, $\mathcal{O}[1/f, s]f^s$ est un $\mathcal{D}_{X,0}[s]$ module pour les actions évidentes.

Théorème 1-1. (Bernstein [1], Björk, Kashiwara [13]) *L'idéal de $\mathbb{C}[s]$:*

$$I = \left\{ e(s) \in \mathbb{C}[s] \; ; \; \exists P \in \mathcal{D}_{X,0}[s] : e(s)f^s = Pf^{s+1} \right\}$$

n'est pas l'idéal $\{0\}$. b_f le générateur unitaire de cet idéal est appelé le polynôme de Bernstein de f.

Faisons quelques remarques d'usage. Si $f(0) \neq 0$, $b_f = 1$. Si $f(0) = 0$ et si 0 n'est pas un point critique de f, alors $b_f = s+1$. Nous donnerons au Section 2 une démonstration de la réciproque. Le polynôme de Bernstein ne dépend que du germe de l'hypersurface $f^{-1}(0)$. Lorsque $f(0) = 0$, b_f est divisible par $s+1$ et on note: $b_f(s) = (s+1)\tilde{b}_f$. Son succès dans la théorie des singularités provient notamment des théorèmes de Malgrange:

Théorème 1-2. (Malgrange [23], [24]) *L'existence d'un entier l tel que $\alpha + l$ est racine de $\tilde{b}_f$ équivaut à l'existence d'un entier naturel k tel que $e^{-2i\pi\alpha}$ est valeur propre de la monodromie de $H^k\big(f^{-1}(t), \mathbb{C}\big)$, $f^{-1}(t)$ désignant la fibre de Milnor de f.*

Il faut noter que sous des versions généralisées, le polynôme de Bernstein est devenu un objet fondamental de la théorie des $\mathcal{D}$ modules ([24], [15], [28], [20]).

Polynôme de Bernstein et problèmes de déformations

Soit $t = (t_1, \ldots, t_p) \in T$ un voisinage de l'origine de $\mathbb{C}^p$, $F(x,t)$ un germe de fonction analytique à l'origine de $\mathbb{C}^n \times \mathbb{C}^p$ tel que $F(x,0) = f(x)$. On notera

$F(x,t) = f_t(x)$. Le problème général est celui de la variation de b_{f_t} en fonction de t. Nous supposerons dans cette sous section que f est à singularité isolée à l'origine. Il résulte des travaux de Le Dung Trang et Ramanujan que (pour $n \neq 3$) si la déformation est à nombre de Milnor constant le long de $\{0\} \times T$, la classe de conjugaison de la monodromie locale est constante. Des théorèmes de Malgrange, on déduit que les racines de b_{f_t} sont constantes modulo $\mathbb{Z}$. Mais b_{f_t} n'est pas constant comme le montrent les exemples les plus simples ([10], [4]). C'est le cas de la déformation:

$$F(x,y,t) = x^5 + y^5 + tx^3y^3,$$

du germe $f(x,y) = x^5 + y^5$ de $\mathbb{C}^2$. Le nombre de Milnor de f_t est égal à 16, mais on a:

$$b_f\left(-\frac{8}{5}\right) = 0 \quad \text{et} \quad b_{f_t}\left(-\frac{8}{5}\right) \neq 0 \quad \text{si} \quad t \neq 0.$$

Mis à part l'invariance du type analytique, on ne sait pas quel autre critère permet d'affirmer que le polynôme de Bernstein est constant. Considérons un germe $f(x)$ de singularité isolée de $\mathbb{C}^n$ et $g(y)$ de $\mathbb{C}^m$, si $f(x)$ est quasi homogène, la déformation

$$F(x,y,t) = f(x) + (1-t)g(y)$$

est à polynôme de Bernstein constant (car à type analytique constant). Mais si l'on ne suppose pas f ou g quasi homogène, on ne sait pas conclure. Pourtant cette déformation garde constant l'idéal jacobien ([2]).

Il y a néanmoins quelques résultats positifs dans le cas d'une déformation à nombre de Milnor constant. F. Geandier ([8], [9]) montre que le polynôme de Bernstein de f_t est constant pour t petit non nul et que

$$b_f \quad \text{divise} \quad b_{f_t}(s)b_{f_t}(s+1)\ldots b_{f_t}(s+k)$$

pour un certain entier k. Elle retrouve de façon algébrique des résultats de semi continuité de certaines racines de b_{f_t}. A. Varchenko montre en utilisant la théorie de Hodge mixte que la plus grande racine de b_{f_t} est constante (conjecture d'Arnold et Malgrange, [30], [31]).

Pour des déformations non nécessairement à type topologique constant, on peut ([5]), en considérant une situation semi locale, stratifier par la condition "b_{f_t} constant" l'espace des paramètres par des strates analytiquement constructibles.

1.2 Polynômes de Bernstein relatifs et génériques

Comme dans la section 1.1, nous considérons T un voisinage de l'origine de $\mathbb{C}^p$, $t = (t_1, \ldots, t_p)$ un point de T. Appelons π_2 la deuxième projection $\pi_2 : X \times T \to T$. Considérons F un germe à l'origine de $X \times T$ de fonction holomorphe. H désignera l'hypersurface d'équation $F = 0$.

Notons par $\mathcal{D}_{X \times T/T}$ l'anneau des opérateurs différentiels relatifs à π_2 et par $\mathcal{D}_{X \times T/T,0}$ sa fibre à l'origine. C'est le sous anneau de $\mathcal{D}_{X \times T}$ formé des opérateurs ne contenant pas de dérivation par rapport aux t_i (ou, intrinsèquement, nuls sur $\pi_2^{-1}(\mathcal{O}_T)$); un opérateur différentiel relatif s'écrit donc:

$$\sum_{\text{fini}} c_\alpha(x,t) \left(\frac{\partial}{\partial x} \right)^\alpha \quad \text{où} \quad c_\alpha(x,t) \in \mathcal{O}_{X \times T}.$$

Par tensorisation par $\mathbb{C}[s]$ au dessus de $\mathbb{C}$, on définit $\mathcal{D}_{X \times T/T}[s]$ et sa fibre à l'origine $\mathcal{D}_{X \times T/T,0}[s]$. On peut alors considérer des équations fonctionnelles relatives et définir:

Définition 1-3. *On dira que F admet un polynôme de Bernstein relatif si l'idéal de $\mathbb{C}[s]$:*

$$I = \left\{ e(s) \in \mathbb{C}[s] \ ; \ \exists P \in \mathcal{D}_{X \times T/T,0}[s] : e(s)F^s = PF^{s+1} \right\}$$

est non réduit à zéro. B_F, le polynôme unitaire engendrant cet idéal, est appelé polynôme de Bernstein relatif.

Définition 1-4. *On dira que F admet un polynôme de Bernstein générique si l'idéal de $\mathbb{C}[s]$:*

$$I = \left\{ e(s) \in \mathbb{C}[s] \ ; \ \exists P \in \mathcal{D}_{X \times T/T,0}[s], \ \exists h \in \mathcal{O}_{T,0} - \{0\} : h(t)e(s)F^s = PF^{s+1} \right\}$$

est non réduit à zéro. $B_{\text{gén},F}$, le polynôme unitaire engendrant cet idéal, est appelé polynôme de Bernstein générique.

Bien sûr si B_F existe, pour tout t: b_{f_t} divise B_F, $B_{\text{gén},F}$ existe et divise B_F. Dans sa thèse [8], F. Geandier montre l'existence de $B_{\text{gén},F}$ lorsque F est une déformation à un paramètre d'une singularité isolée. Elle montre l'existence de B_F lorsque de plus la déformation est à nombre de Milnor constant. On peut à l'aide des méthodes de [4] calculer les premiers exemples dans cette situation. Si F est une déformation à nombre de Milnor constant d'une singularité isolée de courbe plane $(n = 2)$, f semi quasi homogène: $B_F = \text{ppcm}(b_{f_t}) = b_f$. Mais il est faux en géneral que B_F soit le plus petit commun multiple des b_{f_t}, comme le montre l'exemple ([8]):

$$F(x,y,z,t) = x^8 + y^9 + z^{11} + tx^3y^3z^4.$$

Dans [5] est montré que $B_{\text{gén},F}$ existe si F est une déformation d'une singularité isolée et qu'il coincide avec le polynôme de Bernstein (semi local) de f_t pour t générique.

1.3   Déformation non caractéristique

Désormais, T sera un voisinage de l'origine de $\mathbb{C}$ $(p = 1)$. $T^*(X \times T)$ désignera le fibré cotangent à $X \times T$ et

$$(x,t;\xi,\eta) = (x,t;\xi_1,\ldots,\xi_n,\eta)$$

un point de ce fibré cotangent, π la projection canonique du fibré cotangent sur $X \times T$. Si Z est une sous variété lisse de $X \times T$, on notera $T_Z^*(X \times T)$ le conormal à Z, c'est à dire l'ensemble des formes linéaires qui s'annulent sur le fibré tangent à Z. Par exemple $T_{X \times T}^*(X \times T)$ est la section nulle du fibré cotangent. On peut supposer (quitte à diminuer $X \times T$ au voisinage de $(0,0)$) que les points critiques de l'application F sont contenus dans $H = F^{-1}(0)$. C_F le conormal relatif au morphisme F est l'adhérence dans $T^*(X \times T)$ de

$$\big\{(x,t;\xi,\eta) \ ; \ (\xi,\eta) \in T^*_{F^{-1}(F(x,t))}(X \times T), \text{ et } F(x,t) \neq 0\big\}.$$

C_F est donc l'adhérence de

$$\left\{ \left(x,t; \lambda\frac{\partial F}{\partial x_1}, \ldots, \lambda\frac{\partial F}{\partial x_n}, \lambda\frac{\partial F}{\partial t}\right) \ ; \ \lambda \in \mathbb{C}, \text{ et } F(x,t) \neq 0\right\}.$$

C_F est un sous ensemble analytique de $T^*(X \times T)$, conique par rapport aux fibres de la projection π. Au dessus de $(0,0)$, un système de générateurs des germes de fonctions nulles sur C_F est obtenu par

$$\big\{g(x,t,\xi,\eta) \in \mathcal{O}_{X \times T}[\xi,\eta]\big\},$$

homogène en ξ et η tels que

$$g\left(x,t,\frac{\partial F}{\partial x_1}, \ldots, \frac{\partial F}{\partial x_n}, \frac{\partial F}{\partial t}\right) = 0.$$

C_F est une variété involutive de $T^*(X \times T)$. Pour y non nul, il est clair que $C_F \cap \pi^{-1}\big(F^{-1}(y)\big) = T^*_{F^{-1}(y)}(X \times T)$ est une variété lagrangienne de $T^*(X \times T)$. Dans [13] est prouvé que c'est aussi le cas pour $y = 0$.

Notation 1-5. C_F désigne le conormal relatif au morphisme F

Notation 1-6. $W_{0,F} = C_F \cap \pi^{-1}\big(F^{-1}(0)\big)$

Définition 1-7. *On dira que la déformation F est non caractéristique si au voisinage de $(0,0)$:*

$$C_F \cap T^*_{X \times 0}(X \times T) \subset T^*_{X \times T}(X \times T).$$

Proposition 1-8. *La déformation F est non caractéristique si et seulement si une des deux conditions suivantes est vérifiée:*

1. *$(0, \ldots, 0; 0, \ldots, 0; 1)$ n'appartient pas à C_F.*

2. *$\partial F/\partial t$ est entier sur l'idéal jacobien relatif de $\mathcal{O}_{X \times T, 0}$:*

$$\left(\frac{\partial F}{\partial x_1}, \ldots, \frac{\partial F}{\partial x_n}\right).$$

DÉMOSTRATION. La preuve est facile: l'équivalence de 1 et 2 se déduit des équations de C_F données précédemment.

La terminologie non caractéristique provient de la théorie des équations aux dérivées partielles, voir par exemple [16]. Elle s'expliquera un peu plus loin lorsqu'on montrera le lien de ces définitions avec la théorie des $\mathcal{D}$ modules. Dans la preuve du théorème 3.3 de [22], D. T. Lê et Z. Mebkhout donnent un critère topologique pour qu'une déformation soit non caractéristique: Soit j l'inclusion fermée de $H = F^{-1}(0)$ dans $X \times T$, $j_!\mathbb{C}_H$ le prolongement par zéro du faisceau constant sur H de fibre $\mathbb{C}$.

Proposition 1-9. ([22]) *La déformation F est non caractéristique si et seulement si pour tout $g \in \mathcal{O}_{X \times T,0}$ tel que $dg(0,0) = (0,\dots,0,1)$:*

$$\Phi_g(j_!\mathbb{C}_H) = 0.$$

Cette condition s'énonce sous la forme: "g ne crée pas de cycles évanescents sur $j_!\mathbb{C}_H$". Le foncteur Φ_g est défini dans [7]: $\Phi_g(j_!\mathbb{C}_H)$ est un complexe de faisceaux d'espaces vectoriels sur H. Cette condition d'annulation exprime que le morphisme canonique entre $\mathbb{C}_H$ restreint à $g^{-1}(0)$ et $R\Psi_g(\mathbb{C}_H)$ est un quasi isomorphisme. Rappelons que la cohomologie de $R\Psi_g(\mathbb{C}_H)$ en un point x de $g^{-1}(0)$ est la cohomologie de la fibre de Milnor locale en x de la restriction de g à H.

Dans le cas particulier où F est une déformation de f à singularité isolée, la proposition 1-8 s'exprime:

Proposition 1-10. *Une déformation F d'une singularité isolée f est non caractéristique si et seulement si elle est à nombre de Milnor constant le long d'une section de π_2.*

DÉMOSTRATION. Supposons la déformation F non caractéristique; comme F est toujours entier sur l'idéal de ses dérivées, il résulte de la condition 2 de la proposition 1-8 que F s'annule sur les zéros de l'idéal jacobien relatif. Il résulte alors du travail de F. Lazzeri [21] que la déformation F est à nombre de Milnor constant le long d'une section de π_2. Inversement si F est une déformation à nombre de Milnor constant le long d'une section de π_2, il est prouvé par B. Teissier dans [28] que $\partial F/\partial t$ est entier sur l'idéal jacobien relatif.

1.4 L'existence du polynôme de Bernstein pour une déformation non caractéristique

M. Kashiwara montre dans [13] que $\mathcal{D}_{X \times T}[s]F^s$ est un $\mathcal{D}_{X \times T}$ module de type fini de variété caractéristique C_F. On en déduit facilement que $\mathcal{D}_{X \times T,0}F^s$ est un germe de $\mathcal{D}_{X \times T,0}$ module de même variété caractéristique. Ainsi, si $(x_0, t_0, \xi_0, \eta_0)$ n'appartient pas à C_F, il existe un opérateur différentiel P de $\mathcal{D}_{X \times T,0}$ annulant F^s tel que son symbole principal $\sigma(P)$ ne s'annule pas en

$(x_0, t_0, \xi_0, \eta_0)$: Si

$$P = \sum_{\alpha,\beta} p_{\alpha,\beta}(x,t)\left(\frac{\partial}{\partial x_1}\right)^{\alpha_1} \cdots \left(\frac{\partial}{\partial x_n}\right)^{\alpha_n}\left(\frac{\partial}{\partial t}\right)^{\beta}, \qquad \alpha = (\alpha_1, \ldots, \alpha_n),$$

alors $PF^s = 0$ et $\sigma(P)(x_0, t_0, \xi_0, \eta_0) \neq 0$ où

$$\sigma(P) = \sum_{|\alpha|+\beta=d(P)} p_{\alpha,\beta}(x,t)\xi^\alpha \eta^\beta, \quad \text{et} \quad d(P) = \sup\{|\alpha|+\beta : p_{\alpha,\beta} \neq 0\}.$$

Ainsi si F est non caractéristique, $(0, \ldots, 0; 0, \ldots, 0, 1)$ n'appartient pas à C_F. On obtient donc:

Proposition 1-11. *Si F est une déformation non caractéristique, il existe un opérateur $Q \in \mathcal{D}_{X \times T, 0}$ de la forme*

$$Q = \left(\frac{\partial}{\partial t}\right)^k + A_1\left(x, t, \frac{\partial}{\partial x}\right)\left(\frac{\partial}{\partial t}\right)^{k-1} + \cdots + A_n\left(x, t, \frac{\partial}{\partial x}\right),$$

où $A_i \in \mathcal{D}_{X \times T/T, 0}$ de degré $\leq i$ tel que $QF^s = 0$.

Considérons l'équation fonctionnelle réalisant le polynôme de Bernstein absolu:

$$b_F(s)F^s = PF^{s+1}, \qquad P \in \mathcal{D}_{X \times T, 0}[s].$$

Itérons cette relation k fois, il vient:

$$b_F(s)b_F(s+1)\ldots b_F(s+k-1)F^s = TF^{s+k},$$

où $T \in \mathcal{D}_{X \times T, 0}[s]$. Divisons T par Q. On peut écrire: $T = AQ + R$ où $R \in \mathcal{D}_{X \times T, 0}[s]$ et où R a son degré de dérivation par rapport à $\partial/\partial t$ inférieur ou égal à $k - 1$. Ainsi

$$b_F(s)b_F(s+1)\ldots b_F(s+k-1)F^s = RF^{s+k}$$

avec la condition de degré sur R ci-dessus. Nous allons maintenant nous débarrasser complétement de $\partial/\partial t$. Ecrivons:

$$R = R_1\left(\frac{\partial}{\partial t}\right)^{k-1} + \ldots + R_{k-1} \quad \text{où} \quad R_i \in \mathcal{D}_{X \times T/T, 0}[s].$$

Comme pour i plus grand que 1:

$$\left(\frac{\partial}{\partial t}\right)^{k-i} F^{s+k} \in \mathcal{O}_{X \times T, 0}[s]F^{s+1},$$

on en déduit:

$$RF^{s+k} \in \mathcal{D}_{X \times T/T, 0}[s]F^{s+1},$$

et donc:

$$b_F(s)b_F(s+1)\ldots b_F(s+k-1)F^s \in \mathcal{D}_{X\times T/T,0}[s]F^{s+1}.$$

Cela prouve la proposition suivante:

Proposition 1-12. *Si la déformation F est non caractéristique, F admet un polynôme de Bernstein relatif*

1.5   Caractérisation de l'existence du polynôme de Bernstein relatif

Cette caractérisation repose sur le théorème 2-16 de la section suivante

Faisons d'abord quelques rappels sur le $\mathcal{D}_{X\times T,0}$ module $\mathcal{O}_{X\times T,0}[1/F]$. Sa finitude comme $\mathcal{D}_{X\times T,0}$ module résulte de l'existence du polynôme de Bernstein. En effet, soit

$$b_F(s)F^s = PF^{s+1}$$

l'équation fonctionnelle réalisant le polynôme de Bernstein, soit k_0 un entier tel que pour $k \geq k_0$, $-k$ ne soit pas racine de b_F. On a ainsi pour $k \geq k_0$:

$$\frac{1}{F^{k_0}} \in \mathcal{D}_{X\times T,0}\frac{1}{F^{k_0-1}},$$

d'où, par itération pour $k \geq k_0$:

$$\frac{1}{F^k} \in \mathcal{D}_{X\times T,0}\frac{1}{F^{k_0-1}},$$

soit encore:

$$\mathcal{O}_{X\times T,0}[1/F] = \mathcal{D}_{X\times T,0}\frac{1}{F^{k_0-1}}.$$

Dans [14] est montré que $\mathcal{O}_{X\times T,0}[1/F]$ défini un germe de $\mathcal{D}_{X\times T,0}$ module holonome. Dans [18], [22] et [11] est montré par différentes méthodes que $W_{0,F}\cup T^*_{X\times T}(X\times T)$ est la variété caractéristique de $\mathcal{O}_{X\times T,0}[1/F]$. Enfin, $\mathcal{O}_{X\times T,0}[1/F]$ est un germe de $\mathcal{D}_{X\times T,0}$ module holonome régulier [25]

Supposons maintenant que F admette un polynôme de Bernstein relatif B_F, on montre par la même méthode que plus haut que:

$$\mathcal{O}_{X\times T,0}[1/F] = \mathcal{D}_{X\times T/T,0}\frac{1}{F^{k_1-1}},$$

où k_1 est un entier tel que pour $k \geq k_1$, $B_F(-k) \neq 0$. Résumons cela dans la proposition suivante:

Proposition 1-13. *Si F admet un polynôme de Bernstein relatif, le germe de $\mathcal{D}_{X\times T,0}$ module holonome régulier $\mathcal{O}_{X\times T,0}[1/F]$ est de type fini sur $\mathcal{D}_{X\times T/T,0}$.*

Notamment, toute suite croissante de sous $\mathcal{D}_{X\times T/T,0}$- modules de

$$\mathcal{O}_{X\times T,0}[1/F]$$

stationne, par exemple la suite:

$$M_k = \sum_{i=0}^{k} \mathcal{D}_{X \times T/T,0} \left(\frac{\partial}{\partial t}\right)^i \frac{1}{F^{k_1-1}}.$$

On en déduit qu'il existe un opérateur A de la forme

$$A = \left(\frac{\partial}{\partial t}\right)^p + A_1\left(x,t,\frac{\partial}{\partial x}\right)\left(\frac{\partial}{\partial t}\right)^{p-1} + \cdots + A_p\left(x,t,\frac{\partial}{\partial x}\right),$$

où $A_i \in \mathcal{D}_{X \times T/T,0}$ tels que

$$A\frac{1}{F^{k_1-1}} = 0.$$

Mais puisque aucune hypothèse n'est faite sur le degré par rapport aux dérivations des A_i, une telle équation n'entraîne rien à priori sur la variété de $\mathcal{O}_{X \times T,0}[1/F]$ comme $\mathcal{D}_{X \times T,0}$ module. C'est par contre ce que permet de faire le théorème 2-16 démontré plus loin. En effet, si F admet un polynôme de Bernstein relatif, $\mathcal{O}_{X \times T,0}[1/F]$ vérifie les hypothèses de ce théorème. On en déduit donc que l'hypersurface $t = 0$ est non caractéristique pour la variété caractéristique de $\mathcal{O}_{X \times T,0}[1/F]$. Cette condition s'exprime par:

$$(0,\ldots,0)(0,\ldots,0,1) \notin W_{0,F} \cup T^*_{X \times T}(X \times T).$$

Comme $F(0) = 0$ et $(0,\ldots,0)(0,\ldots,0,1) \notin T^*_{X \times T}(X \times T)$, cette condition équivaut donc à

$$(0,\ldots,0)(0,\ldots,0,1) \notin W_{0,F}.$$

On a donc:

Théorème 1-14. *F admet un polynôme de Bernstein relatif si et seulement si la déformation F est non caractéristique.*

Il résulte alors de la proposition 1-10:

Corollaire 1-15. *Si F est une déformation de f à singularité isolée F admet un polynôme de Bernstein relatif si et seulement si la déformation F est à nombre de Milnor constant le long d'une section de π_2.*

2 $\mathcal{D}_{X \times T}$ Module holonome régulier relativement cohérent

2.1 Définition, notation et exposé du problème

Soit X un voisinage de l'origine de $\mathbb{C}^n$ et T un voisinage de l'origine de $\mathbb{C}$; $x = (x_1,\ldots,x_n)$ désigne un point de X et t un point de T. $T^*(X \times T)$ le fibré cotangent, (x,t,ξ,η) un point de ce fibré cotangent.

Rappelons que $\mathcal{D}_{X \times T/T}$ désigne le faisceau des opérateurs relatifs à la projection $\pi_2 : X \times T \to T$. Ce sont les opérateurs dont l'écriture ne contient pas de dérivation par rapport à t.

Soit M un $\mathcal{D}_{X\times T}$ Module cohérent, on note car(M) sa variété caractéristique. Considérons l'hypersurface $X\times\{0\}$ de $X\times T$. Nous rappelons la définition suivante:

Définition 2-1. $X\times\{0\}$ *est non caractéristique pour* car(M) *si* $T^*_{X\times\{0\}}(X\times T)$ *et* car(M) *ne se coupent pas en dehors de la section nulle.*

Supposons que $X\times\{0\}$ soit non caractéristique pour car(M). Soit u une section de M au voisinage de $(0,0)$. L'hypothèse se traduit par le fait que le covecteur $(0,0,\ldots,0;0,\ldots,0,1)\notin$ car(M). Nous en déduisons (comme dans la preuve de la proposition 1-11) qu'il existe un opérateur A de la forme:

$$\left(\frac{\partial}{\partial t}\right)^p + A_1\left(x,\frac{\partial}{\partial x},t\right)\left(\frac{\partial}{\partial t}\right)^{p-1} + \cdots$$

où $A_i\in\mathcal{D}_{X\times T/T,0}$ de degré au plus égal à i par rapport aux dérivations tel que A vérifie:

$$Au = 0.$$

On en déduit que M est un $\mathcal{D}_{X\times T/T}$ module localement de type fini. De la cohérence de M comme $\mathcal{D}_{X\times T}$ module et de la cohérence de $\mathcal{D}_{X\times T/T}$, on déduit que M est cohérent sur l'anneau $\mathcal{D}_{X\times T/T}$. Isolons cela sous forme de remarque:

Remarque 2-2. *Si M est un $\mathcal{D}_{X\times T}$ module cohérent et si $X\times\{0\}$ est non caractéristique pour* car(M), *M est un $\mathcal{D}_{X\times T/T}$ module cohérent.*

Nous dirons qu'un $\mathcal{D}_{X\times T}$ module est relativement cohérent s'il est cohérent comme $\mathcal{D}_{X\times T/T}$ module.

Exemple 2-3. Soit

$$M = \frac{\mathcal{D}_{X\times T}}{\mathcal{D}_{X\times T}P} \quad\text{où}\quad P = \frac{\partial}{\partial t} - \sum_{i=1}^n\left(\frac{\partial}{\partial x_i}\right)^2.$$

M est relativement cohérent. Sa variété caractéristique est

$$\text{car}(M) = \left\{(x,t,\xi,\eta)\in T^*(X\times T):\sigma(P)(x,t,\xi,\eta)=0\right\}.$$

Mais $X\times\{0\}$ n'est pas non caractéristique pour car(M). Nous remarquerons que M n'est pas holonome.

Exemple 2-4. Soit

$$M = \mathcal{D}_{\mathbb{C}\times\mathbb{C}}\exp\left(\frac{x}{t}\right).$$

Notons $u = exp(x/t)$. Soit I l'idéal à gauche de $\mathcal{D}_{\mathbb{C}\times\mathbb{C}}$ des opérateurs différentiels annulant u:

$$I = \mathcal{D}_{\mathbb{C}\times\mathbb{C}}\left(t\frac{\partial}{\partial x}-1,\frac{\partial}{\partial t}+x\left(\frac{\partial}{\partial x}\right)^2\right).$$

Notons aussi que $t\partial/\partial t + x\partial/\partial x$ appartient à I et que $(t\xi, x\xi + t\eta)$ engendrent l'idéal de $\mathcal{O}_{X\times T}[\xi]$ défini par les symboles principaux des éléments de I. M est donc holonome de variété caractéristique:

$$T^*_{\mathbb{C}^2}\mathbb{C}^2 \cup T^*_{\{0,0\}}\mathbb{C}^2 \cup T^*_{\{t=0\}}\mathbb{C}^2.$$

$X \times \{0\}$ n'est pas non caractéristique pour M, mais $\partial/\partial t + x(\partial/\partial x)^2 \in I$, M est relativement cohérent. Indiquons que M n'est pas un $\mathcal{D}$ module régulier.

2.2 Exemple: $\mathcal{O}[*H]$ où H est une hypersurface lisse

On suppose ici que $H = F^{-1}(0)$ est une hypersurface analytique complexe lisse en $0 \in X \times T$. On note $M = \mathcal{O}[1/F]$. La variété caractéristique de M est la réunion de la section nulle de $T^*(X \times T)$ et du conormal à H, au voisinage de zéro. Si H n'est pas tangente à $X \times \{0\}$, $X \times \{0\}$ est non caractéristique pour M; dans l'autre cas:

Proposition 2-5. *Si H est tangente à $X\times\{0\}$, M n'est pas relativement cohérent.*

DÉMONSTRATION. Tout se passe au voisinage de zéro; quitte à multiplier F par une unité, nous pouvons supposer $F = t - f(x)$ où f ainsi que toutes ses dérivées s'annulent en zéro. Nous laissons de coté le cas trivial $f = 0$. Identifions:

$$\frac{\mathcal{O}_{X\times T}[1/t - f]}{\mathcal{O}_{X\times T}} \simeq \mathcal{D}_{x,t}f^s,$$

en envoyant la classe de $1/t - f$ sur f^s. où les actions sont définies par:

$$\begin{aligned}
tg(s)f^s &= g(s+1)f^{s+1}, \\
\frac{\partial}{\partial t}g(s)f^s &= -\frac{1}{t}(s+1)g(s)f^s, \\
&= -sg(s-1)f^{s-1}.
\end{aligned}$$

Si $\mathcal{O}_{X\times T}[1/t-f]$ était relativement cohérent au voisinage de 0, il existerait un opérateur

$$Q = \left(\frac{\partial}{\partial t}\right)^m + A_1\left(\frac{\partial}{\partial t}\right)^{m-1} + \cdots + A_m$$

annulant f^s, où $A_i \in \mathcal{D}_{X,0}$. On en déduit:

$$(-1)^{m-1}s(s-1)\ldots(s-m+1)f^{s-m} =$$
$$A_mf^s - sA_{m-1}f^{s-1} + \cdots + (-1)^{m-1}s(s-1)\ldots(s-m+2)A_1f^{s-m+1}.$$

Faisons $s = 0$, il vient $A_m(1) = 0$, d'où:

$$\begin{aligned}
A_m &= \sum A_{m,i}\frac{\partial}{\partial x_i} \\
A_mf^s &= \sum sA_{m,i}\frac{\partial f}{\partial x_i}f^{s-1}.
\end{aligned}$$

Ainsi:

$$(-1)^{m-1}(s-1)\ldots(s-m+1)f^{s-m} =$$
$$\left(\sum A_{m,i}\frac{\partial f}{\partial x_i}+(-1)^{m-1}A_{m-1}\right)f^{s-1}+\cdots+(s-1)\ldots(s-m+2)A_1f^{s-m+1}.$$

Faisons $s = 1$, $s = 2$, ..., On obtient:

$$(s - m + 1)f^{s-m} \in \mathcal{D}_{X,0}f^{s-m+1}.$$

Ou encore $(s+1)f^s \in \mathcal{D}_{X,0}f^{s+1}$. Ainsi le polynôme de Bernstein de f diviserait $(s + 1)$. Comme $f(0) = 0$, $b_f(s) = s + 1$. Si l'on sait que cela entraine que $f^{-1}(0)$ est lisse en zéro, on aboutit à une contradiction. Ce résultat attendu semble connu des spécialistes. En l'absence de référence, en voilà une preuve relativement simple.

Proposition 2-6. *Soit* $f : \mathbb{C}^n, 0 \to \mathbb{C}, 0$ *un germe de fonction analytique à l'origine de* $\mathbb{C}^n$, $b_f(s)$ *son polynôme de Bernstein. L'hypersurface* $f^{-1}(0)$ *est lisse si et seulement si* $b_f(s) = s + 1$.

DÉMONSTRATION. Si f est à singularité isolée, on sait que $b_f(s) = (s+1)\tilde{b}(s)$, où $\tilde{b}(s)$ est le polynôme minimal d'un endomorphisme sur un espace de dimension μ ([23], [5]). Le résultat provient alors du fait que si f n'est pas lisse, le nombre de Milnor est strictement positif. Supposons maintenant le lieu singulier de $f^{-1}(0)$ de dimension $n' \geq 1$, $n = n' + n''$, nous pouvons trouver une projection linéaire $\rho : \mathbb{C}^n \to \mathbb{C}^{n'}$ de telle sorte que le lieu singulier relatif soit fini sur $\mathbb{C}^{n'}$ (au voisinage de zéro) et égal au lieu singulier absolu ([29]). D'après [B.G.M.], on sait qu'il existe un polynôme de Bernstein générique qui est le polynôme de Bernstein de la fibre générique; donc $\tilde{b}_{\text{gén}}(s)$ ne peut être égal à 1 et donc $\tilde{b}(s) \neq 1$ puisque $\tilde{b}_{\text{gén}}(s)$ divise $\tilde{b}(s)\ldots\tilde{b}(s + l)$ pour un certain entier ([5]).

2.3   Exemple: $\mathcal{O}[*H]$, où H est une hypersurface à singularité isolée

Soit F une fonction holomorphe sur $X \times T$, telle que $F = 0$ soit une équation de l'hypersurface H. Nous supposerons H à singularité isolée en zéro. Nous allons directement montrer que $\mathcal{O}[1/F]$ n'est pas relativement cohérent.

Proposition 2-7. *Soit* H *une hypersurface à singularité isolée en zéro,* $\mathcal{O}[*H]$ *n'est pas relativement cohérent au voisinage de zéro.*

DÉMONSTRATION. Distinguons deux cas: supposons d'abord que la section de H par l'hyperplan $X \times \{0\}$ ne soit pas à singularité isolée en 0; cela signifie qu'il existe une suite de points $(x^{(p)})_{p\in\mathbb{N}}$ de $V \cap (X \times \{0\})$ telle que H est lisse en $x^{(p)}$, de plan tangent en $x^{(p)}$ égal à $X \times \{0\}$. La proposition 2-5 montre alors que $\mathcal{O}[*H]$ n'est pas relativement cohérent au voisinage de $x^{(p)}$; donc que $\mathcal{O}[*H]$ n'est pas relativement cohérent au voisinage de 0.

Plaçons nous dorénavant dans le second cas: π_2 induit un morphisme fini de la courbe polaire Γ définie par l'idéal $J = (F'_{x_1}, \ldots, F'_{x_n})\mathcal{O}_{X\times T}$ sur T au voisinage de 0. On déduit de là que F ne s'annule sur aucune composante irréductible de Γ (si $\gamma : y \in (\mathbb{C}, 0) \mapsto \gamma(y) \in (\Gamma, 0)$ non constant annulait F, on aurait:

$$\frac{\partial}{\partial y}\big(F \circ \gamma(y)\big) = \sum_{i=1}^{n} \frac{\partial x_i}{\partial y} F'_{x_i} \circ \gamma + \frac{\partial t}{\partial y} F'_t \circ \gamma = \frac{\partial t}{\partial y} F'_t \circ \gamma = 0,$$

donc $F'_t \circ \gamma = 0$ ce qui contredirait l'hypothèse de singularité isolée). D'autre part, de manière classique, on déduit du fait que $\big(F'_{x_1}(x;0),\ldots,F'_{x_n}(x,0)\big)$ forme une suite régulière, que $\mathcal{O}_{X \times T}/J^N$ est sans $\mathcal{O}_T$ torsion au voisinage de 0, donc de dimension pure 1. De ces propriétés, il résulte:

$$\lambda F \in J^N \Rightarrow \lambda \in J^N. \tag{2.1}$$

Supposons alors que $\mathcal{O}[1/F]$ soit $\mathcal{D}_{X \times T/T}$ cohérent au voisinage de 0; la suite des $\big(\mathcal{D}_{X \times T/T}(1/F^k)\big)$ stationne et ainsi, il existe un entier k tel que

$$\mathcal{D}_{X \times T/T}(1/F^k) = \mathcal{O}[1/F]$$

au voisinage de 0. En particulier, il existe un opérateur $A(x,t,\partial/\partial x)$ relatif tel que:

$$\frac{1}{F^{k+1}} = A \frac{1}{F^k 3}.$$

Soit $A = \sum_{|\alpha| \le N} a_\alpha(x,t)(\partial/\partial x)^\alpha$ où N est le degré de A et notons son symbole principal: $\sigma_A(\xi_1,\ldots,\xi_n) = \sum_{|\alpha|=N} a_\alpha(x,t)\xi^\alpha$. La relation $1/F^{k+1} = A(1/F^k)$ s'écrit:

$$\frac{1}{F^{k+1}} = (-1)^N k(k+1)\ldots(k+N-1)\frac{\sigma_A(F'_{x_1},\ldots,F'_{x_n})}{F^{k+N}} + \frac{b(x,t)}{F^{k+N-1}}.$$

Donc pour $N \ge 2$, $\sigma_A(F'_{x_1},\ldots,F'_{x_n}) = \lambda F$ dans $\mathcal{O}_{X \times T}$; d'après (2.1) on obtient $\lambda \in J^N$, donc $\lambda = \sum_{|\alpha|=N} \lambda_\alpha(x,t)(F'_x)^\alpha$ puis:

$$\sum_{|\alpha|=N} \big(a_\alpha(x,t) - \lambda_\alpha(x,t)F\big)(F'_x)^\alpha = 0$$

De nouveau en utilisant la régularité de la suite $(F'_{x_1},\ldots,F'_{x_n})$, cette relation implique:

$$\sum_{|\alpha|=N} \big(a_\alpha(x,t) - \lambda_\alpha(x,t)F\big)\xi^\alpha = \sum_{1 \le i < j \le n} c_{i,j}(x,t,\xi)(\xi_i F'_{x_j} - \xi_j F'_{x_i}).$$

dans les polynômes $\mathcal{O}_{X \times T}[\xi]$, homogènes en ξ. L'opérateur

$$Q = \sum_{0 \le i \le j \le n} c_{i,j}\left(x,t,\frac{\partial}{\partial x}\right)\left(\frac{\partial}{\partial x_i}F'_{x_j} - \frac{\partial}{\partial x_j}F'_{x_i}\right)$$

annule $1/F^k$ et a pour symbole principal $\sum_{|\alpha|=N}\bigl(a_\alpha(x,t) - \lambda_\alpha(x,t)F\bigr)\xi^\alpha$; il s'écrit donc:

$$Q = \sum_{|\alpha|=N} \left(a_\alpha(x,t)\left(\frac{\partial}{\partial x}\right)^\alpha - \lambda_\alpha(x,t)\left(\frac{\partial}{\partial x}\right)^\alpha F\right) + R.$$

où R est un opérateur relatif de degré strictement inférieur à N. Nous obtenons finalement:

$$\frac{1}{F^{k+1}} = A\frac{1}{F^k} = (A - Q)\frac{1}{F^k} =$$
$$\sum_{|\alpha|<N}\left(a_\alpha(x,t)\left(\frac{\partial}{\partial x}\right)^\alpha - R\right)\frac{1}{F^k} + \sum_{|\alpha|=N}\lambda_\alpha(x,t)\left(\frac{\partial}{\partial x}\right)^\alpha\frac{1}{F^{k-1}}.$$

En faisant opérer une dérivation sur $1/F^{k-1}$, on arrive à: $1/F^{k+1} = A'(1/F^k)$ où A' est de degré strictement inférieur à N. Par induction, nous sommes ainsi amenés à $1/F^{k+1} = A(1/F^k)$ où A est de degré 1:

$$A = \sum_{i=1}^n a_i(x,t)\frac{\partial}{\partial x_i} + b(x,t);$$

alors $1/F^{k+1} = -k\sum_{i=1}^n a_i(F'_{x_i}/F^{k+1}) + b/F^k$ soit $1 = -k\sum_{i=1}^n a_i F'_{x_i} + bF$ qui donne $1 = 0$ (en "faisant" $x = t = 0$)!

Remarque 2-8. *Cela donne une preuve du fait que si $F = 0$ est l'équation d'une hypersurface à singularité isolée de $\mathbb{C}^n$, $\mathcal{O}[1/F]$ n'est relativement cohérent pour aucune projection. Cela donne en particulier une preuve différente de [18], [22] et [11] du fait que la variété caractéristique de $\mathcal{O}[1/F]$ est la réunion du conormal à $F = 0$, du conormal au point et de la section nulle.*

2.4 Exemple: D Modules holonomes réguliers à support un bouquet de courbes lisses

Soit Γ une réunion de courbes lisses de $X \times T$ passant par l'origine. Nous dirons que Γ est un bouquet de courbes lisses.

Proposition 2-9. *Soit M un $\mathcal{D}_{X\times T}$ module holonome régulier à support Γ un bouquet de courbes lisses. Si M est relativement cohérent, alors $X \times \{0\}$ est non caractéristique pour M.*

DÉMONSTRATION. Soit $\Gamma = \Gamma_1 \cup \Gamma_2 \cup \cdots \Gamma_r$, une réunion de courbes lisses passant par l'origine. Soit M_i le sous module de M formé des sections de M supportées par Γ_i; $M/\sum M_i$ est à support l'origine, donc somme directe de modules isomorphes à $\oplus\mathcal{D}_{X\times T}/\mathcal{D}_{X\times T}(x_1,\ldots,x_n,t)$. Or, un tel module, s'il est non nul, n'est pas $\mathcal{D}_{X\times T/T}$ cohérent. Ainsi $M = \sum M_i$ et pour prouver la proposition, on est ramené à le faire avec l'hypothèse supplémentaire: le support de M est une seule courbe lisse Γ. Il faut distinguer deux cas.

1. Γ n'est pas tangente à $t = 0$

Considérons alors $\big(x_1 = \varphi_1(t), \ldots, x_n = \varphi_n(t), t\big)$ une paramétrisation analytique de Γ. Quitte à faire le changement de coordonnées:

$$(x_1' = x_1 - \varphi_1(t), \ldots, x_n' = x_n - \varphi_n(t), t' = t),$$

on voit que l'on peut supposer que Γ n'est autre que l'axe des t; c'est ce que l'on fait. Ainsi, il existe $u \in M - 0$ tel que:

$$x_i u = 0 \ \ \forall i \in \{1, 2, \ldots, m\}.$$

Comme de plus M est un $\mathcal{D}_{X \times T/T}$ module cohérent, il existe P:

$$P = \left(\frac{\partial}{\partial t}\right)^k + A_1 \left(\frac{\partial}{\partial t}\right)^{k-1} + \cdots + A_k$$

tel que $Pu = 0$ et $A_i = \sum a_{i,\alpha}(\partial/\partial x)^\alpha \in \mathbb{C}\{t\}[\partial/\partial x_1, \ldots, \partial/\partial x_n]$. Soit β un multi indice de longueur maximum (supposée non nulle) parmi ceux intervenant dans la décomposition des A_i:

$$[P, x^\beta]u = \beta! \left(a_{1,\beta} \frac{\partial}{\partial t}^{k-1} + \cdots + a_{k,\beta}\right) u = 0.$$

Alors

$$Q = P - \left(\frac{\partial}{\partial x}\right)^\beta \left(a_{1,\beta} \frac{\partial}{\partial t}^{k-1} + \cdots + a_{k,\beta}\right)$$

annule u. Q est de la forme:

$$Q = \left(\frac{\partial}{\partial t}\right)^k + B_1 \left(\frac{\partial}{\partial t}\right)^{k-1} + \cdots + B_k,$$

où $B_i = \sum b_{i,\alpha}(\partial/\partial x)^\alpha$ avec $\alpha \neq \beta$ et $|\alpha| \leq |\beta|$. En itérant ce procédé, on peut supposer que u est annulé par un opérateur de la forme

$$P = \left(\frac{\partial}{\partial t}\right)^k + A_1 \left(\frac{\partial}{\partial t}\right)^{k-1} + \cdots + A_k$$

où $A_i \in \mathbb{C}\{t\}$. On en déduit $X \times \{0\}$ est non caractéristique pour $\mathcal{D}_{X \times T} u$. Le même résultat s'en déduit pour M par exemple, en faisant une récurrence sur sa longueur.

2. Γ est tangente à $t = 0$

On peut supposer Γ paramétrée par x_1:

$$\Gamma \ : \ t = \varphi(x_1), x_2 = \varphi_2(x_1), \ldots, x_n = \varphi_n(x_1),$$

où $\varphi(0) = \varphi'(0) = 0$. Quitte à faire un changement de coordonnées, on peut supposer: $\varphi_2 = \cdots = \varphi_n = 0$. Du théorème de classification des $\mathcal{D}$ modules réguliers relativement à un croisement normal ([10]), on peut se ramener au cas où M contient un $\mathcal{D}$ module isomorphe à:

$$M' = \frac{\mathcal{D}_{X \times T}}{\mathcal{D}_{X \times T}\left(t - \varphi(x_1), \dfrac{\partial}{\partial x_1} + \varphi'(x_1)\dfrac{\partial}{\partial t}, x_2, \ldots, x_n\right)},$$

ou de la forme:

$$M_\tau = \frac{\mathcal{D}_{X \times T}}{\mathcal{D}_Z\left(t - \varphi(x_1), x_1\left(\dfrac{\partial}{\partial x_1} + \varphi'(x_1)\dfrac{\partial}{\partial t}\right) + \tau, x_2, \ldots, x_n\right)},$$

où $\tau \in \mathbb{C}$. Il reste à prouver que ces deux $\mathcal{D}_{X \times T}$ modules ne sont pas relativement cohérents.

Supposons que le module M' soit relativement cohérent, soit:

$$Q = \left(\frac{\partial}{\partial t}\right)^p + A_1\left(x_1, \frac{\partial}{\partial x_1}, \ldots, \frac{\partial}{\partial x_n}\right)\left(\frac{\partial}{\partial t}\right)^{p-1} + \cdots$$
$$+ A_p\left(x_1, \frac{\partial}{\partial x_1}, \ldots, \frac{\partial}{\partial x_n}\right)$$

de classe nulle dans M', où $p \geq 1$; Q appartient donc à l'idéal à gauche engendré par les générateurs:

$$\left(t - \varphi(x_1), \frac{\partial}{\partial x_1} + \varphi'(x_1)\frac{\partial}{\partial t}, x_2, \ldots, x_n\right),$$

qui commutent entre eux. Les symboles principaux de ces générateurs formant une suite régulière, $\sigma(Q)$ le symbole principal de Q vérifie:

$$\sigma(Q) = \Lambda\big(\xi_1 + \varphi'(x_1)\eta\big),$$

où $\Lambda \in \mathcal{O}_{X \times T}[\xi_1, \ldots, \xi_n, \eta]$. Si $\deg Q > p$ on a: $\deg_\eta\big(\sigma(Q)\big) \leq p-1$, donc: $\deg_\eta \Lambda \leq p - 2$.

Soit $\tilde{\Lambda}$ un opérateur relevant Λ:

$$Q - \tilde{\Lambda}\left(\frac{\partial}{\partial x_1} + \varphi'(x_1)\frac{\partial}{\partial t}\right)$$

vérifie la même propriété que Q, mais son degré est strictement inférieur. On peut ainsi, en itérant, supposer que $\deg Q = p$, d'où:

$$(0, \ldots, 0)(\xi_1 = 0, \ldots, \xi_n = 0, \eta = 1)$$

n'appartient pas à car(M'), ce qui est faux. Ainsi M' n'est pas relativement cohérent. Le même raisonnement s'applique pour M_τ puisque les
symboles principaux de

$$\left(t - \varphi(x_1), x_1\left(\frac{\partial}{\partial x_1} + \varphi'(x_1)\frac{\partial}{\partial t} \right) + \tau, x_2, \ldots, x_n \right)$$

forment une suite régulière.

2.5   Caractérisation géométrique de la cohérence relative

**Nous reprendrons ici la démonstration faite avec Y. Laurent du théorème
fondamental de la note ([6])**
  Soit M un $\mathcal{D}_{X \times T}$ Module cohérent; suivons [20]:

Définition 2-10. *Nous dirons que M est 1 spécialisable (par rapport à $X \times \{0\}$)
si pour toute section u de M définie au voisinage de $X \times \{0\}$, il existe un
opérateur P de la forme:*

$$P = \sum A_i \left(t\frac{\partial}{\partial t} \right)^i,$$

*et un polynôme $b \in \mathbb{C}[s]$ où $A_i \in \mathcal{D}_{X \times T / T}$ et où le degré de P comme élément
de $\mathcal{D}_{X \times T}$ est inférieur ou égal au degré de b et tel que*

$$\big(b(t\partial/\partial t) - tP\big)u = 0.$$

Théorème 2-11. ([17]) *Un $\mathcal{D}_{X \times T}$ Module régulier est 1 spécialisable.*

Remarque 2-12. *Ce théorème relie d'un coté des conditions sur la croissance
les solutions d'un $\mathcal{D}$ module et de l'autre coté des conditions algébriques. On
se reportera aux exposés dans ce volume de Y. Laurent et Z. Mebkhout.*

  Nous allons rappeler maintenant la notion de V filtration ([24], [15], [28]).
Soit $\mathcal{J}$ l'idéal $t\mathcal{O}_{X \times T}$. On pose:

$$V_k(\mathcal{D}_{X \times T}) = \big\{ P \in \mathcal{D}_{X \times T} \; ; \; \forall l \in \mathbb{N} \quad P(\mathcal{J}^l) \subset \mathcal{J}^{l-k} bigr\}.$$

On obtient ainsi une filtration croissante indexée sur $\mathbb{Z}$ de $\mathcal{D}_{X \times T}$:

$$\frac{\partial}{\partial t} \in V_1(\mathcal{D}_{X \times T}) \qquad t \in V_{-1}(\mathcal{D}_{X \times T})$$

$$\frac{\partial}{\partial x_i} \in V_0(\mathcal{D}_{X \times T}) \qquad x_i \in V_0(\mathcal{D}_{X \times T}).$$

Un élément de $V_0(\mathcal{D}_{X \times T})$ s'écrit:

$$\sum a_{\alpha,j}(x,t)\left(t\frac{\partial}{\partial t} \right)^j \left(\frac{\partial}{\partial x} \right)^\alpha.$$

Un élément de $V_0(\mathcal{D}_{X\times T})/V_{-1}(\mathcal{D}_{X\times T})$ admet un représentant qui s'écrit sous la forme:

$$\sum a_{\alpha,j}(x,0)\left(t\frac{\partial}{\partial t}\right)^j\left(\frac{\partial}{\partial x}\right)^\alpha.$$

$\mathrm{gr}_V(\mathcal{D}_{X\times T})$, le gradué pour la V filtration est supporté par $t=0$. Il s'identifie à

$$\pi_*(\mathcal{D}_{[T_X(X\times T)]}),$$

où π est la projection du fibré normal $T_X(X\times T)$ sur X. (x,τ) désignent les coordonnées canoniques de ce fibré normal. $\mathcal{D}_{[T_X(X\times T)]}$ désigne le sous faisceau des opérateurs du fibré normal polynomiaux par rapport aux fibres de π. L'identification:

$$\mathrm{gr}_V(\mathcal{D}_{X\times T})\to\pi_*(\mathcal{D}_{[T_X(X\times T)]})$$

identifie ainsi la multiplication par t et la multiplication par τ, dérivation par rapport à t et dérivation par rapport à τ.

A partir d'une présentation du $\mathcal{D}_{X\times T}$ module M, on définit une bonne V filtration de M et nous noterons $\mathrm{gr}_V(M)$ son gradué V et

$$\tilde{M}=\mathcal{D}_{T_X(X\times T)}\bigotimes_{\mathcal{D}_{[T_X(X\times T)]}\pi^{-1}}\big(\mathrm{gr}_V(M)\big).$$

Si M est spécialisable, $\tilde{M}$ est un $\mathcal{D}_{T_X(X\times T)}$ module cohérent. Notons $\mathrm{car}(M)$ la variété caractéristique de M. Notons:

$$C=C_{T_X^*(X\times T)}\big(\mathrm{car}(M)\big),$$

le cône normal de $\mathrm{car}(M)$ le long de $T_X^*(X\times T)$. C'est un cône de sommet $T_X^*(X\times T)\cap\mathrm{car}(M)$ contenu dans $T_{T_X^*(X\times T)}T^*(X\times T)$. Au moyen de la 2 forme canonique sur $T^*(X\times T)$, on identifie:

$$T^*\big(T^*(X\times T)\big)\quad\text{et}\quad T\big(T^*(X\times T)\big),$$

au dessus de $T^*(X\times T)$. C s'identifie alors à un sous ensemble du fibré conormal $T^*\big(T_X^*(X\times T)\big)$. La transformée de Fourier dans les fibres de

$$T_X^*(X\times T)\to X$$

identifie:

$$T^*\big(T_X^*(X\times T)\big)\quad\text{et}\quad T^*\big(T_X(X\times T)\big).$$

Dans les coordonnées locales canoniques, il s'agit de l'application:

$$(x,\tau;x^*,\tau^*)\to(x,-\tau^*;x^*,\tau).$$

Notation 2-13. $\tilde{C}$ désignera l'image de C par ces transformations

Théorème 2-14. ([20], **Corollary 4.1.2**) *Si M est un $\mathcal{D}_{X \times T}$ Module cohérent 1 spécialisable:*

$$\mathrm{car}(\tilde{M}) = \tilde{C}.$$

Remarque 2-15. *La preuve de [20] se fait par de l'analyse sur les opérateurs 2 microdifférentiels. Lorsque M est holonome régulier, ce théorème résulte également de [15] et [27]. Expliquons brièvement pourquoi. Soit M un module holonome régulier et $\mathcal{F}^{\cdot}$ son complexe des solutions holomorphes. Dans [15] est construit un $\mathcal{D}_{X \times T}$ module dont les solutions sont le "spécialisé" de $\mathcal{F}^{\cdot}$ le long de X. Ce $\mathcal{D}$ module est justement le gradué V de M pour une bonne V filtration de M. Dans [27] est calculé géométriquement le cycle qui correspond (par la correspondance d'Euler-Mac Pherson) à la fonction constructible obtenu en prenant l'indice du spécialisé de $\mathcal{F}^{\cdot}$ le long de X. Ce cycle donne la variété caractéristique de $\mathrm{gr}_V(M)$ (théorème de l'indice de Kashiwara); cette variété caractéristique ne dépend pas de la bonne V filtration.*

Théorème 2-16. *Soit M un $\mathcal{D}_{X \times T}$ Module cohérent 1 spécialisable (ce qui est le cas si M est holonome régulier). Les propriétés suivantes sont équivalentes:*

1. *M est $\mathcal{D}_{X \times T / T}$ cohérent au voisinage de $X \times \{0\}$.*

2. *$X \times \{0\}$ est non caractéristique pour $\mathrm{car}(M)$.*

DÉMOSTRATION. On a vu dans la Remarque 2-2 que $2 \Rightarrow 1$. Montrons que $1 \Rightarrow 2$. Il suffit de faire la preuve au voisinage de 0. Soit (M_r) une bonne V filtration de M. Soit $u \in M_r$, $\mathcal{D}_{X \times T, 0} u$ est de type fini sur $\mathcal{D}_{X \times T / T, 0}$. Ainsi, il existe des opérateurs relatifs $(A_i)_{i \in \{1, \dots, k\}}$ tels que:

$$\left(\left(\frac{\partial}{\partial t} \right)^k + A_1 \left(\frac{\partial}{\partial t} \right)^{k-1} + \cdots + A_k \right) u = 0.$$

D'où $(\partial / \partial t)^k u \in V_{k-1} u \subset M_{k+r-1}$. Donc $\mathrm{gr}_V(M)$ est supporté par $\tau^* = 0$, où (x, τ, x^*, τ^*) sont les coordonnées de $T^*\big(T_X(X \times T)\big)$. Comme M est 1 spécialisable, on déduit du théorème précédent que:

$$\tilde{C} \subset \{\tau^*\} = 0.$$

La base de ce cône, $\mathrm{car}(M) \cap T_X^*(X \times T)$, est donc contenu dans $\{\tau^*\} = 0$, où (x, τ^*) sont les coordonnées de $T_X^*(X \times T)$. Donc $(0, \dots, 0; 0; \dots, 0, 1)$ qui est un covecteur de $T_X^*(X \times T)$ ($x = 0$ et $\tau^* = 1$) n'appartient pas à $\mathrm{car}(M)$. Ainsi $X \times \{0\}$ est non caractéristique pour $\mathrm{car}(M)$ au voisinage de 0.

Remarque 2-17. *Une traduction algébrique de ce théorème est le résultat suivant. Soit $u \in M$ une section d'un $\mathcal{D}_{X \times T}$ module cohérent. S'il existe deux* *équations du type:*

$$\left(\left(\frac{\partial}{\partial t} \right)^k + A_1 \left(x, t, \frac{\partial}{\partial x} \right) \left(\frac{\partial}{\partial t} \right)^{k-1} + \cdots \right) u = 0,$$

$$\left(b \left(t \frac{\partial}{\partial t} \right) + tP \right) u = 0,$$

où $A_i \in \mathcal{D}_{X \times T / T, 0}$, $b(s) \in \mathbb{C}[s]$, $P \in V_0(\mathcal{D}_{X \times T})$ et si le degré de P par rapport aux dérivations est inférieur au degré en s de b, alors il existe des opérateurs relatifs B_i de degré par rapport aux dérivations $\partial/\partial x$ inférieur à i tels que:

$$\left(\left(\frac{\partial}{\partial t} \right)^l + B_1 \left(x, t, \frac{\partial}{\partial x} \right) \left(\frac{\partial}{\partial t} \right)^{l-1} + \cdots + B_l \right) u = 0$$

References

[1] I. N. Bernstein: "The analytic continuation of generalised functions with respect a parameter", Functional. Anal. Appl.6, 26-40, 1972.

[2] J.Briançon: "Quelques remarques sur le joint de deux singularités isolées de fonctions analytiques complexes suivies de deux exemples bizarres", Prépublication de l'Université de Nice 296, Avril 1991.

[3] J. Briançon, Ph. Maisonobe: "Autour d'une conjecture sur les $\mathcal{D}$ Modules holonomes réguliers cohérents relativement à une projection", Prépublication de l'Université de Nice, $n°$ 290, Décembre 1990.

[4] J. Briançon, M. Granger, Ph. Maisonobe, M. Miniconi: "algorithme de calcul du polynôme de Bernstein: cas non dégénéré", Annales de l'Institut Fourier, tome 39, Fascicule 3, 553–610, 1989.

[5] J. Briançon, F. Geandier, Ph. Maisonobe: "Déformation d'une singularité isolée d'hypersurface et polynôme de Bernstein", Bulletin de la Soc. Math. France, 120, 1992, p. 15–49.

[6] J. Briançon, Y. Laurent, Ph. Maisonobe: "Sur les modules différentiels holonomes réguliers, cohérents relativement à une projection", C. R. Acad. Sci. Paris, t. 313, Série 1, p. 285–288, 1991.

[7] P. Deligne, N. Katz: "Groupes de monodromie en Géométrie Algébrique (S.G.A. 7 1/2)", volume 340 of Lect. Notes in Math., Springer-Verlag, Berlin-Heidelberg, 1973.

[8] F. Geandier: "Polynômes de Bernstein et déformations à nombre de Milnor constant", Thèse présentée à l'université de Nice, Juin 1989.

[9] F. Geandier: "Déformations à nombre de Milnor constant: quelques résultats sur les polynômes de Bernstein", Compositio Mathematica 77, n° 2, (1991).

[10] A. Galligo, M. Granger, Ph. Maisonobe: "$\mathcal{D}$ Modules et Faisceaux Pervers dont le Support Singulier Est un Croisement Normal", dans Systèmes Différentiels et Singularités, Astérisque 130, (1985).

[11] V. Ginsburg: "Characteristic varieties and vanishing cycles", Inventiones mathematicae 84, 327–402, (1986)

[12] M. Kato: "The b-function of μ-constant deformation of $x^9 + y^{4}$", Bull. College Sci. Uni. Ryukyus, t 32, (1982), p 5–8.

[13] M. Kashiwara: "B. functions and Holonomic Systems", Inventiones Mathematicae, vol. 38, (1976), p. 33–53.

[14] M. Kashiwara: "On the Holonomic Systems of Linear Differential Equations , II", Inventiones Mathematicae, vol. 49, (1978).

[15] M. Kashiwara: "Vanishing cycles and holonomic systems of differential equations", Lecture Notes in Math. n° 1016, (1983) p. 134–142.

[16] M. Kashiwara: "Systems of microdifferential equations", progress in Math, 34, Birkhauser, Boston, 1983.

[17] M. Kashiwara, T. Kawai: "Second Microlocalisation and Asymptotis Expensions", Lecture Notes in Physics, no 126, Springer, (1980) p. 21–76.

[18] M. Kashiwara, T. Kawai, M. Sato: "Micro local analysis of prehomogeneous vector spaces", inventiones Math, 62, p. 117–179, 1980.

[19] Y. Laurent: "Polygone de Newton et B. Fonctions pour les modules microdifférentiels", Ann. Scient. Ec. Norm. Sup. $4^{\grave{e}me}$ série, t. 20, 1987, p. 391–441.

[20] Y.Laurent: "Vanishing cycles of $\mathcal{D}$ modules", Inventiones Mathematicae, vol. 112 , (1993), p. 491–539.

[21] F. Lazzeri: "A theorem on the monodromy of isolated singularities", Singularités à Cargèse, Astérisque 7 et 8 (1973).

[22] D. T. Lê, Z. Mebkhout: "Variétés caractéristiques et variétés polaires", C. R. Acad. Sc. Paris, t. 296 (17 janvier 1983).

[23] B. Malgrange: "Le polynôme de Bernstein d'une singularité isolée", Lect. Notes Math. vol. 459, Springer-Verlag. p. 98–119, (1975).

[24] B. Malgrange: "Polynôme de Bernstein-Sato et cohomologie évanescente", Astérisque 101–102, 233–267, 1983.

[25] Z. Mebkhout: "Le formalisme des 6 opérations de Grothendieck pour les $\mathcal{D}_X$-modules cohérents", Travaux en cours, n° 35, (1989).

[26] C. Sabbah: "$\mathcal{D}$ modules et cycles évanescents (d'après B. Malgrange et M. Kashiwara)", Conférence de La Rabida 1984, vol 3, Hermann, Paris, (1987), 53–98.

[27] C. Sabbah: "Quelques Remarques sur la Géométrie des Espaces Conormaux", dans Systèmes Différentiels et Singularités, Astérisque 130, 1985.

[28] B. Teissier: "Cycles évanescents, section plane et conditions de Whitney", Singularités à Cargèse, Astérisque 7 et 8, (1973).

[29] B. Teissier: "The Hunting of invariants in the geometry of discriminants", pro-
      ceedings of the Nordic Summer School, P. Holm editor, Sijthoff and Noordhoff
      international Publishers, 1977.

[30] A. N. Varchenko: "Gauss Manin connection of isolated singular point and Bern-
      stein polynomial", Bull. Sc. Math., $2^{\text{éme}}$ série, 104, 205–223, 1980.

[31] A.N. Varchenko: "Asymptotic Hodge structure in the vanishing cohomology",
      Math. USSR Izvestijà, Vol. 18, n 3, 1982.

Addresses of authors:

Prof. J. Briançon
Unité associée au CNRS 168
Université de Nice
Parc Valrose,
F-06108 Nice Cedex 2
France

Prof. Ph. Maisonobe
Unité associée au CNRS 168
Université de Nice
Parc Valrose,
F-06108 Nice Cedex 2
France
Email: phm@math.unice.fr

LE POLYGONE DE NEWTON D'UN $\mathcal{D}_X$-MODULE

Z. Mebkhout

1 Introduction

Dans cet exposé nous définissons l'analogue du polygone de Newton en un point singulier d'une équation différentielle pour un $\mathcal{D}_X$-module holonome le long d'une hypersurface Y.

Le polygone de Newton en un point singulier d'une équation différentielle se définit à l'aide des valuations x-adiques des coefficients de cette équation ([25], [11]). A une variable le cas d'un système se ramène à celui d'une équation grâce, par exemple, au lemme du vecteur cyclique. A plusieurs variables le polygone de Newton d'un opérateur différentiel le long de $\Lambda = T_Y^* X$ se définit à partir des valuations $\mathcal{I}_\Lambda$-adiques des coefficients ([13]). Pour un système différentiel le polygone de Newton n'a plus de sens mais les pentes se définissent à partir des variétés micro-différentielles et qui sont des nombres rationnels ([13]).

A une variable on peut reconstituer, à translation près, le polygone de Newton en un point singulier d'une équation différentielle complexe à partir de la filtration Gevrey de l'espace d'irrégularité de cette équation en ce point singulier ([25], voir aussi [11]).

A plusieurs variables on sait définir le faisceau d'irrégularité $\mathrm{Irr}_Y(\mathcal{M})$ d'un $\mathcal{D}_X$-module holonome $\mathcal{M}$ le long d'une hypersurface Y muni de sa filtration Gevrey $\mathrm{Irr}_Y^{(r)}(\mathcal{M})$ indexée par les nombres réels $r \geq 1$ ([22]). Nous démontrons ([14]) d'une part, le théorème de comparaison des pentes: les pentes transcendantes, les sauts de la filtration Gevrey, sont égales aux pentes algébriques définies dans ([13]), puis d'autre part le théorème d'intégralité: la fonction constructible $(r-1)\big(\chi\big(\mathrm{Irr}_Y^{(r+\epsilon)}(\mathcal{M})\big) - \chi\big(\mathrm{Irr}_Y^{(r-\epsilon)}(\mathcal{M})\big)\big)$ à priori à valeurs rationnelles est à valeurs entières. A partir du théorème de positivité de l'irrégularité et du théorème d'intégralité on peut alors définir le polygone de Newton de $\mathcal{M}$ le long de chaque composante irréductible de la variété caractéristique du faisceau $\mathrm{Irr}_Y(\mathcal{M})$. C'est un polygone du plan réel dont les sommets en nombre fini sont à coordonnées entières.

Le théorème d'intégralité est bien sûr l'analogue en toutes dimensions du théorème d'intégralité de Hasse-Arf de la théorie de la ramification des corps locaux qui permet de même de définir le polygone de Newton d'un module sur le groupe de Galois d'une extension finie d'un corps local (cf. [10]). Ceci est en conformité avec le "principe" de Deligne selon lequel il y a une analogie entre les propriétés de la ramification dite sauvage en caractéristique $p > 0$ et les propriétés de la ramification (l'irrégularité) en caractéristique zéro. On doit donc s'attendre à des résultats analogues pour la ramification en dimension supérieure aussi bien du point de vue l-adique que du point de vue p-adique.

Progress in Mathematics, Vol. 134
© 1996 Birkhäuser Verlag Basel/Switzerland

Dans le cas p-adique, à une variable, on a une définition de l'irrégularité analogue à la définition complexe à l'aide de l'indice [27] des opérateurs différentiels opérant sur l'espace de fonctions analytiques dans un disque ouvert. Mais pour l'instant le problème essentiel demeure de montrer l'existence de l'indice. On dispose d'un substitut de la filtration Gevrey par la filtration à l'aide de la croissance au bord du disque (cf. [26]).

Comme conséquence du théorème de comparaison des pentes on énonce dans le paragraphe 7 une conjecture purement algébrique qui implique l'équivalence entre la nullité du faisceau d'irrégularité, condition générique de la régularité, et l'existence d'une équation fonctionnelle régulière, condition algébrique forte de la régularité. C'est une illustration du fait qu'il est plus simple et plus naturel de déduire les propriétés de la régularité à partir des propriétés de l'irrégularité.

Les résultats des paragraphes 6 et 7 sont le fruit d'une longue collaboration avec Y. Laurent ([14]) sur ces questions de l'irrégularité à plusieurs variables (cf. [22], 6.5, 6.6).

Cet exposé qui est de nature introductif est le prolongement naturel de l'exposé de la deuxième conférence de La Rabida (1984) ([20]). Pour les démonstrations complètes le lecteur est prié de se reporter à l'article ([14]). Les théorèmes de positivité et d'intégralité ont des analogues algébro-géométriques qui peuvent se démontrer par voie purement algébrique ([14]).

Nous voudrions remercier Alberto Arabia pour l'aide considérable qu'il nous a apportée dans la réalisation de ce travail.

2   Le cas d'une variable

(2-1) Nous allons rappeler le cas d'une variable qui sert de modèle à la situation générale.

Soit D un petit disque complexe, voisinage de l'origine muni d'une coordonnée x. Soit

$$P\left(x, \frac{d}{dx}\right) = a_n(x)\frac{d^n}{dx^n} + \cdots + a_0(x)$$

une équation d'ordre n à coefficients holomorphes sur D qui admet l'origine 0 comme unique point singulier. On dit que 0 est un point singulier régulier si $P(x, d/dx)$ admet une base fondamentale formée de fonctions multiformes régulières. Une fonction multiforme de détermination finie fonction multiforme régulière

$$\varphi(x) = \sum_{\alpha, k} c_{\alpha,k}(x) x^\alpha \log(x)^k,$$

où α est un nombre complexe, k est un entier naturel et $c_{\alpha,k}$ est une fonction uniforme, la somme étant finie, est régulière si les coefficients $c_{\alpha,k}$ admettent au plus des singularités polaires.

Définition 2-2. *On appelle nombre de Fuchs ou irrégularité de l'équation* $P(x, d/dx)$ *en 0 le nombre positif*

$$\mathrm{irr}_0\left(P\left(x, \frac{d}{dx}\right)\right) := \sup_{0 \le k \le n}\{k - v_0(a_k)\} - (n - v_0(a_n)).$$

On note v_0 la valuation x-adique attachée à l'origine. Fuchs a démontré le critère numérique suivant (cf. [6]):

Théorème 2-3. *Le point singulier 0 de l'équation* $P(x, d/dx)$ *est régulier si et seulement si le nombre* $\mathrm{irr}_0(P(x, d/dx))$ *est nul.*

(2-4) La définition du nombre de Fuchs est purement algébrique et garde un sens sur un corps de base quelconque. Si $\mathcal{E}$ est un fibré à connexion sur une courbe X non singulière sur un corps de caractéristique nulle, on peut définir à l'aide du théorème du vecteur cyclique le nombre de Fuchs $\mathrm{irr}_\infty(\mathcal{E})$ du fibré $\mathcal{E}$ en chaque point à l'infini ∞ de X. On note $\chi(X; \mathrm{DR}(\mathcal{E}))$ la caractéristique d'Euler-Poincaré de la cohomologie de de Rham de X à valeurs dans $\mathcal{E}$. Le nombre de Fuchs intervient alors dans la formule de Deligne ([2]):

Théorème 2-5. *Si* $\mathcal{E}$ *est un fibré à connexion de rang* r *sur une courbe non singulière connexe* X *sur un corps de caractéristique nulle, on a la formule*

$$\chi\big(X; \mathrm{DR}(\mathcal{E})\big) = r \cdot \chi\big(X; \mathrm{DR}(\mathcal{O}_X)\big) - \sum_\infty \mathrm{irr}_\infty(\mathcal{E}),$$

la somme étant étendue à tous les point à l'infini de X.

(2-6) L'interprétation complexe de Malgrange ([18]) du nombre de Fuchs consiste à faire opérer $P(x, d/dx)$ sur des espaces de séries. Si un opérateur P opère sur un espace fonctionnel $\mathcal{F}$, on dit qu'il est *à indice* si son noyau et son conoyau sont de dimensions finies. On note $\chi(P, \mathcal{F})$ la caractéristique d'Euler-Poincaré de P à valeurs dans $\mathcal{F}$, c'est à dire la dimension complexe du noyau moins la dimension complexe du conoyau. On note $\mathbb{C}\{x\}$ l'espace des séries convergentes à coefficients complexes et $\mathbb{C}[[x]]$ l'espace des séries formelles à coefficients complexes. On a alors le théorème ([18]):

Théorème 2-7.

1. *L'opérateur* $P(x, d/dx)$ *est à indice dans les espaces* $\mathbb{C}\{x\}$ *et* $\mathbb{C}[[x]]$,

$$\chi\big(P, \mathbb{C}\{x\}\big) = n - v_0(a_n) \qquad et \qquad \chi(P, \mathbb{C}[[x]]) = \sup_{0 \le k \le n}\{k - v_0(a_k)\}.$$

2. *Le conoyau de* P *opérant dans le quotient* $\mathbb{C}[[x]]/\mathbb{C}\{x\}$ *est nul et donc*

$$\mathrm{irr}_0 P\left(x, \frac{d}{dx}\right) = \dim_\mathbb{C} \mathrm{Ker}(P, \mathbb{C}[[x]] \,/\, \mathbb{C}\{x\}).$$

Définition 2-8. *On appelle espace d'irrégularité de $P(x, d/dx)$ le long de 0, et on note* $\mathrm{Irr}_0(P)$, *l'espace vectoriel complexe* $\mathrm{Ker}\big(P, \mathbb{C}[\![x]\!] \,/\, \mathbb{C}\{x\}\big)$.

(2-9) Ceci amène à considérer le polygone de Newton de l'opérateur $P(x, d/dx)$ le long de sa singularité 0 (cf. [11], [25]). Pour tout k, $0 \leq k \leq n$, on considère dans le plan réel le quadrant $(\lambda, \mu) \in \mathbb{R}^2$, $\lambda \leq k$, $\mu \geq v_0(a_k) - k$. Le polygone de Newton de l'opérateur $P(x, d/dx)$ le long de sa singularité 0 est l'enveloppe convexe de la réunion de ces quadrants. La frontière de cette enveloppe est constituée d'une ligne horizontale, d'une ligne verticale et de segments de droites dans le cas d'une singularité irrégulière, de pentes strictement positives et finies.

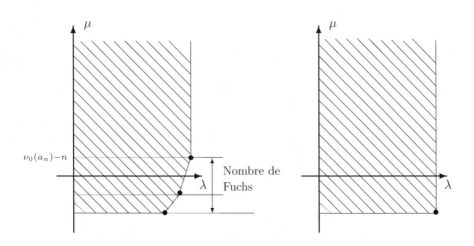

Figure 2.1: Polygone de Newton d'une singularité irregulière (gauche) et d'une singularité regulière (droite).

(2-10) Pour une singularité irrégulière il apparaît, entre les indices formelle et convergent, d'autres indices *critiques* qui sont interprétés par Ramis ([25]) comme les indices de l'opérateur $P(x, d/dx)$ dans des espaces de séries de classe de Gevrey. Auparavant Komatsu ([11]) avait interprété la plus grande pente.

Définition 2-11. *Soit s un nombre réel plus grand ou égal à 1, on dit qu'une série formelle $\sum_k a_k x^k$ est dans la classe de Gevrey d'ordre s si et seulement si la série*

$$\sum_k \frac{a_k}{n!^{s-1}} x^k$$

est convergente.

On note $\mathcal{O}_s$ l'espace vectoriel complexe des séries de classe de Gevrey d'ordre s. En particulier $\mathcal{O}_1$ est l'espace des séries convergentes et $\mathcal{O}_\infty$ est l'espace des séries formelles. L'opérateur $P(x, d/dx)$ opère dans l'espace $\mathcal{O}_s$ pour tout s. On a alors le résultat suivant:

Théorème 2-12.

1. *Pour tout réel $s \geq 1$, l'opérateur $P(x, d/dx)$ est à indice dans l'espace $\mathcal{O}_s$ pour tout s.*

2. *Le conoyau de l'opérateur $P(x, d/dx)$ dans l'espace quotient $\mathcal{O}_s/\mathcal{O}_1$ est nul pour tout s.*

On pose

$$\operatorname{Irr}_Y^{(s)}(P) := \operatorname{Ker}(P, \mathcal{O}_s/\mathcal{O}_1).$$

On a donc $\operatorname{Irr}_0^{(\infty)}(P) = \operatorname{Irr}_0(P)$. L'espace d'irrégularité $\operatorname{Irr}_0(P)$ est muni d'une filtration $\operatorname{Irr}_0^{(s)}(P)$, par des sous-espaces vectoriels complexes, croissante indexée par les nombres réels $s \geq 1$. L'espace vectoriel $\operatorname{Irr}_0(P)$ étant de dimension finie, la filtration admet au plus un nombre fini de sauts. Les sauts et les indices critiques correspondants sont déterminés par le polygone de Newton de P ([25]):

Théorème 2-13.

1. *Un nombre réel $s \geq 1$ est un saut de la filtration $\operatorname{Irr}_0^{(s)}(P)$ si et seulement si $1/(s-1)$ est une pente finie non nulle du polygone de Newton de P.*

2. *Si s est un saut, l'opposé de l'ordonnée du point le plus bas du segment de pente $1/(s-1)$ du polygone de Newton est égal à l'indice $\chi(P, \mathcal{O}_s)$ alors que l'opposé de l'ordonnée du point le plus haut est égal à l'indice $\chi(P, \mathcal{O}_{s-\varepsilon})$ ($\varepsilon > 0$).*

En vertu du théorème 2-13 les sauts de la filtration $\operatorname{Irr}_0^{(s)}(P)$ sont des nombres rationnels. De façon plus précise, si on pose

$$\phi(s) := \lim_{\varepsilon \to 0} \chi(P, \mathcal{O}_{s+\varepsilon}) - \chi(P, \mathcal{O}_{s-\varepsilon}),$$

la fonction ϕ est à valeurs entières et est non nulle pour tous les sauts. Le polygone de Newton de P étant à coordonnées entières, $(s-1)\phi(s)$ est un entier positif pour tout s.

(2-14) Pour une singularité irrégulière la donnée de l'espace $\operatorname{Irr}_0(P)$ muni de sa filtration $\operatorname{Irr}_0^{(s)}(P)$, $s \geq 1$, permet de reconstituer à une translation près, le polygone de Newton de P. En effet

$$\phi(s) = \dim_{\mathbb{C}} \operatorname{Irr}_0^{(s^+)}(P) - \dim_{\mathbb{C}} \operatorname{Irr}_0^{(s^-)}(P).$$

Les points $((s-1)\phi(s), \phi(s))$, à coordonnées entières pour s parcourant les sauts de la filtration de l'espace $\operatorname{Irr}_0(P)$, forment les sommets du polygone de Newton.

(2-15) Si au lieu de considérer un opérateur différentiel P on considère un système différentiel, par exemple l'idéal I à gauche engendré par un nombre fini d'opérateurs différentiels, on ne peut plus construire le polygone de Newton de I directement à partir de I. Par contre l'espace d'irrégularité $\mathrm{Irr}_0(I)$ muni de sa filtration $\mathrm{Irr}_0^{(s)}(I)$, $s \geq 1$ et la fonction saut ϕ, gardent un sens. Les points du plan $\big((s-1)\phi(s), \phi(s)\big)$, en nombre fini, définissent un polygone. Il faut montrer que ce polygone défini par voie *transcendante* est à coordonnées *entières* pour avoir le polygone de Newton de I. Pour cela, on peut montrer que l'espace d'irrégularité de l'idéal I muni de sa filtration est égal à l'espace d'irrégularité muni de sa filtration d'un opérateur de l'idéal et on se ramène aux résultats précédents. Cependant, le fait que le polygone de Newton d'un système se ramène à celui d'un opérateur est caractéristique de la dimension un puisque alors il n'a y qu'une seule dérivation.

3 La catégorie des faisceaux pervers

Si X/k est une variété algébrique non singulière sur un corps k de caractéristique nulle, le polygone de Newton le long d'une sous-variété Y non singulière d'un opérateur différentiel $P(x, \partial_x)$ au voisinage d'un point de Y garde un sens. Il suffit de remplacer la valuation x-adique par la valuation $\mathcal{I}_\Lambda$-adique où Λ est le fibré conormal $T_Y^* X$ de Y dans X ([13]). Pour un idéal d'opérateurs différentiels le polygone de Newton n'a plus de sens mais les pentes gardent encore un sens (loc. cit.). En dimension supérieure, si le corps de base est $\mathbb{C}$ les opérateurs différentiels ne sont pas à indice dans les espaces de séries, le polygone de Newton n'a plus d'interprétation transcendante.

Par contre les systèmes holonomes sont à indice et c'est pour eux que l'on peut définir l'analogue de l'espace d'irrégularité muni de sa filtration. L'idée de base est que la catégorie des faisceaux *pervers* sur une variété complexe joue le rôle de la catégorie des espaces vectoriels complexes de dimension finie à laquelle elle se réduit si la variété en question est un point.

(3-1) Sur un espace analytique complexe X on a la catégorie des complexes constructibles $D_c^b(\mathbb{C}_X)$ qui contient comme sous-catégorie pleine la catégorie des faisceaux pervers $\mathrm{Perv}(\mathbb{C}_X)$. Nous allons supposer pour simplifier que X est sans singularités. Si $\mathcal{F}$ est un complexe constructible nous noterons $h^i(\mathcal{F})$ son i-ème faisceau de cohomologie.

Définition 3-2.

1. *On dit qu'un complexe constructible $\mathcal{F}$ a la propriété de support si ses faisceaux de cohomologie sont concentrés entre le degré 0 et le degré $\dim X$ et si $\dim \mathrm{supp}\big(h^i(\mathcal{F})\big) \leq \dim X - i$, pour tout i.*

2. *On dit qu'un complexe constructible $\mathcal{F}$ est un faisceau pervers s'il a la propriété de support ainsi que son dual $\mathcal{F}^\nu := \mathbb{R}\hom_{\mathbb{C}_X}(\mathcal{F}, \mathbb{C}_X)$.*

La catégorie $\mathrm{Perv}(\mathbb{C}_X)$ est la sous-catégorie pleine de $D^b_c(\mathbb{C}_X)$ dont les objets sont les faisceaux pervers.

Théorème 3-3.

1. La catégorie $\mathrm{Perv}(\mathbb{C}_X)$ est abélienne.

2. La catégorie $\mathrm{Perv}(\mathbb{C}_X)$ est un champ, c'est a dire ses objets ainsi que ses morphismes sont de nature locale.

3. Les faisceaux pervers sont localement de longueur finie.

La démonstration du théorème précédent peut se faire, soit à partir de la théorie des $\mathcal{D}_X$-modules parce que la catégorie des $\mathcal{D}_X$-modules holonomes réguliers sur X, qui a les trois propriétés du théorème, est équivalente par le foncteur de de Rham à la catégorie des faisceaux pervers, soit directement ([1]) à partir de la théorie des catégories dérivées et des catégories triangulées mais dans un cadre beaucoup plus général.

(3-4) Si X est un espace analytique complexe, on a le morphisme d'Euler entre le groupe des cycles de X et le groupe des fonctions constructibles à valeurs entières sur X

$$\mathrm{Eu}\colon \mathbb{Z}(X) \longrightarrow \mathbb{F}(X),$$

qui est un isomorphisme de groupes abéliens en vertu de la partie élémentaire du théorème de MacPherson ([16]). En fait la définition du morphisme d'Euler est algébro-géométrique ([3], [15], [28]) et garde un sens pour les variétés algébriques au-dessus d'un corps de caractéristique nulle. Nous noterons C l'isomorphisme inverse de Eu.

Supposons que X est sans singularités de sorte que l'on peut considérer le fibré cotangent T^*X. L'application qui a une sous-variété irréductible de X associe l'adhérence du conormal de sa partie lisse, se prolonge en un isomorphisme de groupes abéliens entre le groupes des cycles de X et le groupe des cycles lagrangiens homogènes de T^*X:

$$\mathbb{Z}(X) \longrightarrow \mathbb{Z}_\ell(T^*X).$$

Si $\mathcal{F}$ est un complexe constructible on appelle cycle caractéristique et on note $\mathrm{CCh}(\mathcal{F})$ l'unique cycle lagrangien $\sum_\alpha m_\alpha \overline{T^*_{X_\alpha}X}$ tel que

$$\chi(\mathcal{F}) = \mathrm{Eu}\left(\sum_\alpha (-1)^{\mathrm{codim}_X(X_\alpha)} m_\alpha \overline{X_\alpha}\right),$$

où $\chi(\mathcal{F})$ est la caractéristique d'Euler-Poincaré de $\mathcal{F}$.

Théorème 3-5. Si $\mathcal{F}$ est un faisceau pervers son cycle caractéristique est un cycle positif.

En effet si $\mathcal{F}$ est un faisceau pervers il apparaît comme le complexe de de Rham d'un $\mathcal{D}_X$-module holonome $\mathcal{M}$ et le théorème de l'indice de Kashiwara ([7]) dit précisément que les multiplicités m_α sont les multiplicités du gradué associé à une bonne filtration de $\mathcal{M}$ le long de ses composantes irréductibles. Le théorème admet aussi une démonstration directe sans théorie des $\mathcal{D}_X$-modules. La multiplicité m_α apparaît comme la dimension de l'espace de cycles évanescents de $\mathcal{F}$ le long d'une fonction lisse assez générale dont le graphe de sa différentielle coupe transversalement $T^*_{X_\alpha}X$ ([17]).

4 Le faisceau d'irrégularité et le cycle d'irrégularité

(4-1) Soit $(X,\mathcal{O}_X)$ une variété analytique complexe non singulière et $\mathcal{D}_X$ le faisceau des opérateurs différentiels sur X. Si $\mathcal{M}$ est un $\mathcal{D}_X$-module à gauche cohérent, son cycle caractéristique CCh$(\mathcal{M})$ est un cycle positif, involutif et homogène du fibré cotangent T^*X. Un $\mathcal{D}_X$-module holonome est un module cohérent dont le cycle caractéristique est lagrangien. Si $\mathcal{M}$ est holonome alors son complexe des solutions holomorphes $\mathbb{R}\hom_{\mathcal{D}_X}(\mathcal{M},\mathcal{O}_X)$ est un faisceau pervers sur X.

(4-2) Si Y est une sous-variété de X définie par un idéal $\mathcal{I}_Y$, notons

$$\mathcal{O}_{X\hat{|}Y} := \lim_{\leftarrow k} \mathcal{O}_X/\mathcal{I}_Y^k,$$

le complété formel de X le long de Y et $\mathcal{O}_{X|Y}$ la restriction du faisceau des fonctions $\mathcal{O}_X$ à Y. Le faisceau $\mathcal{D}_X$ opère à gauche sur les faisceaux $\mathcal{O}_{X|Y}$ et $\mathcal{O}_{X\hat{|}Y}$. Le faisceau $\mathcal{O}_{X|Y}$ est un sous-$\mathcal{D}_X$-module du faisceau $\mathcal{O}_{X\hat{|}Y}$ et on note $\mathcal{Q}_Y$ le quotient ainsi obtenu, d'où une suite exacte de $\mathcal{D}_X$-modules à gauche

$$0 \to \mathcal{O}_{X|Y} \to \mathcal{O}_{X\hat{|}Y} \to \mathcal{Q}_Y \to 0.$$

On a alors le théorème suivant ([22]):

Théorème 4-3. *Si Y est une hypersurface de X et $\mathcal{M}$ un $\mathcal{D}_X$-module holonome le complexe $\mathbb{R}\hom_{\mathcal{D}_X}(\mathcal{M},\mathcal{Q}_Y)$, décalé d'une unité vers la droite, est un faisceau pervers sur X.*

La démonstration de ce théorème utilise entre autre le théorème de Cauchy et Kowalewski pour les fonctions holomorphes.

Définition 4-4. *Pour un triplet $(X,Y,\mathcal{M})$ comme dans le théorème précédent, on appelle faisceau d'irrégularité de $\mathcal{M}$ le long de Y le complexe*

$$\mathbb{R}\hom_{\mathcal{D}_X}(\mathcal{M},\mathcal{Q}_Y)$$

et on le note Irr$_Y(\mathcal{M})$.

Par construction, le faisceau Irr$_Y(\mathcal{M})$ est porté par Y. Un complexe sur Y est un faisceau pervers sur Y si, en tant que complexe sur X décalé vers la droite de la codimension de Y dans X, c'est un faisceau pervers sur X. Donc, si Y est une hypersurface, $\text{Irr}_Y(\mathcal{M})$ est un faisceau pervers sur Y. Si on note $\text{Mh}(D_X)$ la catégorie des D_X-modules holonomes sur X et $\text{Perv}(\mathbb{C}_Y)$ la catégorie des faisceaux pervers sur Y, on obtient un foncteur

$$\text{Irr}_Y \colon \text{Mh}(D_X) \longrightarrow \text{Perv}(\mathbb{C}_Y).$$

Ce foncteur est un foncteur exact de catégories abéliennes. En effet le foncteur Irr_Y transforme, par construction, une suite exacte de D_X-modules holonomes en un triangle distingué de la catégorie triangulée des complexes constructibles, mais un triangle distingué de faisceaux pervers est, en fait, une suite exacte de la catégorie abélienne des faisceaux pervers (cf. [1]).

Remarque 4-5. *On peut montrer qu'en général l'image essentiel du foncteur* Irr_Y *n'est pas n'est pas la catégorie* $\text{Perv}(\mathbb{C}_Y)$ *toute entière. Autrement dit tout faisceau pervers n'est pas de la forme* $\text{Irr}_Y(\mathcal{M})$. *Donc se pose la question de caractériser l'image essentiel du foncteur* Irr_Y. *Cette question est pour l'instant inaccessible en toute dimension, mais peut être testée en dimension deux.*

(4-6) Si on part d'une variété algébrique complexe non singulière X, un diviseur Y et $\mathcal{M}$ un D_X-module holonome, on définit le faisceau $\text{Irr}_Y(\mathcal{M})$ comme le faisceau d'irrégularité du module transcendant associé. C'est un faisceau pour la topologie transcendante sauf qu'il est algébriquement constructible. Nous allons voir que le cycle caractéristique $\text{CCh}\big(\text{Irr}_Y(\mathcal{M})\big)$ peut se définir par voie algébrique et garde un sens au-dessus d'un corps de base de caractéristique nulle.

Soit U le complémentaire de Y dans X, j l'inclusion de U dans X et i l'inclusion de Y dans X. Si $\mathcal{M}$ est un D_X-module holonome, notons $\mathcal{M}(*Y)$ l'image directe par j de la restriction de $\mathcal{M}$ à U qui est encore un D_X-module holonome. On a alors la proposition (cf. [22]):

Proposition 4-7. *Pour un triplet* $(X, Y, \mathcal{M})$, *le faisceau* $\text{Irr}_Y(\mathcal{M})$ *est isomorphe dans la catégorie dérivée à la restriction du complexe*

$$\mathbb{R}\hom_{D_{X^{an}}}\big((\mathcal{M}(*Y))^{an}, \mathcal{O}_{X^{an}}\big)[+1]$$

à Y.

En vertu de la proposition précédente, on a une suite exacte de faisceaux pervers sur X:

$$0 \longrightarrow j_! j^{-1}\mathbb{R}\hom_{D_{X^{an}}}\big((\mathcal{M}(*Y))^{an}, \mathcal{O}_{X^{an}}\big) \longrightarrow$$
$$\mathbb{R}\hom_{D_{X^{an}}}\big((\mathcal{M}(*Y))^{an}, \mathcal{O}_{X^{an}}\big) \longrightarrow \text{Irr}_Y(\mathcal{M})[-1] \to 0.$$

Le cycle caractéristique du faisceau $\text{Irr}_Y(\mathcal{M})[-1]$ apparaît comme la différence du cycle du faisceau

$$\mathbb{R}\hom_{D_{X^{an}}}\big((\mathcal{M}(*Y))^{an}, \mathcal{O}_{X^{an}}\big)$$

et du cycle du faisceau

$$j_!j^{-1}\,\mathbb{R}\hom_{\mathcal{D}_{X^{an}}}\big((\mathcal{M}(*Y))^{an},\mathcal{O}_{X^{an}}\big),$$

c'est-à-dire

$$\mathrm{CCh}(\mathrm{Irr}_Y(\mathcal{M})[-1]) = \mathrm{CCh}(\mathbb{R}\hom_{\mathcal{D}_{X^{an}}}((\mathcal{M}(*Y))^{an},\mathcal{O}_{X^{an}})$$
$$- \mathrm{CCh}(j_!j^{-1}\,\mathbb{R}\hom_{\mathcal{D}_{X^{an}}}((\mathcal{M}(*Y))^{an},\mathcal{O}_{X^{an}}).$$

Si Y est une sous-variété d'une variété algébrique X non singulière au-dessus d'un corps de caractéristique nulle, notons U le complémentaire de Y dans X et j l'inclusion canonique de U dans X. Au prolongement par zéro à X, $j_!$, des fonctions constructibles sur U à valeurs entières correspond, par l'isomorphisme d'Euler, le prolongement, $j_!$, des cycles lagrangiens homogènes $T^*(U)$ en cycles de $T^*(X)$.

Définition 4-8. *Soit un triplet $(X,Y,\mathcal{M})$ comme ci-dessus, où l'on suppose que Y est un diviseur. On définit le cycle d'irrégularité de $\mathcal{M}$ le long de Y comme le cycle lagrangien homogène*

$$\mathrm{CCh}\big(\mathrm{Irr}_Y(\mathcal{M})\big) = \mathrm{CCh}\big(\mathcal{M}(*Y)\big) - j_!j^{-1}\,\mathrm{CCh}(\mathcal{M}).$$

Pour un triplet $(X,Y,\mathcal{M})$ le cycle $\mathrm{Irr}_Y(\mathcal{M})$ est de la forme $\sum_\alpha m_\alpha \overline{T^*_{Y_\alpha}X}$ pour des sous-variétés Y_α de Y.

Théorème 4-9. *Pour un triplet $(X,Y,\mathcal{M})$ comme dans la définition précédente, le cycle $\mathrm{Irr}_Y(\mathcal{M})$ est positif.*

Pour la démonstration on invoque le principe de Lefschetz pour se ramener au cas où le corps de base est $\mathbb{C}$, où c'est une conséquence du théorème 3-5. Nous ne connaissons pas pour l'instant une démonstration purement algébrique de ce résultat. Si $(X,Y,\mathcal{M})$ est un triplet comme ci-dessus, on obtient une famille (Y_α, m_α) où Y_α est une sous-variété fermée de Y et m_α un entier positif. Nous allons indiquer deux conséquences du théorème de positivité.

Supposons que $\mathcal{M}$ est lisse sur U, supposé connexe, c'est à dire que c'est un fibré à connexion intégrable de rang r. Si la connexion est régulière, la variété caractéristique de $\mathcal{M}(*Y)$ est égale à celle de $\mathcal{O}_X(*Y)$ qui est déterminée par l'éclatement de Nash de Y (cf. [17]).

Mais si la connexion est irrégulière on a

$$\mathrm{CCh}\big(\mathcal{M}(*Y)\big) = r \cdot \mathrm{CCh}\big(\mathcal{O}_X(*Y)\big) + \mathrm{CCh}\big(\mathrm{Irr}_Y(\mathcal{M})\big),$$

et donc, pour r fixé, le cycle $r \cdot \mathrm{CCh}\big(\mathcal{O}_X(*Y)\big)$ est une borne inférieure des cycles caractéristiques des connexions intégrables de rang r. Si de plus X est propre, on obtient la généralisation ([22]) de la formule de Deligne:

$$\chi\big(U;\mathrm{DR}(\mathcal{M}_U)\big) = r \cdot \chi\big(U;\mathrm{DR}(\mathcal{O}_U)\big) + (-1)^{\dim U}\sum_\alpha m_\alpha T^*_X X \cdot \overline{T^*_{Y_\alpha}X},$$

où $T^*_X X \cdot \overline{T^*_{Y_\alpha}X}$ désigne le nombre d'intersection de cycles lagrangiens du fibré cotangent T^*X.

5 La filtration du faisceau d'irrégularité

(5-1) Soit X une variété analytique complexe et Y une hypersurface lisse. Si (t, x) est un système de coordonnées locales tel que Y est définie par $t = 0$, un germe du faisceau $\mathcal{O}_{X\hat{|}Y}$ des fonctions formelles le long de Y est une série $\sum_k a_k(x)t^k$, où $a_k(x)$ est une série de fonctions holomorphes sur Y définies sur le même ouvert de Y. Le faisceau $\mathcal{O}_{X\hat{|}Y}$ admet, comme à une variable, une filtration indexée par les nombres réels $s \geq 1$ par les sous-faisceaux $\mathcal{O}_{X|Y,s}$ des fonctions de classe de Gevrey d'ordre s. Une série $\sum_k a_k(x)t^k$ est de classe de Gevrey d'ordre s si la série

$$\sum_k \frac{a_k(x)}{k!^{s-1}}t^k$$

est convergente. C'est une condition qui ne dépend pas des coordonnées. On a donc $\mathcal{O}_{X|Y} = \mathcal{O}_{X|Y,1}$ et on pose $\mathcal{O}_{X|Y,\infty} := \mathcal{O}_{X\hat{|}Y}$. Le faisceau $\mathcal{D}_X$ opère à gauche sur $\mathcal{O}_{X|Y,s}$ pour tout s et, en fait, la filtration de $\mathcal{O}_{X\hat{|}Y}$ par les sous-faisceaux $\mathcal{O}_{X|Y,s}$ est une filtration de sous-$\mathcal{D}_X$-modules à gauche. Pour tous $s, s' \geq 1, s \leq s'$, on définit le $\mathcal{D}_X$-module à gauche $\mathcal{Q}(s, s')$ par la suite exacte de $\mathcal{D}_X$-modules à gauche

$$0 \to \mathcal{O}_{X|Y,s} \to \mathcal{O}_{X|Y,s'} \to \mathcal{Q}(s, s') \to 0.$$

Théorème 5-2. *Soit un triplet $(X, Y, \mathcal{M})$ comme ci-dessus, pour tous s, s', $1 \leq s \leq s' \leq \infty$, le complexe $\mathbb{R}\hom_{\mathcal{D}_X}(\mathcal{M}, \mathcal{Q}(s, s'))$ est un faisceau pervers sur Y.*

Pour $s = 1$ et $s' = \infty$ le théorème 5-2 se réduit au théorème 4-3. La démonstration de ce théorème ([22]) utilise le théorème de Cauchy pour les fonctions de classe de Gevrey de type r, pour un r fixe ([12]).

Définition 5-3. *Pour tout réel $s \geq 1$, on appelle faisceau d'irrégularité d'ordre s de $\mathcal{M}$ le long de Y, et on note $\mathrm{Irr}_Y^{(s)}(\mathcal{M})$, le complexe $\mathbb{R}\hom_{\mathcal{D}_X}(\mathcal{M}, \mathcal{Q}(1, s))$.*

Pour tous $s, 1 \leq s \leq \infty$, on a donc un foncteur

$$\mathrm{Irr}_Y^{(s)} \colon \mathrm{Mh}(\mathcal{D}_X) \longrightarrow \mathrm{Perv}(\mathbb{C}_Y).$$

Ce foncteur est donc exact.

Corollaire 5-4. *Pour tous $s, s', 1 \leq s \leq s' \leq \infty$ le faisceau $\mathrm{Irr}_Y^{(s)}(\mathcal{M})$ est un sous-faisceau du faisceau $\mathrm{Irr}_Y^{(s')}(\mathcal{M})$.*

En effet on a une suite exacte de $\mathcal{D}_X$-modules à gauche

$$0 \to \mathcal{Q}(1, s) \to \mathcal{Q}(1, s') \to \mathcal{Q}(s, s') \to 0.$$

Mais le foncteur $\mathbb{R}\hom_{\mathcal{D}_X}(\mathcal{M}, ?)$ transforme suite exacte en triangle distingué. Or un triangle distingué de faisceaux pervers est une suite exacte de faisceaux pervers.

En particulier les faisceaux $\operatorname{Irr}_Y^{(s)}(\mathcal{M})$, $1 \le s \le \infty$, constituent une filtration croissante du faisceau d'irrégularité

$$\operatorname{Irr}_Y(\mathcal{M}) := \operatorname{Irr}_Y^{(\infty)}(\mathcal{M}).$$

Un faisceau pervers étant localement de longueur finie, la filtration du faisceau d'irrégularité n'a, localement, qu'un nombre finie de sauts.

Définition 5-5. *Pour un triplet $(X, Y, \mathcal{M})$ comme ci-dessus, pour tout $1 \le s \le \infty$, on définit $\operatorname{Irr}_Y^{(<s)}(\mathcal{M})$ comme la réunion des faisceaux d'irrégularité d'ordre strictement inférieur à s, c'est donc un sous-faisceau pervers de $\operatorname{Irr}_Y^{(s)}(\mathcal{M})$.*

Définition 5-6. *Sous les conditions de la définition précédente, on définit le faisceau $\operatorname{Gr}_s\big(\operatorname{Irr}_Y(\mathcal{M})\big)$ par la suite exacte*

$$0 \to \operatorname{Irr}_Y^{(<s)}(\mathcal{M}) \to \operatorname{Irr}_Y^{(s)}(\mathcal{M}) \to \operatorname{Gr}_s\big(\operatorname{Irr}_Y(\mathcal{M})\big) \to 0.$$

On dit que s est un saut ou un indice critique en un point x de Y, si x appartient au support du faisceau $\operatorname{Gr}_s(\operatorname{Irr}_Y(\mathcal{M}))$. Si s est un saut en x on dit que $1/(s-1)$ est une pente en x du faisceau d'irrégularité.

L'ensemble des pentes est localement fini sur Y. On a

$$\chi\big(\operatorname{Irr}_Y(\mathcal{M})\big) := \sum_s \chi\big(\operatorname{Gr}_s\big(\operatorname{Irr}_Y(\mathcal{M})\big)\big).$$

(5-7) Si Y est une hypersurface éventuellement singulière de X, une fonction formelle le long de Y n'a plus un développement unique en série de puissances par rapport à une équation réduite de Y. Aussi, la filtration de Gevrey de $\mathcal{O}_{X \upharpoonright Y}$ n'a plus à priori de sens de façon évidente (cf. 5-9). Cependant la filtration Gevrey du faisceau d'irrégularité $\operatorname{Irr}_Y(\mathcal{M})$ garde encore un sens dans le cas singulier. En effet, soit $f: X \to \mathbb{C}$ une équation locale de Y, considérons le morphisme $\Delta_f : X \to X \times \mathbb{C}$, graphe de f. Notons $\overline{\mathcal{M}}$ le $\mathcal{D}_{X \times \mathbb{C}}$-module image directe de $\mathcal{M}$ par Δ_f. Le faisceau $\operatorname{Irr}_Y(\mathcal{M})$ est alors isomorphe à $\Delta_f^{-1}\big(\operatorname{Irr}_{X \times 0}(\overline{\mathcal{M}})\big)[1]$. On transpose alors la filtration Gevrey du faisceau $\operatorname{Irr}_{X \times 0}(\overline{\mathcal{M}})$ au faisceau $\operatorname{Irr}_Y(\mathcal{M})$. Cette filtration locale ne dépend pas de l'équation locale de Y ([22]). Comme la catégorie des faisceaux pervers est un champ, cette filtration locale se recolle pour donner naissance à une filtration de Gevrey globale du faisceau $\operatorname{Irr}_Y(\mathcal{M})$. Les pentes du faisceau $\operatorname{Irr}_Y(\mathcal{M})$ sont donc définies.

Remarque 5-8. *On peut aussi montrer que tout faisceau pervers sur Y n'est pas de la forme $\operatorname{Irr}_Y^{(s)}(\mathcal{M})$.*

(5-9) Comme le demande le Referee on peut définir le faisceau des fonctions de classes de Gevrey d'ordre s même si Y est une hypersurface singulière d'une variété analytique complexe.

Définition 5-10. *Soit Y une hypersurface d'une variété analytique complexe X et s une nombre réel ≥ 1, on définit le sous-faisceau $\mathcal{O}_{X|Y,s}$ des fonctions de* *classes de Gevrey d'ordre s du faisceau $\mathcal{O}_{X\hat{\upharpoonright}Y}$ comme le faisceau des germes des fonction formelles qui sont de classe de Gevrey d'ordre s aux points lisses de Y.*

Pour un triplet $(X, Y, \mathcal{M})$ on peut donc définir le complexe $\operatorname{Irr}_Y^s(\mathcal{M})$. Il faut voir maintenant qu'il est isomorphe au complexe obtenu par la méthode du graphe une fois choisie une équation de Y. D'où la filtration Gevrey du faisceau $\operatorname{Irr}_Y(\mathcal{M})$ pour un module holonome obtenue intrinsèquement. Soit $f\colon X \to \mathbb{C}$ une équation locale de Y, considérons le morphisme $\Delta_f\colon X \to X \times \mathbb{C}$, graphe de f. Notons $\overline{\mathcal{M}}$ le $\mathcal{D}_{X\times\mathbb{C}}$-module image directe de $\mathcal{M}$ par Δ_f. On a alors la proposition:

Proposition 5-11. *Pour tout $\mathcal{D}_X$-module cohérent $\mathcal{M}$ et pour tout réel s, $1 \leq s \leq \infty$ on a un isomorphisme*

$$\mathbb{R}\hom_{\mathcal{D}_{X\times\mathbb{C}}}(\overline{\mathcal{M}}, \mathcal{O}_{X\times\mathbb{C}|X\times 0,s}) \to \mathbb{R}\hom_{\mathcal{D}_X}(\mathcal{M}, \mathcal{O}_{X|Y,s})[1].$$

La proposition est une conséquence du cas $\mathcal{M} = \mathcal{D}_X$, c'est à dire d'un isomorphisme

$$f^*\mathcal{O}_{X\times\mathbb{C}|X\times 0,s} \to \mathcal{O}_{X|Y,s}$$

pour tout s, $1 \leq s \leq \infty$. Considérons d'abord le cas $s = \infty$ où on a l'isomorphisme de $\mathcal{D}_X$-modules à gauche

$$\mathcal{O}_{X\hat{\upharpoonright}Y} \to \mathbb{R}\hom_{\mathcal{O}_X}(\mathbb{R}\operatorname{alg}\Gamma_Y(\mathcal{O}_X), \mathcal{O}_X)$$

([22], Section 2). Mais les foncteurs $\mathbb{R}\hom_{\mathcal{O}_X}(-,-)$ et $\mathbb{R}\operatorname{alg}\Gamma_Y(-)$ commutent au foncteur image inverse. On en déduit l'isomorphisme pour $s = \infty$. L'image inverse du faisceau des fonctions de classes de Gevrey d'ordre $s \geq 1$ est contenue dans le faisceau des fonctions de classe de Gevrey d'ordre s en dehors des singularités de Y donc dans le faisceau $\mathcal{O}_{X|Y,s}$. On obtient le morphisme dans la cas $s \geq 1$ qui est alors injectif. Le principe de Hartogs suivant montre que ce morphisme est surjectif:

Proposition 5-12. *Soit Z une hypersurface non singulière d'une variété analytique complexe, alors toute fonction formelle le long de Z qui est de classe de Gevrey d'ordre $s \geq 1$ en dehors d'un ensemble analytique de Y de codimension au moins un est de classe de Gevrey d'ordre s partout sur Z.*

Si l'ensemble analytique est lisse alors la proposition est une conséquence de la formule de Cauchy. Le cas général se réduit à ce cas par récurrence sur la dimension.

(5-13) La filtration Gevrey du faisceau d'irrégularité se transpose en filtration Gevrey de son cycle caractéristique. La filtration Gevrey du cycle caractéristique est purement algébrique et garde un sens au-dessus d'un corps de caractéristique nulle ([22]).

6   Le polygone de Newton d'un $\mathcal{D}_X$-module

Nous avons défini les pentes du faisceaux $\mathrm{Irr}_Y(\mathcal{M})$ par voie transcendante.
Nous allons voir comment on peut montrer que ces pentes sont égales aux
pentes définies par Y. Laurent ([13]) par voie purement algébrique et qui sont
des nombres rationnels.

6.1   Les pentes algébriques de $\mathcal{M}$ le long de Y.

Soit une variété algébrique non singulière X/k au-dessus d'un corps k de caractéristique nulle, Y une sous-variété fermée non singulière de X et Λ le fibré
conormal T_Y^*X de Y dans X. Pour tout nombre rationnel r, $1 \le r \le \infty$, on
définit une filtration $F_{Y_r}^m(\mathcal{D}_X)$ de $\mathcal{D}_X$ indéxée par les entiers m relatifs. Pour
$r = 1$, c'est la filtration par l'ordre des opérateurs différentiels. Pour $r = \infty$
c'est la V-filtration de la théorie des cycles évanescents. On peut construire les
filtrations F_{Y_r} pour $1 < r < \infty$ à partir du cas $r = 1$ et $r = \infty$ (cf. [13], [29]).

Supposons pour simplifier que Y est une hypersurface et soit (t, y); $y =$
$(y_1, \ldots)$ un système de coordonnées locales tel que Y est définie par $t = 0$. Si
$r = p/q$, $1 \le r \le \infty$, où p et q sont des entiers positifs ou nuls premiers entre
eux en convenant que $1 = 1/1$ et $\infty = 1/0$, un monôme $t^\alpha y^l \partial_t^\beta \partial_y^m$ est de degré
$p(\beta - \alpha) + q(|m| + \alpha)$ pour la filtration F_{Y_r}.

Pour $1 < r < \infty$, le gradué $\mathrm{Gr}_{F_r}(\mathcal{D}_X)$ s'identifie au faisceau des fonctions
sur le fibré cotangent $T^*\Lambda$ de Λ. Si $\mathcal{M}$ est un $\mathcal{D}_X$-module cohérent, le gradué de
la filtration induite par la F_r-filtration de $\mathcal{D}_X$ sur $\mathcal{M}$ dans une présentation locale, définit un cycle positif $\widetilde{\sum}_Y^r(\mathcal{M})$ du fibré cotangent $T^*\Lambda$ qui ne dépend pas
de la présentation choisie. La variété micro-caractéristique de type r, $\sum_Y^r(\mathcal{M})$,
est le support du cycle $\widetilde{\sum}_Y^r(\mathcal{M})$.

Le fibré $T^*\Lambda$ admet deux actions du groupe multiplicatif k^*, l'une est
induite par l'action sur les fibres de la projection naturelle $\Lambda \to Y$, l'autre
étant l'action naturelle sur les fibres de $T^*\Lambda \to \Lambda$.

On a les résultats suivants ([13]): localement sur Y il n'y a qu'un nom-
bre fini de rationnels r tels que les variétés $\sum_Y^r(\mathcal{M})$ ne sont pas bi-homogènes
pour tout $\mathcal{D}_X$-module cohérent $\mathcal{M}$; les variétés micro-caractéristiques $\sum_Y^r(\mathcal{M})$
sont lagrangiennes pour tout $\mathcal{D}_X$-module holonome $\mathcal{M}$. Pour tout rationnel
$r, 1 < r < \infty$ et pour tout $\mathcal{D}_X$-module cohérent $\mathcal{M}$, notons $I_Y^{(r)}(\mathcal{M})$ l'adhérence
de la projection sur Y des composantes non bi-homogènes de la variété micro-
caractéristique $\sum_Y^r(\mathcal{M})$. Les variétés fermées $I_Y^{(r)}(\mathcal{M})$ forment donc un ensemble localement fini sur Y.

Définition 6-1. *Pour tout $\mathcal{D}_X$-module cohérent $\mathcal{M}$, on dit qu'un nombre ration-*
nel r, $1 < r < \infty$ est un indice critique, ou que $s = 1/(r - 1)$ est une pente, de
$\mathcal{M}$ le long de Y en un point de Y, si $I_Y^{(r)}(\mathcal{M})$ passe par ce point.

Si on part d'une variété analytique complexe non singulière X, d'une
hypersurface non singulière Y et d'un $\mathcal{D}_X$-module holonome $\mathcal{M}$, on construit

pour tout rationnel $r, 1 < r < \infty$ les variétés micro-caractéristiques $\sum_Y^r(\mathcal{M})$ qui sont alors des variétés analytiques complexes algébriques sur les fibres de la projection $T^*\Lambda \to Y$. Les adhérences des projections $I_Y^{(r)}(\mathcal{M})$ des composantes non bi-homogènes de $\sum_Y^r(\mathcal{M})$ sont donc des ensembles analytiques fermées de Y.

6.2 Le théorème de comparaison des pentes

Soit X une variété complexe non singulière algébrique ou analytique, Y une hypersurface non singulière et $\mathcal{M}$ un D_X-module holonome. Alors on a, d'une part les gradués $\mathrm{Gr}_r\big(\mathrm{Irr}_Y(\mathcal{M})\big)$ associés à la filtration $\mathrm{Irr}_Y^{(r)}(\mathcal{M})$ indéxée par les nombres réels $r \geq 1$ du faisceau $\mathrm{Irr}_Y(\mathcal{M})$ et d'autre part les variétés $I_Y^{(r)}(\mathcal{M})$ indéxées par les nombres rationnels $r \geq 1$. Convenons que pour un nombre réel $r \geq 1$ qui n'est pas rationnel $I_Y^{(r)}(\mathcal{M})$ est vide.

On a le théorème ([14]):

Théorème 6-2. *Soit un triplet $(X, Y, \mathcal{M})$ comme ci-dessus, alors le support du faisceau $\mathrm{Gr}_r\big(\mathrm{Irr}_Y(\mathcal{M})\big)$ est égal à la variété $I_Y^{(r)}(\mathcal{M})$ pour tout $r \geq 1$.*

La démonstration du théorème de comparaison utilise le théorème de Cauchy pour les fonctions de classe de Gevrey de type r, pour tout r simultanément, ce qui n'est valable que pour les modules holonomes.

6.3 Le théorème d'intégralité.

Soit un triplet $(X, Y, \mathcal{M})$ comme ci-dessus, en vertu du théorème de comparaison des pentes, si le faisceau $\mathrm{Gr}_r\big(\mathrm{Irr}_Y(\mathcal{M})\big)$ est non nul, donc si la variété $I_Y^{(r)}(\mathcal{M})$ est non vide, le nombre r est rationnel. En fait, on a le théorème plus précis ([14]):

Théorème 6-3. *Soit un triplet $(X, Y, \mathcal{M})$ comme ci-dessus, alors pour tout $r \geq 1$ la fonction constructible sur Y $(r-1)\chi(\mathrm{Gr}_r\big(\mathrm{Irr}_Y(\mathcal{M})\big))$, à priori à valeurs rationnelles, est à valeurs entières.*

Nous attirons l'attention du lecteur que le théorème d'intégralité ne se déduit pas par spécialisation du point générique au point spécial et que sa démonstration ne se fait pas par récurrence sur la dimension contrairement au théorème de positivité.

6.4 Le polygone de Newton.

Ainsi, en vertu du théorème d'intégralité des pentes pour tout $r \geq 1$, la fonction constructible $(r-1)\chi\big(\mathrm{Gr}_r(\mathrm{Irr}_Y(\mathcal{M}))\big)$ sur Y est à valeurs entières pour un triplet $(X, Y, \mathcal{M})$ comme ci-dessus. En vertu du théorème de MacPherson ([16])

c'est la fonction d'Euler d'un cycle $C\big((r-1)\chi\big(\mathrm{Gr}_r(\mathrm{Irr}_Y(\mathcal{M}))\big)\big)$. On a donc une égalité entre cycles

$$C\big((r-1)\chi\big(\mathrm{Gr}_r(\mathrm{Irr}_Y(\mathcal{M}))\big)\big) = (r-1)\,C(\chi(\mathrm{Gr}_r(\mathrm{Irr}_Y(\mathcal{M}))).$$

De plus en vertu du théorème de positivité, c'est là une égalité entre cycles positifs.

Si on écrit le cycle $C\big(\chi(\mathrm{Irr}_Y(\mathcal{M}))\big)$ de façon unique

$$C\big(\chi\big(\mathrm{Irr}_Y(\mathcal{M})\big)\big) = \sum_\alpha m_\alpha Y_\alpha,$$

pour des sous-variétés irréductibles Y_α de Y et des entiers positifs m_α, le cycle

$$C\big(\chi\big(\mathrm{Gr}_r(\mathrm{Irr}_Y(\mathcal{M}))\big)\big)$$

s'écrit de façon unique

$$C\big(\chi\big(\mathrm{Gr}_r(\mathrm{Irr}_Y(\mathcal{M}))\big)\big) = \sum_\alpha m_{\alpha,r} Y_\alpha,$$

avec des entiers $m_{\alpha,r} \geq 0$. De l'égalité

$$\chi(\mathrm{Irr}_Y(\mathcal{M})) = \sum_r \chi(\mathrm{Gr}_r(\mathrm{Irr}_Y(\mathcal{M}))),$$

on déduit l'égalité

$$m_\alpha = \sum_r m_{\alpha,r}.$$

Si $1 < r_1 < r_2 < \cdots < r_l < \infty$ sont les indices critiques en un point, pour chaque Y_α intervenant dans le cycle d'irrégularité au voisinage de ce point, les points à coordonnées entières du plan réel

$$\big((r_1-1)m_{\alpha,r_1}, m_{\alpha,r_1}\big), \quad \big((r_2-1)m_{\alpha,r_2}, m_{\alpha,r_2}\big), \quad \ldots \quad \big((r_\ell-1)m_{\alpha,r_\ell}, m_{\alpha,r_\ell}\big),$$

sont les sommets du polygone de Newton $N(\mathcal{M}, Y)_\alpha$ de $\mathcal{M}$ le long de Y_α. Ainsi, les polygones de Newton de $\mathcal{M}$ au voisinage d'un point sont indéxés par les composantes irréductibles de la variété caractéristique du faisceau d'irrégularité $\mathrm{Irr}_Y(\mathcal{M})$.

6.5 Semi-continuité du polygone de Newton

Les pentes en un point forment un ensemble semi-continu supérieurement par spécialisation par construction. Mais à priori il n'est pas clair que le polygone de Newton varie de façon semi-continu par spécialisation. Considérons un triplet $(\mathbb{C}^2, Y, \mathcal{M})$ où donc Y est une courbe. En point y_0 de Y il y a le polygone de Newton $N(\mathcal{M}, Y)_0$ au point spécial et le polygone au point générique

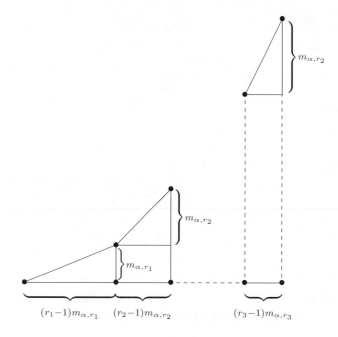

Figure 6.2: Le polygon de Newton $N(\mathcal{M}, Y)_\alpha$.

$N(\mathcal{M}, Y)_1$. Il résulte de la structure des faisceaux pervers portés par une courbe (cf. [23], [24]) que si le point y_0 est singulier les pentes du polygone générique sont nécessairement pentes du polygone spécial. Il est possible que cette propriété soit de nature analytique spéciale aux faisceaux de type $\mathrm{Gr}_r(\mathrm{Irr}_Y(\mathcal{M}))$ et non géométrique et demeure vraie même en un point lisse en ce sens que la variation du polygone de Newton est similaire à la variation des pentes cf. les Remarques 4-5 et 5-8.

7   Sur l'existence d'une équation fonctionnelle régulière

Comme application du théorème de comparaison des pentes, nous allons énoncer une conjecture de nature algébrique qui entraine que la nullité du faisceau d'irrégularité $\mathrm{Irr}_Y(\mathcal{M})$ d'un $\mathcal{D}_X$-modules holonome $\mathcal{M}$ le long d'une hypersurface Y, est équivalente à l'existence d'une équation fonctionnelle régulière le long de Y pour toute section locale de $\mathcal{M}$, ce qui constitue la forme la plus forte de la régularité. Donc par principe de Lefschetz, la nullité du cycle d'irrégularité d'un $\mathcal{D}_X$-module holonome le long d'une hypersurface est équivalente à l'existence d'une équation fonctionnelle régulière le long de cette hypersurface pour toute section locale, c'est à dire à l'existence d'un opérateur de type de Fuchs annulant cette section.

(7-1) Soit X une variété algébrique non singulière, Y une hypersurface non singulière de X et $\mathcal{M}$ un $\mathcal{D}_X$-module holonome. Soit t une équation locale de Y et ∂_t la dérivation correspondante. Rappelons que pour toute section locale non nulle u de $\mathcal{M}$, on a une équation fonctionnelle non triviale

$$b(t\partial_t)u = P(t, x, t\partial_t, \partial_x)tu$$

où b est un polynôme à une variable à coefficients dans le corps de base et P est un opérateur différentiel de $\mathcal{D}_X$ d'ordre zéro pour la $V(= F_\infty)$-filtration. La b-fonction ou le polynôme de Bernstein ou encore le polynôme de Bernstein-Sato, est le polynôme b_u générateur de l'idéal des polynômes réalisant une équation fonctionnelle non nulle. Il ne dépend que de Y et non de son équation t.

Définition 7-2. *Avec les notions précédentes, une équation fonctionnelle de* u

$$b_u(t\partial_t)u = P(t, x, t\partial_t, \partial_x)tu$$

est régulière si le F_1-ordre de P est inférieur ou égal au degré de b_u.

Définition 7-3. *On dit qu'un $\mathcal{D}_X$-module holonome $\mathcal{M}$ est 1-spécialisable le long de Y si toute section u de $\mathcal{M}$ admet une équation fonctionnelle*

$$b(t\partial_t)u = P(t, x, t\partial_t, \partial_x)tu,$$

telle que le F_1-ordre de P est $\leq$ degré de b

On ne demande pas que b soit égal à b_u.

Conjecture 7-4. *Pour un triplet $(X, Y, \mathcal{M})$ comme ci-dessus, les conditions suivantes sont équivalentes:*

a) *toute section locale admet une équation fonctionnelle régulière le long de Y,*

b) *$\mathcal{M}$ est 1-spécialisable,*

c) *$\mathcal{M}$ n'admet pas de pentes le long de Y.*

Remarque 7-5. *La conjecture précédente est vraie pour les modules spécialisables élémentaires.*

(7-6) Supposons que le corps de base est $\mathbb{C}$. Soit X une variété algébrique ou analytique complexe, Y une hypersurface non singulière et $\mathcal{M}$ un $s\mathcal{D}_X$-module holonome. Soit un triplet $(X, Y, \mathcal{M})$ comme ci-dessus, alors la conjecture précédente entraîne l'équivalence entre la nullité du faisceau $\mathrm{Irr}_Y(\mathcal{M})$ et l'existence d'une équation fonctionnelle régulière pour toute section locale de $\mathcal{M}$. En effet, si $\mathrm{Irr}_Y(\mathcal{M})$ est nul, les pentes transcendantes sont nulles, donc les pentes algébriques sont nulles et donc toute section locale admet une équation fonctionnelle régulière. Réciproquement, si toute section locale de $\mathcal{M}$ admet une équation fonctionnelle régulière le long de Y, $\mathcal{M}$ n'admet pas de pentes algébriques, donc n'admet pas de pentes transcendantes, et les faisceaux $\mathrm{Gr}_r\big(\mathrm{Irr}\,_Y(\mathcal{M})\big)$ sont nuls. Il s'ensuit que le faisceau $\mathrm{Irr}\,_Y(\mathcal{M})$ est nul lui aussi.

Kashiwara-Kawai ([9]) ont montré que tout module régulier dans toutes les directions est spécialisable. La démonstration de nature micro-différentiable passe par le théorème d'existence de Riemann et donc par le théorème général de la résolution des singularités.

On aura ramené l'existence d'une équation fonctionnelle régulière à la nullité du faisceau $\mathrm{Irr}\,_Y(\mathcal{M})$ qui est un problème de nature générique et ne repose pas sur le théorème d'existence de Riemann ni sur le théorème de résolution des singularités autrement que pour les courbes planes.

Prenons une fonction non constante $f\colon X \to \mathbb{C}$ et choisissons une coordonnée t sur $\mathbb{C}$, notons Δ_f le graphe de f et Y la fibre $X \times 0$. Considérons le faisceau de cohomologie locale

$$\mathrm{alg}\,H^1_{\Delta_f}(\mathcal{O}_{X\times\mathbb{C}}) := \mathcal{O}_{X\times\mathbb{C}}(*Y)/\mathcal{O}_{X\times\mathbb{C}}$$

de Δ_f à valeur dans $\mathcal{O}_{X\times\mathbb{C}}$ qui est un $\mathcal{D}_{X\times\mathbb{C}}$-module holonome engendré par la distribution de Dirac $\delta\big(t - f(x)\big)$. Le faisceau $\mathrm{Irr}\,_Y(\mathrm{alg}\,H^1_{\Delta_f}\big(\mathcal{O}_{X\times\mathbb{C}}\big))$ est isomorphe au faisceau $\mathrm{Irr}\,_{f^{-1}(0)}(\mathcal{O}_X)[-1]$. Mais le faisceau $\mathrm{Irr}\,_{f^{-1}(0)}(\mathcal{O}_X)[-1]$ est nul ([21]), c'est la forme locale du théorème de comparaison de Grothendieck ([4]). Il existera donc une équation fonctionnelle régulière:

$$b_{\delta(t-f(x))}(t\partial_t)\delta(t - f(x)) = P(t,x,t\partial_t,\partial_x)t\delta(t - f(x)),$$

sauf que dans le cas analytique il faut se restreindre au voisinage d'un polycylindre pour avoir les conditions noethériennes. Si m est l'ordre du polynôme $b_{\delta(t-f(x))}$ alors il existera un opérateur différentiel

$$(t\partial_t)^m - P_{m-1}(t,x,\partial_x)t(t\partial_t)^{m-1} - \cdots - P_0(t,x,\partial_x)t,$$

qui annule la distribution de Dirac $\delta(t - f(x))$, où l'opérateur P_{m-i} est de F_1-ordre au plus i. Autrement dit, par des considérations purement algébro-géométriques, on aura montré l'existence d'un opérateur de type de Fuchs, comme à une variable, dans l'annulateur de la distribution de Dirac $\delta\big(t-f(x)\big)$ et cela quelle que soit la nature de la singularité de l'hypersurface $f(x) = 0$. C'est là une illustration du théorème de type Lefschetz-Zariski ([5]) qui ramène l'étude du groupe fondamental du complémentaire d'une hypersurface dans un voisinage de l'origine de $\mathbb{C}^n$ au groupe fondamental de sa trace dans un deux-plan assez général. De plus, ceci montre que ces questions concernant les singularités, qui sont de nature homotopiques, homologiques ou cohomologiques ne font pas appel aux invariants les plus fins de ces singularités contrairement à la résolution de ces singularités. Ces observations sont particulièrement utiles dans l'étude de ce type de problèmes pour les singularités en caractéristiques positives.

Si on tient compte de la relation $t\delta\big(t - f(x)\big) = f(x)\delta\big(t - f(x)\big)$, de l'équation

$$b_{\delta\big(t-f(x)\big)}(t\partial_t)\delta\big(t - f(x)\big) = P(t,x,t\partial_t,\partial_x)t\delta\big(t - f(x)\big),$$

on obtient une équation de la forme

$$b_{\delta\big(t-f(x)\big)}(t\partial_t)\delta\big(t - f(x)\big) = \tilde{P}(x,t\partial_t,\partial_x)f(x)\delta\big(t - f(x)\big),$$

où l'opérateur $\tilde{P}$ est de F_1-ordre au plus m.

De l'isomorphisme $\mathcal{D}_X$-linéaire

$$\mathcal{D}_X[s]\mathbf{f}^s \simeq \mathrm{Gr}_0^V(\mathcal{D}_{X\times\mathbb{C}})\big(\delta(t - f(x))\big)$$

qui envoie s à $t\partial_t$, $f(x)$ à t, et $\mathbf{f}^s$ à $\delta\big(t - f(x)\big)$, on déduit une équation fonctionnelle

$$b_{\delta\big(t-f(x)\big)}\mathbf{f}^s = \tilde{P}(x,s,\partial_x)f(x)\mathbf{f}^s.$$

Si on écrit l'opérateur $\tilde{P}(x,s,\partial_x)$ sous la forme

$$P_m(x,\partial_x)s^m + P_{m-1}(x,\partial_x)s^{m-1} + \cdots + P_0(x,\partial_x),$$

alors l'opérateur P_{m-i} est de F_1-ordre au plus i. Motivée par une convergence d'une série d'opérateurs différentiels, une telle équation fonctionnelle a été conjecturée dans ([19] page 115). L'opérateur

$$b(s) - P_m(x)f(x)s^m - P_{m-1}(x,\partial_x)f(x)s^{m-1} - \cdots - P_0(x,\partial_x)f(x)$$

annule $\mathbf{f}^s$ et a la forme d'un bon opérateur de Kashiwara ([8]) dont il déduisait l'existence de la structure de la variété caractéristique du $\mathcal{D}_X$-module $\mathcal{D}_X[s]\mathbf{f}^s$ qui repose sur la résolution des singularités.

References

[1] Beilinson A.; Bernstein J.; Deligne P., Faisceaux Pervers, Astérisque **100** (1983).

[2] Deligne P., Equations différentielles à points singuliers réguliers, Lecture Notes in Math. **163** (1970).

[3] Gonzalez-Sprinberg; Verdier J.L., L'obstruction d'Euler locale et le Théorème de MacPherson, Astérisque **82-83** (1981), 7-32.

[4] Grothendieck A., On the De Rham Cohomologie of algebraic varieties, Publ. IHES **29** (1966), 93–103.

[5] Hamm H.; Le D.T., Un théorème de Zariski de type Lefschetz, Ann. Scient. Norm. Sup.,**6** (1973), 317–366.

[6] Ince E.L., Ordinary differential equations, Dover New York (1956).

[7] Kashiwara M., Index theorem for maximally overdetermined systems, Proc. Japan Acad. **49** (1973), 803–804.

[8] Kashiwara M., b-functions and holonomic systems, Inv. Math., **38** (1976), 33–53.

[9] Kashiwara M., Kawai T., Second Microlocalisation and Asymptotic Expansions, Lecture Notes in Physics, Springer **126** (1980), 21–76.

[10] Katz N., Wild ramification and some problems of "independence of ℓ", Amer. J. of Math. **105** (1983), 201–227.

[11] Komatsu H., An introduction to the theory of hyperfunctions, Lecture Notes in Math. **287** (1973), 1–40.

[12] Laurent Y., Théorie de la deuxième micro-localisation dans le domaine complexe, Progress in Math. **53** (1985), Birkhäuser.

[13] Laurent Y., Polygone de Newton et b-fonction pour les modules micro-différentiels, Ann. Scient. Norm. Sup., 4e série **20** (1987), 391–441.

[14] Laurent Y.; Mebkhout Z., Le polygone de Newton d'un $\mathcal{D}_X$-module, à paraître.

[15] Le D.T.; Tessier B., Variétés polaires locales et classe de Chern des variétés singulières, Ann. of Math. (2) **114** (1981), 457–491.

[16] MacPherson R., Chern Class for singular varieties, Ann. of Math. (2) **100** (1974), 423–432.

[17] Le D.T.; Mebkhout Z., Variétés caractéristiques et variétés polaires, C. R. Acad. Sc. Paris t. **296** (17 janvier 1983), 129–132.

[18] Malgrange B., Sur les points singuliers des équations différentielles, Ens. Math **20** (1974), 147–176.

[19] Mebkhout Z., La cohomologie locale d'une hypersurface, Lecture Notes in Math **670** (1977), 89–115.

[20] Mebkhout Z., Sur les cycles évanescents des systèmes différentiels, Proceeding de La Conf. La Rabida 1984, Travaux en Cours **24** (1987) Hermann, 35–47.

[21] Mebkhout Z., Le théorème de comparaison entre cohomologies de de Rham et le théorème d'existence de Riemann, Publ. Math. IHES **69** (1989), 47–89.

[22] Mebkhout Z., Le théoréme de positivité de l'irrégularité pour les $\mathcal{D}_X$-modules, The Grothendieck Festschrift III, Progress in Math. **88** (1990), 84–131.

[23] Narvaez L., "Cycles évanescents et faisceaux pervers: cas des courbes planes irréductibles", Compositio Math. 65, (1988) 321–347.

[24] Narvaez L., "Cycles évanescents et faisceaux pervers II: cas des courbes planes réductibles", à paraître dans London Math. Soc. Lect. Notes Series (Proceedings of the Lille Conference on "Singularities", June 1990).

[25] Ramis J.P., Théorèmes d'indice Gevrey pour les équations différentielles ordinaires, Mem. AMS **296** (1984).

[26] Robba Ph., On the index of p-adic differential operators I, Ann. of Math., 101 (1975), 280–316.

[27] Robba Ph., Indice d'un opérateur différentiel p-adique IV. Cas des Systèmes. Mesure de l'irrégularité dans un disque. Ann. Inst. Fourier, 35, 2 (1985), 13–55.

[28] Sabbah C., Quelques remarques sur la géométrie des espaces conormaux, Astérisque **130** (1985), 161–192.

[29] Sabbah C.; Castro P., Sur les pentes d'un $\mathcal{D}_X$-module le long d'une hypersurface, Prépublication 1989.

Address of author:

UFR de Mathématiques, UA 212,
Université de Paris 7,
2 place Jussieu, F-75251 Paris
France
Email: mebkhout@mathp7.jussieu.fr

How good are real pictures?

David Mond

1 Introduction

This paper is motivated by some very good real pictures, drawn by Victor Goryunov in [10], which we reproduce here. Figure 1.1 shows images of stable perturbations of the $\mathcal{A}_e$-codimension 1 singularities of mappings $\mathbb{C}^2 \to \mathbb{C}^3$. It is known (see [12], [17]) that the image of the complex mapping is homotopy equivalent to a 2-sphere; and Goryunov's real pictures showed a real image with the same homotopy type. Goryunov was able to demonstrate, by elementary, though rather long, proofs (which are omitted in the published paper!), that in each case the inclusion of the real in the complex image induced an isomorphism on the second homology group, so that his real pictures really did show the vanishing cycles associated with the codimension 1 singularities. The pictures in Figure 1.1 show stable perturbations of I: S_1 (birth of two cross-caps); II: Tangential contact of two immersed sheets; III: Cross-cap + immersed sheet; IV: Birth of two triple-points; and V: Quadruple point.

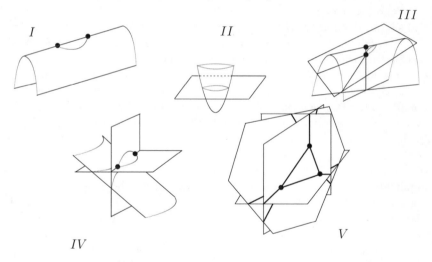

Figure 1.1: Good real pictures: the images of good perturbations of codimension 1 singularities of maps from surfaces to 3-space.

Two questions suggest themselves:

1. Is it possible to avoid the hard work of proving in each case that the vanishing cycles one sees in the real picture really are the complex vanishing cycles?

Progress in Mathematics, Vol. 134
© 1996 Birkhäuser Verlag Basel/Switzerland

2. How often can one find real forms for complex germs and their stable
   perturbations, such that the inclusion of the real image (or discriminant)
   in that of the complexification induces an isomorphism on the vanishing
   homology?

The first question has a straightforward answer:

Theorem 1-1. *Suppose that $f : \mathbb{R}^n, S \to \mathbb{R}^p, 0$ (with $n \geq p-1$ and S a finite set)
is a real-analytic map germ of finite $\mathcal{A}_e$-codimension, and that the complexification $f_{\mathbb{C},t}$ of the real perturbation f_t of f defines, on the domain $U \subseteq \mathbb{C}^n$,
a stable perturbation of the complexification $f_{\mathbb{C}}$ of f. Let $D(f_t)$ and $D(f_{\mathbb{C},t})$
be the discriminants of the mappings f_t and $f_{\mathbb{C},t}$. If rank $H_{p-1}(D(f_t);\mathbb{Z}) =$
rank $H_{p-1}(D(f_{\mathbb{C},t});\mathbb{Z})$ then the inclusion $D(f_t) \hookrightarrow D(f_{\mathbb{C},t})$ induces an isomorphism $H_{p-1}(D(f_t);\mathbb{Z}) \simeq H_{p-1}(D(f_{\mathbb{C},t});\mathbb{Z})$. (Here when $n = p - 1$, we understand discriminant to mean image).*

This is proved in Section 2 . The second question is more difficult. Here,
we limit ourselves to proving

Theorem 1-2. *For all $\mathcal{A}_e$-codimension 1 map-germs $\mathbb{C}^n, 0 \to \mathbb{C}^p, 0$ ($n \geq p$, (n,p)
nice dimensions), of corank 1 (i.e., such that the cokernel of $df(0) : \mathbb{C}^n \to \mathbb{C}^p$ is
1-dimensional) there exists a real form with a real stable perturbation to which
Theorem 1-1 applies.*

The proof makes use of Looijenga's description of the discriminant of a
real simple singularity ([13]), which we recall at some length in Section 4.

We call a real stable perturbation to which Theorem 1-1 applies a "good
real perturbation", and its image or discriminant a "good real picture": such
pictures faithfully illustrate, at least homologically, what is going on in complex
space. However, even bad real pictures are interesting. As is well known, given
a real 1-parameter deformation of an unstable map-germ, what we usually see
over the reals, as the parameter passes through 0, is a "catastrophic" change
in the topology of the image or discriminant. The most familiar example of
this is provided by the family of functions $f_t(x) = x^3 + tx$. When $t < 0$, the
discriminant of f_t consists of two points; as t passes through 0 they fuse and
annihilate one another. In Section 5 we investigate the generalisation of this
behaviour, to the real versal deformations of other germs of codimension 1 (or
more precisely, of discriminant Milnor number 1: see below). We show first,
by elementary Morse theory, that if f_t is a 1-parameter real deformation of
a real germ f_0 of codimension 1, with f_t stable for $t \neq 0$, then the real part
of the discriminant of the complexification $f_{\mathbb{C},t}$ of f_t is, for $t \neq 0$, homotopy
equivalent to a k-sphere, for some k between -1 and $p-1$ (a (-1)-sphere is by
definition empty). In the example above, the real part of the discriminant of
$f_{\mathbb{C},t}$ coincides with the discriminant of f_t, and so $k = 0$ for $t < 0$ and $k = -1$
for $t > 0$.

Now, the discriminants $D(f_{\mathbb{C},t})$, for $t \neq 0$, fit together into a locally trivial
fibration over $\mathbb{C} - 0$, analogous to the Milnor fibration of a hypersurface singularity. In the main theorem of this section, we show that, modulo some auxil-

iary hypotheses, the homological monodromy of this fibre bundle, acting on the $(p-1)$-st homology of the complex discriminant (in fact $H_{p-1}\big(D(f_{\mathbb{C},t};\mathbb{Z})\big) = \mathbb{Z}$) is simply multiplication by $(-1)^{\delta k}$, where δk is the change in the homotopy dimension k of the real part of $D(f_{\mathbb{C},t})$ as t passes through 0 on the real axis.

Similar statements hold for the images of stable perturbations of codimension 1 map-germs $\mathbb{R}^n, S \to \mathbb{R}^{n+1}, 0$.

Let us make the situation clearer. In a recent paper [8] J. Damon and the author have shown the following:

Theorem 1-3. *Let $f : \mathbb{C}^n, S \to \mathbb{C}^p, 0$, with $n \geq p$, (n,p) nice dimensions and S a finite set, have finite $\mathcal{A}_e$-codimension. If f_t is a stable perturbation of f, then the discriminant $D(f_t)$ of f_t has the homotopy type of a wedge of $(p-1)$-spheres, whose number, the "discriminant Milnor number" μ_Δ, satisfies*

$$\mu_\Delta \geq \mathcal{A}_e - \mathrm{codim}(f)$$

with equality if f is quasihomogeneous.

The proof in [8] is given only for the case where S consists of a single point, but is valid for any finite set S.

In similar vein, in [17] it is shown that if $f : \mathbb{C}^n, S \to \mathbb{C}^{n+1}, 0$ is of finite $\mathcal{A}_e$-codimension, then the image of a stable perturbation has the homotopy type of a wedge of n-spheres; from results of [12] it follows that if $n = 2$, the number of 2-spheres in the image is greater than or equal to $\mathcal{A}_e - \mathrm{codim}(f)$, with equality if f is quasihomogeneous. The same relation holds if $n = 1$ ([18]).

Thus, it is only the $(p-1)$-st homology of the discriminant (or image) of f_t that is of interest. If f is of $\mathcal{A}_e$-codimension 1, then the discriminant of a stable perturbation is homotopy equivalent to a wedge of one or more $(p-1)$-spheres; however, all of the examples dealt with in this paper are quasihomogeneous, and so in all these cases $D(f_t)$ is a homotopy-sphere.

The proof of Theorem 1-2 is based on the following classification theorem of V. V. Goryunov:

Theorem 1-4. (V.Goryunov, [9]) *i) Let $f_0 : \mathbb{C}^n, 0 \to \mathbb{C}, 0$ be a simple singularity (i.e., of type A_k, D_k, E_6, E_7 or E_8), and let $\varphi_1 = 1$, $\varphi_2, \ldots, \varphi_\mu$ be a homogeneous basis for $\mathcal{O}_{\mathbb{C}^n,0}/J_{f_0}$, with φ_μ the (unique) term of highest weight. Then the map-germ $f : \mathbb{C}^n \times \mathbb{C}^{\mu-2} \times \mathbb{C}^q \to \mathbb{C} \times \mathbb{C}^{\mu-2} \times \mathbb{C}^q$ defined by*

$$f(x, u, y) = \left(f_0(x) + \sum_{i=2}^{\mu-1} u_i \varphi_i + \sum_{j=1}^{q} y_j^2 \varphi_\mu, u, y \right)$$

has $\mathcal{A}_e$-codimension equal to 1. ii) All $\mathcal{A}_e$-codimension 1 map-germs $\mathbb{C}^n, 0 \to \mathbb{C}^p, 0$ ($n \geq p$, (n,p) nice dimensions) of corank 1 are of the type described in i).

We give a proof of part i) of this theorem in Section 3 below.

In [15] it is shown that the only map-germs $\mathbb{C}^2, 0 \to \mathbb{C}^3, 0$ beyond codimension 1 having good real perturbations are the members of the series H_k;

that is, the existence of good real pictures is relatively uncommon. However, we adventure the following conjecture:

Conjecture. Every codimension 1 germ in the nice dimensions has a real form with a good real perturbation.

The wording of the conjecture is deliberately vague, in order to include more phenomena (rather than to exclude the possibility of refutation!).

Convention. Throughout this paper we will be dealing with discriminants of maps $k^n \to k^p$ (where $k = \mathbb{R}$ or $\mathbb{C}$, and $n \geq p$), and with images of maps $k^n \to k^{n+1}$. Of course, the latter are also discriminants (sets of critical values), and so in Sections 2 and 5, where our arguments apply to both images and discriminants, we shall refer to both as discriminants, and denote them by $D(f)$.

The author is grateful to Victor Goryunov and C. T. C. Wall for helpful conversations, and to the Department of Algebra, Computation, Geometry and Topology of the University of Seville, for its hospitality and financial support during the period in which this paper was written.

2 Comparison of real and complex discriminants and images

In order to compare $D(f_t)$ and $D(f_{\mathbb{C},t})$, we have to introduce an intermediate object, $D(f_{\mathbb{C},t}) \cap \mathbb{R}^p$. Evidently we have $D(f_t) \subseteq D(f_{\mathbb{C},t}) \cap \mathbb{R}^p \subseteq D(f_{\mathbb{C},t})$, and theorem 1-1 is proved by showing first

Lemma 2-1. The inclusion $D(f_t) \hookrightarrow D(f_{\mathbb{C},t}) \cap \mathbb{R}^p$ induces an isomorphism of p-1'st integral homology groups.

And then

Lemma 2-2. Under the hypothesis of theorem 1-1, the inclusion $D(f_{\mathbb{C},t}) \cap \mathbb{R}^p \hookrightarrow D(f_{\mathbb{C},t})$ is a homotopy equivalence.

Before beginning the proofs of these lemmas, let us fix some notation: since $f_{\mathbb{C},t}$ is finite on its critical set, $D(f_{\mathbb{C},t})$ is an analytic hypersurface, and as $f_{\mathbb{C},t}$ is defined by real analytic functions, we may choose for $D(f_{\mathbb{C},t})$ a real defining equation g_t. In fact we may suppose that the family g_t varies analytically with t (and thus that the discriminant of the mapping $F(x,t) = (f_{\mathbb{C},t}(x),t)$ is defined by the equation $G(z,t) = g_t(z)$).

The proof of Lemma 2-1 is elementary: first, $D(f_{\mathbb{C},t}) \cap \mathbb{R}^p - D(f_t)$ is of dimension less than $p - 1$. To prove this, observe first that if $y \in \mathbb{R}^p$ is a regular point of g_t, then y can have only one preimage in the critical set of $f_{\mathbb{C},t}$; this critical point must therefore be real, and hence y lies in $D(f_t)$. Thus, $D(f_t) \supseteq D(f_{\mathbb{C},t})_{\text{reg}} \cap \mathbb{R}^p$. It follows that $D(f_{\mathbb{C},t}) \cap \mathbb{R}^p - D(f_t)$ is contained in the critical locus of g_t, and hence has dimension less than $p - 1$. (The reader will recognise this argument from [13]). Now a Mayer-Vietoris argument shows that the top homology groups of $D(f_{\mathbb{C},t}) \cap \mathbb{R}^p$ and $D(f_t)$ are the same.

The proof of Lemma 2-2 is a fairly straightforward application of Morse theory. In [21] D. Siersma shows that $D(f_{\mathbb{C},t}) = g_t^{-1}(0)$ has the homotopy type of a wedge of $(p-1)$-spheres, whose number μ_Δ satisfies

$$\mu_\Delta = \sum_{y \notin D(f_{\mathbb{C},t})} \mu(g_t, y).$$

This is proved simply by observing that up to homotopy, one obtains the complex epsilon ball from $D(f_{\mathbb{C},t})$ by gluing in p-cells to kill the homotopy of the Milnor fibres of the isolated singular points of g_t. Suppose now that the hypothesis of theorem 1-1 holds. By Lemma 2-1, rank $H^{p-1}\big(D(f_{\mathbb{C},t}) \cap \mathbb{R}^p\big) = \mu_\Delta$. Now $D(f_{\mathbb{C},t}) \cap \mathbb{R}^p$ is the 0-set of the function $g_t \colon \mathbb{R}^p \to \mathbb{R}$, and we can apply a similar Morse-theoretic analysis to Siersma's, only now in the real context. That is, the real ϵ-ball $B_{\mathbb{R},\epsilon}$ is obtained, up to homotopy, from the set $D(f_{\mathbb{C},t}) \cap \mathbb{R}^p$ by gluing in a cell as we pass each critical level of g_t, whose dimension is determined by the rank of the Hessian of g_t at its singular point. As the rank of $H^{p-1}\big(D(f_{\mathbb{C},t}) \cap \mathbb{R}^p\big)$ is equal to μ_Δ, we conclude that all of the complex critical points of g_t are in fact real, and, moreover, that all those whose critical value is less than 0 are local minima, and all those whose critical value is greater than 0 are local maxima. Now we return to the complex discriminant: its homotopy is killed by gluing in p-cells corresponding to the critical points of g_t; but because at each critical point of the function g_t it is equivalent by a real coordinate change (i.e., one that preserves the real subspace $\mathbb{R}^n$) to a sum of squares, or to a negative sum of squares, the p-cell that we glue in may be taken to be the same real p-cell that we glue in to kill the homotopy of $D(f_{\mathbb{C},t}) \cap \mathbb{R}^p$. Let us denote the union of these cells by E, and write $D(f_{\mathbb{C},t}) \cap \mathbb{R}^p = X$, $D(f_{\mathbb{C},t}) = Y$. Both $X \cup E$ and $Y \cup E$ are contractible, and so $H^q(Y \cup E, X \cup E; \mathbb{Z}) = 0$ for all q. Applying excision, we find $H^q(Y, X; \mathbb{Z}) = 0$ for all q, and hence the inclusion $X \hookrightarrow Y$ induces isomorphisms on all homology groups. If $p \geq 3$, then both X and Y are simply connected, and so by Whitehead's theorem, [22], IV.7.15, the inclusion is a weak homotopy equivalence, and thus a homotopy equivalence. When $p = 2$, it's easy to see that X has as deformation retract a wedge of μ_Δ circles, the boundaries of the bounded components of the complement of X in $\mathbb{R}^2$, to which Y also retracts, by Siersma's argument. This completes the proof of Lemma 2-2.

Lemmas 2-1 and 2-2 together of course prove theorem 1-1.

Remark 2-3. *It seems to be the case that under the hypothesis of theorem 1-1, the inclusion $D(f_t) \hookrightarrow D(f_{\mathbb{C},t}) \cap \mathbb{R}^p$ should induce a homotopy equivalence also. Unfortunately I have been unable to prove this. It is not enough simply to observe that $D(f_t)$ is the purely $(p-1)$-dimensional part of $D(f_{\mathbb{C},t}) \cap \mathbb{R}^p$; it is easy to construct examples of spaces showing that this hypothesis is not sufficient. It seems that in order to prove that $D(f_t) \hookrightarrow D(f_{\mathbb{C},t}) \cap \mathbb{R}^p$ is a homotopy equivalence, it will be necessary to use somehow a description of the normalisation of $D(f_t)$.*

Remark 2-4. *Theorem 1-1 of course applies to the case of the Milnor fibre of a hypersurface singularity defined by a real analytic function; indeed, its proof* consists essentially in reducing the general result to the case of the simplest non-trivial Milnor fibre, with $\mu = 1$. However, the range of hypersurface singularities for which it is possible to find a real Milnor fibre satisfying the hypotheses is very limited; for if the real fibre $X_{\mathbb{R}}$ is of positive dimension n, then the rank of $H_n(X_{\mathbb{R}})$ is equal to the number of its closed connected components (each of these is of course an compact oriented n-manifold without boundary), and so the sum of the Betti numbers is at least 2μ. However, the sum of the Betti numbers of the complex Milnor fibre is $\mu + 1$, and so by Smith theory ([3]; $X_{\mathbb{R}}$ is the fixed set of the $\mathbb{Z}_2$ action generated by complex conjugation) we have $2\mu \leq 1 + \mu$, and hence $\mu = 1$. Of course, if the fibre dimension is zero, then the singularity is of type A_k, and it is always possible to find a real Milnor fibre whose inclusion in the complex Milnor fibre is actually an equality.

Generalisation of the preceding observations to the case of an ICIS of embedding codimension greater than 1 can actually be proved very easily by Smith theory. That is, if $X_{\mathbb{R}}$ and $X_{\mathbb{C}}$ are n-dimensional real and complex Milnor fibres, of an n-dimensional ICIS X_0 defined by a real analytic map-germ, and the rank of $H_n(X_R; \mathbb{Z})$ is equal to μ, then by the universal coefficient theorem for homology, the n-th $\mathbb{Z}_2$-Betti number of $X_{\mathbb{R}}$ is greater than or equal to μ, and just as before the 0-th $\mathbb{Z}_2$ Betti number is also greater than or equal to μ. Hence, unless $n = 0$ we have $\mu = 1$. In the case $n = 0$, then provided the embedding dimension is 2 or the algebra $\mathcal{O}_{X_O}$ is of "discrete algebra type" one can again always find a complex Milnor fibre which is in fact real: this follows from a theorem of Damon and Galligo in [7]. For a versal deformation F of the ICIS X_0 is in particular a stable map-germ, whose local algebra has length equal to $1 + \mu$, where μ is the Milnor number of X_0. From 3.8 of [7], it follows that there are real points in the image of F whose real preimage consists of $1 + \mu$ points; such a set is a real Milnor fibre equal to the complex Milnor fibre.

3   Codimension 1 germs

We begin by proving the Theorem of Goryunov (Theorem 1-4, i).) In order to calculate the $\mathcal{A}_e$-codimension of the map-germs listed in Theorem 1-4, we make use not of the standard formula

$$\mathcal{A}_e - \mathrm{codim}(f) = \dim_{\mathbb{C}} N\mathcal{A}_e f,$$

where

$$N\mathcal{A}_e f = \frac{f^*\theta_{\mathbb{C}^p}}{df(\theta_{\mathbb{C}^n}) + f^{-1}(\theta_{\mathbb{C}^p})},$$

but of an auxiliary result of J. Damon ([6], but see also [19]) which we now describe: for every map-germ $f : \mathbb{C}^n, 0 \to \mathbb{C}^p, 0$ of finite singularity type, there exists (by results of Mather, [16]) a stable unfolding $F: \mathbb{C}^n \times \mathbb{C}^d, 0 \to \mathbb{C}^p \times \mathbb{C}^d, 0$, from which f may be recovered, by fibre product, via a germ $\gamma: \mathbb{C}^p, 0 \to \mathbb{C}^p \times \mathbb{C}^d, 0$. If we take F to be level-preserving, as we may, then γ takes the especially simple form $\gamma(y) = (y, 0)$. Let $\text{Derlog}(D(F))$ denote the module of germs of vector fields on $\mathbb{C}^p \times \mathbb{C}^d, 0$ which are tangent to $D(F)$. It is a free $\mathcal{O}_{\mathbb{C}^p \times \mathbb{C}^d, 0}$–module ([14], Corollary 6.13 page 108). Damon's result states that there is an isomorphism of $\mathcal{O}_{\mathbb{C}^p, 0}$-modules

$$\frac{f^* \theta_{\mathbb{C}^p}}{(\theta_{\mathbb{C}^n}) + f^{-1}(\theta_{\mathbb{C}^p})} \simeq \frac{\gamma^* \theta_{\mathbb{C}^p \times \mathbb{C}^d}}{d\gamma(\theta_{\mathbb{C}^p}) + \gamma^* \left(\text{Derlog}(D(F))\right)}.$$

The module on the right is known as $N\mathcal{K}_{D(F),e}\gamma$, since it is the (extended) normal space to the orbit of γ under a certain subgroup $\mathcal{K}_{D(F)}$ of the contact group $\mathcal{K}$. Moreover, it is not necessary to use an unfolding of f: if $F :: \mathbb{C}^m, 0 \to \mathbb{C}^q, 0$ is any stable map-germ from which f can be obtained by fibre product via some map-germ $\gamma : \mathbb{C}^p, 0 \to \mathbb{C}^q, 0$, (not necessarily an immersion) then $N\mathcal{A}_e f$ and $N\mathcal{K}_{D(F),e}\gamma$ are isomorphic.

One advantage of this means of calculating $\mathcal{A}_e$-codimension (f) is that both the numerator and the denominator in $N\mathcal{K}_{D(F),e}\gamma$ are finitely generated $\mathcal{O}_{\mathbb{C}^p}$-modules, and indeed the numerator is free. This is in sharp distinction to the previously mentioned formula, where the numerator is finite and free over $\mathcal{O}_{\mathbb{C}^p}$ only when $n = p$. The principal disadvantage in computations is that it is necessary to have an explicit list of the generators of $\text{Derlog}(D(F))$.

We now use Damon's Theorem to calculate $N\mathcal{K}_{D(F),e}\gamma$ in the case of the map-germs f listed in Theorem 1-4. In what follows, $\mathcal{O}$ denotes $\mathcal{O}_{\mathbb{C}^p}$. As F we take the standard miniversal deformation of the ICIS $f_0^{-1}(0)$, $F : \mathbb{C}^n \times \mathbb{C}^{\mu-1} \to \mathbb{C} \times \mathbb{C}^{\mu-1}, 0$ given by

$$F(x, u) = \left(f_0(x) + \sum_{i=2}^{\mu} u_i \varphi_i, u \right).$$

The map $\gamma : \mathbb{C} \times \mathbb{C}^{\mu-2} \times \mathbb{C}^q, 0 \to \mathbb{C} \times \mathbb{C}^{\mu-1}, 0$ inducing f from F is now given by

$$\gamma(z, u, y) = \left(z, u, \sum_{j=1}^{q} y_j^2 \right)$$

and thus we have

$$N\mathcal{K}_{D(F),e}\gamma = \frac{\mathcal{O}\left\{ \dfrac{\partial}{\partial z}, \dfrac{\partial}{\partial u_i} \right\}_{\{i=2,\ldots,\mu\}}}{\mathcal{O}\left\{ \dfrac{\partial}{\partial z}, \dfrac{\partial}{\partial u_i}, y_j \dfrac{\partial}{\partial u_\mu} \right\}_{\substack{\{j=1,\ldots,q\} \\ \{i=2,\ldots,\mu-1\}}} + \gamma^* \left(\text{Derlog}(D(F))\right)}$$

which is clearly isomorphic to

$$\frac{\mathcal{O}\left\{ \dfrac{\partial}{\partial u_\mu} \right\}}{\mathcal{O}\{(y_1, \ldots, y_q) + \mathcal{I}\} \dfrac{\partial}{\partial u_\mu}}$$

where $\mathcal{I}$ is the ideal in $\mathcal{O}$ generated by the coefficients of $\partial/\partial u_\mu$ in the members of $\mathrm{Derlog}\big(D(F)\big)$. It is clear that $N\mathcal{K}_{D(F),e}\gamma = 1$ if and only if this ideal is equal to $(z, u_2, \ldots, u_{\mu-1})$.

Lemma 3-1. *Provided that f_0 is a simple singularity, $\mathcal{I}$ is equal to*

$$(z, u_2, \ldots, u_{\mu-1}).$$

PROOF. This depends on a calculation of a set of generators of the module $\mathrm{Derlog}\big(D(F)\big)$, due to Zakalyukin [23] and, earlier, to V.I. Arnold, in [2]. We do not enter into details here; the important point is that one may choose a generating set $\xi_j, j = 1, \ldots, \mu$ such that the matrix of coefficients is symmetric. The symmetry is due to the fact that one can view the matrix as the matrix of the $\mathcal{O}_{\mathbb{C}\times\mathbb{C}^{\mu-1}}$-module $\mathcal{O}_{\Sigma(F)}$ induced by multiplication by $f_0 + \sum_{i=2}^{\mu} u_i\varphi_i$, with respect to appropriate $\mathcal{O}_{\mathbb{C}\times\mathbb{C}^{\mu-1}}$-bases (see for example [4] or [24] Sections 1.2 and 1.3) . Since $\mathcal{O}_{\Sigma(F)}$ is Gorenstein, one can choose dual bases (with respect to the perfect $\mathcal{O}_{\mathbb{C}\times\mathbb{C}^{\mu-1}}$-bilinear pairing $\mathcal{O}_{\Sigma(F)} \times \mathcal{O}_{\Sigma(F)} \to \mathcal{O}_{\mathbb{C}\times\mathbb{C}^{\mu-1}}$ given by projection to the socle (see [20])), and the matrix becomes symmetric. Because the pairing respects the grading, the lowest weight element of one basis (the element 1) is paired with the highest weight element of the other (the Hessian determinant φ_μ) ; in our situation this translates into the fact that the column of the matrix consisting of the coefficients of $\partial/\partial u_\mu$ in the generators of $\mathrm{Derlog}\big(D(F)\big)$ is equal to the row consisting of the coefficients of the lowest weight vector field, namely the Euler vector field

$$\deg(f_0)z\frac{\partial}{\partial z} + \sum_{i=2}^{\mu} w_i u_i \frac{\partial}{\partial u_i},$$

where the w_i are weights of the unfolding parameters u_i, t. From this, the equality we want is immediate. $\qquad\square$

Remark 3-2. *Once one knows the result of Damon's we have used above, the theorem also follows immediately from Arnold's theorem ([2], Theorem 1.1) that a generic smooth section of the discriminant $D(F)$ is isomorphic, by a diffeomorphism preserving $D(F)$, to the section by the hyperplane $\{u_\mu = 0\}$.*

4   Good real forms and their perturbations

We now prove Theorem 1-2, that all of the germs described in the preceding section have real forms with stable perturbations whose discriminant has a (non-trivial) vanishing cycle in dimension $p - 1$. The argument depends on the description of the discriminant of the simple singularities in terms of reflection groups, to which we now turn. We give rather more detail than is strictly required for our proof, since we found the existing accounts a little terse.

As is well known, the discriminant $D_{\mathbb{C}}$ of a miniversal unfolding of a simple singularity (i.e., one of type A_k, D_k, E_6, E_7 or E_8) is the discriminant of the corresponding Weyl group, which acts as a reflection group on a Cartan subalgebra $V_{\mathbb{C}}$ of the Lie algebra of the same name. That is, it is the union of critical orbits, and thus the image of any one of the reflecting hyperplanes.

The discriminant of a real miniversal deformation has a somewhat more complicated description, due to Looijenga ([13]); the Weyl group action on the Cartan subalgebra described above is the complexification of its action as a reflection group on a real vector space V of the same dimension, but the real discriminant is not simply the image under the quotient map σ of any one of the reflecting hyperplanes H_i, or even of their union. This is clear: W acts transitively on its Weyl chambers, and on its reflecting hyperplanes, and thus the image under σ of $V - \cup H_i$ is connected. That is, $V - \cup H_i$ is mapped to a single connected component $\mathcal{B}$ of the complement of the real discriminant $D_{\mathbb{R}}$, and each reflecting hyperplane H_i is mapped to the boundary of this component. The remainder of $D_{\mathbb{R}}$, and the remaining connected components of its complement, arise from other "real forms" of $V_{\mathbb{C}}$, which are defined as follows: let $F \colon X \to S$ be a (complex) miniversal deformation of the surface singularity defined by a simple function of three variables[1] of type A_k, D_k, E_6, E_7 or E_8, with F defined by real weighted homogeneous polynomials. By the monodromy representation, the fundamental group of the complement of the discriminant D_C of F acts on the cohomology $H^2(X_s; \mathbb{R})$ of the (complex) Milnor fibre X_s as a reflection group, W_s. By means of the extended period mapping, ([13] Section 3), the base S and the discriminant D_C of the deformation are identified with the quotient of $H^2(X_s; \mathbb{C}) = H^2(X_s; \mathbb{R}) \otimes \mathbb{C}$ by the complexification of this action and with the discriminant of this quotient, respectively. For $s \in S_{\mathbb{R}}$, complex conjugation induces an involution $\mathbf{u}$ on $H^2(X_s; \mathbb{R})$, which normalises W_s. The quotient N_s/W_s (where N_s is the normaliser of W_s) and the class of $\mathbf{u}_s$ in this quotient, are independent of s, and we may refer simply to $\mathbf{u}$ in N/W. Write $V = H^2(X_s; \mathbb{R})$, $V_{\mathbb{C}} = H^2(X_s; \mathbb{C})$.

Now it is easy to see (cf. [13], Proposition 2.2) that the real part of S is the image under the quotient morphism q of the the fixed point spaces of involutions $u \otimes \mathbf{c}$ of $V_{\mathbb{C}}$ (where $\mathbf{c}$ is complex conjugation), for u an involution in the coset $\mathbf{u}W$. Evidently the fixed point spaces for conjugate involutions u are identified by W; Looijenga shows that the conjugacy classes of involutions in $\mathbf{u}W$ are in bijection with the connected components of the complement of the discriminant $D_{\mathbb{R}}$. Note that each point of $S_{\mathbb{R}} - D_{\mathbb{R}}$ has $|W|$ preimages in $V_{\mathbb{C}}$, and so the topological multiplicity of σ on each space $\mathrm{Fix}(u)$ is equal to $|W|/|$conjugacy class of u in $N/W|$. This number is in general strictly less than $|W|$, which explains in some sense why $\sigma(V)$ is the smallest component of $S - D_{\mathbb{R}}$. Since $\sigma(V)$ is the homeomorphic image of the closure of a single Weyl chamber, it has the aspect of a "curvilinear simplicial cone"; except for its vertex, at 0, it lies entirely in the half-space $\{u_\mu > 0\}$, since $u_\mu \circ \sigma$ is the

---

[1]Since the discriminant is invariant under stable equivalence, there is no loss of generality here

square of the Euclidean norm on V (this of course is subject to the appropriate choice of sign for the unfolding term φ_μ, which we assume made). The map σ itself is homogeneous, since its component functions are the (homogeneous) generators of the algebra of W-invariant polynomials on V; the images of the lines through 0 in V are $\mathbb{R}^*$ orbits in $S_\mathbb{R}$, and because $u_\mu \circ \sigma$ is the unique lowest-degree invariant, the limiting tangent to each of these orbits at 0 is the u_μ-axis.

For example, in the case of A_2, we have $V = \{(x_0, x_1, x_2) \in \mathbb{R}^3 : \sum x_i = 0\}$, on which $W = S_3$ acts by permuting the coordinates. Under the quotient mapping $\sigma(x) = \big(x_1 x_2 x_3, -(x_1 x_2 + x_1 x_3 + x_2 x_3)\big)$, V is mapped to the closure of the component $\{(z, u_2) : 4u_2^3 - 27z^2 > 0\}$ of the complement of the discriminant $D_\mathbb{R}$. In this case $\mathbf{u}$ is $\pm$ the identity ([13] page 54), and so to find subspaces of $V_\mathbb{C}$ parametrising the other component, we look for involutions $u \neq$ identity in W itself. These are just $(1,2)$, $(1,3)$ and $(2,3)$; the fixed point spaces of $(1,2) \otimes \mathbf{c}$, $(1,3) \otimes \mathbf{c}$ and $(2,3) \otimes \mathbf{c}$ are $\{(a + ib, a - ib, -2a) : a, b \in \mathbb{R}\}$, $\{(a + ib, -2a, a - ib) : a, b \in \mathbb{R}\}$ and $\{(-2a, a + ib, a - ib) : a, b \in \mathbb{R}\}$. All three involutions are conjugate, and one checks easily that under the mapping σ all three fixed point spaces are mapped onto the closure of the component $\{(z, u_2) : 4u_2^3 - 27z^2 < 0\}$ of the complement of $D_\mathbb{R}$.

Now Looijenga also proves ([13], Theorem 1.7(iv),) that the Euler characteristic of the real fibre $X_{\mathbb{R},s}$ for $s \in S_\mathbb{R} - D_\mathbb{R}$ in the image under σ of $\mathrm{Fix}(u)$ is equal to 1 plus the trace of u (as element of N). This has special significance for us: for in the cases where $\mathbf{u}$ is trivial, it follows that the Euler characteristic of the real fibre is equal to $1 + \mu$. This means that this fibre is an M-space - that is, it is the real part of a complex space (in this case the complex Milnor fibre) whose Betti numbers have the same sum as its Betti numbers; for if $\beta_{\mathbb{R},i}$ is the $i - th$ Betti number of the real fibre $X_{\mathbb{R},s}$ and $\beta_{\mathbb{C},i}$ that of $X_{\mathbb{C},s}$, on one hand we have $\beta_{\mathbb{R},0} - \beta_{\mathbb{R},1} + \beta_{\mathbb{R},2} = 1 + \mu$, as just mentioned, while on the other hand $\beta_{\mathbb{R},0} + \beta_{\mathbb{R},1} + \beta_{\mathbb{R},2} \leq \beta_{\mathbb{C},0} + \beta_{\mathbb{C},1} + \beta_{\mathbb{C},2} = 1 + \mu$. The inequality follows from Smith theory: $X_{\mathbb{R},s}$ is the fixed point set of the $\mathbb{Z}_2$ action on $X_{\mathbb{C},s}$ defined by complex conjugation (see e.g., [3]). Thus, $\beta_{\mathbb{R},1} = 0$ and $\beta_{\mathbb{R},0} + \beta_{\mathbb{R},2} = \beta_{\mathbb{C},0} + \beta_{\mathbb{C},2}$. Note that the fibres in question are precisely those over points of $S_\mathbb{R} - D(F_\mathbb{R})$ coming from V itself: for since W acts orthogonally on V, the trace of u is equal to μ only when u is the identity, in which case $\mathrm{Fix}(u) = V$.

Let us remark further that one can obtain these M-space real Milnor fibres from the constructions of A'Campo ([1]) and Gussein-Sade ([11]) of real Morsifications of simple curve singularities with only two critical values. It is easy to prove that the real level curve for a level strictly between these two critical values is indeed an M-space: we observe in each of the pictures that all saddles lie on a single connected level set, and that in each of the bounded regions of the complement of this level set, the function is positive. The real level set X_ϵ for a value ϵ strictly betwen the two critical values consists of an oval inside each of the bounded components, together with r non-compact components, where r is the number of branches of the original (unperturbed) curve singularity. One checks easily that in each of the pictures, there are then $(1/2)\{\mu - r + 1\}$ ovals; thus $H_1(X_\epsilon; \mathbb{Z}) = \mathbb{Z}^{(1/2)\{\mu - r + 1\}}$ and $H_0(X_\epsilon; \mathbb{Z}) = \mathbb{Z}^{(1/2)\{\mu + r + 1\}}$, so that X_ϵ is indeed an M-space. Now after adding the sum of squares of k new variables to the defining equation, the real level hypersurface (for the same value of ϵ) consists of the disjoint union of a k-fold suspension of each connected component of the level curve X_ϵ, and is thus once again an M-space.

PROOF. (OF THEOREM 1-2) To get a good real picture we take the real form of the simple singularity f_0 described in table 4.1. In each of these cases, when

A_k	$x_1^{k+1} + \sum_{i \geq 2} x_i^2$
D_{2m}	$x_1^{2m-1} - x_1 x_2^2 + \sum_{i \geq 3} x_i^2$
D_{2m+1}	$x_1^{2m1} + x_1 x_2^2 + \sum_{i \geq 3} x_i^2$
E_6	$x_1^4 + x_2^3 + \sum_{i \geq 3} x_i^2$
E_7	$x_1^3 x_2 + x_2^3 + \sum_{i \geq 3} x_i^2$
E_8	$x_1^5 + x_2^3 + \sum_{i \geq 3} x_i^2$

Table 4.1: Real forms of simple singularities.

$n = 3$ the involution $\mathbf{u}$ is trivial ([13] page 54).

Now let $f: \mathbb{C}^n \times \mathbb{C}^{\mu-2} \times \mathbb{C}^q \to \mathbb{C} \times \mathbb{C}^{\mu-2} \times \mathbb{C}^q = \mathbb{C}^p$ be defined by

$$f(x, u, y) = \left(f_0(x) + \sum_{i=2}^{\mu-1} u_i \varphi_i + \sum_{j=1}^{q} y_j^2 \varphi_\mu, u, y \right).$$

As described in Section 2, f is induced from the miniversal deformation F of the singularity f_0 by a map $\gamma: \mathbb{C} \times \mathbb{C}^{\mu-2} \times \mathbb{C}^q \to \mathbb{C} \times \mathbb{C}^{\mu-1}$. The same, of course, is true with $\mathbb{C}$ replaced by $\mathbb{R}$. One obtains a versal deformation of f by adding to the first component $t\varphi_\mu$, where t is the unfolding parameter; this unfolding is induced from F by the deformation $\Gamma = \gamma_t$ of γ defined by adding t to the μ-th component of γ. Thus, since the formation of discriminants commutes with transverse fibre product (over both $\mathbb{R}$ and $\mathbb{C}$), we have $D(f_t) = \gamma_t^{-1}(D(F))$. There are now two cases to consider.

First, where the number q of y-variables is equal to 0. Then γ_t is an immersion for each t, and $D(f_t)$ is simply a slice of $D(F)$. We find the $(p-1)$-cycle we want in $D(f_t)$ when $t > 0$, when it is simply the intersection of $\gamma_t(\mathbb{R}^p)$ with the boundary of the region $\mathcal{B}$. This is the homeomorphic image under σ of the intersection of the sphere $\{\| v \|^2 = t\}$ with the boundary of a Weyl chamber in V, and hence homeomorphic to S^{p-1}.

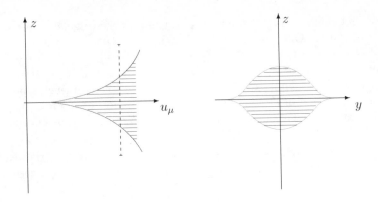

Figure 4.2: Discriminant of the lips singularity

Second, where $q > 0$, we obtain the real form we want by taking the sign of the sum $\sum_{i=1}^{q} y_i^2$ in f to be negative. Now when t is negative, $\gamma_t(\mathbb{R}^p)$ is contained in $\{u_\mu < 0\}$, and thus does not meet $\sigma(V)$. In order to describe what happens when $t > 0$, a diagram is helpful.

In figure 4.2(a) we have taken f_0 of type A_2; the discriminant is the cuspidal curve, equal in this case to $\delta\mathcal{B}$; $\mathcal{B}$ itself is the shaded region of the complement of the cusp. The vertical dotted line represents the boundary of the image of $\gamma_t(\mathbb{R}^p)$ for $t > 0$, with the image of $\gamma_t(\mathbb{R}^p)$ to the left of it. Figure 4.2(b) shows $D(f_t)$, for $t > 0$, in the case $q = 1$; the shaded region enclosed by $D(f_t)$ is $\gamma_t^{-1}(\mathcal{B})$. This is of course the well-known "lips" singularity, given by $f(x, y) = (x^3 + y^2 x, y)$, with stable perturbation $f_t(x, y) = (x^3 + y^2 x - tx, y)$, for $t \neq 0$.

In the general case, the argument is as follows: we have $\gamma_t(z, u, y) \in \delta\mathcal{B} \Leftrightarrow$ $(z, u, 0) \in \delta\mathcal{B} \cap \{u_\mu = t - \sum y_i^2\}$; when $\sum y_i^2 < t$, $\delta\mathcal{B} \cap \{u_\mu = t - \sum y_i^2\}$ is homeomorphic to $\delta\mathcal{B} \cap \{u_\mu = t\}$, while when $\sum y_i^2 = t$, this intersection is just a point (the vertex of the cone, 0). Hence $\gamma_t^{-1}(\delta\mathcal{B})$ is homeomorphic to the q-fold suspension of the intersection of $\{u_\mu = t\}$ with $\delta\mathcal{B}$, and thus to a sphere of dimension $p - 1$. □

Remark 4-1. We obtain an alternative proof of theorem 1-2 by using the characterisation of the interior of $\mathcal{B}$ as the only component of $S_\mathbb{R} - D(F_\mathbb{R})$ over which the fibres of $F_\mathbb{R}$ are M-spaces. For since (in the notation of the preceding proof) $\gamma(\mathbb{R}^p) \cap \text{int}(\mathcal{B})$ is empty, and $\gamma(\mathbb{R}^p) \cap \text{int}(\mathcal{B})$ is not, the real map f_t has local Milnor fibres of a type not present for f. The open set in the target over which these fibres occur must be enclosed by the discriminant $D(f_t)$, and thus by Alexander duality $H^{p-1}(D(f_t), \mathbb{Z})$ has rank at least 1.

5   Bad real pictures

If f_t is a 1-parameter family of real maps, versally unfolding a singularity $f_0 \colon \mathbb{C}^n, 0 \to \mathbb{C}^p, 0$, $(p \leq n+1)$ with $\mu_\Delta = 1$, then as the parameter t passes through 0 the discriminant of f_t undergoes a topological change: over the complex numbers for $t \neq 0$, $D(f_t)$ acquires a non-trivial vanishing cycle in middle dimension $p - 1$. Over the real numbers, the situation is more complicated: as we have seen, frequently it is possible to choose a real form within a given complex orbit so that for the (real) parameter t either greater than 0 or less than 0, the real discriminant carries the same vanishing cycle. However, as t changes sign, examples show that there may still be a vanishing cycle, but that in general it is no longer in dimension $p-1$. It turns out that if we pass from the real discriminant to the real part of the complex discriminant then we always observe a vanishing cycle:

Lemma 5-1. *In the situation just described, the real part of the complex discriminant has the homotopy type of a sphere S^{k-1} for some k with $0 \leq k \leq p$ (this can be improved to $1 \leq k \leq p$ in the case of images). When $k > 0$, the real part of the complex discriminant (or image) is an M-space.*

PROOF. As in the proof of theorem 1-1, we may choose a real family g_t of defining equations for the complex discriminant. The real part of the complex discriminant is then the zero-locus of g_t in $\mathbb{R}^p$. As $\mu_\Delta = 1$, for $t \neq 0$ g_t must have a single (Morse) critical point off its zero-locus (see the proof of 2-2); as g_t is real, this critical point must lie in $\mathbb{R}^p$. If it has index k and positive critical value, then the contractible space B_ϵ is obtained from $g_t^{-1}(0)$ by gluing in a k-cell, by the standard argument of Morse theory. It follows that $g_t^{-1}(0)$ is homotopy equivalent to the boundary S^{k-1} of this k-cell. If the critical value is less than zero, then the same argument applies after reversing all of the signs, and $g_t^{-1}(0)$ is homotopy equivalent to S^{p-k-1}.

The last assertion is obvious.   □

In the situation of the lemma, we will denote the homotopy dimension of the real part of $D(f_t)$ for $t > 0$, by k_+, and for $t < 0$ by k_-.

The drawings in Figure 5.3 are obtained from those in Figure 1.1 by letting the unfolding parameter cross 0, and show in each case the real part of the complex image. Points lying in this and not in the real image are indicated by dotted lines. In each case one observes a non-trivial vanishing cycle; in (II) and (IV) it is actually carried by the real image. We omit the corresponding drawing for case (V) (the quadruple point), since in this case the drawing is the same as in Figure 1.1; and only in this case is the vanishing cycle in dimension $p - 1 = 2$. In the remaining cases its dimension is, respectively, $0, 0, 1$ and 1. The vanishing 1-cycle in (III) is carried by the triangle whose vertices are the special point of the Whitney umbrella (= cross-cap), the point of intersection of the handle of the Whitney umbrella with the immersed plane, and the point

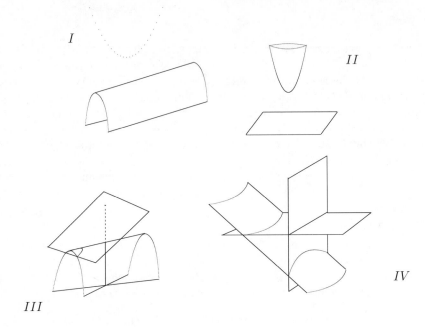

I

II

III

IV

Figure 5.3: Bad real pictures

where the plane cuts the upper apparent profile of the umbrella. We leave the reader to find the vanishing 1-cycle in (IV).

The drawings in Figure 5.4 show stable perturbations of other real forms of the singularities depicted in (I) and (II) of Figures 1.1 and 5.3. In the lower right hand picture we have indicated the vanishing 1-cycle (carried by the real image in this case) by a curve bearing two arrows.

Although these bad pictures do not show the vanishing cycle carried by the complex image, under certain mild hypotheses which are in fact satisfied by all of the germs mentioned in this paper, they do in fact convey information on the complex discriminant:

Theorem 5-2. *In the situation of Lemma 5-1, provided that i) the germ f_0 is quasihomogeneous; ii) the quasihomogeneous monodromy of the fibration of the discriminant of the family over the complement of 0 in the parameter space $\mathbb{C}$ is real, and iii) either f_t for $t > 0$, or f_t for $t < 0$, is a good perturbation of f_0, then the monodromy transformation, acting on $H_{p-1}\big(D(f_{\mathbb{C},t})\big)$, consists of multiplication by $(-1)^{k_+ - k_-}$.*

An important element in the proof is the fact that the geometric monodromy on the complex discriminant $g_1^{-1}(0)$ can be represented by the quasihomogeneous monodromy, which is defined as follows: as f_0 is quasihomogeneous, we may take the family $f_{\mathbb{C},t}$ to be quasihomogeneous also; note that the parameter t cannot have weight equal to zero, as by a result of Damon ([5])

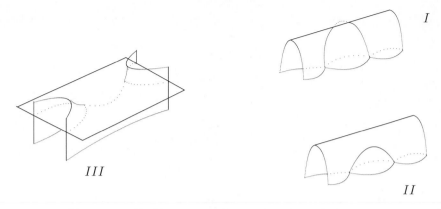

Figure 5.4: Stable perturbations of bad real forms

this would imply that the family $f_{\mathbb{C},t}$ was topologically trivial. As in the proof of 4-1 we take a real family of defining equations g_t for the complex discriminant, which again we may take to be quasihomogeneous, with the weights w_1, ..., w_p of the coordinates z_1, ..., z_p equal to the degrees of the p component functions of f, and the weight of the parameter t equal to its weight in the family $f_{\mathbb{C},t}$, which, after renormalising, we take to be equal to 1. Now since $g_{\lambda t}(\lambda^{w_1} z_1, \ldots, \lambda^{w_p} z_p) = \lambda^d g_t(z_1, \ldots, z_p)$, (where d is the degree of g_t), the linear map $h_\lambda(z_1, \ldots, z_p) = (\lambda^{w_1} z_1, \ldots, \lambda^{w_p} z_p)$ maps $g_t^{-1}(0)$ to $g_{\lambda t}^{-1}(0)$; taking $t = 1$, $\lambda = e^{2\pi i \theta}$ and letting θ vary from 0 to 1, we have a covering homotopy, in the fibration of the discriminant of the family $f_{\mathbb{C},t}$ over $\mathbb{C} - \{0\}$, of the path $e^{2\pi i\theta}$ in $\mathbb{C} - 0$, and $h_1 \colon g_1^{-1}(0) \to g_1^{-1}(0)$ thus represents the geometric monodromy.

We single out one consequence of this, and of the supposition that f_t is a good real perturbation for some nonzero t:

Lemma 5-3. *In the situation just described, then providing that i) f_1 is a good real perturbation, and ii) the quasihomogeneous monodromy is real, then the homological mondodromy, acting on $H_{p-1}(D(f_{\mathbb{C},1}; \mathbb{Z}))$, is just multiplication by $\det(h_1)$.*

PROOF. Because f_1 is a good real perturbation, by Theorem 1-1 the vanishing homology of the complex discriminant is carried by the real discriminant $g_1^{-1}(0)$; by ii), h_1 is a real linear map, and so maps the real discriminant to itself; since this discriminant is the boundary of an open p-cell it follows that the monodromy transformation h_{1*} on $H_{p-1}(D(f_{\mathbb{C},1}); \mathbb{Z}) = \mathbb{Z}$ consists of multiplication by 1 if h_1 preserves orientation and -1 otherwise. Therefore it is just multiplication by $\det(h_1)$, since $\det(h_1)$ is a root of unity. $\square$

PROOF. (OF THEOREM 5-2) Continuing with the notation used above, let P_1 and P_{-1} be the unique critical points, with non-zero critical values, of g_1 and g_{-1} respectively; because $g_{e^{2\pi i\theta}} \circ h_{e^{2\pi i\theta}} = e^{2d\pi i\theta} g_1$, we have $P_{-1} = h_{-1}(P_1)$, and as h_{-1} is linear , and indeed its matrix $[h_{-1}]$ is diagonal, we have

$$[h_{-1}]\left[\frac{\partial^2 g_{-1}}{\partial z_i \partial z_j}(P_{-1})\right][h_{-1}] = (-1)^d\left[\frac{\partial^2 g_1}{\partial z_i \partial z_j}(P_1)\right]$$

and hence taking determinants and noting that $(h_{-1})^2 = h_1$, we have

$$\det(h_1)\det\left[\frac{\partial^2 g_{-1}}{\partial z_i \partial z_j}(P_{-1})\right] = (-1)^{pd}\det\left[\frac{\partial^2 g_1}{\partial z_i \partial z_j}(P_1)\right]$$

There are now two cases:

First, if $(-1)^d = 1$, then the critical values $g_1(P_1)$ and $g_{-1}(P_{-1})$ are equal; the dimension of the vanishing cycles in the real discriminants $D(f_1)$ and $D(f_{-1})$ are then each one less than the indices of

$$\left[\frac{\partial^2 g_1}{\partial z_i \partial z_j}(P_1)\right] \quad \text{and} \quad \left[\frac{\partial^2 g_{-11}}{\partial z_i \partial z_j}(P_{-1})\right]$$

respectively (the former is of course equal to p, since we are asuming that f_1 is a good perturbation of f_0, but this has no bearing on the argument at this point). This difference is thus even if the two determinants have the same sign, and odd if the sign is different. But the two determinants have the same sign if and only if the sign of $\det(h_1)$ is positive, and thus if and only if the monodromy transformation h_{1*} is multiplication by 1.

In the second case, $(-1)^d = -1$; now $g_1(P_1)$ and $g_{-1}(P_{-1})$ are of opposite sign, and so (assuming, say, that $g_1(P_1) > 0$), the dimension of the vanishing cycle in $g_1^{-1}(0)$ is equal to $\text{index}\left[\frac{\partial^2 g_1}{\partial z_i \partial z_j}(P_1)\right] - 1$ while the dimension of the vanishing cycle in $g_{-1}^{-1}(0)$ is equal to $p - \text{index}\left[\frac{\partial^2 g_{-1}}{\partial z_i \partial z_j}(P_{-1})\right] - 1$. Once again, the difference of the two dimensions is even if h_{1*} is the identity, and is odd if h_{1*} is multiplication by -1. □

Remark 5-4. *It is clear from the proof that the theorem is valid even without the hypothesis iii), provided that there does exist a real form for the complex singularity f_0, with a good perturbation, and that this good real form is quasi-homogeneous of the same weights and degrees as the real form considered in the theorem. Thus, from the 1-parameter bifurcation sequence of fusion and annihilation of two cross-caps, of which we show the "before" and "after" in Figure 5.4 I, one can deduce that the monodromy of the stable complex image over the complement of 0 in the parameter space is trivial; for here the real form of f_0 differs from the good real form $(x_1, x_2) \to (x_1, x_2^2, x_2^3 + x_1^2 x_2)$ only in a change of sign of one of its monomials, and therefore has the same weights and degrees; and the vanishing homology in both real images is in the same dimension.*

References

[1] N. A'Campo, Sur la monodromie des singularités isolées d'hypersurfaces complexes, Invent. Math. 20, 147–169

[2] V.I. Arnold, Wave front evolution and the equivariant Morse lemma, Comm. Pure and Appl. Math. 29 (1976) 557–582

[3] G. Bredon, Introduction to compact transformation groups, Pure and applied mathematics volume 46, Academic Press, New York, 1972

[4] J.W. Bruce, Functions on discriminants, J. London Math. Soc. 30 (1984) 551–567

[5] J. Damon, Finite determinacy and topological triviality I, Invent. Math. 62 (1980), 299–324

[6] J. Damon, $\mathcal{A}$-equivalence and equivalence of sections of images and discriminants, Singularity Theory and its Applications, Warwick 1989, Part 1 , D. Mond and J. Montaldi eds., Lecture Notes in Math. 1462, Springer-Verlag, 1991, 93–121

[7] J. Damon and A. Galligo, A topological invariant for stable map-germs, Invent. Math. 32 (1976), 103–132

[8] J. Damon and D. Mond, $\mathcal{A}$-codimension and the vanishing topology of discriminants, Invent. Math. 106 (1991), 217–242

[9] V.V. Goryunov, Singularities of projections of full intersections, Journal of Soviet Mathematics 27, (1984) 2785–2811

[10] V.V. Goryunov, The monodromy of the image of a mapping from $\mathbb{C}^2$ to $\mathbb{C}^3$, Functional Analysis and Applications, Vol. 25 No. 3 (1991), 174–180

[11] S.M. Gusein Sade, Dynkin diagrams for certain singularities of functions of two variables, Functional Analysis and Appl. 8, (1974), 295–300

[12] T. de Jong and D. van Straten, Disentanglements, Singularity Theory and its Applications, Warwick 1989, Part 1, D. Mond and J. Montaldi eds., Lecture Notes in Math. 1462, Springer-Verlag (1991), 199–211

[13] E.J.N. Looijenga, The discriminant of a real simple singularity, Compositio Math. 37 (1978), 51–62

[14] E.J.N. Looijenga, Isolated singular points of complete intersections, London Math. Soc. Lecture Notes 77, 1984

[15] W.L. Marar and D. Mond, Real germs with good perturbations, in preparation

[16] J.N. Mather, Stability of C^∞ mappings IV, Classification of stable germs by R-algebras, Pub. Math. I.H.E.S. 37 (1969), 223–248

[17] D. Mond, Vanishing cycles for analytic maps, Singularity Theory and its Applications, Warwick 1989, Part 1, D. Mond and J. Montaldi eds., Lecture Notes in Math. 1462, Springer-Verlag (1991), 221–234

[18] D. Mond, Looking at bent wires, in preparation

[19] D. Mond and J. Montaldi, Deformations of maps on complete intersections, Damon's $\mathcal{K}_V$-equivalence and bifurcations, preprint, University of Warwick, 1991

[20] G. Scheja and U. Storch, Über Spurfunktionen bei Vollständigen Durchschnitten, J. Reine und Angew. Math. 278/279 (1975), 174–189

[21] D. Siersma, Vanishing cycles and special fibres, Singularity Theory and its Applications, Warwick 1989, Part 1, D. Mond and J. Montaldi eds., Lecture Notes in Math. 1462, Springer-Verlag (1991), 292–301

[22] G.W. Whitehead, Elements of homotopy theory, Graduate texts in Maths. 61, Springer-Verlag, Berlin, Heidelberg, 1978

[23] V.M. Zakalyukin, Reconstruction of wavefronts depending on one parameter, Functional Analysis and Applications, Vol. 10 No.2 (1976) 69–70

[24] V.M. Zakalyukin, Reconstruction of fronts and caustics depending on a parameter and versality of mappings, Itogi Nauki i Tekhni, Seriya Sovremennye Problemy Matematiki, Vol. 22 (1983), 56–93, translated in Journal of Soviet Mathematics Vol. 27 (1984), 2713–2735

Address of author:

Mathematics Institute
Univ. of Warwick
Coventry CV 4 7AL
United Kingdom
Email: mond@maths.warwick.ac.uk

Weighted homogeneous complete intersections

C. T. C. Wall

Abstract

Suppose given a set of weights and degrees defining $\mathbb{C}^{\times}$ actions on $\mathbb{C}^n$ and $\mathbb{C}^p$, with $n \geq p$. Necessary and sufficient conditions are obtained for the existence of an equivariant map $f : \mathbb{C}^n \to \mathbb{C}^p$ such that $f^{-1}(\mathbf{0})$ has an isolated singularity at $\mathbf{0}$. These are somewhat complicated, but simplify if $n - p = 0$ or 1 or if $p = 1$. The former case gives conditions for (weighted) homogeneously generated ideals of finite codimension in the ring $\mathcal{O}_n$ of germs of holomorphic functions; these are generalised to submodules of finite codimension in free $\mathcal{O}_n$-modules. For maps f as above, there are known formulae for the Poincaré series of the Jacobian algebra and the $\mathcal{K}$-cotangent space; we also have a corresponding formula for the quotient in the submodule case.

For the case of $\mathcal{A}$- (right-left-) equivalence of maps f, the above results can be used to give an algorithm for the Poincaré series of the $\mathcal{A}$-cotangent space (of a finitely $\mathcal{A}$-determined germ) in terms of the weights and degrees. The method yields necessary conditions for existence of a finitely $\mathcal{A}$-determined germ which are not, however, sufficient.

To express the condition for $\mathcal{K}$-finite maps, write the source as $\mathbb{C}^I$ with weights $\{w_i \mid i \in I\}$ and the target as $\mathbb{C}^J$ with degrees $\{d_j \mid j \in J\}$. For $A \subseteq I$ write $\mathbb{N}(A)$ for the additive monoid generated by $\{w_i \mid i \in A\}$ and $J(A) = \{j \in J \mid d_j \in \mathbb{N}(A)\}$; write $\#$ to denote cardinality. The condition is that for all $A \subseteq I$ such that $\#A > \#J(A)$ and all non-empty $B \subseteq J \setminus J(A)$,

$$\#\{i \in I \setminus A \mid \exists j \in B \text{ with } d_j - w_i \in \mathbb{N}(A)\} \geq \#A + \#B - \#J(A) - 1.$$

1 Introduction

A map-germ $f : \mathbb{C}^n \to \mathbb{C}$ is said to be weighted homogeneous with weights $(w_1, \ldots, w_n)$ and degree d if for all $t \neq 0$,

$$f(t^{w_1}x_1, \ldots, t^{w_n}x_n) = t^d f(x_1, \ldots, x_n).$$

It is a well known result of Milnor and Orlik [12] that if f has an isolated singularity at $\mathbf{0}$, its Milnor number is given by

$$\mu = \frac{\prod(d - w_i)}{\prod w_i},$$

Progress in Mathematics, Vol. 134
© 1996 Birkhäuser Verlag Basel/Switzerland

and this formula can be refined to describe the Poincaré series of natural vector spaces with dimension μ. Thus for $\theta(f)/T_e\mathcal{R}(f)$ we obtain

$$P(z) = z^{-d}\frac{\prod(1 - z^{d-w_i})}{\prod(1 - z^{w_i})}.$$

This series then reduces to a polynomial (in z and z^{-1}).

If d and the w_i are given, then it is known (Arnol'd [2]) that there exists such an f if and only if $(n = 2)$ for each i, $1 \le i \le n$ there exists j, $1 \le j \le n$ such that $d - w_j$ is divisible by w_i or $(n = 3)$ this holds and for each $i \ne j$, d is a nonnegative integer linear combination of w_i and w_j. When this holds, a generic function of degree d with respect to the weights w_i has an isolated singular point at $\mathbf{0}$. Moreover, Saito has shown [16] that when $n = 3$, if $P(z)$ is a polynomial there exist corresponding functions with an isolated singular point.

In this paper we investigate analogous results for other properties of map-germs which are determined by the weights and degrees, related to the equivalence relations $\mathcal{C}, \mathcal{K}$ and $\mathcal{A}$ of Mather [11] in the same way as the above is related to $\mathcal{R}$.

More precisely, using the terminology and notation of [18], we seek, for $\mathcal{B}$ one of the above equivalence relations, necessary and sufficient conditions for the existence of a $\mathcal{B}$-finite germ with the given weights and degrees, and when it exists a formula for the $\mathcal{B}_e$-codimension $d_e(f, \mathcal{B})$ of such a germ in terms of the weights and degrees, and more precisely, a formula for the Poincaré polynomial of $\theta(f)/T_e(f, \mathcal{B})$.

I am indebted to Kyoji Saito for discussions (during my visit to Japan in 1987 sponsored by the JSPS) stimulating my interest in certain aspects of these problems; to Chris Gibson for asking for a formula for $\mathcal{A}$-codimensions in terms of weights and degrees; to Rodney Sharp and John Pym for helpful correspondence; to Terry Gaffney for useful discussions especially concerning the module case (Section 4 below); to Victor Goryunov for inspiration, and telling me of the codimension formulae in Section 5; to Sasha Aleksandrov for drawing my attention to the work [17] of O.P.Sherbak; and to Andy du Plessis for collaboration giving me the insight into the geometry in the $\mathcal{A}$-case, and for the counterexample at the end.

I now introduce notation and terminology; after that I give the plan of the paper and outline of the main results.

2   Notation

We recall some results about map-germs, following the notation of [18]. We have the equivalence relations $\mathcal{R}$ (right-equivalence), $\mathcal{L}$ (left-equivalence), $\mathcal{A}$ ($\mathcal{R} \times \mathcal{L}$), $\mathcal{C}$ and $\mathcal{K} = \mathcal{R} \cdot \mathcal{C}$ (contact-equivalence); $\mathcal{B}$ may denote any of these. For a map-germ f, the tangent space $\theta(f)$ to the space of map-germs is identified with the space of vector fields along f; it contains subspaces $TB(f)$ of tangent spaces to the orbits of f under the above groups and 'extended tangent spaces' $T_e(B, f)$ related to deformations which do not preserve base points. f is said to be B-stable if $T_e(B, f) = \theta(f)$. In the complex case, R-stable is equivalent to being a submersion, C-stable to $f(\mathbf{0}) \neq \mathbf{0}$, K-stable to being either R-stable or C-stable.

The map-germ f is finitely determined for B-equivalence, or, as we will say, B-finite, if and only if the codimension $d_e(B, f)$ of $T_e(B, f)$ in $\theta(f)$ is finite. In the complex case, for R, C and K this is equivalent to $\mathbf{0}$ being an isolated point of instability. This is the form of B-stability which will be used below: in particular, f is C-finite if and only if $\mathbf{0}$ is isolated in $f^{-1}(\mathbf{0})$ and K-finite if and only if $\mathbf{0}$ is isolated in the intersection $f^{-1}(\mathbf{0}) \cap \Sigma(f)$, where the singular set $\Sigma(f)$ is the set of points $\mathbf{x}$ with $Tf_{\mathbf{x}}$ not surjective.

We now introduce notations for weighted homogeneous maps. We suppose we have an action of the multiplicative group $\mathbb{C}^\times$ on $\mathbb{C}^n$ given by weights w_i, $1 \leq i \leq n$ by

$$t \cdot (x_1, \dots, x_n) = (t^{w_1} x_1, \dots, t^{w_n} x_n).$$

It will be convenient to denote the index set $\{1, 2, \dots, n\}$ by I, and hence write $\mathbb{C}^I$ instead of $\mathbb{C}^n$ and have weights w_i for $i \in I$. We use $\#$ to denote cardinalities, so that $\#I = n$. Similarly, we have an action on $\mathbb{C}^J$ given correspondingly by degrees d_j $(j \in J)$. We consider holomorphic maps $f : \mathbb{C}^I \to \mathbb{C}^J$ which are equivariant with respect to these actions: we call them weighted homogeneous, or homogeneous for short. Write $\mathcal{O}_I$ for the ring of germs at $\mathbf{0}$ of holomorphic functions on $\mathbb{C}^I$: this ring has a grading given by the weights w_i. We often suppose the w_i and d_j to be positive: this implies f polynomial.

For any $A \subseteq I$, write $\mathbb{N}(A)$ (resp. $\mathbb{Z}(A)$) for the additive monoid (i.e. semigroup with 0) (resp. additive group) generated by the w_i with $i \in A$. The elements of $\mathbb{N}(A)$ are the degrees of the monomials in the x_i with $i \in A$.

Suppose given a set R (in the first instance, finite) of monomial maps $\mathbb{C}^I \to \mathbb{C}^J$; say $\{\mathbf{x}^{\alpha(r)} \epsilon_{\sigma(r)} \mid r \in R\}$. Here ϵ_j denotes the j-th basis vector in $\mathbb{C}^J$.

Write U for the vector space of maps $\mathbb{C}^I \to \mathbb{C}^J$ which are linear combinations of the above: we will call these H-maps. Write t_r for the coordinate functions on U corresponding to the basis R. Write $F : U \times \mathbb{C}^I \to \mathbb{C}^J$ for the evaluation map: this is given by $F(\mathbf{t}, \mathbf{x}) = \sum_r t_r \mathbf{x}^{\alpha(r)} \epsilon_{\sigma(r)}$. Set $X = F^{-1}(\mathbf{0})$ and $X^* = X \setminus (U \times \{\mathbf{0}\})$.

For each subset $A \subseteq I$ introduce the sets

$$
\begin{aligned}
\mathbb{C}^A &= \{\mathbf{x} \in \mathbb{C}^I \mid x_i = 0 \text{ for } i \notin A\}, \\
\mathbb{C}^{\times A} &= \{\mathbf{x} \in \mathbb{C}^I \mid x_i = 0 \Leftrightarrow i \notin A\}, \\
V^A &= U \times \mathbb{C}^{\times A}, \\
X^A &= X \cap V^A.
\end{aligned}
$$

Thus the $\mathbb{C}^{\times A}$ partition $\mathbb{C}^I$; the V^A partition $U \times \mathbb{C}^I$; and the X^A partition X, while those with $A \neq \emptyset$ partition X^*. Write

$$J(A) = \{j \in J \mid \exists r \in R \text{ with } \sigma(r) = j \text{ and } \mathbf{x}^{\alpha(r)} \mid \mathbb{C}^A \not\equiv \mathbf{0}\}.$$

We will be particularly interested in the case when R consists of all the monomial maps which are homogeneous with respect to the weights w_i and degrees d_j. It is convenient to assign ϵ_j the weight $-d_j$: then each $\mathbf{x}^{\alpha(r)}\epsilon_{\sigma(r)}$ has weight 0. Also, $J(A) = \{j \in J \mid d_j \in N(A)\}$.

Finally, we need notation for Poincaré series. If $V = \oplus_d V_d$ is a graded vector space, its *Poincaré series* is defined by $P(V, z) = \sum (\dim V_d) z^d$. This is a rational function for important examples: of course it is a polynomial if $\dim V < \infty$.

If Φ is a formal Laurent series $\Phi(z, t) = \sum_{i,j} a_{i,j} z^i t^j$, its *residue* is $\operatorname{Res}_t \Phi = \sum_i a_{i,-1} z^i$.

Now let $f : \mathbb{C}^n \to \mathbb{C}^p$, where $n > p$, be a germ, homogeneous with respect to weights $(w_1, \ldots, w_n)$ and degrees $(d_1, \ldots, d_p)$. Set

$$m = n - p, \quad W = \sum w_i, \quad D = \sum d_j,$$

and

$$\Phi(z, t) = \frac{\prod_j (1 + t z^{d_j})}{\prod_i (1 + t z^{w_i})}.$$

Thus we have for example

$$P(\mathcal{O}_n, z) = \frac{1}{\prod_i (1 - z^{w_i})}$$

and

$$P(Q(f), z) = P(\mathcal{O}_n / f^* \mathfrak{m}_p \cdot \mathcal{O}_n, z) = \Phi(z, -1),$$

provided the components f_j form an $\mathcal{O}_n$-sequence.

In Section 3 we give necessary and sufficient conditions for the existence of $\mathcal{C}$-finite H-maps when $\#I = \#J$. This is equivalent to having an ideal of finite codimension: the corresponding question for modules is treated in Section 4. In Section 5 we obtain corresponding results for $\mathcal{K}$-finite maps when $\#I \geq \#J$. This problem is more complicated, and necessary combinatorial results are discussed in Section 6. The main results of Section 5 were announced by O.P.Sherbak in [17]. I have been unable to find a published proof of these results, so feel the present account is still of some interest. In an earlier version of this paper I considered only the weighted homogeneous case, but the arguments sufficed for the more general situation.

In the same cases we also present formulae for the Poincaré series of the corresponding deformation spaces $\theta(f)/T_e(\mathcal{B}, f)$. These reduce to polynomials when f is $\mathcal{B}$-finite, but we see that only in rather low dimensions is this condition sufficient to ensure existence of $\mathcal{B}$-finite maps. We also argue that simple results are not to be expected for $\mathcal{C}$ and $\mathcal{K}$ in other dimensions.

In Section 7 we obtain some necessary conditions on weights and degrees for existence of homogeneous $\mathcal{A}$-finite maps when $\#I > \#J$; an example shows that these are not sufficient, and we give some discussion of this problem. However we do succeed in obtaining a formula for the Poincaré series of $\theta(f)/T_e(\mathcal{A}, f)$ valid whenever f is $\mathcal{A}$-finite.

We observe that corresponding results for $\mathcal{A}$-equivalence are known when $\#I = \#J = 2$ [7] and when $\#I = 2, \#J = 3$ [14].

3 Ideals and $\mathcal{C}$-equivalence

In this section, we restrict to the case $\#I = \#J$, and will at first suppose R finite.

Theorem 3-1. *The following are equivalent:*

(a) *there exists a $\mathcal{C}$-finite H map $\mathbb{C}^I \to \mathbb{C}^J$;*

(b) *a generic H map $\mathbb{C}^I \to \mathbb{C}^J$ is $\mathcal{C}$-finite;*

(e) $\dim X^* \le \dim U$;

(f) *for each non-empty subset $A \subseteq I$, $\#J(A) \ge \#A$.*

PROOF. Clearly, (b) implies (a).

(a) $\Rightarrow$ (f): if (f) fails there is a coordinate subspace $\mathbb{C}^A$ such that there are fewer than $\#A$ components F_j of F which do not vanish identically on V^A. Thus for any homogeneous map f_t, $f_t^{-1}(\mathbf{0})$ meets $\{\mathbf{t}\} \times \mathbb{C}^A$ in a subset of codimension $< \#A$, hence of positive dimension. Thus f_t is not $\mathcal{C}$-finite.

(f) $\Rightarrow$ (e): suppose (f) holds. For each $A \ne \emptyset$, we have at least $\#A$ components which do not vanish identically. Since these involve disjoint sets of coefficients u_k, their vanishing imposes independent conditions, so $\dim X^A \le \dim U$. Indeed, for each fixed $\mathbf{x} \in \mathbb{C}^{\times A}$, $\{\mathbf{t} \mid (\mathbf{t}, \mathbf{x}) \in X\}$ is a linear subspace of codimension $\#J(A) \ge \#A$. As this holds for all nonempty A, $\dim X^* \le \dim U$.

(e) $\Rightarrow$ (b): suppose (e) holds. As X^* is constructible, to obtain its closure $\overline{X}^*$ we add a set of strictly lower dimension. Thus $\overline{X}^* \cap (U \times \{\mathbf{0}\}) = W \times \{\mathbf{0}\}$, with W a proper algebraic subset of U. If $\mathbf{t} \notin W$, the corresponding map f_t has $\{\mathbf{0}\}$ isolated in $f_t^{-1}(\mathbf{0})$, so the germ of f_t is $\mathcal{C}$-finite. $\qquad\square$

Now apply this to the weighted homogeneous case with $\#I = \#J$, and all the $w_i > 0$ (and hence $d_j > 0$).

Theorem 3-2. *The following are equivalent:*

(a) *there exists a $\mathcal{C}$-finite homogeneous map $\mathbb{C}^I \to \mathbb{C}^J$;*

(b) *a generic homogeneous map $\mathbb{C}^I \to \mathbb{C}^J$ is $\mathcal{C}$-finite;*

(c) *generic elements f_j of $\mathcal{O}_I$ of weights d_j generate an ideal of finite codimension;*

(d) *generic elements f_j of $\mathcal{O}_I$ of weights d_j form an $\mathcal{O}_I$-sequence;*

(e) $\dim X^* \leq \dim U$.

(f) *for each non-empty subset $A \subseteq I$, $\#J(A) \geq \#A$.*

PROOF. It is standard that $\{f_1, \dots, f_n\}$ form an $\mathcal{O}_I$-sequence if and only if they generate an ideal of finite codimension; also (see e.g. [18]) that this is equivalent to the map with these as components being $\mathcal{C}$-finite. Thus (b), (c) and (d) are equivalent. The rest follows from 3-1.   □

When f is $\mathcal{C}$-finite, $f_1, \dots, f_n$ is an $\mathcal{O}_n$-sequence so it follows from exactness of the Koszul complex that the quotient Q of $\mathcal{O}_n$ by the ideal I generated by the f_j has Poincaré series

$$P(Q, z) = P(z, \mathbf{d}, \mathbf{w}) \equiv \frac{\prod(1 - z^{d_j})}{\prod(1 - z^{w_i})}.$$

Thus in particular we have

(g). $P(z, \mathbf{d}, \mathbf{w})$ is a polynomial with nonnegative coefficients.

We next enquire whether (g) implies the conditions of 3-2. Were this true, it would give a condition much quicker to verify (e.g. in a computer search) than (f), which requires separate consideration of $2^n - 1$ different cases.

First observe that if d is the greatest common divisor of the natural numbers a and b then $(1 - z^d)$ is the greatest common divisor of the polynomials $(1 - z^a)$ and $(1 - z^b)$. Hence a necessary and sufficient condition for $P(z, \mathbf{d}, \mathbf{w})$ to be a polynomial is that any natural number m which divides r of the w_i must divide at least r of the d_j. This is equivalent to a condition similar to (f):

(f'). *for each non-empty subset $A \subseteq I$, $\#\{j \in J \mid d_j \in \mathbb{Z}(A)\} \geq \#A$.*

But (for $\#I > 1$) (f') is strictly weaker than (f).

Lemma 3-3. *If $\#I = \#J = 2$ then (g) implies (f).*

PROOF. We have just seen that (g) implies (f'), so that (after perhaps reordering) either w_1 divides d_1 and w_2 divides d_2 (when (f) certainly holds) or we have $w_1 = pq$, $w_2 = pr$, say, with p the greatest common divisor of w_1 and w_2, and p divides d_1 and pqr divides d_2. Now all the exponents occurring in the power series expansion of $(1 - z^{d_2})/((1 - z^{pq})(1 - z^{pr}))$ belong to the semigroup generated by pq and pr. Thus if d_1 does not belong to this semigroup, when we multiply by $(1 - z^{d_1})$ there can be nothing to cancel the negative coefficient of z^{d_1}.   □

This result does not extend to the case $\#I = \#J = 3$. For example, take $\mathbf{w} = (1,3,5)$, $\mathbf{d} = (4,7,15)$. This gives

$$P(z,\mathbf{w},\mathbf{d}) = 1 + z + z^2 + 2z^3 + z^4 + 2z^5 + 3z^6 + z^7 + 2z^8 + 2z^9 + z^{10}$$

$$+3z^{11} + 2z^{12} + z^{13} + 2z^{14} + z^{15} + z^{16} + z^{17}.$$

However, only 15 belongs to the semigroup generated by 3 and 5 (4 and 7 do not) so (f) fails.

A map-germ f can only be finitely $\mathcal{C}$-determined if $n \leq p$. However if $p > n$ it is easy to construct examples where one of the coordinates is 0, so can be replaced by homogeneous elements of arbitrarily high weight; thus it seems unlikely that any simple results can be obtained in terms of the w_i and d_j.

We now return to 3-1, but abandon the restriction that R is finite. In this context we define *generic* in (b) to mean that any finite subset R' of R is contained in a finite subset R'' such that the conclusion holds for a generic member of the corresponding U''. Similarly replace clause (e) of 3-1 by: any U' is contained in a U'' for which the conclusion holds. With these understandings, we have

(3-4) ADDENDUM The conditions of 3-1 remain equivalent when R is infinite.

PROOF. Our proof of (f) $\Rightarrow$ (e) required certain coefficients u_k which are provided by the hypothesis. We enlarge U' to U'' so that these coefficients are independent on U''. The proof then proceeds as before. The proofs of the other implications are unaltered. $\square$

Applying this to the homogeneous case now shows that 3-2 remains true when the weights (and degrees) are allowed to be negative or zero. The subsemigroups $\mathbb{N}(A)$ of $\mathbb{Z}$ may now contain negative numbers. Now any subsemigroup B of $\mathbb{Z}$ containing elements of both signs is a subgroup. For if, say, B contains $a > 0$ and $-b < 0$ then $-a = (b-1)a + a(-b)$ and $b = (a-1)(-b) + ba$ belong to B. This will make the conditions easier to verify in practice.

4   Submodules

We next consider a generalisation of the problem studied in Section 3 which does not directly concern map-germs, but will be needed below. Let $\mathcal{O}_I$ be as above, let F be the free $\mathcal{O}_I$-module with basis $\{\epsilon_j \mid j \in J\}$, let K be a set with $\#K = \#I + \#J - 1$, and for each $k \in K$ let R_k be a finite set of monomials $\{\mathbf{x}^{\alpha(r)}\epsilon_{\sigma(r)}\}$. Consider the submodules M of F generated by linear combinations γ_k ($k \in K$) of elements of R_k. We seek the condition that the γ_k can be chosen so that $\dim_{\mathbb{C}}(F/M) < \infty$. Observe (this will follow from the arguments below) that except in the special case $M = F$, this is the smallest number of generators required to generate an ideal of finite codimension.

We may consider F as a sheaf (indeed, a trivial vector bundle) over $\mathbb{C}^I$, and M as the image of a map of sheaves. The codimension is then finite if and only if the quotient sheaf has support $\{\mathbf{0}\}$. Now a point $\mathbf{x}$ belongs to this support if the matrix $(\gamma_{k,j}(\mathbf{x}))$ (where $\gamma_k = \sum \gamma_{k,j}\epsilon_j$) has rank $< \#I$, or equivalently, if there exist numbers t_j, not all 0, such that $\sum \gamma_{k,j}t_j = 0$ for each k. We thus introduce an extended ideal as follows (I am grateful to Terry Gaffney for this suggestion).

Introduce new variables $\mathbf{t} = \{t_j \mid j \in J\}$, set $\tilde{\gamma}_k = \sum \gamma_{k,j}t_j$, and write $\tilde{I}$ for the ideal in $\mathcal{O}_I[\mathbf{t}]$ generated by the $\tilde{\gamma}_k$. The zero locus Y_γ of $\tilde{I}$ clearly contains the subspaces $\mathbf{x} = \mathbf{0}$ and $\mathbf{t} = \mathbf{0}$; by the preceding paragraph, we see that we need the condition that it is no larger. Since the $\tilde{\gamma}_k$ are all linear in the t's, we may also regard this zero locus as a subvariety $P(Y_\gamma)$ of $\mathbb{C}^I \times \mathbb{P}(\mathbb{C}^J)$.

Let $A \subseteq I$, $B \subseteq J$; write

$$V^{A,B} = \{(\mathbf{x},\mathbf{t}) \mid x_i = 0 \Leftrightarrow i \notin A, \; t_j = 0 \Leftrightarrow j \notin B\},$$

and write $\mathbb{P}V^{A,B}$ for the corresponding subset of $\mathbb{C}^I \times \mathbb{P}(\mathbb{C}^J)$ and $Y_\gamma^{A,B}$, $\mathbb{P}Y_\gamma^{A,B}$ for their intersections with Y_γ and $\mathbb{P}Y_\gamma$.

Proposition 4-1. *There exists a submodule M, as described above, such that* $\dim_{\mathbb{C}}(F/M) < \infty$ *if and only if, for any non-empty subsets $A \subseteq I$, $B \subseteq J$, there are at least $\#A + \#B - 1$ values of k for which there exists $r \in R_k$ with* $\mathbf{x}^{\alpha(r)} \mid \mathbb{C}^A \not\equiv \mathbf{0}$ *and $\sigma(r) \in B$.*

PROOF. If the condition fails, choose A and B such that all but at most $\#A + \#B - 2$ of the generators of M vanish identically on $V^{A,B}$. Then $\dim Y_\gamma^{A,B} \geq 2$, so $\dim \mathbb{P}Y_\gamma^{A,B} \geq 1$ and M cannot have finite codimension.

The proof of the converse proceeds as for 3-1: we consider the universal family (with parameter space U) of modules M, observe that if our condition holds the dimension of the support in $U \times \mathbb{C}^I \times \mathbb{P}(\mathbb{C}^J)$ is at most that of U, and hence (using the $\mathbb{C}^\times$-action) that its projection in U is nowhere dense. Choosing a point $\mathbf{u}$ not in this image now provides the required module M. $\square$

Again we can apply this to the weighted homogeneous situation. Assign weights w_i to x_i, b_j to ϵ_j and associate c_k to k; let R_k consist of all monomial elements of weight c_k. Assume the weights w_i in $\mathcal{O}_I$ positive (the signs of the b_j and c_k are immaterial).

Corollary 4-2. *There exists a submodule M as above with $\dim_{\mathbb{C}}(F/M)$ finite if and only if, for any non-empty $A \subseteq I$, $B \subseteq J$, there are at least $\#A + \#B - 1$ values of k with $c_k - b_j \in \mathbb{N}(A)$.*

The above assumes the R_k finite. However, just as in Section 3, the argument extends with virtually no change to the general case.

We remark that an alternative approach to the geometry is to consider the γ_k as defining a linear map $\Gamma : \mathcal{O}_I^{r+s-1} \to \mathcal{O}_I^s$; and reinterpret the matrix $\Gamma \in M_{r+s-1,s}(\mathcal{O}_I)$ as a map-germ $H : \mathbb{C}^r \to M_{r+s-1,s}(\mathbb{C})$. If X denotes the subspace of matrices of rank $< s$, F/M is a sheaf with support $H^{-1}(X)$, so we seek the condition that $H^{-1}(X) = \{0\}$, and interpret $\dim_{\mathbb{C}}(F/M)$ as the local intersection number of $H(\mathbb{C}^r)$ with X. Since X is determinantal, this can be well defined (cf. [4]).

When M has finite codimension, the Eagon-Northcott complex ([15], p.253, with $t = 1$; see also [4]) provides a resolution of F/M which allows us to infer its Poincaré series.

We now have a commutative ring $R = \mathcal{O}_I$, free modules G of rank $\#I + \#J - 1$, F of rank $t = \#J$, and a linear map $\phi : G \to F$. Write Λ^k for the k-th exterior power of G, S^k for the k-th symmetric power of the dual $F^\vee$ of F, and $\tilde{R} = \Lambda^t F^\vee$ for the top exterior power of $F^\vee$. Then set

$$K_0 = F, \quad K_1 = G, \quad K_r = \Lambda^{t+r-1} \otimes S^{r-2} \otimes \tilde{R} \quad (r \geq 2),$$

and define a differential d_* by: $d_1 = \phi$; $d_2 : \Lambda^{t+1}G \to \Lambda^t F^\vee$ is induced by $\Lambda^t \phi$ and inner product; and for $r \geq 3$,

$$d_{r+1} : \Lambda^{t+r} \otimes S^{r-1} \otimes \tilde{R} \to \Lambda^{t+r-1} \otimes S^{r-2} \otimes \tilde{R}$$

is induced by ϕ and inner product.

We write $m = \#I - 1$ and J for the ideal generated in R by the $t \times t$ minors of the matrix of ϕ.

Theorem 4-3. ([15], p.259) *The complex*

$$0 \to K_{m+1} \to K_m \to \ldots \to K_1 \to K_0$$

is exact, provided that $\mathrm{Gr}_R(J, R) \geq m + 1$.

The condition $\mathrm{Gr}_R(J, R) \geq m + 1$ is satisfied if J contains an R-sequence of length $m + 1$; i.e. if the subset of $\mathbb{C}^I$ where the matrix has rank $< \#J$ has codimension $m + 1$, i.e. is reduced to $\{0\}$. This is just the condition above.

Since we have explicit bases for all the modules, and the sequence is exact, the desired Poincaré series is given by an alternating sum, which reduces to

Proposition 4-4. *When the conditions of 4-2 hold, the Poincaré series of F/M is given by*

$$P(z) = \prod(1 - z^{w_i})^{-1} \left(\sum z^{b_i} - \sum z^{c_j} + z^{C-B} \, \mathrm{Res}_t \left(t^{1-r} \frac{\prod(1 + tz^{-c_j})}{\prod(1 + tz^{-b_i})} \right) \right).$$

As in Section 3, we cannot expect that $P(z)$ being a polynomial with nonnegative coefficients will guarantee the conditions of 4-2. This does hold, however, if $n = 1$: the proof is straightforward.

5 $\mathcal{K}$-equivalence

As in the case of $\mathcal{C}$-equivalence, we cannot expect meaningful results if $n < p$. Moreover, if $n = p$, $\mathcal{K}$-finiteness is equivalent to $\mathcal{C}$-finiteness. So we may suppose $n > p$.

Introduce U, F, X and $\mathbb{C}^A$ etc. as before; recall that

$$\Sigma F = \{(\mathbf{t}, \mathbf{x}) \mid TF_{(t,x)} \text{ is not surjective}\},$$

write $\Sigma X^* = X^* \cap \Sigma F$ for the singular set of X^*, and $\Sigma^A = \Sigma X^* \cap V^A$. Define

$$\alpha(A) = \{r \mid (\mathbf{x}^{\alpha(r)} | \mathbb{C}^A) \neq 0\},$$

and recall that

$$J(A) = \{\sigma(r) \mid r \in \alpha(A)\}.$$

Thus F_j vanishes on V^A if and only if $j \notin J(A)$.

Theorem 5-1. *The following are equivalent:*

(a) *there exists a $\mathcal{K}$-finite H map $\mathbb{C}^I \to \mathbb{C}^J$;*

(b) *a generic H map $\mathbb{C}^I \to \mathbb{C}^J$ is $\mathcal{K}$-finite;*

(e) *$\dim \Sigma X^* \leq \dim U$;*

(f) *for $A \subseteq I$ such that $\#A > \#J(A)$ and B a non-empty subset of $J \setminus J(A)$, if $C(A, B)$ denotes the set of those $i \in I \setminus A$ such that for some $r \in R$ with $\sigma(r) \in B$, we have $\mathbf{x}^{\alpha(r)} = x_i m$, where $m|\mathbb{C}^A \not\equiv 0$, then we have $\#C(A, B) \geq \#A + \#B - \#J(A) - 1$.*

PROOF. Consider the jacobian matrix $TF(\mathbf{t}, \mathbf{x})$ for $\mathbf{x} \in \mathbb{C}^A$. This has columns indexed by J and rows indexed by $I \cup R$. If $r \in \alpha(A)$, then $\partial F/\partial t_r = \mathbf{x}^{\alpha(r)} \epsilon_{\sigma(r)}$ has just one nonzero entry, and this is nowhere zero on V^A. Thus the rows of $TF(\mathbf{t}, \mathbf{x})$ corresponding to the $j \in J(A)$ are linearly independent everywhere on V^A. The other rows have zero entries in the t-columns.

Next, if $i \in A$, the only terms in $\partial F/\partial x_i$ not zero on V^A are those with $r \in \alpha(A)$, and these occur only in the rows $J(A)$. It thus remains to study that part of the matrix with rows $I \setminus A$ and columns $J \setminus J(A)$.

Write $E(A)$ for the set of pairs (j, i) with $i \notin A$, $j \notin J(A)$ and $\partial F_j/\partial x_i$ not identically zero on V^A. This is equivalent to the existence of r with $\sigma(r) = j$, $\mathbf{x}^{\alpha(r)} = x_i m$, $m|\mathbb{C}^A \not\equiv 0$, so that

$$C(A, B) = \{i \in I \setminus A \mid \exists j \in B \text{ with } (j, i) \in E(A)\}.$$

Write $M(A)$ for the space of $(I \setminus A) \times (J \setminus J(A))$ matrices with zero entries in all positions $(j, i) \notin E(A)$, and $c(A)$ for the codimension in $M(A)$ of the subset of matrices in $M(A)$ of rank $< \#J - \#J(A)$.

We will show that the conditions of the theorem are also equivalent to

(c). for all non-empty $A \subseteq I$, $\#J(A) + c(A) \geq \#A$.

As before, we begin by observing that (b) $\Rightarrow$ (a) is trivial.

(a) $\Rightarrow$ (c) Suppose (c) fails. For any H map f, Tf is obtained from TF by substitution, so on $\mathbb{C}^{\times A}$ its rank falls on a subset of codimension $\leq c(A)$. Similarly, $f^{-1}(0)$ has codimension at most $\#J(A)$. Thus $\mathbb{C}^{\times A} \cap \Sigma f \cap f^{-1}(0)$ has dimension > 0, so f cannot be $\mathcal{K}$-finite.

(c) $\Rightarrow$ (e) It follows since all the sets involved are constructible that we have $\dim \Sigma X^* = \max\{\dim \Sigma^A\}$. Now Σ^A is defined in V^A by two conditions: it is the intersection of $F^{-1}(0)$ with the subset where the rank of TF falls. As to $F^{-1}(0)$, it is defined by $\#J(A)$ conditions, which are independent since they involve distinct coordinates t_r $(r \in \alpha(A))$. The rank of TF falls on a subset of codimension $c(A)$ since the entries in the residual part of TF again involve distinct coordinates t_r: moreover, as these $r \notin \alpha(A)$, these two conditions are independent. (As before, this independence holds in a strong sense: for each $\mathbf{x} \in \mathbb{C}^{\times A}$, the two conditions on $\mathbf{t}$ each imply the codimension conditions above, and apply to disjoint sets of variables). Thus Σ^A has codimension $\#J(A)+c(A)$ in N^A, so its dimension is $\leq \dim U$ if and only if $\#J(A) + c(A) \geq \#A$.

(e) $\Rightarrow$ (b) The proof is similar to that of 3-1. Since $\dim \Sigma X^* \leq \dim U$, the set B of points $\mathbf{u}$ of U with $(\mathbf{u}, 0)$ in the closure of ΣX^* has codimension > 0 in U. Now if $\mathbf{u}$ is any regular value of the projection $X^* \cap \pi^{-1}(U \setminus B) \to (U \setminus B)$, the corresponding map $f_{\mathbf{u}}$ has $f^{-1}(0)$ smooth in a punctured neighbourhood of 0, so has a $\mathcal{K}$-finite germ.

We have thus proved equivalence of (a), (b), (c) and (e); it remains to show that (c) $\Leftrightarrow$ (f). This is the step the author found the most difficult, and it requires some further ideas. We will thus state the result required, and show how the theorem follows, deferring the proof of the auxiliary theorem to the next section. □

Let M be the space of matrices with rows (resp. columns) indexed by I' resp. J'. For any subset $E \subseteq J' \times I'$, write $M(E)$ for the subspace of M of matrices with zero entries in all positions $(j, i) \notin E$, Δ for the subvariety of M of matrices with rank $< \#J'$ and $\Delta(E) = \Delta \cap M(E)$. Write $c(E)$ for the codimension in $M(E)$ of $\Delta(E)$.

Theorem 5-2. *If $\delta \geq 1$, we have $c(E) \geq \delta$ if and only if, for all nonempty $B \subseteq J'$,*

$$\#\{i \in I' \mid \exists j \in B \text{ with } (j, i) \in E\} \geq \#B + \delta - 1$$

PROOF. (5-1) We now complete the proof of 5-1. We apply 5-2 with $I' = I \setminus A$, $J' = J \setminus J(A)$, $E = E(A)$. Recall that

$$\{i \in I' \mid \exists j \in B \text{ with } (j, i) \in E\}$$

coincides with the set $C(A, B)$ defined in (f). So 5-2 becomes:

$$c(A) \geq \delta \Leftrightarrow \text{ for all nonempty } B \subseteq J \setminus J(A), \#C(A, B) \geq \#B + \delta - 1.$$

Taking $\delta = \#A - \#J(A)$, this yields the desired equivalence of (c) and (f). □

We now apply this result to the homogeneous situation. Now if any d_j equals any w_i, a generic homogeneous map will have nonzero linear part, and thus be an unfolding of an $\mathcal{E}\mathcal{K}$-equivalent map with w_i and d_j deleted. Removing all such pairs (w_i, d_j) we may suppose no such pairs exist. Then according to (3.3) of [19], for a $\mathcal{K}$-finite map to exist, all weights (and hence all degrees) must be positive (if $p > 1$, and also when $p = 1$ if we use the splitting theorem to reduce to the case when f has no quadratic terms). From now on in this section we assume these normalisations made.

Theorem 5-3. *The following are equivalent:*

(a) *there exists a $\mathcal{K}$-finite homogeneous map $\mathbb{C}^I \to \mathbb{C}^J$;*

(b) *a generic homogeneous map $\mathbb{C}^I \to \mathbb{C}^J$ is $\mathcal{K}$-finite;*

(e) $\dim \Sigma X^* \leq \dim U$;

(f) *for $A \subseteq I$ such that $\#A > \#J(A)$ and $B \subseteq J \setminus J(A)$ nonempty, if*

$$C(A, B) = \{i \in I \setminus A \mid \exists j \in B \text{ with } d_j - w_i \in \mathbb{N}(A)\},$$

we have $\#C(A, B) \geq \#A + \#B - \#J(A) - 1$.

Since condition (f) is somewhat complicated, we next show how it simplifies in some special cases. We begin with one useful general remark.

(5-4) If (f) holds, then for all nonempty $A \subseteq I$,

$$2(\#A - \#J(A)) \leq \#I - \#J + 1.$$

For whenever $\#A > \#J(A)$ we apply (f) with $B = J \setminus J(A)$, and note $C(A, B) \subseteq I \setminus A$. Thus $\#I - \#A \geq \#C(A, B) \geq \#A + \#J - 2\#J(A) - 1$, and the result follows.

(5-5) *Equidimensional case $\#I = \#J$:*

Although this case was excluded from our discussion, we can now see that condition (f) reduces to the condition (f) of 3-1 for $\mathcal{C}$-finiteness. For by 5-4, it implies $\#A \leq \#J(A)$ in all cases, and (f) then yields no further condition.

(5-6) *Curve case $\#J = \#I - 1$ and surface case $\#J = \#I - 2$:*

By 5-4 we must have $\#A \leq \#J(A) + 1$ in all cases, and when equality holds then for all nonempty $B \subseteq J \setminus J(A)$, $\#C(A, B) \geq \#B$. By the theorem of Hall recalled in Section 6 below, this is equivalent to saying that the sets $C(A, \{j\})$, $(j \in J \setminus J(A))$ have disjoint representatives in $I \setminus A$. Our condition thus reduces to:

For all nonempty $A \subseteq I$, either $\#J(A) \geq \#A$ or $\#J(A) = \#A - 1$ and we can find distinct elements $r(j) \in I \setminus A$ for each $j \in J \setminus J(A)$ such that $d_j - w_{r(j)} \in \mathbb{N}(A)$.

(5-7) *Hypersurface case $\#J = 1$*:

For (f) to apply we must have $B = J$, so $J(A) = \emptyset$. Setting

$$C_A = C(A, J) = \{i \in I \setminus A \mid (d - w_i) \in \mathbb{N}(A)\},$$

the condition reduces to $\#C_A \geq \#A$. By 5-4, this implies $2\#A \leq \#I$. We can thus reformulate (f) as follows.

If $\#A > (1/2)\#I$, then $d \in \mathbb{N}(A)$. If $A \neq \emptyset$ and $d \notin \mathbb{N}(A)$, then

$$\#\{i \in I \setminus A \mid (d - w_i) \in \mathbb{N}(A)\} \geq \#A.$$

We observe that for $\#I = 2$ or 3 this recovers the conditions recalled in the introduction to this paper.

(5-8) *Case $\#J = 2$*:

Condition (f) is vacuous if $\#J(A) = 2$. If $\#J(A) = 1$ the same holds unless $\#A \geq 2$. In this case, we require that $\#C(A, J \setminus J(A)) \geq \#A - 1$.

If $J(A) = \emptyset$ we can take B with 1 or 2 elements and need $\#C(A, B) \geq \#A + \#B - 1$ in all three cases. When $\#A = 1$, this is equivalent to saying that if $J = \{1, 2\}$ we can find $r \in C(A, \{1\})$, $s \in C(A, \{2\})$ with $r \neq s$.

We next give formulae for the relevant Poincaré series. Let $f : \mathbb{C}^n \to \mathbb{C}^p$ be a $\mathcal{K}$-finite germ $(n > p)$, homogeneous with respect to (positive) weights $(w_1, \ldots, w_n)$ and degrees $(d_1, \ldots, d_p)$. Write $M(f) = \theta(f)/tf(\theta_n)$,

$$N(f) = \frac{\theta(f)}{tf(\theta_n) + f^*\mathfrak{m}_p \cdot \theta(f)} = \frac{\theta(f)}{T_e\mathcal{K}(f)},$$

$$\overline{Q}(f) = \frac{\mathcal{O}_n}{f^*\mathfrak{m}_p \cdot \mathcal{O}_n + J(f)},$$

where $J(f)$ is the Jacobian ideal.

Theorem 5-9.

$$P(M(f), z) = \frac{\sum_j z^{-d_j} - \sum_i z^{-w_i} + z^{D-W} \operatorname{Res}_t(t^{-m}/\Phi)}{\prod_i (1 - z^{w_i})},$$

$$P(N(f), z) = (-1)^{m-1} z^{D-W} + \Phi(z, -1) \cdot B,$$

where

$$B = 1 + \sum_j z^{-d_j} - \sum_i z^{-w_i} + z^{D-W} \operatorname{Res}_t\left(\frac{t^{1-m}}{(1+t)\Phi}\right),$$

and

$$P(\overline{Q}(f), z) = (-1)^{m-1} z^{D-W} + \Phi(z, -1) \cdot \left(1 - z^{D-W} \operatorname{Res}_t\left(\frac{t^{-m}}{(1+t)\Phi}\right)\right).$$

PROOF. The formula for $M(f)$ follows from 4-3. That for $\overline{Q}(f)$ is due to Goryunov [9] and that for $N(f)$ to Aleksandrov [1]. See also [8]. □

The formulae for $\overline{Q}(f)$ will be needed below. Note that neither the above, nor indeed any formula for $N(f)$, is valid when $n = p$ ($\#I = \#J$). Consider for example the map-germs

$$f(x, y, z) = (x^2 + ayz, y^2 + bzx, z^2 + cxy),$$

with $abc \neq -1, 8$. Then $d_e(f, \mathcal{K}) = 9 + m$, where m ($= 0, 1, 2$ or 3) is the number of a, b and c which vanish. But all these germs are $\mathcal{K}$-finite, and homogeneous with the same degrees $(2, 2, 2)$.

The formula for $P(\overline{Q}, z)$ continues to hold when $n = p$. For since the residue term vanishes, it reduces to $\Phi(z, -1) - z^{D-W}$. But by the remark following Theorem 3-2, $\Phi(z, -1)$ is the Poincaré series $P(Q, z)$, and $\overline{Q}(f)$ is obtained in this case from $Q(f) = \mathcal{O}_n / f^* \mathfrak{m}_p \cdot \mathcal{O}_n$ by factoring out the ideal generated by the Jacobian determinant J of f. It is well known (see e.g. [3], ch.5) that J generates the 1-dimensional socle of $Q(f)$, and it clearly has degree $D - W$.

The formula for $P(N(f), z)$ simplifies as follows:
$m = 1$:

$$P(N(f), z) = z^{D-W} + \Phi(z, -1) \left(1 + \sum_j z^{-d_j} - \sum_i z^{-w_i} \right),$$

$m = 2$:

$$P(N(f), z) = -z^{D-W} + \Phi(z, -1) \left(1 + \sum_j z^{-d_j} - \sum_i z^{-w_i} + z^{D-W} \right).$$

We deduce that, at least in these cases, a sufficient condition for $P(N(f), z)$ to be a polynomial is obtained from (c) of 5-3 by replacing $\mathbb{N}$ by $\mathbb{Z}$. For in the reformulation 5-6 in the case $\#J(A) = \#A - 1$ we have distinct elements $r(j) \in I \setminus A$, $(j \in J \setminus J(A))$ with $d_j - w_{r(j)} \in \mathbb{N}(A)$, so can pair all the terms z^{-d_j} above with terms z^{-w_i} so that the difference is divisible by $1 - z^c$, where c generates $\mathbb{Z}(A)$: it then follows that the term in parentheses above is divisible by $1 - z^c$.

It is not clear that this condition is necessary, since the term in question may well be divisible by the c-th cyclotomic polynomial without being divisible by $1 - z^c$. It is even less likely that the condition that $P(N(f), z)$ is a polynomial with nonnegative coefficients will suffice for the conditions of 5-3. However I tentatively conjecture that these implications do hold in the case $\#I = 4, \#J = 2$: they have survived at least a few checks in this case.

For the case $\#I = 4, \#J = 1$ consider the example (communicated to me by Victor Goryunov) with weights $1, 24, 33, 58$ and degree 265. Here it may be verified that P is a polynomial with non-negative coefficients. Our condition holds for all A *except* $\{2, 3\}$ (corresponding to the weights $24, 33$). Thus there is no $\mathcal{K}$-finite singularity of this type.

6   Combinatorial arguments

We now prove Theorem 5-2 as announced above.

We recall that M is the space of matrices with rows (resp. columns) indexed by I' (resp. J'); $E \subseteq J' \times I'$ and $M(E)$ is the subspace of M of matrices with zero entries in all positions $(j, i) \notin E$, Δ the subvariety of M of matrices with rank $< \#J'$, $\Delta(E) = \Delta \cap M(E)$, and $c(E)$ the codimension in $M(E)$ of $\Delta(E)$.

Theorem 6-1.   *If $\delta \geq 1$, we have $c(E) \geq \delta$ if and only if, for all nonempty $B \subseteq J'$,*

$$\#\{i \in I' \mid \exists j \in B \text{ with } (j, i) \in E\} \geq \#B + \delta - 1.$$

PROOF. Suppose $B \subseteq J$ a non-empty set and let

$$C(B) = \{i \in I' \mid \exists j \in B \text{ with } (j, i) \in E\}.$$

Now if the rows in B are linearly dependent, then certainly the set of all rows is so. For these rows, we have effectively only $\#C(B)$ columns, so even if all $(\#B) \cdot (\#C(B))$ entries were independent, the codimension of the subset where the rank of this submatrix dropped would be $\#C(B) - \#B + 1$; in fact it must be at most this. Hence if $c(E) \geq \delta$, we must have $\#C(B) \geq \#B + \delta - 1$.

For the converse we proceed by induction on $\#J'$: the result is clear if $\#J' = 1$. Write Δ_0 for the set of those $\mathbf{t} \in M$ such that some proper subset of the rows of $\mathbf{t}$ is dependent, and $\Delta_0(E) = \Delta_0 \cap M(E)$. The induction hypothesis shows that $\mathrm{codim}(\Delta_0(E)) \geq i$. Since $\Delta(E)$ is an algebraic set and $\Delta_0(E)$ an algebraic subset, it will now suffice to show that $\mathrm{codim}(\Delta(E) \setminus \Delta_0(E)) \geq i$. Observe that for $\mathbf{t} \in \Delta \setminus \Delta_0$, the rank of $\mathbf{t}$ is $\#J' - 1$. Thus there are uniquely determined coefficients $\lambda_i(\mathbf{t})$, all nonzero, with $\sum \lambda_i(\mathbf{t})(i\text{-th row of } \mathbf{t}) = \mathbf{0}$.

We claim that the tangent space $T_t\Delta$ at $\mathbf{t}$ to Δ has codimension

$$\mathrm{codim}_{M(E)}(M(E) \cap T_t\Delta) \geq i.$$

It follows that $\mathbf{t}$ has a neighbourhood U in $M(E)$ such that the codimension of $U \cap \Delta(E)$ in U is $\geq i$; since this holds for all $\mathbf{t}$, the desired conclusion will follow.

Since the space of matrices of rank $\#J' - 1$ is a single orbit under the product of the general linear groups of source and target, $T_t\Delta$ is the sum $\mathcal{L}\mathbf{t} + \mathbf{t}\mathcal{R}$ of the subspaces obtained by multiplying $\mathbf{t}$ on the left or on the right by arbitrary square matrices. Now $\mathbf{t}\mathcal{R}$ is the space of matrices $\mathbf{u}$ such that $\sum \lambda_i(\mathbf{t})(i\text{-th row of } \mathbf{u}) = \mathbf{0}$, so we see that $\mathcal{L}\mathbf{t} + \mathbf{t}\mathcal{R}$ is the space of matrices $\mathbf{u}$ such that $\sum \lambda_i(\mathbf{t})(i\text{-th row of } \mathbf{u})$ is a linear combination of the rows of $\mathbf{t}$.

Now by hypothesis, the matrices in $M(E)$ involve at least $\#J' + \delta - 1$ columns. Forming $\sum \lambda_i(\mathbf{t})(i\text{-th row})$ leads to a space of row vectors of dimension at least $\#J' + \delta - 1$, since the nonzero entries in $M(E)$ are independent and the constants $\lambda_i(\mathbf{t})$ are all nonzero. The space of linear combinations of rows of

t has dimension $\#J' - 1$ since this is the rank of **t**. This has codimension $\geq \delta$ in the above, proving the theorem.   □

Before discovering the above argument using a partial transversality, the author had obtained partial results using more combinatorial arguments, and some of the ideas seem worth rehearsing here.

Consideration of the patterns of subsets E led to the formulation of the following generalisation of Hall's selection theorem. (For combinatorial results of this kind see e.g. [13]).

Theorem 6-2. *Let $\{A_i \mid i \in K\}$ be subsets of a finite set F such that for all nonempty $I \subseteq K$, if $A_I = \cup\{A_i \mid i \in I\}$ then $\#A_I \geq \#I + r - 1$. Then there exist subsets $B_i \subseteq A_i$, with $\#B_i = r$ for each i, such that each $\#B_I \geq \#I + r - 1$.*

PROOF. By induction on $\sum \#A_i$, it will suffice to show that if (say) $\#A_1 > r$ then we can find $x \in A_1$ such that replacing A_1 by $A_1' = A_1 \setminus \{x\}$ does not destroy the hypothesis.

If $x \in A_1$ does not satisfy this, we can find $I \subseteq K$, with $1 \notin I$, and $\#(A_I \cup A_1') < \#I + r$ and so, since we have only removed one element, is equal to $\#I + r - 1$. As $\#A_I \geq \#I + r - 1$, these two sets are equal; thus $A_1' \subseteq A_I$. Call such a subset I *x-critical*. Also, call I *critical* if $\#A_I = \#I + r - 1$. We now interrupt the proof for a lemma.   □

Lemma 6-3.

  (a) I *is x-critical if and only if I is critical and $A_I \cap A_1 = A_1 - \{x\}$.*

  (b) *If I and J are critical, then either*

   (i) $I \cap J = \emptyset$ or

   (ii) $I \cap J$ *and $I \cup J$ are critical, and $A_{I \cap J} = A_I \cap A_J$.*

  (c) *If $x \neq y \in A_1$, I is x-critical and J is y-critical, then $I \cap J = \emptyset$.*

PROOF. (a) follows from the discussion above. As to (b), we have

 • $\#(I \cup J) + r - 1 \leq \#A_{I \cup J} = \#(A_I \cup A_J)$ and, if $I \cap J \neq \emptyset$,

 • $\#(I \cap J) + r - 1 \leq \#A_{I \cap J} \leq \#(A_I \cap A_J)$.

Since

 • $\#(I \cup J) + \#(I \cap J) = \#I + \#J$,

 • $\#(A_I \cup A_J) + \#(A_I \cap A_J) = \#A_I + \#A_J$, and

 • $\#A_I = \#I + r - 1$ and $\#A_J = \#J + r - 1$,

equality must hold throughout.

For (c), observe that by (a), $A_I \cap A_1 = A_1 \setminus \{x\}$ and $A_J \cap A_1 = A_1 \setminus \{y\}$, so $A_1 \subseteq A_I \cup A_J$. If $I \cap J \neq \emptyset$, then by (b) $I \cup J$ would be critical, so $\#A_{I \cup J \cup \{1\}} = \#A_{I \cup J} = \#(I \cup J) + r - 1$, contradicting our hypothesis.   □

PROOF. (6-2) We now complete the proof of 6-2. Suppose if possible that $\#A_1 > r$ but we cannot delete any member of A_1 without destroying our hypothesis. Choose distinct elements x_i, $(0 \le i \le r)$ in A_1, and choose x_i-critical subsets I_i of K. By 6-3, (c), these are disjoint. Hence

$$\# \cup_0^r A_{I_i} \le \# \cup_0^r (A_{I_i} - A_1) + \#A_1 \le \sum_0^r \#(A_{I_i} - A_1) + \#A_1$$

$$= \sum_0^r (\#I_i + r - \#A_1) + \#A_1$$

by 6-3, (a),

$$= \sum_0^r \#I_i + r(r + 1 - \#A_1) \le \sum_0^r \#I_i = \# \cup_0^r I_i$$

since the I_i are disjoint. But this violates our hypothesis.   □

Hall's theorem is the case $r = 1$ of 6-2, showing that the sets A_i admit disjoint representatives.

It seems an attractive idea to find a 'linear algebra version' of this result, and the author was led to conjecture the following, which would have included 6-1 as a special case:

Let V, W and X be vector spaces, $\phi : X \otimes V \to W$ a linear map such that for all subspaces A of V, $\dim \phi(X \otimes A) \ge \dim A + r - 1$. Then there is an r-dimensional subspace Y of X such that $\psi = \phi|(Y \otimes V)$ has the same property.

Unfortunately this conjecture is false, as the simple example

$$\phi((x_1, x_2, x_3) \otimes (v_1, v_2)) = (x_1 v_1, x_2 v_1 + x_3 v_2, x_1 v_2)$$

(with $r = 2$) demonstrates. It would be very interesting if reasonably simple conditions on ϕ could be found which imply the above conclusion. The conjecture is true if $r = 1$.

Proposition 6-4. *Let V, W and X be vector spaces, $\phi : X \otimes V \to W$ a linear map such that for all subspaces A of V, $\dim \phi(X \otimes A) \ge \dim A$. Then there is a 1-dimensional subspace Y of X such that $\psi = \phi|(Y \otimes V)$ is injective.*

PROOF. By induction on $\dim W$. First suppose V has a proper subspace V_0 such that $W_0 = \phi(X \otimes V_0)$ has $\dim W_0 = \dim V_0$. Then if $V_1 = V/V_0$ and $W_1 = W/W_0$, ϕ induces $\phi_0 : X \otimes V_0 \to W_0$ and $\phi_1 : X \otimes V_1 \to W_1$. Clearly ϕ_0 satisfies the hypothesis; so does ϕ_1 since a subspace of V_1 has the form B/V_0 with $V_0 \subseteq B \subseteq V$, and

$$\dim \phi_1(X \otimes B/V_0) = \dim \phi(X \otimes B)/W_0 \ge \dim B - \dim W_0 = \dim B/V_0.$$

By the induction hypothesis, the set of $x \in X$ such that $\phi_0(x \otimes ?)$ is injective is non-empty; it thus contains a dense and open subset, since its complement is constructible. The same argument also applies to ϕ_1. Thus for a dense open set of x we have injective maps $V_0 \to W_0$ and $V_1 \to W_1$ and hence also $V \to W$.

If, however, for every proper subspace V_0 of V we have $\dim \phi(X \otimes V_0) > \dim V_0$, choose any projection $\pi : W \to \overline{W}$ with $\dim \ker \pi = 1$. Then $\pi \circ \phi$ satisfies the hypothesis, so by induction hypothesis we can choose x with $\pi \phi(x \otimes ?)$ injective. But then it follows that $\phi(x \otimes ?)$ is injective.   □

7 $\mathcal{A}$-equivalence

For the case of $\mathcal{A}$-equivalence we take a rather different approach to the above. First, we wish to consider our map F as an unfolding of a map f with zero 1-jet at $\mathbf{0}$. Now F is a generic map of given weights and degrees, so if any equality $d_j = w_i$ holds, TF will be nonzero, and F an unfolding of a homogeneous map with fewer components in source and target. We thus begin by cancelling as many such pairs as possible to obtain weights and degrees for f.

Next, $\mathcal{A}$-finiteness of F implies $\mathcal{K}$-finiteness of F, and hence of f. For f, if $\#J > 1$ we know by the discussion at the beginning of Section 5 that $\mathcal{K}$-finiteness implies that all weights and degrees are positive, and then 5-1 gives the precise conditions required for $\mathcal{K}$-finiteness. If $\#J = 1$, and if we have $r, s \in I$ with $w_r + w_s = d$ then by the splitting theorem we can reduce to the case with r and s deleted. We may either delete such pairs (noting that if this reduces to $\#I \leq 1$ we may have to reinstate a pair) or replace the weights by $d/2$ and thus suppose all weights > 0. We refer to the process defined in these two paragraphs as *reduction*. Notationally, we consider $f : \mathbb{C}^{I_0} \to \mathbb{C}^{J_0}$ as unfolded by $F : \mathbb{C}^I \to \mathbb{C}^J$, and abbreviate $\mathcal{O}_{J_0}$ to $\mathcal{O}_0$. We assume $\#I > \#J$.

We next consider $M(f) = \theta(f)/\mathrm{Im}(tf)$ as a module over the ring $\mathcal{O}_0$. It follows from the preparation theorem that this is finitely generated, and a set of generators may be obtained by lifting a set $\{\psi_r\}_{1 \leq r \leq g}$ of generators of $N(f)$. These may be supposed homogeneous; their weights are provided by the formula 5-9 for the Poincaré series of $N(f)$. Writing $\{\eta_r\}$ for the standard basis of $\mathcal{O}_0^g$, we define a surjection $\phi f : \mathcal{O}_0^g \to M(f) = \theta(f)/\mathrm{Im}(tf)$ by $\phi f(\eta_r) = \psi_r$.

It is also known (this is essentially due to Looijenga [10]; discussions closer to our viewpoint will be found in [6], [8]) that $\ker \phi f$ is a free $\mathcal{O}_0$-module. More precisely, a homogeneous base for $\ker \phi f$ may be obtained by applying the Euler derivation (of weight 0) to a basis of $\overline{Q}(f)$ as follows (this is due to Goryunov [8]; an alternative account appears in Chapter 10 of our book [20]).

Theorem 7-1. (10.5.12, loc.cit.) Let $\{\chi_i\}$ be a $\mathbb{C}$-basis for $\overline{Q}(f)$; acting on the basis $\{\partial/\partial y_j\}$, $(j \in J_0)$ for $\theta(f)$ yields elements projecting to $[\chi_i \partial/\partial y_j] \in M(f)$, say. Choose lifts

$$\phi f \left(\sum a_{ijr} \eta_r \right) = \left[\chi_i \frac{\partial}{\partial y_j} \right],$$

with $a_{ijr} \in \mathcal{O}_0$. Then the elements $\Xi_i = \sum_j d_j y_j a_{ijr} \eta_r$ form a free $\mathcal{O}_0$-base of $\ker \phi f$.

Thus we have an exact sequence

$$0 \longrightarrow \mathcal{O}_0^g \xrightarrow{\alpha f} \mathcal{O}_0^g \xrightarrow{\phi f} M(f) \longrightarrow 0.$$

We could indeed have given this discussion for F itself, rather than for f. Since inclusion induces isomorphisms $N(f) \to N(F)$, $\overline{Q}(f) \to \overline{Q}(F)$ this leads to a sequence of the same form with $\mathcal{O}_0$ replaced by the ring $\mathcal{O}_J$. The point of working over f first is (a) usually many fewer variables, so easier calculations, and (b) positive weights.

We wish to study the deformation space

$$N\mathcal{A}(F) = \theta(F)/T_e\mathcal{A}(F) = \theta(F)/(\operatorname{Im} tF + \operatorname{Im} \omega F).$$

Now ωF is a map to $\theta(F)$ of the free R-module θ_J with basis the coordinate vectors ϵ_j. Here $j \in J$, the index set for coordinates in the target of F, and hence for generators of R. We thus have a presentation of $N\mathcal{A}(F)$ as a quotient of free R-modules with homogeneous generators of known degrees:

- Module: same as $\mathbb{C}$-generators for $N(F)$,

- Submodule: the $\mathbb{C}$-generators for $\overline{Q}(F)$, together with the elements $\{-d_j \mid j \in J\}$.

We wish to apply 4-1 to this situation. At present, the submodule appears to have one generator more than is needed for that result to be applicable. But the generator 1 of $\overline{Q}(F)$ corresponds to the Euler relation which (modulo $\operatorname{Im} tF$) belongs to the image of ωF. It may thus be omitted.

Theorem 7-2. *Suppose given weights and degrees as usual, with $\#I > \#J$. Then necessary conditions that a generic homogeneous map be $\mathcal{A}$-finite are: (a) after reduction, condition (f) of 5-3 holds. We then use 5-9 to determine the sets of degrees $\{\alpha_l \mid l \in L\}$, say, occurring in a basis of $N(F)$ and $\{\beta_l \mid l \in L\}$ in a basis of $\overline{Q}(F)$. Let $\beta_c = 0$ and $L' = L \setminus \{c\}$. Then the condition of 4-2 must hold, where notation is changed as follows:*

4-2	I	$\mathcal{O}_I$	w_i	J	b_j	K	c_k
Here	J	$\mathcal{O}_J$	d_j	L	α_l	$L' \cup J$	$\beta_l, -d_j$

We recall that the condition in 4-2 was: for any non-empty subsets $A \subseteq I$, $B \subseteq J$, there are at least $\#A + \#B - 1$ of the c_k for which there exists $j \in B$ with $c_k - b_j \in \mathbb{N}(A)$. So here we get:

- for any non-empty subsets $A \subseteq J$, $B \subseteq L$, set

$$C_1 = \{l \in L' \mid \exists m \in L, \ \beta_l - \alpha_m \in \mathbb{N}(A)\},$$

$$C_2 = \{j \in J \mid \exists m \in L, \ -d_j - \alpha_m \in \mathbb{N}(A)\},$$

then $\#C_1 + \#C_2 \geq \#A + \#B - 1$.

Here the β_l (as well as the w_i, d_j) are positive, but the α_l may have either sign. From 4-3 we have, writing $AA = \sum \alpha_m$, $BB = \sum \beta_l$:

Proposition 7-3. *When the conditions of 7-2 hold, the Poincaré series of $N\mathcal{A}(F)$ is given by*

$$P(z) = \prod (1 - z^{d_j})^{-1} \left(\sum z^{\alpha_l} - \sum z^{\beta_l} - \sum z^{-d_j} + z^{BB-D-AA} \tilde{R} \right)$$

with

$$\tilde{R} = \operatorname{Res}_t \left(t^{1-r} \frac{\prod (1 + tz^{-\beta_l}) \prod (1 + tz^{d_j})}{\prod (1 + tz^{-\alpha_l})} \right).$$

We will give an example to show that the necessary conditions 7-2 for existence of $\mathcal{A}$-finite maps are insufficient. We now give a fuller discussion of the problem of finding sufficient conditions, to put it in context and indicate the nature of the difficulty.

We start, as above, with a map $f : \mathbb{C}^s \to \mathbb{C}^t$ having zero 1-jet at $\mathbf{0}$. Let $\widehat{F} : \mathbb{C}^{s+a} \to \mathbb{C}^{t+a}$ be a miniversal unfolding of f. Thus any $\mathcal{EK}$-equivalent germ $F : \mathbb{C}^n \to \mathbb{C}^p$ is induced from $\widehat{F}$ by some map $g : \mathbb{C}^p \to \mathbb{C}^{t+a}$. The map g may be considered arbitrary subject to the requirement that it is transversal at $\mathbf{0}$ to $\widehat{F}$; if F is weighted homogeneous, we may choose g also to be.

Since $\widehat{F}$ is stable, the $t + a = \tau$ coordinate vectors $\partial/\partial y_j$ provide a basis for $N(\widehat{F})$. The same discussion as above shows that, setting $\widehat{\mathcal{O}}$ for the ring $\mathcal{O}_{t+a}$, these induce a surjective map

$$\phi\widehat{F} : \widehat{\mathcal{O}}^\tau \to M(\widehat{F}) = \theta(\widehat{F})/\operatorname{Im}(t\widehat{F}).$$

Again the kernel is free on τ generators α_i whose degrees are those of the elements of a $\mathbb{C}$-basis of $\overline{Q}(\widehat{F})$.

Now $M(\widehat{F})$ behaves naturally under base-change: $M(F)$ is the (completed) tensor product over $\widehat{\mathcal{O}}$ of $M(\widehat{F})$ and $\mathcal{O}_p$, where the module structure on the latter is that induced by g. It is thus the quotient of $\mathcal{O}_p^\tau$ by the free submodule generated by the images of the α_i. To be completely explicit, if α_i has components a_{ij}, this image $g^*\alpha_i$ has corresponding components $a_{ij} \circ g = g^* a_{ij}$. Note that the larger module $\mathcal{O}_p^\tau$ can be identified with $\theta(g)$.

We must also consider the image of

$$\theta_p \xrightarrow{\omega F} \theta(F) \to \theta(F)/\operatorname{Im}(tF) = M(F).$$

But θ_p is freely generated by the coordinate vector fields $\partial/\partial v_i$, say, and this corresponds under g to $\sum (\partial y_j/\partial v_i)(\partial/\partial y_j)$. Since we have identified the $\partial/\partial y_j$ with the free generators of our module, we see that these same vectors are the natural generators of the image of tg.

We have thus identified the modules

$$\theta(F)/(tF(\theta_n) + \omega F(\theta_p)) \text{ and } \theta(g)/(tg(\theta_p) + \omega g(A)),$$

where A is the module spanned by the α_i. (Note in passing that tg corresponds to ωF and ωg to tF!). This is essentially the same as the identification made by Damon [5] of $\theta(F)/T_e\mathcal{A}(F)$ with $\theta(g)/T_e\mathcal{K}_V(g)$, where V is identified with the discriminant $\Delta(\widehat{F})$: it is well known [8] that the α_i span the module of vector fields tangent to $\Delta(\widehat{F})$.

It can now be seen why it is not straightforward to apply the methods of the preceding sections of this article to the problem of finding sufficient conditions for the existence of $\mathcal{A}$-finite maps: the structure of the module A is not closely tied to the coordinate system, and g^*A does not depend linearly on the map g.

We now present the example mentioned above. Take

  weights w_i as $\{2, 2, 2, 2, 1, 1, 1, 1, 1, 1\}$ and

  degrees d_j as $\{3, 3, 2, 2, 2, 1, 1, 1, 1\}$.

Then F is an unfolding of a map f with weights $\{2, 1, 1\}$ and degrees $\{3, 3\}$. Put f in the normal form $\langle xz + y^3, xy + z^3 + vzy^2 \rangle$, where $v^2 \neq -4$ for $\mathcal{K}$-finiteness. The versal unfolding of f is

$$\widehat{F}(x, y, z, u_1, \ldots, u_8) = \langle u_1, \ldots, u_8, xz + y^3 + u_1 y,$$

$$xy + z^3 + (v + u_8)y^2 z + u_2 y + u_3 z + u_4 x + u_5 y^2 + u_6 yz + u_7 z^2 \rangle$$

which is $\mathcal{A}$-stable. Any unfolding with the given weights is induced from this by a map j which may be assumed weight-preserving. For a generic such j, the matrix of partial derivatives of the u_i, $(1 \leq i \leq 7)$ with respect to the given unfolding parameters will be nonsingular. The unfolding is thus equivalent to the map F obtained from the above by omitting u_8. Thus F represents the generic map with the given weights and degrees.

F is not, however, $\mathcal{A}$-finite. For if $\alpha^4 + v\alpha^2 - 1 = 0$, the map-germ

$$(x, y, z) \to \langle xz + y^3 + (6 - 9v\alpha^2)t^2 y + (18v\alpha^2 - 20)t^3,$$

$$x(y + 4t) + \frac{(z + 3\alpha t)^4 - (3\alpha t)^4}{z} + v(y + t)^2(z + 6\alpha t) \rangle$$

has an $\tilde{E}_7$ singularity at $(0, -4t, -3\alpha t)$ as we can see since

$$zf_2 - (y + 4t)f_1 = (z + 3\alpha t)^4 + v(z + 3\alpha t)^2(y + t)^2 - (y + t)^4.$$

This germ is obtained from F by substituting for the u_i:

u_1	u_2	u_3	u_4	u_5	u_6	u_7
$6 - 9v\alpha^2$	$12v\alpha t^2$	$54\alpha^2 + v$	$4t$	$6v\alpha t$	$2vt$	$12\alpha t$,

and adding a constant. We have thus identified a curve of points at which the germ of F belongs to the stratum $\tilde{E}_7$. These are the only $\tilde{E}_7$ points in the source of F; they are not versally unfolded by F since the modulus v is constant. Thus we have a curve of points at which the germ of F is not $\mathcal{A}$-stable, so the germ of F at $\mathbf{0}$ is not $\mathcal{A}$-finite.

On the other hand, we calculate

$$P(N(f), z) = 2z^{-3} + 3z^{-2} + 4z^{-1} + 1;$$

$$P(\overline{Q}(f), z) = 1 + 2z + 4z^2 + 2z^3 + z^4;$$

Hence the weights in the ring R are $\{3, 3, 2, 2, 2, 1, 1, 1, 1\}$; the weights in the module are $\{-3, -3, -2, -2, -2, -1, -1, -1, -1, 0\}$; and those in the submodule are $\{-3, -3, -2, -2, -2, -1, -1, -1, -1, 1, 1, 2, 2, 2, 2, 3, 3, 4\}$. Most of these cancel with generators of the module; we are left with a module with 1 generator of degree 0 and a submodule (ideal) with generators of weights $\{1, 1, 2, 2, 2, 2, 3, 3, 4\}$. To check the conditions of 3-1, (f) for this case we consider the possible semigroups $\mathbb{N}(A)$:

- $\mathbb{N}(A) = \mathbb{N}$: then $J(A) = J$, with 9 elements while $\#A \leq \#I = 9$;

- $\mathbb{N}(A) = 2\mathbb{N}$: then $\#J(A) = 5$, $\#A \leq 3$;

- $\mathbb{N}(A) = 3\mathbb{N}$: then $\#J(A) = 2$, $\#A \leq 2$;

- $\mathbb{N}(A) = \langle 2, 3 \rangle$: then $\#J(A) = 7$, $\#A \leq 5$.

Thus the condition holds in all cases.

Alternatively, if we write the generators of the ring as a_1, b_1, c_1, d_1, a_2, b_2, c_2, a_3, b_3 (the subscript denoting the weight) we may consider the ideal generated by a_1, b_1, a_2, b_2, c_2, $c_1 d_1$, a_3, b_3, and $c_1^4 + d_1^4$, which clearly has finite codimension (viz. 8).

It seems likely that to make effective progress on sufficient conditions for the existence of homogeneous $\mathcal{A}$-finite maps we will have to use the characterisation as maps $\mathcal{A}$-stable outside the origin, and thus use the classification of $\mathcal{A}$-stable germs. This is likely to be easier in the nice dimensions where, indeed, it is still plausible that the necessary conditions given above may be sufficient.

8   Other ground fields

Several of the arguments in this article remain valid over fields $\mathfrak{k}$ other than $\mathbb{C}$.

In the proof of 5-1, the arguments for (a) $\Rightarrow$ (f) and for (f) $\Rightarrow$ (e) are valid over any algebraically closed field, so the implications hold for any field. For (e) $\Rightarrow$ (b) we require the existence of a point in the complement of an algebraic subset of positive codimension: this holds over any infinite field. Similar considerations apply to 4-1.

In 5-1 however, though most arguments (and in particular, those involving the Jacobian matrix of f) remain valid over any $\mathfrak{k}$, the existence of regular values of the projection $\pi : X^* \to U$ breaks down (this map may be inseparable). The conclusion does follow for any infinite field of characteristic 0; in particular, for $\mathbb{R}$.

References

[1] A.G. Aleksandrov, *Cohomology of a quasihomogeneous complete intersection*, Math. USSR Izvestiya **26**iii (1986) 437–477.

[2] V.I. Arnol'd, *Normal forms of functions in neighbourhoods of degenerate critical points*, Russian Math. Surveys **29**ii (1974) 10–50.

[3] V.I. Arnol'd, S.M. Gusein-Zade & A.N. Varchenko, *Singularities of differentiable maps I*, Birkhäuser 1985.

[4] D.A. Buchsbaum & D.S. Rim, *A generalised Koszul complex II: depth and multiplicity*, Trans. Amer. Math. Soc. **3** (1964) 197–224.

[5] J.N. Damon, *A-equivalence and the equivalence of sections of images and discriminants*, pp 93–121 in *Singularity theory and its applications I*, Springer lecture notes in math. **1462** (1991).

[6] A.A. du Plessis, T. Gaffney & L.C. Wilson, *Map-germs determined by their discriminants*, to appear.

[7] T. Gaffney & D.M.Q. Mond, *Weighted homogeneous maps from the plane to the plane*, Math. Proc. Camb. Phil. Soc. **109** (1991) 451–470.

[8] V.V. Goryunov, *Vector fields and functions on discriminants of complete intersections and bifurcation diagrams of projections*, pp 31–54 in Current problems in math. (Itogi Nauki i Tekhniki) **33**, VINITI, 1988. Translated in Journal of Soviet Math. **52** (1990) 3231–3245.

[9] V.V. Goryunov, *Poincaré polynomial of the space of residue forms on a quasihomogeneous complete intersection*, Russian Math. Surveys **35**ii (1980) 241–242.

[10] E.J.N. Looijenga, *Isolated singular points on complete intersections*, LMS lecture note series **77**, Cambridge University Press, 1984.

[11] J.N. Mather, *Stability of C^∞-mappings III: finitely determined map-germs*, Publ. Math. IHES **35** (1969) 127–156.

[12] J.W. Milnor & P.Orlik, *Isolated singularities defined by weighted homogeneous polynomials*, Topology **9** (1970) 385–393.

[13] L. Mirsky, *Transversal theory*, Academic Press, 1971.

[14] D.M.Q. Mond, *The number of vanishing cycles for a quasihomogeneous mapping from $\mathbb{C}^2$ to $\mathbb{C}^3$*, Quart. J. Math. Oxford **42** (1991) 335–345.

[15] D.G. Northcott, *Finite free resolutions*, Cambridge University Press, 1976.

[16] K. Saito, *Regular systems of weights and associated singularities*, pp 479–526 in *Advanced studies in Math.* **8**, Kinokuniya & North Holland, 1986.

[17] O.P. Sherbak, *Conditions for the existence of a nondegenerate mapping with a given support*, Func. Anal Appl. **13** (1979) 154–155.

[18] C.T.C. Wall, *Finite determinacy of smooth mappings*, Bull. London Math. Soc. **13** (1981) 481–539.

[19] C.T.C. Wall, *A second note on symmetry of singularities*, Bull. London Math. Soc. **12** (1980) 347–354.

[20] A.A. du Plessis & C.T.C. Wall, *The geometry of topological stability*, Oxford University Press, 1995.

Address of author:

Department of Mathematics
University of Liverpool
PO. Box 147
Liverpool L69 3BX
United Kingdom
Email: c.t.c.wall@liverpool.ac.uk

Part III

Curves and Surfaces

Degree 8 and genus 5 curves in $\mathbb{P}^3$ and the Horrocks-Mumford bundle.

M. R. González-Dorrego

Introduction

We construct curves of degree 8 and genus 5 on a Kummer surface S (Proposition 1-6). We show that the exact sequence of normal bundles associated with a generic such curve splits (Theorem 3-4).

In [1, p. 43] W. Barth gave a construction of the Horrocks-Mumford bundle assuming the existence of a nonsingular irreducible curve C with certain properties (cf. Theorem 2-3); one of them is that, set-theoretically, C is the complete intersection of two surfaces S_1, $S_2 \subset \mathbb{P}^3$, both of degree n. For $n = 4$, the existence of such a curve ($d = 8$, $g = 5$) was proved using computational methods (Macaulay), by F. O. Schreyer and W. Decker. To my knowledge this is the only existence result for such curves so far. In this paper, we explain why a generic irreducible nonsingular curve of degree 8 and genus 5 on a Kummer surface satisfies all but one of Barth's conditions (Proposition 3-6); thus, such a curve cannot be used to do Barth's construction of the Horrocks-Mumford bundle. This result sheds new light on the search of an irreducible nonsingular curve of degree 8 and genus 5 in $\mathbb{P}^3$ satisfying all of Barth's conditions; it must be "special". It will be nice to give a conceptual proof of the existence of such a curve and to describe its locus.

I would like to thank L. Ein, K. Hulek and A. Sommese for discussions which were a great help to me in this work.

1 Construction of curves of degree 8 and genus 5 on a Kummer surface $S \in \mathbb{P}^3$

Let k be an algebraically closed field of characteristic different from 2.

Definition 1-1. *A Kummer surface S in $\mathbb{P}^3$ is a reduced, irreducible surface of degree 4 having 16 nodes and no other singularities.*

Remark 1-2. *The nodes of S cannot be 16 arbitrary points of $\mathbb{P}^3$. They form a special configuration called a (16,6) configuration (i.e., there exist 16 planes such that every plane contains exactly 6 nodes and every node lies on exactly 6 planes). These planes are called special planes; they correspond to the nodes of the dual Kummer surface S^*. The section of S by one of the special planes is a non-singular conic, counted twice; we call this conic a special conic. [2, Proposition 2.16, Corollary 2.18, Corollary 4.27].*

Progress in Mathematics, Vol. 134
© 1996 Birkhäuser Verlag Basel/Switzerland

A Rosenhain tetrahedron is a set of 4 special planes forming a tetrahedron whose vertices are nodes. There exist 80 Rosenhain tetrahedra in S [2, Corollary 3.21].

The minimal desingularization $\pi \colon X \to S$ of the Kummer surface S is obtained by blowing-up its 16 nodes $\{P_i\}_{1 \leq i \leq 16}$. Let $E_i = \pi^{-1}(P_i)$ be the exceptional divisor associated to P_i, $1 \leq i \leq 16$. Let J denote the set of subindices of the 16 nodes of S. Let H be a hyperplane section of S. Let $\tilde{H} = \pi^*(H)$ be the total transform of the hyperplane section of S.

It is known that X is a K3 surface [2, Theorem 3.4].

Proposition 1-3. *For any Rosenhain tetrahedron, let D be the divisor on X formed by the sum of the proper transforms of the 4 conics in which the planes meet S and the four exceptional curves, inverse images of the 4 nodes:*

$$D = \tilde{C}_0 + \tilde{C}_1 + \tilde{C}_2 + \tilde{C}_3 + E_0 + E_1 + E_2 + E_3.$$

Then the linear equivalence class of D is independent of the Rosenhain tetrahedron chosen. In fact,

$$\sum_{i=1}^{16} E_i \sim 4\tilde{H} - 2D.$$

PROOF. [2, Proposition 3.22]. □

Theorem 1-4. C *be an irreducible curve on a K3 surface X such that $C^2 > 0$. Then $\mathcal{O}_X(C)$ is base-point free.*

PROOF. [6, Theorem 3.1] □

Corollary 1-5. *Let L be an invertible sheaf on a K3 surface such that $|L|$ has no fixed components and $L^2 \geq 2$. The generic member of $|L|$ is an irreducible nonsingular curve.*

PROOF. [6, (4.1), (5.8), (6.1)]. □

Proposition 1-6. *There exists an irreducible nonsingular curve of degree 8 and genus 5 on S, passing through the 16 nodes of S.*

PROOF. Let H be a hyperplane section of S. We define the linear system on X

$$|\tilde{C}| = \left| 2\tilde{H} - \tfrac{1}{2} \sum_{i \in J} E_i \right|.$$

By Proposition 1-3 $|\tilde{C}| = |D|$, which has no fixed components,

$$\tilde{C} = 8 \quad \text{and} \quad p_a(\tilde{C}) = 5.$$

Let char $k \neq 2$. By Corollary 1-5 the generic member of $|\tilde{C}|$ is an irreducible nonsingular curve. Consider its image in S. Since $E_i \cdot \tilde{C} = 1$, $1 \leq i \leq 16$, we obtain an irreducible nonsingular curve of degree 8 and genus 5; it passes through the 16 nodes of S.

See [3] for another, more classical, construction of this linear system. □

2 Barth's Construction

Let char $k = 0$. Let C be an irreducible nonsingular curve in $\mathbb{P}^3$ of degree d and genus g.

Let $\mathcal{N}_C = \mathcal{N}_C|_{\mathbb{P}^3}$ denote the normal sheaf of C in $\mathbb{P}^3$. Using the exact sequences of [4, p. 182],

$$0 \to \mathcal{T}_C \to \mathcal{T}_{\mathbb{P}^3} \otimes \mathcal{O}_C \to \mathcal{N}_C \to 0, \qquad 0 \to \mathcal{O}_C \to \mathcal{O}_C(1)^4 \to \mathcal{T}_C|_{\mathbb{P}^3} \to o,$$

we can compute the degree of $\mathcal{N}_C$: $\deg \mathcal{N}_C = 4d + 2g - 2$. Besides,

$$\begin{aligned}
\chi(\mathcal{N}_C) &= \chi(\mathcal{T}_C|_{\mathbb{P}^3}) - \chi(\mathcal{T}_C) \\
&= \chi(\mathcal{O}_C(1)^4) - \chi(\mathcal{O}_C) - \chi(\mathcal{T}_C) = 4d \\
&= h^0(\mathcal{N}_C) - h^1(\mathcal{N}_C),
\end{aligned}$$

so, $h^0(\mathcal{N}_C) = 4d + h^1(\mathcal{N}_C)$.

Proposition 2-1. Let C be a nonsingular curve on a surface S of degree s.

1. If no singular points of S lie on C, we have the following exact sequence:

$$0 \to \omega_C(4 - s) \to \mathcal{N}_C \to \mathcal{O}_C(s) \to 0.$$

2. If S has n nodes $e_1, \cdots, e_n$ on C, we have:

$$0 \to \omega_C(4 - s)(e_1 + \cdots + e_n) \to \mathcal{N}_C \to \mathcal{O}_C(s)(-e_1 - \cdots - e_n) \to 0.$$

PROOF. [5]. □

Remark 2-2. Let C and S be as in Proposition 2-1, 2. Then

$$\mathcal{N}_C|_S \simeq \omega_C(4 - s)(e_1 + \cdots + e_n).$$

Barth's Construction [1, p. 43–46]

Let C be a nonsingular curve in $\mathbb{P}^3$ having the following properties:

1. Set-theoretically, C is the complete intersection of two surfaces $S_1, S_2 \subset \mathbb{P}^3$, both of degree n.

2. C is the curve of contact of these two surfaces, i.e., no point $z \in C$ is singular for both S_1 and S_2, and the intersection multiplicity $i_C(S_1, S_2)$ equals 2.

   By 2, the tangent planes to S_1 and/or S_2 define a line bundle $\mathcal{L} \subset \mathcal{N}_C$. Let $\mathcal{M} = \mathcal{N}_C/\mathcal{L}$.

3. The exact sequence

$$0 \to \mathcal{L} \to \mathcal{N}_C \to \mathcal{M} \to 0,$$

  splits.

4. $\mathcal{M} = \mathcal{L}(-1)$.

5. C is linearly normal, i.e., the restriction map $H^0(\mathcal{O}_{\mathbb{P}^3}(1)) \to H^0(\mathcal{O}_C(1))$ is surjective.

   This curve C has degree $(n^2)/2$. Note that n has to be even.

   If S_1 has only ordinary double points (nodes) as singularities, one finds that the number of nodes is $\lambda = (n^2(n+2))/6$. If $n = 4$, the number of nodes is 16 and the degree of the curve C is 8 (of genus 5); we are dealing then with Kummer surfaces in $\mathbb{P}^3$.

   Assume that a curve C satisfying conditions 1–5 exists. Additionally, fix some point $x_0 \in \mathbb{P}^4$, and let $\sigma \colon \tilde{\mathbb{P}} \to \mathbb{P}^4$ be the blowing up of x_0. Let $q \colon \tilde{\mathbb{P}} \to \mathbb{P}^3$ be the canonical projection sending a point $y \neq x_0$ to the line joining it with x_0. Denote by $T = \sigma(q^{-1}C)$ the cone over C with vertex x_0.

Theorem 2-3. ([1, p. 45]) *Given C with the above properties, there exists a stable rank 2 vector bundle F on $\mathbb{P}^4$. This bundle admits a section $f \in H^0(F)$ vanishing with multiplicity 2 on T. Its Chern classes are:*

$$c_1(F) = 2n - 1, \qquad c_2(F) = n^2.$$

3 A generic curve of degree 8 and genus 5 in $\mathbb{P}^3$

Let char $k = 0$. Let C be an irreducible nonsingular curve of degree 8 and genus 5 on a Kummer surface S (see Proposition 1-6).

Proposition 3-1. *Let X be a nonsingular surface, not ruled. Let L be a very ample line bundle, with $g(L) \neq h^1(\mathcal{O}_X)$. If $\mathcal{L} \in \mathrm{Pic}(X)$ is a line bundle on X such that $\mathcal{L}|_C \simeq \mathcal{O}_C$, for C in an open set of $|L|$, then $\mathcal{L}$ is trivial.*

PROOF. [7, Proposition (0.9)].   □

Proposition 3-2. *A generic irreducible nonsingular curve C of degree 8 and genus 5 on a Kummer surface $S \subset \mathbb{P}^3$ is not canonical.*

PROOF. Consider $C \subset S$ and $\tilde{C} \subset X$. $\tilde{C}$ on the K3 surface X is canonical since it is given as a complete intersection of three quadrics in $\mathbb{P}^4$. If $C \subset \mathbb{P}^3$ were canonical, then its hyperplane section $\mathcal{O}_C(1)$ in $\mathbb{P}^3$ would be linearly equivalent on C to a hyperplane section of $\tilde{C}$ in $\mathbb{P}^4$, since both are canonical (we identify C with $\tilde{C}$ via π).

Pulling back the hyperplane section H of $C \subset \mathbb{P}^3$ to the linear system $|\tilde{H}|$ in $\mathbb{P}^5$, restricted to $\tilde{C}$, we obtain that the following two linear systems coincide, restricted to $\tilde{C}$:

$$2\tilde{H} - \tfrac{1}{2}\sum_{i \in J} E_i \sim \tilde{H},$$

that is to say that $\tilde{H} \sim \tfrac{1}{2}\sum_{i \in J} E_i$. If this were true for the generic curve $\tilde{C}$, by Proposition 3-1 we would have that $\tilde{H} \sim \tfrac{1}{2}\sum_{i \in J} E_i$ on X. This is a contradiction since $\tilde{H} \cdot E_i = 0$, $i \in J$. $\square$

Lemme 3-3. *A generic irreducible nonsingular curve C of degree 8 and genus 5 on a Kummer surface S in $\mathbb{P}^3$ is not projectively normal, but it is linearly normal.*

PROOF. Consider the exact sequence

$$0 \to \mathcal{I}_C \to \mathcal{O}_{\mathbb{P}^3} \to \mathcal{O}_C \to 0. \tag{3.1}$$

Twisting (3.1) by $\mathcal{O}_{\mathbb{P}^3}(2)$ and taking cohomology, we obtain the exact sequence

$$0 \to H^0\big(C, \mathcal{I}_C(2)\big) \to H^0\big(\mathbb{P}^3, \mathcal{O}_{\mathbb{P}^3}(2)\big) \to H^0\big(C, \mathcal{O}_C(2)\big) \to H^1\big(C, \mathcal{I}_C(2)\big) \to 0,$$

so $h^0\big(C, \mathcal{I}_C(2)\big) + 12 = h^1\big(C, \mathcal{I}_C(2)\big) + 10$. Therefore, $h^1\big(C, \mathcal{I}_C(2)\big) \geq 2$, and C is not projectively normal. It is linearly normal, since a curve of degree 8 and genus 5 is not a plane curve, so $h^0\big(\mathcal{I}_C(1)\big) = 0$, and $h^0\big(\mathcal{O}_{\mathbb{P}^3}(1)\big) = 4 = h^0\big(\mathcal{O}_C(1)\big)$ since C is not canonical by Proposition 3-1 , hence $h^1\big(\mathcal{I}_C(1)\big) = 0$. $\square$

Theorem 3-4. *Let C be a generic irreducible nonsingular curve of degree 8 and genus 5 on a Kummer surface S passing through its 16 nodes P_i, $1 \leq i \leq 16$. Then the exact sequence*

$$0 \to \omega_C\left(\sum_{i=1}^{16} P_i\right) \to \mathcal{N}_C \to \mathcal{O}_C(4)\left(-\sum_{i=1}^{16} P_i\right) \to 0$$

splits.

PROOF. By Proposition 2-1, 2, with $s=4$, $n=16$, $e_i = P_i$, $1 \leq i \leq 16$, we have

$$0 \to \omega_C\left(\sum_{i=1}^{16} P_i\right) \to \mathcal{N}_C \to \mathcal{O}_C(4)\left(-\sum_{i=1}^{16} P_i\right) \to 0. \tag{3.2}$$

Tensoring (3.2) with $\mathcal{O}_C(-4)$ and dualizing, we obtain

$$0 \to \mathcal{O}_C\left(\sum_{i=1}^{16} P_i\right) \to \mathcal{N}_C^*(4) \to \omega_C^*(4)\left(-\sum_{i=1}^{16} P_i\right) \to 0. \tag{3.3}$$

By Remark 2-2, $(\mathcal{N}_C|_S)^*(4) = \omega^*(4)\left(-\sum_{i=1}^{16} P_i\right)$. Let us denote by $\mathcal{F}$, $\mathcal{O}_C\left(\sum_{i=1}^{16} P_i\right)$, and $\omega^*(4)\left(-\sum_{i=1}^{16} P_i\right)$ by $\mathcal{E}$. Then $\deg\mathcal{F} = 16$, $\deg\mathcal{N}_C^*(4)=24$, $\deg\mathcal{E}=8$. To see that the above sequence splits we shall prove that

$$0 \to \mathcal{F} \to \mathcal{N}_C^*(4) \to \mathcal{E} \to 0$$

splits. For this we shall study $H^1(\mathcal{F} \otimes \mathcal{E}^*)$. By Serre duality,

$$h^1(\mathcal{F} \otimes \mathcal{E}^*) = h^0(\mathcal{F}^* \otimes \mathcal{E} \otimes \omega_C).$$

We have $\mathcal{F} \otimes \mathcal{E} = \mathcal{O}_C(4) \otimes \omega_C{}^*$. Thus,

$$\mathcal{F}^* \otimes \mathcal{O}_C(4) \simeq \mathcal{E} \otimes \omega_C.$$

On the other hand, $\deg(\mathcal{F}^* \otimes \mathcal{E} \otimes \omega_C)=0$. We are going to prove that $\mathcal{F}^* \otimes \mathcal{E} \otimes \omega_C \not\simeq \mathcal{O}_C$. Suppose that $\mathcal{F}^* \otimes \mathcal{E} \otimes \omega_C \simeq \mathcal{O}_C$. This would imply that $\mathcal{F}^2 \simeq \mathcal{O}_C(4)$, $\mathcal{O}_C\left(2\sum_{i=1}^{16} P_i\right) \simeq \mathcal{O}_C(4)$, $\mathcal{O}_C(4)\left(-2\sum_{i=1}^{16} P_i\right) \simeq 0$. Consider $\tilde{C} \subset X$. If this were true for the generic curve $\tilde{C}$, by Proposition 3-1 we would have that $4\tilde{H} \sim 2\sum_{i=1}^{16} E_i$, where H is a hyperplane section of S. This is a contradiction since $4\tilde{H} \cdot E_i = 0$, $1 \leq i \leq 16$. Thus, $\mathcal{F}^* \otimes \mathcal{E} \otimes \omega_C \not\simeq \mathcal{O}_C$, but $\deg(\mathcal{F}^* \otimes \mathcal{E} \otimes \omega_C)=0$. Hence, $h^1(\mathcal{F} \otimes \mathcal{E}^*) = h^0(\mathcal{F}^* \otimes \mathcal{E} \otimes \omega_C)=0$, which implies that

$$0 \to \mathcal{F} \to \mathcal{N}_C^*(4) \to \mathcal{E} \to 0$$

splits, and so does (3.3). $\square$

Proposition 3-5. Let C be a generic irreducible nonsingular curve of degree 8 and genus 5 on a Kummer surface S. If $\mathcal{L} = \omega_C\left(\sum_{i=1}^{16} P_i\right)$ and $\mathcal{M} = \mathcal{O}_C(4)\left(-\sum_{i=1}^{16} P_i\right)$, then $\mathcal{M} \not\simeq \mathcal{L}(-1)$.

PROOF. We have the exact sequence

$$0 \to \mathcal{L} \to N_C|_{\mathbb{P}^3} \to \mathcal{M} \to 0.$$

Since $2C$ is the complete intersection of two surfaces of degree 4,

$$\mathcal{L} \otimes \omega_C = \omega_{2C} = \mathcal{O}_C(4).$$

Thus,

$$\omega_C{}^2 \simeq \mathcal{O}_C(4)\left(-\sum_{i=1}^{16} P_i\right) = \mathcal{M}.$$

Suppose that $\mathcal{M} \simeq \mathcal{L}(-1)$. Then,

$$\omega_C{}^2 \simeq \omega_C(-1)\left(\sum_{i=1}^{16} P_i\right),$$

which implies that

$$\omega_C \simeq \mathcal{O}_C(-1)\left(\sum_{i=1}^{16} P_i\right).$$

On the other hand, we have that $\omega_C \simeq \mathcal{O}_C(5)\left(-2\sum_{i=1}^{16} P_i\right)$, and so, $\mathcal{O}_C(6)\left(-3\sum_{i=1}^{16} P_i\right) = 0$. Lifting C to the K3 surface X, minimal desingularization of S, we have that $6\tilde{H} - 3\sum_{i=1}^{16} E_i \sim 0$, where H is a hyperplane section of S. This is a contradiction, since $\tilde{H} \cdot E_i = 0$, $1 \le i \le 16$. Therefore, $\mathcal{M} \not\simeq \mathcal{L}(-1)$. $\qquad\square$

Proposition 3-6. *A generic irreducible nonsingular curve C of degree 8 and genus 5 on a Kummer surface S in $\mathbb{P}^3$ satisfies all but one of Barth's conditions. In particular, such a curve of degree 8 and genus 5 cannot be used to do Barth's construction of a stable 2-bundle F in $\mathbb{P}^4$, through a fixed point $P \in \mathbb{P}^4$, with $c_1 = -1$, $c_2 = 4$.*

PROOF. According to Barth's construction, (see Section 2), it has to satisfy 5 properties; we can prove that it satisfies the following four:

1. Set-theoretically, C is the complete intersection of a Kummer surface S_1 and a quartic surface S_2 in $\mathbb{P}^3$, since $2C \sim 4H$.

2. C is the curve of contact of these surfaces ([1], [8, Proposition 2.1]).

3. It is Theorem 3-4.

5. C is linearly normal (Lemma 3-3).

But C does not satisfy the required fourth property as we show in Proposition 3-5. $\qquad\square$

References

[1] W. Barth. Kummer surfaces associated with the Horrocks-Mumford bundle. Journées de géometrie algébrique d'Angers. Sijthoff and Noordhoff, Alphen aan den Rijn (1980). 1979 pages 29–48

[2] M. R. Gonzalez-Dorrego. (16,6) configurations and geometry of Kummer surfaces in $\mathbb{P}^3$. To appear in Memoirs of the A. M. S.

[3] M. R. Gonzalez-Dorrego. Curves on a Kummer surface in $\mathbb{P}^3$, I. To appear in Mathematische Nachrichten Vol 164. 1993

[4] R. Hartshorne. Algebraic Geometry. Springer-Verlag. Graduate Texts in Mathematics, 52 1977

[5] D. Perrin. Courbes passant par m points généraux de $\mathbb{P}^3$. Mém. Soc. Math. de France No:28-29, 138 pp. 1987

[6] B. Saint-Donat. Projective models of K3 surfaces. Amer. J. Math.Vol 96, 1974 pages 602–639

[7] A. J. Sommese. Hyperplane sections of projective varieties I. The adjunction mapping. Duke Math.J. vol 46. 1979 pages 377–401

[8] A. Verra. Contact curves of Two Kummer surfaces. Lecture Notes in Pure and Applied Mathematics Series vol 132 pages 397–407. 1991. Marcel Dekker Inc

Address of author:

Department of Mathematics
University of Toronto
Toronto, M5S 1A1 CANADA
Email: dorrego@ccuam3.sdi.uam.es

IRREDUCIBLE POLYNOMIALS OF $k((X))[Y]$

A. Granja

1 Introduction

Let k be a field and let X, Y, Z be indeterminates over k. We will denote by Ord_X the usual valuation on the meromorphic function field $k((X))$.

The main objective of this paper is to study when a polynomial $P(Y) \in k((X))[Y]$ is irreducible.

Let k be an algebraically closed field and let $P(Y) \in k((X))[Y]$ be a monic polynomial of degree N. If N is nondivisible by the characteristic of k, Abhyankar has given in [2] a criterion to determine when $P(Y)$ is irreducible. Here we will give a criterion to see when $P(Y)$ is irreducible without any assumptions on k or $P(Y)$. In particular, this will answer a question of S. S. Abhyankar (see [1]) relative to give irreducibility criterions for elements of $k[[X, Y]]$. (Note that if $Q(X, Y) \in k[[X, Y]]$ then by the Weierstrass Preparation Theorem we can write $Q(X, Y) = X^\alpha Q'(X, Y)P(X, Y)$, where $Q'(X, Y) \in k[[X, Y]]$ is a unit, $P(X, Y) \in k[[X]][Y]$ is a monic polynomial and α is a nonnegative integer. So $Q(X, Y)$ is irreducible on $k[[X, Y]]$ if and only if $P(X, Y)$ is irreducible on $k[[X]][Y]$ and $\alpha = 0$).

We study the irreducibility of the polynomial $P(Y) \in k((X))[Y]$ in two steps.

First, in Section 2 we note that we must only consider polynomials

$$P(Y) = \sum_{i=0}^{n} a_i(X)Y^{n-i} \in k[[X]][Y],$$

such that there is $0 \le i \le n$ with $a_i(0) \ne 0$. In this situation we have:

1. If $a_i(0) = 0$, $0 \le i \le n-1$ and $a_n(0) \ne 0$, then $P(Y)$ is irreducible if and only if $P^*(Z) = Z^n P(1/Z) = \sum_{i=0}^{n} a_i(X)Z^i \in k[[X]][Z]$ is irreducible.

2. If $a_0(0) = 0$ and there is $0 \le i \le n-1$ such that $a_i(0) \ne 0$, then $P(Y)$ is reduced. (Corollary 2-4).

So we must only consider polynomials

$$P(Y) = \sum_{i=0}^{n} a_i(X)Y^{n-i} \in k[[X]][Y]$$

such that $a_0(X) = 1$, that is, monic polynomials.

In section 4, we will give a criterion to see when a monic polynomial of $k[[X]][Y]$ is irreducible.

In fact, we have (see Theorem 4-7): For a monic polynomial $P \in k[[X]][Y]$ of degree n, the following statements are equivalent:

Progress in Mathematics, Vol. 134
© 1996 Birkhäuser Verlag Basel/Switzerland

1. P is irreducible.

2. $V_i(P)$ is finite, $1 \leq i \leq n-1$.

3. $V_{[n/2]}(P)$ is finite.

Here $V_i(P) = \{\mathrm{Ord}_X(r(P,Q)); Q \in k[X][Y]$ is monic of degree i and $r(P,Q) \neq 0\}$, $1 \leq i < n$, $r(P,Q)$ denotes the usual resultant and $[n/2]$ is the greatest integer s such that $s \leq n/2$.

In the rest of the paper we give some results about the computation of $V_{[n/2]}(P)$.

2   Reduction of the Problem

First we note that $P(Y) \in k((X))[Y]$ is irreducible if and only if $X^\alpha P(Y)$ is irreducible for any integer α. So we can assume that $P(Y) \in k[X][Y]$ and $P(Y) \notin X(k[X][Y])$.

On the other hand, if $P(Y) = \sum_{i=0}^n a_i(X)Y^{n-i} \in k[X][Y]$, with $a_i(0) = 0$ for $0 \leq i \leq n-1$ and $a_n(0) \neq 0$, then we can consider $P^*(Z) = Z^n P(1/Z) = \sum_{i=0}^n a_i(X)Z^i$ and we have that $P(Y)$ is irreducible if and only if $P^*(Z)$ is irreducible.

So we can also assume that $P(Y) = \sum_{i=0}^n a_i(X)Y^{n-i} \in k[X][Y]$ and $a_i(0) \neq 0$ for some $0 \leq i \leq n-1$.

Definition 2-1. *Let $P(X,Y) \in k[X,Y]$ be such that $P = \sum_{i=0}^\infty a_i(X)Y^i$, with $a_i(X) \in k[X]$. We denote by $j(P) = \min\{j \geq 0; \mathrm{Ord}_X(a_j) \leq \mathrm{Ord}_X(a_i)$, for each $0 \leq i\}$.*

Lemma 2-2. *Let $P, Q, S \in K[X,Y]$ be such that $P = QS$. Then $j(P) = j(Q) + j(S)$.*

PROOF. It is an easy computation.   □

Lemma 2-3. *Let $P(Y) = \sum_{i=0}^n a_i(X)Y^{n-i} \in k[X][Y]$ be such that $a_n(0) = 0$ and $a_s(0) \neq 0$ for some $0 < s < n$. Then $P(Y)$ is reduced.*

PROOF. Let s be such that $a_s(0) \neq 0$ and $a_j(0) = 0$ for $0 \leq j < s$. By the Weierstrass Preparation Theorem we can write $P = QP'$, with $P' \in k[X][Y]$ a monic polynomial of degree s and $Q \in k[X,Y]$ such that $Q(0,0) \neq 0$.

By Lemma 2-2, if $P' = Y^s + c_1(X)Y^{s-1} + \cdots + c_s(X)$, we have that $c_j(0) = 0$, $1 \leq j \leq s$.

On the other hand, we can write $P = Q^*P' + A$ with $Q^*, A \in k[X][Y]$, Q^* of degree $n - s$ and A of degree less or equal to $s - 1$. Also by Lemma 2-2 we have that $A = 0$, so $Q = Q^*$ and P is reduced.   □

Corollary 2-4. *If $P(Y) = \sum_{i=0}^n a_i Y^{n-i} \in k[X][Y]$ with $a_0(0) = 0$ and there is $0 \leq i \leq n-1$ such that $a_i(0) \neq 0$ then P is reduced.*

PROOF. Apply Lemma 2-3 to $P^*(Z) = Z^n P(1/Z)$.   □

Remark 2-5. *We have reduced the irreducibility problem of polynomials of $k((X))[Y]$ to study the monic irreducible polynomials of $k[\![X]\!][Y]$.*

3   Some Maximal Ideals of $k[\![X]\!][Y]$

Let $P(Y) = Y^n + a_1 Y^{n-1} + \cdots + a_n \in k[\![X]\!][Y]$ be a monic polynomial and consider the ring $R = k[\![X]\!][Y]/P(k[\![X]\!][Y])$.

Lemma 3-1. *We have that:*

1. *R is a finitely generated $k[\![X]\!]$-module.*

2. *The $Xk[\![X]\!]$-topology of R as a $k[\![X]\!]$-module is the $(X + P)R$-topology as a ring.*

3. *R is complete for the $Xk[\![X]\!]$-topology.*

4. *R is a semilocal ring and $\sqrt{(X + P)R}$ is the intersection of the maximal ideals of R.*

PROOF. See [6, Th. 15, p. 276].   □

Remark 3-2. *We note that if a maximal ideal M of $k[\![X]\!][Y]$ contains a monic polynomial P then $Xk[\![X]\!] \subseteq M$.*

Proposition 3-3. *M is a maximal ideal of $k[\![X]\!][Y]$ with $Xk[\![X]\!] \subseteq M$ if and only if $M = Xk[\![X]\!][Y] + \left(Y^s + a_1(X)Y^{s-1} + \cdots + a_s(X)\right)k[\![X]\!][Y]$, where $Q^*(Y) = Y^s + a_1(0)Y^{s-1} + \cdots + a_s(0)$ is an irreducible polynomial of $k[Y]$.*

PROOF. Let $Q(Y) = Y^s + a_1(X)Y^{s-1} + \cdots + a_s(X)$ be such that $Q^*(Y) = Y^s + a_1(0)Y^{s-1} + \cdots + a_s(0)$ is an irreducible polynomial of $k[Y]$. Let

$$\Phi : k[Y] \longrightarrow k[Y]/Q^*(k[Y])$$

be the canonical epimorphism and let

$$\eta_X : k[\![X]\!][Y] \longrightarrow k[Y]$$

be the morphism defined by

$$\eta_X\left(Y^r + b_1(X)Y^{r-1} + \cdots + b_r(x)\right) = Y^r + b_1(0)Y^{r-1} + \cdots + b_r(0).$$

If we write $j = \Phi \circ \eta_X$, then we have that $\ker(j)$ is a maximal ideal of $k[\![X]\!][Y]$ and $Xk[\![X]\!][Y] + (Y^s + a_1 Y^{s-1} + \cdots + a_s)Xk[\![X]\!][Y] \subseteq \ker(j)$.

On the other hand, if $j(Y^r + b_1(X)Y^{r-1} + \cdots + b_r(X)) = 0$, then we can write $Y^r + b_1(X)Y^{r-1} + \cdots + b_r(X) = \left(Y^s + a_1(X)Y^{s-1} + \cdots + a_s(X)\right)S(Y) + c_0(X)Y^s + c_1(X)Y^{s-1} + \cdots + c_s(X)$, with $c_i(0) = 0$, $0 \le i \le s$. So $\ker(j) = Xk[\![X]\!][Y] + \left(Y^s + a_1(X)Y^{s-1} + \cdots + a_s(X)\right)k[\![X]\!][Y]$.

To finish, if M is a maximal ideal of $k[\![X]\!][Y]$ with $Xk[\![X]\!] \subseteq M$ then $\eta_X(M)$ is a maximal ideal of $k[Y]$ and we have $\eta_X(M) = \eta_X\left(Y^s + a_1(X)Y^{s-1} + \cdots + a_s(X)\right)k[Y]$, being $\eta_X\left(Y^s + a_1(X)Y^{s-1} + \cdots + a_s(X)\right)$ an irreducible polynomial. So $M = Xk[\![X]\!][Y] + \left(Y^s + a_1(X)Y^{s-1} + \cdots + a_s(X)\right)k[\![X]\!][Y]$. □

Corollary 3-4. With the notations as in Proposition 3-3, assume that $P_i(Y) \in k[\![X]\!][Y]$, $1 \leq i \leq r$ are monic polynomials such that

$$\eta_X(P) = \left(\eta_X(P_1)\right)^{s_1} \ldots \left(\eta_X(P_r)\right)^{s_r}$$

is the factorization of $\eta_X(P)$ in irreducible polynomials of $k[Y]$. In this situation if M is a maximal ideal of $k[\![X]\!][Y]$ such that $P \in M$ then $M = Xk[\![X]\!][Y] + P_i k[\![X]\!][Y]$ for some $1 \leq i \leq r$.

PROOF. It is an easy consequence of Proposition 3-3. $\qquad\square$

Remark 3-5. *Note that, if R is a semilocal ring but not a local ring, then R is not a domain and P is reduced.*

4   Irreducibility Criterion for Monic Polynomials of $k[\![X]\!][Y]$

For any pair of polynomials $u_0 Y^n + \cdots + u_n, v_0 Y^m + \cdots + v_m \in k[\![X]\!][Y]$, we denote by

$$r(u_0 Y^n + \cdots + u_n, v_0 Y^m + \cdots + v_m) =$$

$$\det \begin{pmatrix} u_0 & u_1 & \cdots & u_n & 0 & 0 & \cdots & 0 \\ 0 & u_0 & \cdots & u_{n-1} & u_n & 0 & \cdots & 0 \\ & & \cdots\cdots\cdots\cdots\cdots\cdots\cdots\cdots\cdots\cdots\cdots & & & \\ v_0 & v_1 & \cdots\cdots\cdots\cdots & v_m & 0 & \cdots & 0 \\ 0 & v_0 & \cdots\cdots\cdots\cdots & v_{m-1} & v_n & \cdots & 0 \\ & & \cdots\cdots\cdots\cdots\cdots\cdots\cdots\cdots\cdots\cdots\cdots & & & \end{pmatrix}$$

the usual resultant.

Let $P(Y) = Y^n + \sum_{i=1}^{n} a_i(X) Y^{n-i} \in k[\![X]\!][Y]$ be a monic polynomial of degree n.

Definition 4-1. For $1 \leq i \leq n-1$ we denote by $M_i(P) = \{r(P,Q) \in k[\![X]\!]; Q$ is monic of degree $i\}$ and by $V_i(P) = \{\mathrm{Ord}_X(b); b \in M_i(P), b \neq 0\}$.

Remark 4-2. *Note that P is reduced if and only if $0 \in M_i(P)$ for some $1 \leq i \leq n-1$. We also note that $\mathrm{Ord}_X\left(r(P,Y)\right) V_i(P) \subseteq V_{i+1}(P)$. Thus if $V_i(P)$ is infinite then $V_j(P)$ is also infinite for $j > i$.*

Lemma 4-3. *If $P = ST$ with $S, T \in k[\![X]\!][Y]$ and S a monic polynomial of degree $0 < s < n$, then $V_s(P)$ is infinite.*

PROOF. For each $h > 0$ consider $S_h = S + X^h$; we have that $r(P, S_h) = r(T, S_h) r(S, X^h)$, so $\mathrm{Ord}_X\left(r(P, S_h)\right) \geq h$. To finish, we note that S_h and S are coprime polynomials and that there is h_0 such that S_h and T are also coprime polynomials for $h \geq h_0$. $\qquad\square$

Corollary 4-4. *If $V_{[n/2]}(P)$ is finite then P is irreducible, where $[n/2]$ is the greatest integer s such that $s \leq n/2$.*

PROOF. It is an easy consequence of Lemma 4-3. (Note that we can assume that $s \leq [n/2]$ in Lemma 4-3). $\qquad\square$

Lemma 4-5. *Let R be a complete noetherian local domain of Krull dimension one and let R' be the integral closure of R in its quotient field. Then R' is a finitely generated R-module and R' is a discrete valuation ring. Moreover there is $j_0 \geq 1$ such that $(M(R'))^{j_0} \subseteq M(R)$, where $M(R)$ (resp. $M(R')$) is the maximal ideal of R (resp. of R').*

PROOF. By [3, p. AC IX 33] R' is a finitely generated R-module. By the Krull-Akizuki Theorem ([3, Cap. VII p. 224]) R' is noetherian and has Krull dimension one. By [6, Th. 15 pag. 276] R' is a complete semilocal domain, so we have that R' is a local ring and as R' is an integrally closed ring, then R' is a discrete valuation ring.

To finish the proof, we can write $R' = \sum_{i=1}^{n}(x_i/y_i)R + R$, with $y_i \in M(R) - \{0\}$, $1 \leq i \leq n$. We have that $\prod_{i=1}^{n} y_i R' \subseteq M(R)$ and, as R' is a discrete valuation ring, there is $j_0 \geq 1$ with $(M(R'))^{j_0} \subseteq \prod_{i=1}^{n} y_i R' \subseteq M(R)$. $\qquad\square$

Lemma 4-6. (Teissier) *Consider $P, Q \in k[[X]][Y]$, then*

$$L_{k[[X]]}(k[[X]][Y]/(P,Q)) = \mathrm{Ord}_X(r(P,Q)).$$

Where $L_{k[[X]]}$ denotes the usual length function of $k[[X]]$-modules.

PROOF. On can adapte the proof given in [4, Proposition 5.2.2., p. 88]. $\qquad\square$

Theorem 4-7. *With the above notations, if $P(Y) \in k[[X]][Y]$ is a monic polynomial of degree n then the following statements are equivalent:*

1. *P is irreducible.*

2. *$V_i(P)$ is finite, $1 \leq i \leq n-1$.*

3. *$V_{[n/2]}(P)$ is finite.*

PROOF. We must only see that a)$\Rightarrow$ b). Assume that P is an irreducible polynomial and let R be the ring $R = k[[X]][Y]/P(k[[X]][Y])$. Then R is a complete noetherian local domain of Krull dimension one. (See Lemma 3-1). If R' is the integral closure of R in its quotient field, by Lemma 4-3 there is $j_0 \geq 1$ such that $(M(R'))^{j_0} \subseteq M(R)$, where $M(R)$ and $M(R')$ are the respective maximal ideals of R and R'.

Assume that $V_i(P)$ is infinite for some $1 \leq i \leq n-1$, in particular $V_{n-1}(P)$ is infinite. Consider a sequence $\{Q_i\}_{i \geq 1}$ such that Q_i is monic of degree $n-1$, $r(P,Q) \neq 0$ and $\mathrm{Ord}_X(r(P,Q_i)) = L_{k[[X]]}(k[[X]][Y]/(P,Q_i)) \geq i$, for each $i \geq 1$. Let q_i be the class of Q_i in R, $i \geq 1$. By [5, Th. 13 p. 168] and by Lemma 4-6 we have that

$$L_{k[[X]]}(k[[X]][Y]/(P,Q_i)) = [(R/M(R)){:}k]\, L_R(R/q_iR)$$

and

$$L_R(R/q_iR) = [(R'/M(R')){:}(R/M(R))]\, L_{R'}(R'/q_iR').$$

Thus $q_i \in (M(R'))^{j_0 n} \subseteq (M(R))^n$, for $i > [(R'/M(R')){:}k]\, j_0 n$.

On the other hand, as P is irreducible, by Hensel's Lemma we have that
$P(0,Y) = Y^n + \sum_{i=1}^n a_i(0)Y^{n-i} = (Y^e + \sum_{i=1}^e \alpha_i Y^{e-i})^{n/e}$, with $Q(Y) = Y^e + \sum_{i=1}^e \alpha_i Y^{e-i} \in k[Y]$ irreducible. So the maximal ideal of R is generated by the classes of X and $Q(Y)$. (Corollary 3-4).

Now we have that $Q_i \in (X,Q)^n$, so we can write

$$Q_i = S(X,Y)\big(Q(Y)\big)^n + XT(X,Y),$$

with $S,T \in k[X][Y]$. In particular $Q_i(0,Y) = S(0,Y)\big(Q(0,Y)\big)^n$ but $Q_i(0,Y)$ is monic of degree $n-1$ and $S(0,Y)\big(Q(0,Y)\big)^n$ has degree at least n. □

5 Some Ideas to Compute $V_{[n/2]}(P)$

Theorem 4-7 gives a criterion to see when a monic polynomial $P \in k[X][Y]$ is irreducible, that is, see if $V_{[n/2]}(P)$ is finite or not. In this section we give some results that can be used to compute $V_{[n/2]}(P)$.

Let us write

$$\begin{aligned} Q_{s,U}(Y) &= Y^s + U_1 Y^{n-1} + \cdots + U_s, \\ r(P,s,U_1,\ldots,U_s) &= r(P,Q_{s,U}), \end{aligned}$$

where $U_1,\ldots,U_s$ are indeterminates over $k[X]$. To see if $V_s(P)$ is finite we need to bound the values of $\mathrm{Ord}_R\big(r(P,s,u_1,\ldots,u_s)\big)$, when $u_1,\ldots,u_s \in k[X]$.

Lemma 5-1. Let $P,Q,Q' \in k[X][Y]$ be such that $Q - Q' \in X^\alpha k[X][Y]$, with $\alpha > \mathrm{Ord}_X\big(r(P,Q)\big)$. If Q and Q' have the same degree then $\mathrm{Ord}_X\big(r(P,Q)\big) = \mathrm{Ord}_X\big(r(P,Q')\big)$.

PROOF. If h is the degree of both Q and Q', consider

$$r(P,U_0,\ldots,U_h) = r(P,U_0 Y^h + U_1 Y^{h-1} + \cdots + U_h),$$

with $U_0,\ldots,U_h$ indeterminates over $k[X]$. If $n = \deg P$, then $r(P,U_0,\ldots,U_h)$ is a homogeneous polynomial of degree n. We can write $Q2 = Q + a_0(X)Y^h + \cdots + a_h(X)$ with $a_i(X) \in X^s k[X]$. $0 \le i \le h$, so $r(P,Q') = r(P,Q) + J$, with $J \in X^s k[X]$. □

Corollary 5-2. Let $P \in k[X][Y]$ be such that

$$P = Y^n + \sum_{j=1}^n \left(\sum_{i=0}^\infty a_i^j X^i\right) Y^{n-j}.$$

For each nonnegative integer α we denote by

$$P_\alpha = Y^n + \sum_{j=1}^n \left(\sum_{i=0}^\alpha a_i^j X^i\right) Y^{n-j}.$$

Assume that there is α such that $\mathrm{Ord}_X(r(P_\alpha, Q')) \le \alpha$ for each $Q' \in k[X][Y]$ with

$$Q' = Y^{[n/2]} + \sum_{j=1}^{[n/2]} \left(\sum_{i=0}^{\alpha} b_i^j X^i \right) Y^{[n/2]-j}.$$

Then $V_{[n/2]}(P)$ is finite and P is irreducible.

PROOF. It is a easy consequence of Lemma 5-1. □

References

[1] S.S. Abhyankar, Desingularization of plane curves. Proceedings, Symposia Pure Mathematics, **40**. Part 1, 1983, pp. 1–45.

[2] S.S. Abhyankar, Irreducibility Criterion for Germs of Analytic Functions of Two Complex Variables. Advances in Mathematics, Vol. 74, 1989, pp. 190–257.

[3] N. Bourbaki, Algèbre Commutative. Masson. 1985.

[4] A. Chenciner, Courbes Algébriques Planes. Publications Mathématiques de l'Université Paris VII, **4**. U.E.R. de Mathématiques. Paris 1978.

[5] B.G. Northcott, Lessons on Rings Modules and Multiplicities. Cambridge University Press. 1968.

[6] O. Zariski, P. Samuel, Commutative Algebra. Vol. I and II. Springer-Verlag. 1986

Address of author:

Dpto. Matemáticas
Universidad de León
24071-León
Spain
Email: granja@eleule11.bitnet

Examples of Abelian Surfaces with Polarization Type $(1,3)$

Isidro Nieto

1 Abstract

In the family

$$A(z_0^4 + z_1^4 + z_2^4 + z_3^4) + 2B(z_0^2 z_1^2 + z_2^2 z_3^2) + 2C(z_0^2 z_2^2 + z_1^2 z_3^2)+$$
$$2D(z_0^2 z_3^2 + z_1^2 z_2^2) + 4E z_0 z_1 z_2 z_3 = 0$$

of quartic surfaces in 4 variables invariant under the level $(2,2)$-Heisenberg Group $H_{2,2}$ we study explicitly two subfamilies,

$$\mathcal{F}_{AB} := \{A = B = 0\} \quad \text{and} \quad \mathcal{F}_{AE} := \{A = E = 0\},$$

and show that

1. To every point in $\mathcal{F}_{AE}$ there corresponds an abelian surface which is a product of elliptic curves; it carries a polarization of type $(2,2)$ and $(2,6)$.

2. $\mathcal{F}_{AB}$ is a $\mathbb{P}_1$-bundle over an elliptic curve C_{AB}.

Both 1 and 2 follow by analysing the configuration of lines lying on each element of $\mathcal{F}_{AE}$ and $\mathcal{F}_{AB}$.

Our main motivation is to study the singular $H_{2,2}$ quartic surfaces arising in the study of the moduli space of abelian surfaces of type $(1,3)$.

2 Introduction

The main objects of study will be two subfamilies of the family of quartic surfaces in $\mathbb{P}_3$ given by

$$A(z_0^4 + z_1^4 + z_2^4 + z_3^4) + 4E z_0 z_1 z_2 z_3 + 2B(z_0^2 z_1^2 + z_2^2 z_3^2)+$$
$$2C(z_3^2 z_1^2 + z_0^2 z_2^2) + 2D(z_0^2 z_3^2 + z_1^2 z_2^2),$$

and the two subfamilies are given by $\mathcal{F}_{AB}$ and $\mathcal{F}_{AE}$. The family of quartic surfaces is invariant under the Heisenberg Group $T := H_{2,2}$ of level $(2,2)$ generated by

$$\sigma_1 := (01)(23), \quad \tau_1 := \text{diag}(1\!:\!-1\!:\!1\!:\!-1),$$
$$\sigma_2 := (02)(13), \quad \tau_2 := \text{diag}(1\!:\!1\!:\!-1\!:\!-1),$$

which can be realized as the central extension

$$1 \longrightarrow \mu_2 \longrightarrow T \longrightarrow (\mathbb{Z}_2 \oplus \mathbb{Z}_2)^2 \longrightarrow 0.$$

Progress in Mathematics, Vol. 134
© 1996 Birkhäuser Verlag Basel/Switzerland

The family $\mathcal{F}_{AB}$ as well as the group T are well known, both appear ([3]) in connection with the moduli space of abelian surfaces with a level $(2,2)$-structure. Both subfamilies also appear in connection with the moduli space of abelian surfaces of type $(1,3)$ ([7]).

The main results concerning these two families are contained in Section 5.2.

Our main contribution is that starting with the basic examples (Section 4.1) of products of elliptic curves, the geometric properties are well reflected by the symmetries of the Heisenberg Group T, e.g. from equivariance of the embedding given by the linear system (e.g. Section 4.1), then to incidence properties of lines lying e.g. on the surfaces in a subfamily of $\mathcal{F}_{AE}$ studied in Section 7.

Acknowledgments Part of this work was done in the fall of 1990, the author is thankful to W. Barth for profitable discussions and the DAAD and CINVESTAV for financial support.

3   Preliminaries

We introduce the main definitions and facts used throughout. Our main reference will be [5]. Fix an abelian surface A and an ample line bundle $\mathcal{L}$ on A. If $p \in A$,

$$
\begin{array}{rccc}
t_p & : & A & \longrightarrow & A \\
 & & x & \longmapsto & x + p
\end{array}
$$

is the translation at p. By definition,

$$
G(\mathcal{L}) := \{(p, \varphi) \mid t_p \mathcal{L} \overset{\varphi}{\simeq} \mathcal{L}\}, \qquad H(\mathcal{L}) := \{p \in A \mid t_p \mathcal{L} \overset{\varphi_p}{\simeq} \mathcal{L}\}.
$$

$G(\mathcal{L})$ can be given group structure ([6]), and $H(\mathcal{L})$ is an abelian group. Both groups are related by

$$
1 \longrightarrow \mathbb{C}^* \longrightarrow G(\mathcal{L}) \longrightarrow H(\mathcal{L}) \longrightarrow 0. \tag{3.1}
$$

$\mathcal{L}$ ample implies that equation (3.1) is a central extension. Let's put

$$
\begin{array}{rccc}
e^{\mathcal{L}} & : & H(\mathcal{L}) \times H(\mathcal{L}) & \longrightarrow & \mathbb{C}^* \\
 & & (p, q) & \longmapsto & \varphi_p \varphi_q \varphi_p^{-1} \varphi_q^{-1},
\end{array}
$$

a skew-symmetric bilinear map.

For $\delta = (\delta_1, \delta_2)$ with δ_1, δ_2 positive integers with $\delta_1 | \delta_2$,

$$
G(\delta) := \mu_{\delta_2} \times \mathbb{Z}_{\delta_1} \oplus \mathbb{Z}_{\delta_2} \times \operatorname{Hom}(\mathbb{Z}_{\delta_1} \oplus \mathbb{Z}_{\delta_2}, \mu_{\delta_2})
$$

is a set with multiplication law :

$$
(\alpha, t, l) \cdot (\alpha', t', l') = \big(\alpha \alpha' l'(t), t + t', l \cdot l'\big).
$$

This is the Heisenberg Group $G(\delta)$ of level δ. Set also

$$H(\delta) := \mathbb{Z}_{\delta_1} \oplus \mathbb{Z}_{\delta_2} \times \operatorname{Hom}(\mathbb{Z}_{\delta_1} \oplus \mathbb{Z}_{\delta_2}, \mu_{\delta_2}).$$

Both groups are related by

$$0 \longrightarrow \mu_{\delta_2} \longrightarrow G(\delta) \longrightarrow H(\delta) \longrightarrow 0, \qquad (3.2)$$

which is a central extension with center $Z\big(G(\delta)\big) = \mu_{\delta_2}$.

The bilinear form

$$
\begin{array}{rcl}
e^\delta \quad : \quad H(\delta) \times H(\delta) & \longrightarrow & \mathbb{C}^* \\
\big((t,l),(t',l')\big) & \longmapsto & l'(t)/l(t'),
\end{array}
$$

is anti-symmetric.

Recall by Riemann-Roch for abelian varieties with $\delta = (\delta_1, \dots, \delta_g)$, $g := \dim(A)$ that

$$h^0(\mathcal{L}) = \delta_1 \dots \delta_g.$$

A level δ-structure on $(A, \mathcal{L})$ consists of a group isomorphism

$$\alpha \colon H(\mathcal{L}) \simeq H(\delta),$$

such that $\alpha|_{\mu_{\delta_2}} = \operatorname{id}_{\mu_{\delta_2}}$, preserving $e^{\mathcal{L}}$ and e^δ.

For $g = 2$, let $V(\delta) = \operatorname{Map}(\mathbb{Z}_{\delta_1} \oplus \mathbb{Z}_{\delta_2}, \mathbb{C})$. The group $G(\mathcal{L})$ admits a unique (up to a constant) irreducible representation isomorphic to

$$
\begin{array}{rcl}
\varrho \quad : \quad G(\delta) & \longrightarrow & \operatorname{Aut}\big(V(\delta)\big) \\
(\alpha, t, l) & \longmapsto & U_{(\alpha,t,l)}(f(x)) = \alpha l(x) f(x + t).
\end{array}
$$

By a theorem of Stone-Von Neumann-Mackey [6, Prop. 3], this is the unique irreducible representation of $G(\delta)$ by which $Z\big(G(\delta)\big) = \mu_{\delta_2}$ as scalar multiplication operates. This is the *Schrödinger representation* of $G(\delta)$. Therefore,

$$H^0(A, \mathcal{L}) \simeq V(\delta),$$

as $G(\mathcal{L})$ (resp. $G(\delta)$)-modules.

4   First examples: products of elliptic curves

In this section we collect some examples of abelian surfaces decomposing as a product of elliptic curves with both a polarization of type $(2,2)$ and $(2,6)$.

We introduce our notation for this section.

For the vector space

$$V_n := \operatorname{Map}(\mathbb{Z}_n, \mathbb{C}),$$

with basis $\{u_i\}_{i=0,\ldots,n-1}$, $V^{\pm}$ denotes its ι-eigenspace decomposition under

$$\iota \ : \ \begin{array}{ccc} V_n & \longrightarrow & V_n \\ u_k & \longmapsto & u_{-k}. \end{array}$$

The Heisenberg Group $G(2,2n)$, $n \geq 1$ is a central extension

$$1 \longrightarrow \mu_{2n} \longrightarrow G(2,2n) \longrightarrow (\mathbb{Z}_2 \oplus \mathbb{Z}_{2n})^2 \longrightarrow 0,$$

and $G_2(2,2n)$ its subgroup of 2-torsion elements.

Let

$$\varrho \colon G(2,2n) \longrightarrow \mathrm{Aut}\big(V(2,2n)\big), \qquad \rho_m \colon G(m) \longrightarrow \mathrm{Aut}(V_m),$$

denote the Schrödinger representations of degree $4n$, m of ϱ (resp. ρ_m).

Proposition 4-1.

1. *There is a group isomorphism*

$$G(2) \times G(2n) \simeq G(2,2n),$$

   in particular

$$G(2) \times G_2(2n) \simeq G_2(2,2n).$$

2. *If $\delta_{i,j}$ (resp. δ_i) are the (Kronecker) delta functions on $V(2,2n)$ (resp. on V_2), then we identify*

$$\begin{array}{ccc} V(2,2n) & \simeq & V_2 \otimes V_{2n} \\ \delta_{i,j} & \rightarrow & \delta_i \otimes \delta_j, \end{array}$$

   and ϱ is equivalent to

$$\rho \ : \ \begin{array}{ccc} G(2) \times G(2n) & \longrightarrow & \mathrm{Aut}(V_2 \otimes V_{2n}) \\ (g,h) & \longmapsto & \rho_2(g)(u_i) \otimes \rho_{2n}(h)(v_j). \end{array}$$

3. *$V_{2n}^{\pm}$ are $G_2(2,2n)$-modules.*

PROOF. 1. It is enough to give the isomorphism with respect to the generators of both groups. For $G(2,2n)$:

$$\sigma \colon \xi_{i,j} \mapsto \xi_{i,j+1}, \quad \sigma' \colon \xi_{i,j} \mapsto \xi_{i+1,j}, \quad \tau \colon \xi_{i,j} \mapsto \omega^j \xi_{i,j}, \quad \tau' \colon \xi_{i,j} \mapsto (-1)^i \xi_{i,j},$$

for a fixed $\omega \in \mu_{2n}$ and $i = 0$, 1; $j = 0, \ldots, 2n-1$.

For $G(2)$, generators are given by:

$$\eta \colon u_i \mapsto u_{i+1}, \qquad \epsilon \colon u_i \mapsto (-1)^i u_i,$$

with $i = 0$, 1.

For $G(2n)$:

$$\alpha: v_i \mapsto v_{i+1}, \qquad \beta: v_i \mapsto \kappa^i v_i,$$

for a fixed $\kappa \in \mu_{2n}$, $i = 0, \ldots, 2n - 1$.

The mapping

$$
\begin{array}{ccc}
G(2) \times G(2n) & \longrightarrow & G(2, 2n) \\
(\eta, 1) & \longmapsto & \sigma' \\
(\epsilon, 1) & \longmapsto & \tau' \\
(1, \alpha) & \longmapsto & \sigma \\
(1, \beta) & \longmapsto & \tau,
\end{array}
$$

gives the desired isomorphism.

2. Given basis elements u_0, u_1 of V_2 (resp. v_0, $\ldots$, v_{2n-1} of V_{2n}) then

$$V_2 \otimes V_{2n} = \langle u_i \otimes v_j \rangle_{i=0,1;\, j=0,\ldots,2n-1}$$

is a basis for this vector-space and one checks directly that one obtains the same generators of the ϱ-representation as those given by ρ.

3. Recall that if $V_{2n} = \langle \omega_k \rangle_{k=0,\ldots,2n-1}$ then

$$
\begin{aligned}
V_{2n}^+ &= \langle \omega_0, \omega_1 + \omega_{2n-1}, \ldots, \omega_{n-1} + \omega_{n+1}, \omega_n \rangle, \\
V_{2n}^- &= \langle \omega_1 - \omega_{2n-1}, \ldots, \omega_{n-1} - \omega_{n+1} \rangle,
\end{aligned}
$$

and

$$G_2(2, 2n) = \langle \sigma^n, \tau^n \rangle,$$

with generators acting on the representation space V_{2n} as:

$$
\begin{aligned}
\sigma^n : w_i &\mapsto w_{i+n}, \qquad i = 0, \ldots, 2n - 1, \\
\tau^n : w_i &\mapsto (-1)^i w_i
\end{aligned}
$$

gives

$$
\begin{aligned}
\sigma^n(\omega_0) &= \omega_n, & \sigma^n(\omega_m \pm \omega_{2n-m}) &= \omega_{m+n} \pm \omega_{n-m}, \\
\tau^n(\omega_0) &= \omega_0, & \tau^n(\omega_m \pm \omega_{2n-m}) &= (-1)^m(\omega_m \pm \omega_{2n-m}). \qquad \square
\end{aligned}
$$

4.1 Projective embeddings of products of elliptic curves with both a polarization of type $(2, 2)$ and $(2, 6)$

For this section we will fix points P_0 (resp. P_1) over the elliptic curves E_0 (resp. E_1) and $\{e_i^k\}_{k=0,1;\, i=0,\ldots,3}$ the 2-torsion points on $\{E_k\}_{k=0,1}$. For $m, n \geq 1$ positive integers, $\mathbb{P}_{m-1}^j$ will denote the projectivization of the vector spaces $H^0(E_j, \mathcal{O}_{E_j}(mP_j))$ for $j = 0, 1$. Finally, $s: \mathbb{P}_r \times \mathbb{P}_s \hookrightarrow \mathbb{P}_{r+s+rs}$ is the Segre Embedding.

Proposition 4-2.

1. The image of $A = E_0 \times E_1$ under $\mathcal{O}_A(2P_0 + 2P_1)$ is a $4:1$ cover over a T-invariant quadric ramified along

$$\bigcup_{i,j \in \{0,1,2,3\}} (e_i^0 \times e_j^1).$$

   Moreover, the mapping φ given as the composition of

   $$E_0 \times E_1 \xrightarrow{\;\left|\mathcal{O}_A(2P_0 + 2P_1)\right|\;} \mathbb{P}_1^0 \times \mathbb{P}_1^1 \overset{s}{\hookrightarrow} \mathbb{P}_3$$

   is $T = G(2,2)$-equivariant.

2. Let $n \geq 2$ be a positive integer. The image of $A = E_0 \times E_1$ under the linear system $\mathcal{O}_A(2P_0 + 2nP_1)$ composed with s is a surface of degree $4n$ in $\mathbb{P}_{4n-1}$, such that A is a $2:1$ cover of it, ramified along

$$\bigcup_{j=0,\dots,3} \{e_j^0 \times \mathbb{P}_{2n-1}^1\}.$$

   Moreover, the mapping ϕ given as the composition of

   $$A \xrightarrow{\;\left|\mathcal{O}_A(2P_0 + 2nP_1)\right|\;} \mathbb{P}_1^0 \times \mathbb{P}_{2n-1}^1 \overset{s}{\hookrightarrow} \mathbb{P}_{4n-1}$$

   is $G_2(2,2n)$-equivariant.

PROOF. 1. By Riemann-Roch,

$$\deg\bigl(\varphi(E_0 \times E_1)\bigr) = 4 \cdot 2/4 = 2.$$

The $G(2,2)$-equivariance follows from 4-1, 1 above.

   2. The mapping given by the linear system $\left|\mathcal{O}_{E_1}(2nP_1)\right|$,

$$E_1 \longrightarrow \mathbb{P}_{2n-1}^1,$$

is very ample for $n \geq 2$. This implies that

$$\deg\bigl(\varphi(A)\bigr) = 2 \cdot 4n/2 = 4n.$$

The $G_2(2,2n)$-equivariance follows from 4-1, 1 above.   □

Remark 4-3.

1. *The family of T-invariant quadrics is generated by*

$$H^0\big(\mathbb{P}_3, \mathcal{O}_{\mathbb{P}_3}(2)\big) = H^0\big(\mathbb{P}_3, \mathcal{O}_{\mathbb{P}_3}(2)\big)^T.$$

2. *We claim that the image of $A = E_0 \times E_1$ under the odd linear system defined by $|\mathcal{O}_A(2P_0 + 6P_1)|^-$ is the image of A under $|\mathcal{O}_A(2P_0 + 2P_1)|$. Let $v_0, \ldots, v_5$ be a basis for V_6; u_0, u_1 a basis for V_2 and t the section which generates V_3^-.*

PROOF. The group G generated by

$$\sigma': t \otimes u_i \mapsto t \otimes u_{i+1}, \qquad \tau': t \otimes u_i \mapsto (-1)^i t \otimes u_i,$$

for $i = 0, 1$, is naturally isomorphic to $G_2(6)$. This group isomorphism induces an isomorphism between the $G_2(6)$-module V_6^- and the G-module $V_3^- \otimes V_2$ given as:

$$\phi \; : \; \begin{array}{ccc} V_6^- & \longrightarrow & V_3^- \otimes V_2 \\ v_1 - v_5 & \longmapsto & t \otimes u_0, \\ v_4 - v_2 & \longmapsto & t \otimes u_1, \end{array}$$

and is equivariant with respect to the action of both groups. We also define

$$m_t \; : \; \begin{array}{ccc} V_2 & \longrightarrow & V_3^- \otimes V_2 \\ u_i & \longmapsto & t \otimes u_i, \end{array}$$

for $i = 0, 1$, inducing

$$[m_t] \colon \mathbb{P}(V_2) \to \mathbb{P}(V_3^- \otimes V_2).$$

We denote by id the identity morphism. Using the Künneth identification we have

$$\begin{aligned} H^0\big(E_0 \times E_1, \mathcal{O}(2P_0 + 6P_1)^-\big) &= H^0\big(E_0, \mathcal{O}(2P_0)\big) \otimes H^0\big(E_1, \mathcal{O}(6P_1)^-\big) \\ &= V_2 \otimes V_6^-. \end{aligned}$$

Thus

$$\big(\mathrm{id} \times [\phi]\big) \circ \big|\mathcal{O}_A(2P_0 + 6P_1)^-\big| = \big(\mathrm{id} \times [m_t]\big) \circ \big|\mathcal{O}_A(2P_0 + 2P_1)\big|,$$

and hence

$$s \circ \big|\mathcal{O}_A(2P_0 + 6P_1)^-\big| = s \circ \big(\mathrm{id} \times [\phi^{-1}]\big) \circ \big(\mathrm{id} \times [m_t]\big) \circ \big|\mathcal{O}_A(2P_0 + 2P_1)\big|.$$

From $t(e_0^1) = t(e_1^1) = t(e_2^1) = t(e_3^1) = 0$ it follows that the above composition has as base locus the elliptic curves $\{E_0 \times e_k^1\}_{k=0,\ldots,3}$. $\qquad\square$

5   The two-dimensional families of T-invariant quartic surfaces

In this section we study the two subfamilies $\mathcal{F}_{AB}$, $\mathcal{F}_{AE}$.

5.1 T-invariant quartic surfaces

Consider the level $(2,2)$-Heisenberg Group $T = G(2,2)$ given as the central extension

$$1 \longrightarrow \mu_2 \longrightarrow T \longrightarrow (\mathbb{Z}_2 \oplus \mathbb{Z}_2)^2 \longrightarrow 0.$$

It is well known ([6, Prop. 3]) that

$$\varrho\colon T \longrightarrow \mathrm{Aut}(V_2)$$

is the unique irreducible representation of T such that μ_2 operates as scalar multiplication. It induces

$$\mathrm{Symm}^4 \varrho\colon T \longrightarrow \mathrm{Aut}\big(\mathrm{Symm}^4(V_2)\big).$$

One can show (e.g., [6]) that the subspace W_0^T of T-invariant quartic polynomials in 4 variables is given by

$$W_0^T := \mathrm{Symm}^4(V_2)^T = \langle g_0, \ldots, g_4 \rangle,$$

with

$$\begin{aligned} g_0 &:= z_0^4 + z_1^4 + z_2^4 + z_3^4, & g_3 &:= 2(z_0^2 z_3^2 + z_1^2 z_2^2), \\ g_1 &:= 2(z_0^2 z_1^2 + z_2^2 z_3^2), & g_4 &:= 4z_0 z_1 z_2 z_3, \\ g_2 &:= 2(z_0^2 z_2^2 + z_1^2 z_3^2), \end{aligned}$$

and

$$f_\omega \in W_0^T \iff f_\omega = Ag_0 + Bg_1 + Cg_2 + Dg_3 + Eg_4.$$

Henceforth we write $f_\omega := f$ with $\omega := (A\colon B\colon C\colon D\colon E)$ to express that f_ω is the formula above.

5.2   The family $\mathcal{F}_{AB}$

Recall that the Segre-Primal ([1]) is the hypersurface of degree 3 in $\mathbb{P}_4$ singular in 10 ordinary double points. These double points lie in a set of planes *the Segre planes* contained in the primal. Both the double points and the Segre planes form a configuration of type $(15_4, 10_6)$ meaning that there are 15 Segre planes each containing 4 double points and 10 double points each lying on 6 such planes. The dual of the Segre-Primal is a hypersurface of degree 4 in another $\mathbb{P}_4$ which is singular in 15 double lines. By dualizing these lines one obtains the 15 Segre-Planes. One such plane in the coordinates $A, \ldots, E$ of Section 5 is given by

$$\mathcal{F}_{AB} := \{A = B = 0\}.$$

A 5-dimensional representation of the symmetric group in 6 letters operates transitively on the Segre planes (cf. [8]), hence it is enough to study one $\mathcal{F}_{AB}$.

Proposition 5-1. *For each* $\omega \in W_0^T$ *let*

$$X_\omega := \{z \in \mathbb{P}_3 \mid f_\omega(z) = 0\}, \qquad l_{i,j} := \{z_i = z_j = 0\},$$

hence

$$\mathrm{Sing}(\mathcal{F}_{AB}) = l_{01} \cup l_{23}.$$

For each $\omega \in \mathcal{F}_{AB}$, l_{01}, l_{23} *are double lines for the surface* X_ω *and it is a* $\mathbb{P}_1$*-bundle over an elliptic curve* C_ω *lying on the quadric* $l_{01} \times l_{23} \hookrightarrow \mathbb{P}_3$.

PROOF. $A = B = 0$ if and only if $g_2 C + g_3 D + g_4 E = 0$. The line spanned by $(0\!:\!0\!:\!\lambda_2\!:\!\lambda_3)$ and $(\mu_0\!:\!\mu_1\!:\!0\!:\!0)$ lies on X_ω if and only if for all t

$$(1-t)^2 t^2 \big(C(\mu_0^2 \lambda_2^2 + \mu_1^2 \lambda_3^2) + D(\mu_0^2 \lambda_3^2 + \mu_1^2 \lambda_2^2) + 2E(\mu_0 \mu_1 \lambda_2 \lambda_3) \big) = 0.$$

This implies that

$$C_\omega := \big\{ (\mu_0\!:\!\mu_1); (\lambda_2\!:\!\lambda_3) \in \mathbb{P}_1 \times \mathbb{P}_1 \mid f_w(\mu_0, \mu_1, \lambda_0, \lambda_1) = 0 \big\}$$

is a curve of bidegree $(2,2)$ on $\mathbb{P}_1 \times \mathbb{P}_1$ and hence elliptic. The line spanned by the plane in $\mathbb{C}^4$

$$\begin{pmatrix} \mu_0 & \mu_1 & 0 & 0 \\ 0 & 0 & \lambda_2 & \lambda_3 \end{pmatrix},$$

is the fibre of the mapping

$$\begin{array}{ccc} C_\omega & \longrightarrow & \mathbb{P}_3 \\ (\mu; \lambda) & \longmapsto & (\mu\!:\!\lambda) \end{array}$$

at the point $(\mu\!:\!\lambda)$. $\qquad\qquad\qquad\qquad\qquad\qquad\qquad\qquad\qquad$ □

For the computations in the next proposition it will be useful to introduce the Klein-coordinates (x_i) from the Plücker coordinates (p_{ij}) in $\mathbb{P}_5$ given as

$$x_0 := p_{01} - p_{23}, \qquad x_1 := i(p_{01} + p_{23}), \qquad x_2 := p_{02} + p_{13},$$
$$x_3 := i(p_{02} - p_{13}), \qquad x_4 := p_{03} - p_{12}, \qquad x_5 := i(p_{03} + p_{12}).$$

We will also adapt coordinates A, B, C, D, E for a $\mathbb{P}_4$. Choose also coordinates u_0, u_1, u_2, u_3, u_4, u_5 in $\mathbb{P}_5$ and let $U := \{u_0 + \cdots + u_5 = 0\}$. Define the following imbedding of $\mathbb{P}_4$ into U,

$$\begin{array}{ccc} t \; : & \mathbb{P}_4 & \longrightarrow & U \subset \mathbb{P}_5 \\ & (A\!:\!B\!:\!C\!:\!D\!:\!E) & \longmapsto & (u_0\!:\!u_1\!:\!u_2\!:\!u_3\!:\!u_4\!:\!u_5) \end{array}$$

with

$$\begin{array}{ll} u_0 = -B - C - D + A, & u_3 = B + C - D + A, \\ u_1 = -B + C + D + A, & u_4 = -2A + E, \\ u_2 = B - C + D + A, & u_5 = -2A - E. \end{array}$$

In particular for

$$L_{i,j} := \{u_i = u_j = 0\}, \qquad L_{i,j,k} := \{u_i = u_j = u_k = 0\},$$

we obtain

$$t(\{A = E = 0\}) = L_{0,1},$$
$$t(\{A = E = B + C + D = 0\}) = L_{0,1,2}.$$

Proposition 5-2. *For $\omega \in \mathbb{P}_4$ satisfying one of the equations of the left hand side, it is singular in $l_{01} \cup l_{23}$ and in the corresponding points (in Klein coordinates) of the right hand side:*

$$A = B = C + D - E = 0 \longleftrightarrow (0:0:0:\pm\iota:1:0)$$
$$A = B = C - D - E = 0 \longleftrightarrow (0:0:0:\pm1:0:\iota)$$
$$A = B = C + D + E = 0 \longleftrightarrow (0:0:\pm1:0:0:\iota)$$
$$A = B = C - D + E = 0 \longleftrightarrow (0:0:\pm\iota:0:1:0)$$

Each $\mathcal{F}_{AB}$ contains four double points. Every line in the $\mathbb{P}_4$ above contains 2 double points and every double point is contained in two lines in $\mathbb{P}_4$ forming a configuration of type $(4_2, 4_2)$.

PROOF. Let $p := (z_0 : z_1 : z_2 : z_3)$ be a singular point of X_ω and $\partial_i(f_\omega) := \partial f_\omega / \partial z_i$ for $i = 0, \ldots, 3$. Then for $f_\omega := Cg_2 + Dg_3 + Eg_4$

$$\partial_0 f_\omega(p) = Cz_0 z_2^2 + Dz_0 z_3^2 + Ez_1 z_2 z_3 = 0,$$
$$\partial_1 f_\omega(p) = Cz_1 z_3^2 + Dz_1 z_2^2 + Ez_0 z_2 z_3 = 0,$$
$$\partial_2 f_\omega(p) = Cz_0^2 z_2 + Dz_1^2 z_2 + Ez_0 z_1 z_3 = 0,$$
$$\partial_3 f_\omega(p) = Cz_1^2 z_3 + Dz_0^2 z_3 + Ez_0 z_1 z_2 = 0.$$

Assume that (i) $\prod_{i=0}^{5} z_i \neq 0$. Then

$$z_0 \partial_0 f_\omega - z_2 \partial_2 f_\omega = z_0^2 z_3^2 - z_1^2 z_2^2 = z_0^2 z_2^2 - z_1^2 z_3^3 = 0.$$

The possible values for the lines are then $z_1/z_0 = \pm 1, \pm i$, $z_2/z_3 = \pm 1, \pm i$. Substituting these values in the expressions for the partials of f_ω we obtain

1. $z_1/z_0 = 1$, $z_2/z_3 = -1$ or $z_1/z_0 = -1$, $z_2/z_3 = 1$; giving $C + D - E = 0$.

2. $z_1/z_0 = z_2/z_3 = 1$ or $z_1/z_0 = z_2/z_3 = -1$; giving $C + D + E = 0$.

3. $z_1/z_0 = z_2/z_3 = \iota$ or $z_1/z_0 = z_2/z_3 = -\iota$ giving $C - D + E = 0$.

4. $z_1/z_0 = \iota$, $z_2/z_3 = -\iota$ or $z_1/z_0 = -\iota$ or $z_2/z_3 = \iota$; giving $C - D - E = 0$.

To see the configuration use the coordinates $u_0, \ldots, u_5$ in U. For example, one of the lines in the above table is written as,

$$u_0 + u_1 = u_2 + u_3 = u_4 + u_5 = u_0 + u_5 = 0.$$

The double points $(1: -1: 1: -1: 1: -1)$, $(1: -1: -1: 1: 1: -1)$, are contained in this line. Conversely, given e.g., the first double point above, it is contained in the lines

$$u_0 + u_1 = u_2 + u_3 = u_4 + u_5 = u_0 + u_5 = 0,$$
$$u_0 + u_1 = u_2 + u_3 = u_4 + u_5 = u_2 + u_5 = 0.$$

If (ii) $\prod_{i=0}^{5} z_i = 0$, it follows from the expressions for the partials of f_ω that $z_1 = z_0 = 0$ or $z_2 = z_3 = 0$; otherwise we have $z_2 = z_0 = 0$, implying $z_1 z_3 = 0$; or $z_3 = z_0 = 0m$, implying $z_1 z_2 = 0$; or $z_2 = z_1 = 0$, implying $z_0 z_3 = 0$; or $z_1 = z_3 = 0$, implying $z_0 z_2 = 0$.

In any case, $\operatorname{Sing}(X_\omega) = l_{01} \cup l_{23}$.   □

6   The Family $\mathcal{F}_{AE}$

Let

1. $\mathrm{NS}(A)$ be the Neron-Severi group of A. This is the free abelian group generated by the equivalence classes of divisors of A modulo algebraic equivalence, and

   $$\rho := \operatorname{rank} \mathrm{NS}(A) \otimes \mathbb{Q}$$

   will denote its $\mathbb{Q}$-rank.

2. NC be the Nikulin construction for $K3$ surfaces [9, Theorem 1]:

   Given X a $K3$ surface and $\{L_i\}_{i=1,\ldots,16}$ 16 disjoint rational curves on X, there exists a unique torus A and a $2:1$ cover $f: \widetilde{X} \to X$ ramified along $\mathbf{L} := \cup_{i=1}^{16} \overline{L_i}$ such that A is obtained from $\widetilde{X}$ by contracting $\{f^*(\overline{L_i})\}_{L_i \in \mathbf{L}}$ to smooth points as in the following diagram:

   $$\widetilde{X} \xrightarrow{f} X$$
   $$\downarrow \sigma$$
   $$A$$

   with σ the contracting morphism.

   Our basic assumption in this section is that $\rho = 2$.

Proposition 6-1. For $\omega \in L_{i,j} - \cup_{i \neq j \neq k} L_{i,j,k}$ there exists an abelian surface

$$A_\omega \simeq E_1' \times E_2'/(\mathbb{Z}_2 \times \mathbb{Z}_2)$$

(where $\simeq$ is isomorphism) such that the following polarizations, by means of NC, are constructed on A_ω

1. There exists $\Theta \subset A_w$ hyperelliptic such that $2 \cdot \Theta$ gives a $(2,2)$-polarization on A_ω.

2. There exists $E \subset A_\omega$ elliptic such that $2 \cdot (E+\Theta)$ gives a $(2,6)$-polarization on A_ω.

PROOF. For points R, $S \in X_\omega$, $\langle R, S \rangle$ is the line spanned by both points. Fix z_1, z_2, $z_3 \in \mathbb{C}$ such that $z_1 z_2 z_3 \neq 0$ and fix the points

$$Q_1 := (0 : z_1 : z_2 : z_3), \qquad Q_1' := (0 : 1/z_1 : 1/z_2 : 1/z_3).$$

For the group T we write the generators of the representation ϱ of T as in the introduction. Let also

$$\sigma_3 := \sigma_1 \sigma_2, \qquad \tau_3 := \tau_1 \tau_2.$$

Let P_i be the i-th unit point, namely the point whose coordinates are all zero except on the i-th coordinate whose value is equal to 1 for $i = 1, \ldots, 4$. Let also

$$B \colon X \to X_\omega$$

be the blowing-up of X_ω with centre at $\{P_i\}_{i=1,\ldots,4}$.

Let

$$l_{i,j} := \langle \sigma_{i-1} P_1, \sigma_{i-1} \tau_{j-1} Q_1 \rangle.$$

$L_{i,j}$ will denote the strict transform of $l_{i,j}$ under B (resp. $L_{i,j}'$, replacing $l_{i,j}$ by $l_{i,j}'$). Choose

$$\mathbf{L}_0 := \{L_{i,k}\}_{i,k \in \{1,\ldots,4\}}$$

as the set of 16 disjoint rational lines and let f be the associated double cover as in NC. We have also

$$E_i := B^*(p_i), \qquad \overline{E_i} := f^*(E_i),$$

for $i = 1, \ldots, 4$. We fix once and for all $i_0 \in \{1, \cdots, 4\}$ and choose the following curves on A_ω,

$$\begin{aligned}
\Theta_{i_0,k}' &:= \sigma_* f^*(L_{i_0,k}^{*}{}') \\
\Theta_{i,k} &:= \sigma_* f^*(L_{i,k})
\end{aligned}$$

with $k \in \{1, \cdots, 4\}$, and $\sigma \colon \tilde{X} \dashrightarrow X$ the contracting morphism on $\mathbf{L}_0$ as in NC. The situation can be pictured as:

$$\tilde{X} \xrightarrow{\ f\ } X \xrightarrow{\ B\ } X_\omega$$
$$\downarrow{\scriptstyle\sigma}$$
$$A$$

$\square$

Due to its technical nature we shall omit the proof of the next lemma for the next section, but shall use it in the course of the proof of Proposition 6-1.

Lemma 6-2. *If* $L' \in \{L'_{i_0,1}, \dots, L'_{i_0,4}\}$ *then*

$$L' \cdot \sum_{L \in \mathbf{L}_0} L = 6.$$

We shall also use the following lemma proved after Proposition 6-1.

Lemma 6-3.

1. $g(\overline{\overline{E}}_i) = 1$, *for* $i = 1, \dots, 4$.

2. *If* $\overline{\overline{E}}_i$ *is complementary to* $\overline{\overline{E}}'_i$ *i.e, it induces a decomposition* $A \sim \overline{\overline{E}}'_i \times \overline{\overline{E}}_i$ *then*

$$\overline{\overline{E}}_i \cdot \overline{\overline{E}}_i' = 4, \qquad for\ i = 1, \dots, 4.$$

3. $g(\Theta'_{i_0,k}) = 2$ *for* $k \in \{1, \cdots, 4\}$.

PROOF. (OF PROPOSITION 6-1) 1. Given E'_0 elliptic in A, by Poincaré's irreducibility theorem there exists E'_1 such that by Lemma 6-3 2,

$$E'_0 \cdot E'_1 = 4, \tag{6.3}$$

but together with the assumption $\rho = 2$ and that $\Theta_{i,k}$ can not be a multiple of E'_0 or E'_1,

$$E'_1 = n_0 E'_0 + n_1 \Theta'_{i_0,k} \qquad \text{for some } n_0,\ n_1 \in \mathbb{Q}.$$

Fixing the pair (i_0, k) we write $\Theta := \Theta'_{i_0,k}$ and it follows from the proof of Lemma 6-3 2, that $n_0 = -1$ and $n_1 = 2$. Thus

$$\Theta = 1/2(E'_0 + E'_1)$$

induces a principal polarization on A. It remains to prove the claimed isomorphism. Let

$$\varphi \colon \check{E}_0 \times \check{E}_1 \longrightarrow A$$

be a morphism of degree 4 and $E'_i \subset A$ the image of $\{\check{E}_i\}_{i=0,1}$ and $\mathcal{L} \in \mathrm{Pic}(A)$ such that $\mathcal{L}^{\otimes 2} = \mathcal{O}_A(E'_0 + E'_1)$ (which exists by equation (6.3)).

We claim that $\ker \varphi \equiv \mathbb{Z}_2 \oplus \mathbb{Z}_2$. It is enough to show that there is no cyclic group $\mathbb{Z}_4$ acting on $\check{E}_0 \times \check{E}_1$ leaving $\varphi^*(\mathcal{L})$ invariant. Indeed, since

$$\varphi^*(E'_i) \simeq 4\check{E}_i \Rightarrow \varphi^* \mathcal{L} \sim 2(\check{E}_0 + \check{E}_1)$$

and since $\check{E}_0 + \check{E}_1$ induces a principal polarization $H\big(\varphi^*(\mathcal{L})\big) \simeq \mathbb{Z}_2^4$, i.e., $\varphi^*(\mathcal{L})$ induces a $(2, 2)$-polarization.

2. By part 1, we have constructed $\Theta_1, E_1 \subset \tilde{X}$ such that Θ_1 is hyperelliptic, E_1 is elliptic, and

$$E_1 \cdot \Theta_1 = 2 = \Theta_1^2,$$

denoting their images in A under σ by the same letters. $2(E_1+\Theta_1)$ is a $(2,6)$ polarization, for

$$2^2(E_1 + \Theta_1)^2 = 24$$

and $E_1 + \Theta_1$ symmetric implies $2(E_1 + \Theta_1)$ totally symmetric. Hence the latter is a $(2,6)$ polarization on A. $\square$

A preliminary observation to the proof of lemma 6-3 is that if $\mathcal{L}$ is the line bundle associated to the cyclic double covering (e.g., [2, Lemma (17.1)]),

$$\omega_{\tilde{X}} = f^*(\omega_X \otimes \mathcal{L}),$$

in particular

$$K_{\tilde{X}} = f^*(K_X) + \frac{1}{2} \sum_{L \in L_0} f^*(L).$$

PROOF. (OF LEMMA 6-3) 1. Since $K_{\tilde{X}} = B^*(K_{\mathbb{P}_3} + X_\omega) = B^*(-4H + 4H) = 0$ where H is a hyperplane in $\mathbb{P}_3$. Applying the adjuction formula to $E_i = B^*(P_i)$ for $i = 1, \ldots, 4$, we obtain

$$-2 = E_i \cdot (E_i + K_X) = E_i^2.$$

The covering f is given by

$$\mathcal{L}^{\otimes 2} = \mathcal{O}\left(\sum_{L \in L_0} L \right)$$

and

$$K_{\tilde{X}} = f^*(K_X) + \frac{1}{2} f^*\left(\mathcal{O}\left(\sum_{L \in L_0} L \right) \right) = \frac{1}{2} f^*\left(\mathcal{O}\left(\sum_{L \in L_0} L \right) \right).$$

We use again the adjuction formula applied to $\overline{E}_i$, but first we compute

$$\begin{aligned}
\overline{E}_i \cdot \overline{E}_i &= -4, \\
\overline{E}_i \cdot K_{\tilde{X}} &= \frac{1}{2} \overline{E}_i \cdot f^*\left(\mathcal{O}\left(\sum_{L \in L_0} L \right) \right) \\
&= \frac{1}{2} \sum_{L \in L_0} f^*(E_i) \cdot f^*(L) = \sum_{L \in L_0} E_i \cdot L \\
&= 4.
\end{aligned}$$

Hence

$$2g(\overline{E}_i) - 2 = -4 + 4 = 0,$$

which implies $g(\overline{E}_i) = 1$.

3. Fix a pair (i_0, k) with $k \in \{1, \ldots, 4\}$ and let

$$L' = L_{i_0, k}'.$$

Since $g(L') = g(l') = 0$ using again the adjunction formula

$$-2 = 2g(L') - 2 = L' \cdot (L' + K_X) = L'^2,$$

then

$$\overline{L}' \cdot \overline{L}' = f^*(L') \cdot f^*(L') = -4.$$

Computing the canonical class of $\tilde{X}$ in the case of a cylic covering gives

$$K_{\tilde{X}} = f^*(K_X + \mathcal{L}) = \frac{1}{2} f^* \left(\sum_{L \in L_0} L \right).$$

Also

$$\overline{L}' \cdot K_{\tilde{X}} = \frac{1}{2} \overline{L}' \cdot \sum_{L \in L_0} f^*(L) = \frac{1}{2} \cdot 2 \cdot L' \cdot \sum_{L \in L_0} L = 6.$$

Another application of the adjunction formula gives $g(\overline{L}') = 2$. By passing to the image of L' under σ, we have $g(\Theta') = 2$.

2. Using the same notation as introduced in 3,

$$\Theta' := \sigma_*\big(f^*(L')\big) = \sigma_*(\overline{L}'), \qquad \overline{\overline{E}}_0 := \sigma_*\big(f^*(E_0)\big) = \sigma_*(\overline{E}_0).$$

By the projection formula

$$\overline{\overline{E}}_0 \cdot \overline{L}' = f^*(E_0) \cdot f^*(L') = 2E_0 \cdot L' = 2.$$

The hypothesis on ρ implies that there exist $n_0, n_1 \in \mathbb{Q}$ with

$$\overline{\overline{E}}_0{}' = n_0 \overline{\overline{E}}_0 + n_1 \Theta'$$

or

$$0 = \overline{\overline{E}}_0{}'^2 = 2n_0 n_1 \overline{\overline{E}}_0 \cdot \Theta' + n_1^2 \Theta'^2 = 2n_1(n_1 + 2n_0).$$

The $\{\overline{\overline{E}}_0, \overline{\overline{E}}_0'\}$ are integral elliptic curves. Then $n_1 = -2n_0$ implies $n_0 = \pm 1$, but

$$0 \le n := \overline{\overline{E}}_0 \cdot \overline{\overline{E}}_0{}'$$
$$= \overline{\overline{E}}_0 \cdot \big(n_0 \overline{\overline{E}}_0 + n_1 \Theta'\big)$$
$$= n_1 \overline{\overline{E}}_0 \cdot \Theta' = 2n_1$$
$$= -4n_0,$$

hence $n_0 = -1$, $n = 4$, $n_1 = 2$, giving $\Theta' = 1/2(\overline{\overline{E}}_0 + \overline{\overline{E}}_0')$. $\qquad \square$

Remark 6-4. *The following curves lie on A*

1. *4 Elliptic curves* $\overline{\overline{E}}_1', \ldots, \overline{\overline{E}}_4'$ *(and their complements* $\overline{\overline{E}}_1, \cdots, \overline{\overline{E}}_4$*).*

2. *16 Hyperelliptic curves* $\{\Theta_{i,k}'\}_{k,i \in \{1,\ldots,4\}}$.

6.1   The proof of Lemma 6-2

PROOF. The elements of the group T are linear automorphisms hence preserve
the incidence relation of intersection. Moreover, the sum $\sum_{L \in L_0} L$ is invariant
under the group T. This implies that it is enough to prove the claim for a fixed
$i_0 \in \{1, \cdots, 4\}$. The Klein coordinates are very appropiate for this computation
because if (x_i), (x_i') are two lines in $\mathbb{P}_3$ then (x_i) will intersect (x_i') if and only
if $\sum_{i=0}^5 x_i \cdot x_i' = 0$.

Choose L the line spanned by $(1:0:0:0)$ and $(0:z_1:z_2:z_3)$, and L' the line
spanned by $(1:0:0:0)$ and $(0:1/z_1:1/z_2:1/z_3)$. In Klein coordinates

$$
\begin{aligned}
L &= (z_1 : \iota z_1 : z_2 : \iota z_2 : z_3 : \iota z_3), \\
L' &= \left(\frac{1}{z_1} : \frac{\iota}{z_1} : \frac{1}{z_2} : \frac{\iota}{z_2} : \frac{1}{z_3} : \frac{\iota}{z_3} \right), \\
\sigma_1(L) &= (z_1 : \iota z_1 : -z_2 : \iota z_2 : z_3 : -\iota z_3), \\
\sigma_2(L) &= (z_1 : -\iota z_1 : z_2 : \iota z_2 : -z_3 : \iota z_3), \\
\sigma_3(L) &= (z_1 : -\iota z_1 : -z_2 : \iota z_2 : -z_3 : -\iota z_3), \\
\tau_2 \sigma_1(L) &= (z_1 : \iota z_1 : z_2 : -\iota z_2 : -z_3 : \iota z_3), \\
\tau_1 \sigma_2(L) &= (-z_1 : \iota z_1 : z_2 : \iota z_2 : z_3 : -\iota z_3), \\
\tau_3 \sigma_3(L) &= (z_1 : -\iota z_1 : -z_2 : \iota z_2 : z_3 : \iota z_3).
\end{aligned}
$$

Clearly, the following pairs of lines are transversal to each other

$$
\begin{array}{ccc}
\big(\sigma_1(L), L'\big), & \big(\sigma_2(L), L'\big), & \big(\sigma_3(L), L'\big), \\
\big(\tau_2\sigma_1(L), L'\big), & \big(\tau_1\sigma_2(L), L'\big), & \big(\tau_3\sigma_3(L), L'\big).
\end{array}
$$

$\square$

7   The Family $t^{-1}(L_{0,1,2})$

Proposition 7-1. *Fix points*

$$
R_0 := (1:1:1:1), \qquad Q_0 := (1:1:1:-1),
$$

and define for $j = 0, \ldots, 3$,

$$
R_j := \tau_j Q_0, \qquad Q_j := \tau_j Q_0.
$$

Let $P_1, \ldots, P_4$ be the unit points in $\mathbb{P}_3$. For $\omega \in t^{-1}(L_{0,\,1,\,2})$,

$$\mathrm{Sing}(X_\omega) = \bigcup_{i=1}^{4} P_i \bigcup_{j=0}^{3}(R_j \cup Q_j).$$

Each line $\{l_{i,j}\}_{i,j=0,\ldots,3}$ through the coordinate points p_j in X_ω passes through one R_j and Q_j and a line containing R_j and Q_j is contained in each X_ω of $\omega \in t^{-1}(L_{0,\,1,\,2})$. Therefore, the configuration of singular points and lines $\{l_{i,j}\}$ is $(12_4, 16_3)$.

PROOF. We will compute the singular locus of X_ω as follows: Write the gradient of f_ω as $\partial_i f_\omega := \partial_i f_\omega / \partial z_i$ and let $p := (z_0 : z_1 : z_2 : z_3)$ be a singular point of X_ω. Then

$$\partial_0 f_\omega(p) = z_0\big(C(z_2^2 - z_1^2) - D(z_1^2 - z_3^2)\big) = 0$$
$$\partial_1 f_\omega(p) = z_1\big(C(z_0^2 - z_3^2) + D(z_0^2 - z_2^2)\big) = 0$$
$$\partial_2 f_\omega(p) = z_2\big(C(z_0^2 - z_3^2) - D(z_3^2 - z_1^2)\big) = 0,$$
$$\partial_3 f_\omega(p) = z_3\big(C(z_2^2 - z_1^2) - D(z_0^2 - z_2^2)\big) = 0.$$

We consider two cases.

Suppose (i) $z_0 z_1 z_2 z_3 \neq 0$. From $(\partial_1/z_1 - \partial_2/z_2)f_\omega = (\partial_0/z_0 - \partial_3/z_3)f_\omega = 0$ it follows

$$z_1^2 - z_3^2 = z_0^2 - z_2^2 = 0.$$

Analogously from $\partial_0 f_\omega = 0$ (resp. $\partial_2 f_\omega = 0$) it follows $z_1^2 - z_2^2 = 0$ (resp. $z_0^2 - z_3^2 = 0$). This implies $z_0^2 = z_1^2 = z_3^2 = z_2^2$ giving the solutions $(1 : \pm 1 : \pm 1 : \pm 1)$.

Otherwise assume that (ii) $z_0 z_1 z_2 z_3 = 0$. By $H_{2,2}$-symmetry, we assume $z_0 = 0$ then writing

$$\partial_1 f_\omega(p) = \quad\ z_1(C z_3^2 + D z_2^2) \qquad = 0$$
$$\partial_2 f_\omega(p) = z_2\big(-C z_3^2 + D(z_3^2 - z_1^2)\big) = 0$$
$$\partial_3 f_\omega(p) = \quad z_3\big(C(z_2^2 - z_1^2) + D z_2^2\big) \ = 0.$$

If (a) $z_1 z_2 z_3 \neq 0$, from $C z_3^2 = -D z_2^2$, substituting in the equations:

$$\partial_2 f_\omega(p) = D(z_2^2 + z_3^2 - z_1^2) = 0,$$
$$\partial_3 f_\omega(p) = C(z_2^2 - z_1^2 - z_3^2) = 0.$$

$CD \neq 0$ implies $z_2^2 - z_1^2 = z_3^2 = 0$ but e.g. from $\partial_1 f_\omega = 0$ implies $z_2 = 0$ a contradiction.

Otherwise (b) $z_2 z_3 z_1 = 0$. From e.g. $z_1 = 0$ we have

$$\partial_2 f_\omega : z_2 z_3^2(-C + D) \ = \ 0,$$
$$\partial_3 f_\omega : z_2^2 z_3(C + D) \ = \ 0.$$

Since $(D - C)(C + D) \neq 0$ then $z_2 z_3 = 0$ in particular

$$(0 : 0 : 1 : 0), \qquad (0 : 0 : 0 : 1)$$

are singular points.

To determine the configuration, it is enough by $H_{2,2}$-symmetry to show that there is a line in X_ω containing P_0, R_0, Q_0. But if L_0 is the line spanned by P_4 and R_0 clearly Q_0 is contained in the lines $L_0, \tau_2\sigma_1 L_0, \tau_1\sigma_2 L_0, \tau_3\sigma_3 L_0$. Finally, the line spanned by P_1 and R_0 denoted by l_0 is contained in X_ω and clearly contains the additional point Q_3. $\qquad\square$

Remark 7-2. *Let*

$$d_{ijkl} := \{u_i = u_j = u_k = u_l = 0\} \bigcap \left\{ \sum u_i = 0 \right\}.$$

There are 15 such points. It is easy to see that these correspond up to permutation of coordinates to the $H_{2,2}$-invariant tetrahedra

$$z_0 z_1 z_2 z_3 = 0.$$

Each plane L_{ij} contains four lines $L_{ijk}, L_{ijl}, L_{ijm}, L_{ijn}$. Each line L_{ijk} contains $d_{ijkl}, d_{ijkm}, d_{ijkn}$. Conversely, every d_{ijkl} is contained in the lines $L_{ijk}, L_{ijl}, L_{jkl}, L_{ikl}$. The planes L_{ij} intersect as

$$\begin{aligned} L_{ij} \cap L_{kl} &= d_{ijkl} \quad \text{or} \\ L_{ij} \cap L_{il} &= L_{ijl}. \end{aligned}$$

The lines L_{ijk}, L_{ijl} intersect in d_{ijkl}.

8 The Family $\mathcal{F}_{AB} \cap \mathcal{F}_{AE}$

If we let

$$S_{(ij)(kl)} := \{u_i + u_j = u_k + u_l = u_m + u_n = 0\}$$

we have two cases up to permutation of coordinates for the intersection of a plane $L_{i,j}$ and $S_{(kl)(mn)}$

1. $L_{45} \cap S_{(01)(23)} = \{(u_0 : -u_0 : u_2 : -u_2 : 0 : 0)\}$,

2. $L_{02} \cap S_{(01)(23)} = \{(0 : 0 : 0 : 0 : 1 : -1)\} = d_{0123}$.

For the first case this is $A = B = E = 0$ in the A, B, C, D, E coordinates. This is not one of the lines in Proposition 5-2 hence it is singular along $l_{01} \cup l_{23}$. As for the second case d_{0123} is up to permutation of coordinates an $H_{2,2}$-invariant tetrahedra.

Remark 8-1. *Let N be the closure of $H_{2,2}$-invariant quartic surfaces in $\mathbb{P}_3$ containing lines. This is a threefold in U with coordinates $u_0, \ldots, u_5$. It is singular along the 10 double points of the Segre-primal and the lines L_{ijk} we have described in 5.2 (cf. [7]). We have shown that every X in N lying outside the Segre-planes and L_{ij} is smooth. The surfaces parametrised by the L_{ijk} have been described under the name of "desmic surfaces" (cf. [4, ch. II]). If we fix*

Δ_1, Δ_2, Δ_3, *three $H_{2,2}$-invariant tetrahedra in $\mathbb{P}_3$ the equation of a desmic surface is*

$$\lambda\Delta_1 + \mu\Delta_2 + \nu\Delta_3 = 0, \quad \lambda, \mu, \nu \in \mathbb{C},$$

and where an identity exists of the form

$$\alpha\Delta_1 + \beta\Delta_2 + \gamma\Delta_3 \equiv 0.$$

For the surfaces introduced in Section 7,

$$\begin{aligned}
\Delta_1 &:= (z_1^2 - z_0^2)(z_2^2 - z_3^2) \\
\Delta_2 &:= (z_2^2 - z_1^2)(z_0^2 - z_3^2) \\
\Delta_3 &:= (z_3^2 - z_1^2)(z_0^2 - z_2^2),
\end{aligned}$$

and $\lambda + \mu + \nu = 0$ with $\alpha = \beta := 1$, $\gamma := -1$. The Segre-planes and the planes L_{ij} arise as follows: if $u_0, \ldots, u_5$ are the coordinates of a $\mathbb{P}_5$ as in Section 5.2 then $\{u_i + u_j = 0\} \cap N$ is the union of three Segre-planes and a residual quadric, a plane L_{ij} with multiplicity two.

References

[1] Baker,H.: Principles of Geometry, Cambridge University Press, 1925.

[2] Barth, W. Peters C., Van de Ven A.: Compact Complex Surfaces: Springer Verlag, 1984.

[3] Hudson, R.: Kummer's Quartic Surface: Cambridge University Press, 1905.

[4] Jessop,C.M.: Quartic Surfaces with Singular Points: Cambridge University Press, 1916.

[5] Mumford, D.: Abelian Varieties: Oxford University Press, 1970.

[6] Mumford, D.: On the Equations Defining Abelian Varieties. Invent. Math, **1**, 287–354, 1966.

[7] Nieto, I. and Barth, W.: Abelian surfaces of type (1,3) and quartic surfaces with 16 skew lines. Journal of Algebraic Geometry, vol. 14, 1994.

[8] Nieto, I.: The Normalizer of the Level (2,2)-Heisenberg Group. Manuscripta math. 76, 257–267, 1992.

[9] Nikulin, V. V.: On Kummer Surfaces, Math. USSR Izvestija, **9** 1975, No. 2.

Address of author:

CIMAT A.C.
Apartado Postal 402
3600 Guanajato, Gto.
México.
Email: nieto@buzon.main.conacyt.mx

SEMIGROUPS AND CLUSTERS AT INFINITY

Ana-José Reguera López

1 Introduction

Let k be a field of arbitrary characteristic $p \geq 0$ and C a curve defined over k with only one branch at infinity, that is, a projective absolutely irreducible plane curve with resolution over k for which there is a line (line at infinity) intersecting C in only one point P, and C has only one analytic branch at P.

The only branch at infinity corresponds to a valuation v of $K(C)$, the field of rational functions over k. We define

$$\Gamma_1(P) = \{-v(g) \mid g \in R\},$$

where R is the affine k-algebra of C, that is, obtained taking away the infinity line. Let $\{x, y\}$ be coordinates of R and $f(x, y) = 0$ the expression of C in the affine part. Then, if p does not divide $n = -v(x)$ and $m = -v(y)$ at the same time, approximate roots can be defined for the polynomial $f(x, y)$ (see [5], chapter II, sections 6, 7) and it can be proved that there exist a positive integer h and a sequence of positive integers $\delta_0, \ldots, \delta_h$ such that they generate $\Gamma_1(P)$ and

1. If $d_i = \gcd(\delta_0, \ldots, \delta_{i-1})$, for $1 \leq i \leq h+1$ and $n_i = d_i/d_{i+1}$, $1 \leq i \leq h$, then $d_{h+1} = 1$ and $n_i > 1$ for $1 \leq i \leq h$.

2. For $1 \leq i \leq h$, $n_i \delta_i$ belongs to the semigroup generated by $\delta_0, \ldots, \delta_{i-1}$.

3. $\delta_i < \delta_{i-1} n_{i-1}$ for $i = 2, \ldots, h$.

(see [5], section 7 and [17], section 2). From now on, we will refer to these properties as "1, 2 and 3", meaning "1, 2 and 3 as in the Introduction".

What interests us is the converse problem, that is, given a semigroup Γ generated by a sequence of natural numbers $\delta_0, \delta_1, \ldots, \delta_h$ satisfying 1, 2 and 3 and fixed a field k, try to find curves with only one branch at infinity and such that $\Gamma_1(P) = \Gamma$.

This problem has been studied in the literature with some restrictions. Thus, if p does not divide $d_2 = \gcd(\delta_0, \delta_1)$, then there exists a projective plane curve defined over k with only one branch at infinity and such that $\Gamma_1(P) = \Gamma$ ([17], Section 2).

In this paper we study curves with the above property, that is, only one branch at infinity and such that $\Gamma_1(P) = \langle \delta_0, \delta_1, \ldots, \delta_h \rangle$, using as main tool Newton's polygons. Precisely, to each branch of curve centered at a point P one can associate a sequence of polygons which give information about the singularity of the branch at P (see 2.1).

Progress in Mathematics, Vol. 134
© 1996 Birkhäuser Verlag Basel/Switzerland

We also introduce in section 2 a weaker notion than approximate roots: the notion of approximants (definition 2-3). The main property they satisfy is that given a sequence of approximants for the polynomial $f(x,y)$ defining a curve with only one branch at infinity (no matter about the characteristic of the field), then there is a sequence of natural numbers generating $\Gamma_1(P)$ and satisfying 1, 2 and 3 as above (see result 2-7).

In section 3 we introduce the concept of cluster in order to make more precise our approach. A cluster with origin at a point P in a regular surface S is a finite set of points infinitely near P, each with an assigned integral multiplicity called virtual multiplicity (definition 3-2). If the cluster K satisfies the proximity relations (definition 3-4), then the set of curves going through it (definition 3-3) defines a complete ideal I_K of the local ring $\mathcal{O}_{S,P}$ and generic element of I_K defines a curve which goes through K with effective multiplicities equal to the virtual ones (see [8], section 3; [14], Section 2 and [18], appendix 5). If the cluster has only one branch of infinitely near points, then the sequence of Newton's polygons and the transformations in the algoritm in Section 2.1 are the same for the generic element of I_K.

Therefore, we say that a cluster K satisfying the proximity relations is compatible with the semigroup Γ generated by $\delta_0, \delta_1, \ldots, \delta_h$ if the sequence of Newton's polygonons for a generic element is the one in figure 3.5.

The concept of cluster allows us to classify the set of curves with only one branch at infinity and such that $\Gamma_1(P)$ is the semigroup $\Gamma = \langle \delta_0, \delta_1, \ldots, \delta_h \rangle$. In fact, given a cluster K compatible with Γ, we can calculate explicitly a basis of the k-vector space $W_{\delta_0}(K)$ consisting of the polynomials in $k[X_0, X_1, X_2]$ of degree than δ_0 and whose germ at P belons to I_K (see proposition 3-10).

Now, if one of the elements in $W_{\delta_0}(K)$ defines a curve with only branch at infinity and having a sequence of approximants, then the generic element in $W_{\delta_0}(K)$ has the same property (i.e., only one branch at infinity and approximants) and for it $\Gamma_1(P) = \langle \delta_0, \delta_1, \ldots, \delta_h \rangle$ and the semigroup of values at P is $S(P) = \langle r_0, \ldots r_h \rangle$ where the r's can be arithmetically described in terms of the δ's (theorem 3-13).

In section 4 we give explicit equations of plane curves defined over any field k, with only one branch at infinity and satisfying $\Gamma_1(P) = \langle \delta_0, \delta_1, \ldots, \delta_h \rangle$ and $S(P) = \langle r_0, \ldots r_h \rangle$. When p does not divide d_2, it is possible to find a curve with the above properties and smooth in the affine part. For these curves the Weierstrass semigroup of P coincides with $\Gamma_1(P) = \langle \delta_0, \delta_1, \ldots, \delta_h \rangle$ and we can also compute the genus in terms of the δ's (theorem 4-6).

When p divides d_2, we study necessary conditions for the existence of such curves (theorem 4-8) reaching the same result for the Weierstrass semigroup of infinity and the genus of the curves so obtained. Finally, we show with an example how the method described in this work can be modified in order to obtain such curves when the field k and the semigroup Γ do not satisfy the hypothesis of theorem 4-8.

2   The concept of approximant

2.1   Newton polygon

We describe an algorithm that allows us to compute the semigroup of values of a branch from the Newton diagram of this branch. The construction we show here is studied with detail in [7], algorithm 3.4.14.

Let k be any field and $f(u,\omega) = \sum_{\alpha,\beta \geq 0} A_{\alpha\beta} u^\alpha \omega^\beta$ be a polymonial in $k[u,\omega]$ defining a germ of curve at $P = (0,0)$, and suppose it has only one branch which is rational over k. Let us consider the Newton diagram of f:

$$D(f) = \{(\alpha,\beta) \mid A_{\alpha,\beta} \neq 0\}.$$

The bounded segments of the convex hull of $D(f) + \mathbb{R}_+^2$ is called the Newton polygon of f and denoted by $P(f)$.

Let us suppose we have chosen coordinates $\{u,w\}$ in such a way that $v(u) < v(\omega)$, where v is the valuation associated to the branch defined by f at P. Then, one of the following situations must happen:

a) The polygon is the point $(0,1)$.

b) The polygon is the segment with extremes in $(m,0)$ and $(0,n)$ (if $v(\omega) < \infty$).

In the case a) the algorithm ends. In the case b) it may happen:

b.1) $m = en$ with $e \in \mathbb{Z}_+$. Then, there exist $a, \lambda \in k^*$ so that

$$\sum_{(\alpha,\beta) \in P(f)} A_{\alpha\beta} u^\alpha \omega^\beta = a\,(\omega - \lambda u^e)^n.$$

In this case, after a change of coordinates $u' = u$, $\omega' = \omega - \lambda u'^e$ the polygon is transformed into a segment of extremes $(m',0)$ and $(0,n)$ with $m' > m$. The preceding transformation will be called b1-transformation and is univocally determined by the equation of f.

b.2) If b.1) does not happen, let $e_1 = \gcd(m,n) < n$ and let us take (σ,τ) the only integral solution of the equation

$$|\tau m - \sigma n| = e_1,$$

satisfying the conditions $\sigma \geq 0$, $\tau \geq 0$, $2\tau \leq n/e_1$, $2\sigma \leq m/e_1$. In this case, we make the transformation

$$u \longmapsto u^{\frac{n}{e_1}} \omega^\tau,$$
$$\omega \longmapsto u^{\frac{m}{e_1}} \omega^\sigma,$$

(that is, a finite sequence of blow ups) and we consider the strict transform of f. In this way, the Newton polygon is transformed into a vertical segment with extremes $(0,0)$ and $(0,e_1)$. The preceding transformation will be called b2-transformation and is also univocally determined by the equation of f.

If f is irreducible as a power series centered at P, then we will carry out the following algorithm.

We make sucesive b1-transformations whenever n divides m (m being the suitable value obtained after the preceding transformation). It may happen:

1.1) The process is infinite.

2.1) The process is finite, that is, after a finite number of b1-transformations we obtain a Newton polygon whose projections on the axes, m_0 and $n = e_0$, are such that $e_0 < m_0$ and $e_1 = \gcd(e_0, m_0) < e_0$.

In the case 1.1 the algorithm ends (that happens if and only if $n = 1$). In the case 2.1 we make the suitable b2-transformation. After that, we make another sequence of b1-transformations and we reach again one of the two cases

1.2) The process is infinite.

2.2) The process is finite.

In the case 1.2 the algorithm ends ($e_1 = 1$). In the case 2.2 we denote $e_2 = \gcd(m_1, e_1)$ and we continue in the same way.

Since we have $e_0 > e_1 > \cdots$, there must exist an integer h such that $e_h = 1$ and so the algorithm has an end. After this algorithm, we obtain h polygons $(P_\nu)_{\nu=0}^{h-1}$ which are straight lines with vertices $(m_\nu, 0)$ and $(0, e_\nu)$, where e_ν and m_ν satisfy

  i) $\gcd(e_\nu, m_\nu) < e_\nu$ for $0 \leq \nu \leq h - 1$.

 ii) $e_{\nu+1} = \gcd(e_\nu, m_\nu)$ for $0 \leq \nu \leq h - 1$.

(see figure 2.1).

In this situation, we have the following result.

Result 2-1. [[7], chapter 4, section 3] The semigroup of values of the curve defined by f at P is generated by

$$S(P) = \langle \overline{\beta}_0, \overline{\beta}_1, \ldots, \overline{\beta}_h \rangle ,$$

where $\overline{\beta}_0 = e_0$, $\overline{\beta}_\nu = (1/e_{\nu-1})(e_0 m_0 + \cdots + e_{\nu-1} m_{\nu-1})$, for $1 \leq \nu \leq h$.

The preceding study is done for curves with only one branch. Conversely, let $F \in k[X_0, X_1, X_2]$ be a reduced homogeneous polynomial defining a projective curve having only one point P in the infinity line and let $\widetilde{f}(u, w) \in k[u, \omega]$

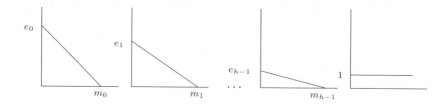

Figure 2.1

a local equation for the germ of curve at P. If it is possible to carry out all the steps of the preceding algorithm for $\tilde{f}(u, w)$ and the Newton polygons so obtained are segments, then $\tilde{f}$ is an irreducible power series, that is, the curve $F = 0$ has only one branch at P.

2.2   Approximants

Definition 2-2. Let k be a field of arbitrary characteristic $p \geq 0$. A curve C defined over k with only one branch at infinity is a projective plane curve defined over k, absolutely irreducible (that is, irreducible as a curve defined over the algebraic closure $\bar{k}$ of k) and with resolution defined over k (that is, all infinitely near points obtained in a resolution of C as a curve over $\bar{k}$ have to be defined over k) such that there exists a line L for which $L \cap C$ is only one point P and C has only one analytic branch at P. For the sake of simplicity, we will choose coordinates $(X_0 \colon X_1 \colon X_2)$ in such a way that L is the line at infinity $X_2 = 0$ and $P = (1 \colon 0 \colon 0)$.

Definition 2-3. Let $f(x, y)$ be a monic polynomial of y over $k[x]$ defining a curve C with only one branch at infinity and let v be its associated valuation. Let $\tilde{f}(u, w)$ be the local expression of C around the infinity point (where $u = y/x, w = 1/x$) and let us consider its sequence of Newton polygons (figure 2.1). We call approximants of $f(x, y)$ to a sequence of polynomials $g_0(x, y)$, $g_1(x, y)$, $\ldots, g_{h+1}(x, y)$ such that

i) $g_0(x, y) = x$.

ii) For $r = 1, \ldots, h + 1$, $g_r(x, y)$ is a monic polynomial of y of degree $\bar{\delta}_0/d_r$ where
$$\bar{\delta}_0 = -v\big(g_0(x, y)\big) = \deg_y f(x, y),$$
$$\bar{\delta}_i = -v\big(g_i(x, y)\big),$$
$$d_r = \gcd\big(\bar{\delta}_0, \ldots, \bar{\delta}_{r-1}\big).$$

iii) For $r = 1, \ldots, h + 1$ the sequence of Newton polygons for the local expression $\tilde{g}_r(u, w)$ of the curve $g_r(x, y) = 0$ around infinity is shown in figure 2.2, where $\gamma_r > m_r/d_r$ and the transformations from one polygon

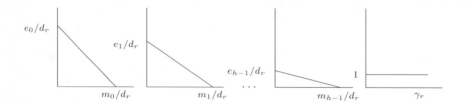

Figure 2.2

to the next one are the same as for $\widetilde{f}(u,w)$. That is, $\widetilde{g}_r(u,w)$ has maximal contact of genus r with $\widetilde{f}(u,w)$ (see [7], chapter 4).

Remark 2-4. Note that the i-th approximant $g_i(x,y)$ defines a curve which is the projection of a germ of curve transversal to the exceptional divisor placed at the level of the i-th terminal free point (see [7], chapter 3 and [8], section 9).

Remark 2-5. When $p = \operatorname{char} k$ does not divide $\overline{\delta}_0 = \deg_y f(x,y)$, approximate roots can be defined for $f(x,y)$ (see [5], chapter II, sections 6, 7). Approximate roots are approximants for $f(x,y)$ in the sense of definition 2-3. Therefore, the existence of approximants is guaranteed when p does not divide $\overline{\delta}_0$.

Let us consider a curve C with only one branch at infinity and take co-ordinates as in definition 2-2. In this situation, we will denote by R the affine k-algebra for the chart $X_2 \neq 0$ of C, by $\widetilde{R}$ the normalization of R, by $\mathcal{O} = \mathcal{O}_{C,P}$, the local ring of C at P and by $K(C)$ the field of rational functions over k.

The only branch at infinity corresponds to a valuation v of $K(C)$, that is, the valuation associated to the only valuation ring R_v satisfying $R \not\subset R_v$ and R_v dominates $\mathcal{O}$. From this situation, three semigroups can be considered: the semigroup of values of C at P given by

$$S(P) = v\big(\mathcal{O} - \{0\}\big),$$

the Weierstrass semigroup of P

$$\Gamma(P) = \big\{-v(g) \mid g \in \widetilde{R}\big\},$$

that is, Γ is the set of orders in P of the rational functions defined on the normalization $\widetilde{C}$ of the curve which are regular outside P, and

$$\Gamma_1(P) = \big\{-v(g) \mid g \in R\big\}.$$

Obviously $\Gamma_1(P)$ is contained in $\Gamma(P)$ and they are equal if and only if $R = \widetilde{R}$, that is, if and only if $C - P$ is an smooth affine curve. Properties of these semigroups are extensively studied in the literature ([2], [5], [6], [7], [16], [17]).

Now, suppose $f(x, y)$ has a sequence of approximants $\{g_i(x, y)\}_{i=0,\dots,h+1}$, and assume the notation in 2-3. We can calculate $\Gamma_1(P)$ for the curve C in terms of the g_i's. In fact, the following properties are deduced from [5]:

Result 2-6. [[5], section 7.2] Let

$$\mathcal{M} = \left\{ (b_0, \dots, b_h) \;\middle|\; 0 \le b_i \le \frac{d_i}{d_{i+1}} \text{ for } 1 \le i \le h \text{ and } b_0 \ge 0 \right\}.$$

Then, every element of $R = k[x, y]/(f)$ can be uniquely expressed as k-linear combination of

$$g^{\underline{b}} = g_0{}^{b_0} \cdot g_1{}^{b_1} \cdots g_h{}^{b_h},$$

where $\underline{b} \in \mathcal{M}$.

Result 2-7. We have

$$\Gamma_1(P) = \langle \bar{\delta}_0, \bar{\delta}_1, \dots, \bar{\delta}_h \rangle.$$

Moreover, the sequence of Newton polygons for the curve $f = 0$ can be obtained from the $\bar{\delta}$'s as follows.

(a) If $\bar{\delta}_0 - \bar{\delta}_1$ does not divide $\bar{\delta}_0$ (see figure 2.3).

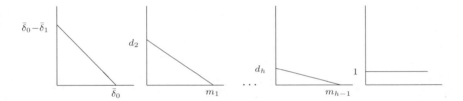

Figure 2.3

(b) If $\bar{\delta}_0 - \bar{\delta}_1$ divides $\bar{\delta}_0$ (see figure 2.4),

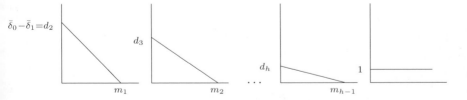

Figure 2.4

where

$$m_i = \frac{d_i \bar{\delta}_i - d_{i+1} \bar{\delta}_{i+1}}{d_{i+1}}, \qquad 1 \le i \le h - 1,$$

$$m_1' = \frac{\bar{\delta}_0^{\,2} - d_2 \bar{\delta}_2}{d_2}.$$

PROOF. [16], Section 1, Lemma 2, assures that if $\underline{b}, \underline{b}' \in \mathcal{M}$ and $\underline{b} \neq \underline{b}'$ then $b_0 \delta_0 + \cdots + b_h \delta_h \neq b_0' \delta_0 + \cdots + b_h' \delta_h$ and hence $v\left(g^{\underline{b}}\right) \neq v\left(g^{\underline{b}'}\right)$. Therefore,

$$\Gamma_1(P) = \left\{ -v\left(g^{\underline{b}}\right) \mid \underline{b} \in \mathcal{M} \right\},$$

and it is generated by $\left\{ -v\left(g_i(x,y)\right) \right\}_{i=0,\ldots,h} = \left\{ \bar{\delta}_i \right\}_{i=0,\ldots,h}$.

Now, let us suppose $\bar{\delta}_0 - \bar{\delta}_1$ does not divide $\bar{\delta}_0$. Then, from the definition of $\bar{\delta}_0$ and $\bar{\delta}_1$, it follows that for the first Newton polygon we have $m_0 = \bar{\delta}_0$ and $e_0 = \bar{\delta}_0 - \bar{\delta}_1$. Suppose we have proved the $(r-1)$-th first polygons are the desired ones then, for the r-th, we have

$$e_{r-1} = \gcd\left(e_{r-2}, m_{r-2}\right) = \gcd\left(d_{r-1}, \bar{\delta}_r\right) = d_r.$$

In a neighbourhood of P we have $x = 1/w$, $y = u/w$ and

$$g_r(x, y) = w^{-\frac{\bar{\delta}_0}{d_i}} \tilde{g}_r(u, w).$$

Since we know the sequence of Newton polygons for $g_r(x, y)$, it follows

$$\begin{aligned}
v\left(\tilde{g}_r(u, w)\right) &= \frac{m_0 e_0}{e_1 e_{r-1}} e_1 + \frac{m_1 e_1}{e_2 e_{r-1}} e_2 + \cdots + \frac{m_{r-2} e_{r-2}}{e_{r-1} e_{r-1}} e_{r-1} + m_{r-1} \\
&= \frac{1}{e_{r-1}} \left(m_0 e_0 + \cdots + m_{r-2} e_{r-2}\right) + m_{r-1} \\
&= \frac{\bar{\delta}_0^{\,2} - d_{r-1} \bar{\delta}_{r-1}}{d_r} + m_{r-1},
\end{aligned}$$

and

$$\bar{\delta}_r = -v\left(g_r(x, y)\right) = \frac{\bar{\delta}_0^{\,2}}{d_r} - v\left(\tilde{g}_r(u, w)\right).$$

Therefore

$$m_{r-1} = \frac{d_{r-1} \bar{\delta}_{r-1}}{d_r} - \bar{\delta}_r$$

and (a) is proved. The reasoning is the same when $\bar{\delta}_0 - \bar{\delta}_1$ divides $\bar{\delta}_0$. $\qquad\square$

3 Curves associated to a semigroup

Let us fix an arbitrary field k and a semigroup $\Gamma = \langle \delta_0, \delta_1, \dots, \delta_h \rangle$ satisfying 1, 2 and 3 (see introduction). We want to study all plane curves of degree δ_0 having only one branch at infinity, such that $\Gamma_1(P)$ is exactly Γ and having approximants. For $p = \operatorname{char} k$ not dividing d_2, the existence of such curves is assured ([17], section 2 and [16], section 4).

We are going to classify these curves by asking their sequence of Newton's polygons and the transformations of the algoritm in Section 2.1 to the be same. This notion can be expressed in terms of clusters. Therefore, we start defining the notion of cluster and specifying some properties on the theory of clusters and complete ideals in regular surfaces. For more information on this subject see [8],[10], [11], [12], [13], [14], [18].

Definition 3-1. *Let S be a regular surface, let P be a point in S and $\pi\colon S' \to S$ the blowing up of P. The exceptional locus $E = \pi^{-1}(P)$ of π will be called the first infinitesimal neighbourhood of P and its points will be called points in the first neighbourhood of P.*

Since S' is also a regular surface, we may define inductively: for $i > 1$, the points in the i-th infinitesimal neighbourhood of P are the points in the $(i-1)$-th infinitesimal neighbourhood of some point in the first infinitesimal neighbourhood of P. The points in some neighbourhood of P are called points infinitely near P (see [9], libro IV, chapters I, II and [8], Section 1). The points infinitely near P are partially ordered in a natural way: We say that Q precedes Q', if and only if Q' is infinitely near Q.

Given a point P on a regular surface S, we define the points proximate to P to be the points infinitely near P which belong, as ordinary or infinitely near points, to the first neigbourhood E of P. Thus, the points proximate to P are all points in E and furthermore, all points in the i-th neigbourhood of P $(i > 1)$ lying on the corresponding strict transform of E.

Definition 3-2. *Let P be a point in a regular surface S. A cluster with origin at P is a finite set of points P_τ infinitely near P, P itself included, each with an assigned integral multiplicity ν_τ. The multiplicity ν_τ is called virtual multiplicity of P_τ and we may assume, by adding points with multiplicity zero if necessary, that all points preceding a point in the cluster are also in the cluster. The depth of the cluster is the greatest i for which the i-th neighbourbood of the origin contains points of the cluster.*

Let S be a regular surface and $\pi\colon S' \to S$ be the blowing up of a point $P \in S$, let E be the exceptional locus of π and ν an integer. Given an algebroid curve C with origin at P we say that C goes through the depth-zero cluster $K = (P, \nu)$ if and only if $e_P(C) \geq \nu$, being $e_P(C)$ the multiplicity of C at P, and in this case we define the virtual transform of C when P is taken with virtual multiplicity ν to be the curve $\pi^*C - \nu E$.

Now, let K be a cluster with origin at P. Assume that K has positive depth and denote by $P_{11}, \ldots, P_{1s}$ the points of K in the first neighbourhood of P. For each $j = 1, \ldots, s$ denote by K_j the cluster with origin at P_{1j} constructed by taking the points of K infinitely near P_{1j}, each one with the same virtual multiplicity as in K. We call these clusters the clusters induced by K with origin at the points of K in the first neighbourhood of P.

Definition 3-3. *We say that a curve C goes through the cluster K if and only if:*

i) *C goes through the cluster (P, ν_P), that is, $e_P(C) \geq \nu_P$ where ν_P is the virtual multiplicity of P.*

ii) *The virtual transform of C when P is taken with virtual multiplicity ν_P goes through the clusters $K_1, \ldots, K_s$ induced by K with origin at the points of K in the first neigbourhood of P.*

To avoid confusions, the multiplicities of a curve C at the points of K are called effective multiplicities. The curves going through K may have at some points effective multiplicities smaller than the virtual ones.

Definition 3-4. *A cluster K satisfies the proximity relations if and only if for each point P_τ in K, if $P_{\tau_1}, \ldots, P_{\tau_r}$ are the points of K proximate to P_τ, then we have $\nu_\tau \geq \sum_{i=1}^r \nu_{\tau_i}$ (proximity relation at P_τ), empty sums being taken equal to zero.*

Definition 3-5. *Related to a cluster K with origin at P in a regular surface S, we define a $\mathfrak{m}_P$-primary ideal I_K of $\mathcal{O}_{S,P}$ (where $\mathfrak{m}_P$ is the maximal ideal of $\mathcal{O}_{S,P}$) consisting of the equations of all the curves which go through K, that is,*

$$I_K = \{0\} \cup \{h \in \mathcal{O}_{S,P} \mid \text{ the curve } h = 0 \text{ goes through } K\}.$$

Now, the cluster K satisfies the proximity relations if and only if the generic element of the ideal I_K defines a curve which goes through K with effective multiplicities equal to the virtual ones. The elements of I_K satisfying the above property are called general.

The correspondence $K \to I_K$ defines a bijection between clusters with origin at P satisfying the proximity relations and $\mathfrak{m}_P$-primary complete ideals of $\mathcal{O}_{S,P}$, or equivalently, $\mathfrak{m}_P$-primary integrally closed ideals (see [18]). The inverse map associates to an $\mathfrak{m}_P$-primary complete ideal I the unique cluster K_I such that a generic element of I goes through K_I with effective multiplicities equal to the virtual ones.

We can calculate the colength of a complete ideal in terms of its associated cluster.

Result 3-6. [[10], Theorem 5.2] *Let S be a germ of a regular surface with origin at P. Let $\mathfrak{m}_P$ be its maximal ideal and I a complete $\mathfrak{m}_P$-primary ideal with associated cluster $\{\nu_Q\}_{\mathcal{O}_{S,Q}>\mathcal{O}_{S,P}}$. Then $\nu_Q = 0$ for all but finitely many Q*

and

$$\ell_{\mathcal{O}_{S,P}}(\mathcal{O}_{S,P}/I) = \sum_Q [Q\colon P]\,\frac{\nu_Q(\nu_Q+1)}{2},$$

where $\ell_{\mathcal{O}_{S,P}}$ denotes the lenth as $\mathcal{O}_{S,P}$-module and $[Q\colon P]$ denotes the (finite) degree of the residue field extension $\mathcal{O}_{S,Q}/\mathfrak{m}_Q \supset \mathcal{O}_{S,P}/\mathfrak{m}_P$.

(3-7) Now, let C be one of the curves we are interested in. That is, C is a plane curve of degree δ_0, having only one branch at infinity P, such that $\Gamma_1(P)$ is exactly Γ and having approximants. Let us suppose $f(x,y) = 0$ is an equation for C. Whith no loss of generallity we can suppose f is a monic polynomial in y of degree δ_0, where $x = X_0/X_2$, $y = X_1/X_2$ and $(X_0\colon X_1\colon X_2)$ are projective coordinates in $\mathbb{P}_k^2$, so that $P = (1\colon 0\colon 0)$. Let $\{g_i(x,y)\}_{i=0,\dots,\bar{h}}$ be a sequence of approximants of $f(x,y)$. Then, Result 2-7 gives a set of generators $\{\bar{\delta}_i\}_{i=0,\dots,\bar{h}}$ of $\Gamma_1(P)$ satisfying 1, 2 and 3, and such that $\bar{\delta}_0 = \delta_0$. By an inductive reasoning we see that for $\Gamma_1(P)$ to be equal Γ it is necessary that $\bar{h} = h$ and $\bar{\delta}_i = \delta_i$ ($0 \le i \le h$). Therefore, the sequence of Newton polygons for the germ of C at P is as follows.

(a) If $\delta_0 - \delta_1$ does not divide δ_0, figure 3.5.

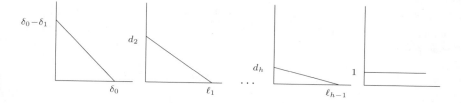

Figure 3.5

(b) If $\delta_0 - \delta_1$ divides δ_0, figure 3.6, where

$$\ell_i = \frac{d_i\delta_i - d_{i+1}\delta_{i+1}}{d_{i+1}}, \qquad i = 1,\dots,h-1.$$

This implies (Result 2-1) that the semigroup of values $S(P)$ of C at P is generated by $r_0,\dots,r_h$ where

$$r_0 = \delta_0 \qquad \text{and} \qquad r_i = \frac{\delta_0^2}{d_i} - \delta_i, \qquad i = 1,\dots,h.$$

(3-8) Let us consider the cluster K with origin at P and satisfying the proximity relations defined by the minimal resolution of the germ of C at P, each point

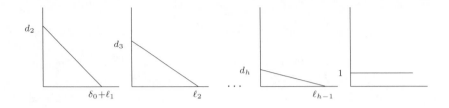

Figure 3.6

with virtual multiplicity equal to the effective multiplicity of C. That is, K is the cluster consisting of all the points infinitely near P which are singular for the strict transform of the germ of C at P, each point with assigned multiplicity the multiplicity of the corresponding strict transform of C, together with some points which are nonsingular for the strict transform of C, with assigned multiplicity 1 (these last points added in order to obtain a cluster satisfying the proximity relations). K is then the unique cluster with origin at P and satisfying the proximity relations for which all general curves going through it have the same sequence of Newton polygons as $\tilde{f}(u, w)$ and the same sequence of transformations from the one polygon to the following one.

Let us note that, since C has only one branch at P, the cluster K consists of exactly one point in the i-th neighbourhood of P for each i less or equal the depth of K. That is why "the sequence of virtual multiplicities of K" has sense in a natural way. Moreover, by [7], proposition 2.2.10 and Chapter 3, this sequence of virtual multiplicities can be calculated as follows: Let

$$
\begin{cases}
\delta_0 = c_0^0(\delta_0 - \delta_1) + n_1^0 \\
\delta_0 - \delta_1 = c_1^0 n_1^0 + n_2^0 \\
\dotfill \\
n_{s_0-1}^0 = c_{s_0}^0 n_{s_0}^0
\end{cases}
\qquad
\begin{cases}
\ell_i = c_0^i d_{i+1} + n_1^i \\
d_{i+1} = c_1^i n_1^i + n_2^i \\
\dotfill \\
n_{s_i-1}^i = c_{s_i}^i n_{s_i}^i
\end{cases}
\tag{3.1}
$$

be the Euclidean algorithms for $\gcd(\delta_0, \delta_0 - \delta_1)$ and $\gcd(\ell_i, d_{i+1})$, $1 \le i \le h-1$ respectively. Then, the c_0^0 first points in K have virtual multiplicity equal to $\delta_0 - \delta_1$, after which there are c_1^0 points with multiplicity $n_1^0, \ldots, c_{s_0}^0$ points with multiplicity $n_{s_0}^0$, c_0^1 points with multiplicity d_2, c_1^1 points with multiplicity $n_1^1, \ldots$ and finally $c_{s_{h-1}}^{h-1}$ points with

Now, let us denote by I the complete ideal I_K associated to K (note that I is an ideal of $\mathcal{O}_{k^2,P}$). Let $k_{\delta_0}[X_0, X_1, X_2]$ be the k-vector space of homogeneous polynomials of degree δ_0 and define the k-vector space

$$
W_{\delta_0}(K) = \{H \in k_{\delta_0}[X_0, X_1, X_2] \mid \text{ the germ of } H \text{ at } P \text{ belongs to } I\}. \tag{3.2}
$$

There is an injective morphism of k-vector spaces

$$
k_{\delta_0}[X_0, X_1, X_2]/W_{\delta_0}(K) \longrightarrow \mathcal{O}_{k^2,P}/I
$$

defined as follows: For $H(X_0, X_1, X_2) \in k_{\delta_0}[X_0, X_1, X_2]$, the image of the element $H(X_0, X_1, X_2) + W_{\delta_0}(K)$ by the above morphism is the (unique) element of $\mathcal{O}_{k^2,P}/I$ defined by the localization of $H(1, u, w)$ at the maximal ideal of $\mathcal{O}_{k^2,P}$.

Therefore, we have

$$\dim W_{\delta_0}(K) \geq \dim k_{\delta_0}[X_0, X_1, X_2] - \ell\left(\mathcal{O}_{k^2,P}/I\right).$$

Using Result 3-6, for $\delta_0 - \delta_1$ not dividing δ_0, we obtain

$$2\ell(\mathcal{O}_{k^2,P}/I) = (\delta_0 - \delta_1)(\delta_0 - \delta_1 + 1)c_0^0 + n_1^0(n_1^0 + 1)c_1^0 + \cdots + n_{s_0}^0(n_{s_0}^0 + 1)c_{s_0}^0$$
$$+ d_2(d_2 + 1)c_0^1 + n_1^1(n_1^1 + 1)c_1^1 + \cdots + n_{s_1}^1(n_{s_1}^1 + 1)c_{s_1}^1$$
$$\cdots\cdots\cdots\cdots\cdots\cdots\cdots\cdots\cdots\cdots\cdots\cdots\cdots\cdots$$
$$+ d_h(d_h + 1)c_0^{h-1} + \cdots + n_{s_{h-1}}^{h-1}\left(n_{s_{h-1}}^{h-1} + 1\right)c_{s_{h-1}}^{h-1}$$
$$= (2\delta_0 - \delta_1 - 1) + (\delta_0^2 - d_h\delta_h) + \sum_{i=1}^{h-1}\ell_i$$

where we set $n_0 = 1$, and hence,

$$\dim W_{\delta_0}(K) \geq \frac{(\delta_0 + 1)(\delta_0 + 2)}{2} - \ell(\mathcal{O}_{k^2,P}/I)$$
$$= \frac{1}{2}\left[3 + 2\delta_0 + d_h\delta_h - \sum_{i=0}^{h-1}(n_i\delta_i - \delta_{i+1})\right]. \tag{3.3}$$

In fact, we can estimate a basis of $W_{\delta_0}(K)$ using the ideas of result 2-6.

Lemma 3-9. Let $f(x,y) = 0$ a plane curve having only one branch at infinity, such that $\Gamma_1(P) = \langle \delta_0, \ldots, \delta_h \rangle$ and having a sequence of approximants $\{g_i(x,y)\}_{i=0,\ldots,h}$. Let us consider the set $B = \{(b_0, \ldots, b_h, 0) \in \mathbb{N}^{h+2}$ so that $0 \leq b_i < n_i$, for $1 \leq i \leq h$, and $b_0 \leq \delta_0 - b_1 - b_2(\delta_0)/d_2 - \cdots - b_h(\delta_0/d_h)\} \cup \{(0, \ldots, 0, 1)\}$ and for any $(h+2)$-uple $\underline{b} \in B$ let

$$g^{\underline{b}} = g_0{}^{b_0} g_1{}^{b_1} \cdots g_h{}^{b_h} g_{h+1}{}^{b_{h+1}} \in k[x,y].$$

Then $\{g^{\underline{b}}/\underline{b} \in B\}$ is a basis of the k-vector space $k_{\delta_0}[x,y]$ of polynomials in x, y of degree smaller or equal δ_0.

Moreover, if $G_i(X_0, X_1, X_2) \in k_{\delta_0}[X_0, X_1, X_2]$ is the homogenization of $g_i(x,y)$ and we consider the set

$$B' = \left\{(\underline{b}, b_{h+2}) \in \mathbb{N}^{h+3} \,\middle|\, \underline{b} \in B, \, b_0 + b_1\frac{\delta_0}{d_1} + \cdots \right.$$
$$\left. + b_h\frac{\delta_0}{d_h} + b_{h+1}\delta_0 + b_{h+2} = \delta_0\right\}$$

then, $\{G^{\underline{b}}/\underline{b} \in B'\}$ is a basis of the k-vector space $k_{\delta_0}[X_0, X_1, X_2]$.

PROOF. Obviously, the second assertion follows from the first one. Besides, since the polynomials $g_i(x, y)$ are monic in y of degree δ_0/d_i, Proposition 1 in [16], Section 2, assures that $\{g^{\underline{b}} \mid \underline{b} = (0, b_1, \ldots, b_{h+1}) \in \mathbb{N}^{h+2}, 0 \leq b_i \leq n_i\}$ is a basis of the $k[x]$-module $k[x, y]$. By the other hand, $g_0(x, y) = x$ and hence $k\{g^{\underline{b}} \mid \underline{b} = (b_0, b_1, \ldots, b_{h+1}) \in \mathbb{N}^{h+2}, 0 \leq b_i \leq n_i \text{ for } 1 \leq i \leq h\}$ is a basis of $k[x, y]$ as k-vector space. Now, for each $\underline{b} = (b_0, \ldots, b_{h+1}) \in \mathbb{N}^{h+2}$, the degree of $g^{\underline{b}}$ is

$$\deg g^{\underline{b}} = b_0 + b_1 \frac{\delta_0}{d_1} + \cdots + b_h \frac{\delta_0}{d_h} + b_{h+1}\delta_0 \tag{3.4}$$

(see condition (iii) in definition 2-3) and therefore, lemma 3-9 follows. □

Proposition 3-10. *We keep on the notation above. Then, a basis of $W_{\delta_0}(K)$ is defined by the polynomials $G^{\underline{b}}$ where $\underline{b} = (b_0, \ldots, b_{h+2}) \in B'$ satisfies*

$$b_0\delta_0 + b_1\delta_1 + \cdots + b_h\delta_h + b_{h+1}d_h\delta_h \leq d_h\delta_h. \tag{3.5}$$

Moreover, suppose we have

$$d_h\delta_h = a_{h0}\delta_0 + a_{h1}\delta_1 + \cdots + a_{h\ h-1}\delta_{h-1},$$

where $a_{h0} \geq 0$, $0 \leq a_{hj} < n_j$ $(1 \leq j \leq h-1)$, and let $G^{\underline{a}} = G_0^{a_{h0}} \ldots G_{h-1}^{a_{h\ h-1}} X_2^{\beta}$, where

$$\beta = \delta_0 - a_{h0} - a_{h1}\frac{\delta_0}{d_1} - \cdots - a_{h\ h-1}\frac{\delta_0}{d_{h-1}}.$$

Then, for a dense set of $(\alpha, \beta) \in k^* \times k$ and for any linear combination H' of those and $G^{\underline{b}}$ satisfying (3.5) with $G^{\underline{b}} \neq F$ and $G^{\underline{b}} \neq G^{\underline{a}}$, the germ at P of $H = \alpha F + \beta G^{\underline{a}} + H' \in W_{\delta_0}(K)$ defines a general element of the ideal I_K.

PROOF. Suppose $\delta_0 - \delta_1$ does not divide δ_0. In the other case the reasoning is the same. Fist, we will prove that (3.5) is a necessary and sufficient condition for $G^{\underline{b}}$ to belong to $W_{\delta_0}(K)$. Take one of them, say $M = G^{\underline{b}}$ with $\underline{b} \in B'$ and $\underline{b} \neq 0$, and let $\widetilde{M}(u, w)$ be the polynomial defined by M in $k[u, w]$, being $u = X_1/X_0$, $w = X_2/X_0$. Let $\widetilde{M}^{(i)}$ be the i-th transform of M by the transformations defined to reach the i-th Newton polygon of $\tilde{f}$ from the $(i-1)$-th one, and let us denote by $P_i(M)$ the Newton polygon of $\widetilde{M}^{(i)}$. In the same way we define $\widetilde{G}_j = \tilde{g}_j, \tilde{g}_j^{(i)}$ and $P_i(\tilde{g}_j)$.

If $b_{h+1} = 1$, then $\widetilde{M} = \tilde{g}_{h+1} = \tilde{f}$ and $M = G_{h+1}$, and hence $P_i(\widetilde{M}) = P_i(\tilde{f})$ for $i = 1, \ldots, h+1$. So, let us suppose $b_{h+1} = 0$. Then, we have

$$\widetilde{M} = u^{b_1} \tilde{g}_2^{b_2} \cdots \tilde{g}_h^{b_h} w^{\beta_1},$$

where

$$\beta_1 = \delta_0 - b_0 - b_1 \frac{\delta_0}{d_1} - \cdots - b_h \frac{\delta_0}{d_h},$$

and the virtual transform of $\widetilde{M}$ after applying the corresponding b1-transformations is

$$\widetilde{M}^{(1)} = u^{b_1} \widetilde{g}_2^{(1)^{b_2}} \cdots \widetilde{g}_h^{(1)^{b_h}} \left(w + \phi_1(u) \right)^{\beta_1},$$

where $\phi_1(u) \in k[u]$ and $\phi_1(0) = 0$.

Besides, for $i \geq 2$, the Newton polygon $P(\widetilde{g}_i^{(1)})$ is the segment with extremes $(\delta_0/d_i, 0)$ and $(0, \delta_0 - \delta_1/d_i)$ and hence $P(\widetilde{g}_i^{(1)})$ is parallel to $P_1(\widetilde{f})$. Therefore, the polygon $P_1(\widetilde{M})$ is above $P_1(\widetilde{f})$ if and only if the point $(b_1 + b_2(\delta_0/d_2) + \cdots + b_h(\delta_0/d_h), \beta_1)$ is above $P_1(\widetilde{f})$, that is, if and only if

$$\frac{b_1}{d_1} + \frac{b_2}{d_2} + \cdots + \frac{b_h}{d_h} + \frac{\beta_1}{\delta_0 - \delta_1} \geq 1,$$

or equivalently

$$b_0 \delta_0 + b_1 \delta_1 + d_1 \delta_1 \left(\frac{b_2}{d_2} + \cdots + \frac{b_h}{d_h} \right) \leq d_1 \delta_1. \tag{3.6}$$

Now, the transformation applied to reach the second Newton polygon of $\widetilde{f}$ from the first one is defined by

$$u \longmapsto u^{\frac{\delta_0 - \delta_1}{d_2}} \left(w + \varphi_2(u) \right)^{\tau},$$

$$w \longmapsto u^{\frac{\delta_0}{d_2}} \left(w + \varphi_2(u) \right)^{\sigma},$$

where $|\tau \delta_0 - \sigma(\delta_0 - \delta_1)| = d_2$, $0 \leq 2\tau \leq \delta_0 - \delta_1/d_2$, $0 \leq 2\sigma \leq \delta_0/d_2$ and $\varphi_2(u) \in k[u]$ satisfies $\varphi_2(0) \neq 0$. Therefore, the virtual transform $\widetilde{M}^{(2)}$ of M after applying this first transformation is

$$\widetilde{M}^{(2)} = u^{\beta_2} w^{b_2} \widetilde{g}_3^{(2)^{b_3}} \cdots \widetilde{g}_h^{(2)^{b_h}} + R_2,$$

where

$$\beta_2 = \frac{\delta_0}{d_2} \left[(\delta_0 - \delta_1) \left(\frac{b_1}{d_1} + \frac{b_2}{d_2} + \cdots + \frac{b_h}{d_h} - 1 \right) + \beta_1 \right],$$

and $P(R_2)$ is above $P(\widetilde{M}^{(2)} - R_2)$. Again, for $i \geq 3$, the Newton polygon $P(\widetilde{g}_i^{(2)})$ is parallel to $P_2(\widetilde{f})$ and hence the Newton polygon $P_2(\widetilde{M})$ is above $P_2(\widetilde{f})$ if and only if

$$\frac{\beta_2}{\ell_1} + \frac{b_2}{d_2} + \frac{b_3}{d_3} + \cdots + \frac{b_h}{d_h} \geq 1,$$

or equivalently

$$b_0 \delta_0 + b_1 \delta_1 + b_2 \delta_2 + d_2 \delta_2 \left(\frac{b_3}{d_3} + \cdots + \frac{b_h}{d_h} \right) \leq d_2 \delta_2. \tag{3.7}$$

Let us suppose we have imposed $k - 1$ conditions relative to $P_1(\widetilde{M})$, $\ldots$, $P_{k-1}(\widetilde{M})$, then we have

$$\widetilde{M}^{(k)} = u^{\beta_k} w^{b_k} \widetilde{g}_{k+1}^{(k)^{b_{k+1}}} \cdots g_h^{(k)^{b_h}} + R_k,$$

where

$$\beta_k = n_{k-1} \left[\ell_{k-2} \left(\frac{b_{k-1}}{d_{k-1}} + \cdots + \frac{b_h}{d_h} - 1 \right) + \beta_{k-1} \right], \qquad (3.8)$$

and $P(R_k)$ is above $P(M^{(k)} - R_k)$. Since $P(\widetilde{g}_i^{(k)})$ is parallel to $P_k(\widetilde{f})$ for $i > k$, and

$$d_k \beta_k = -b_0 \delta_0 - b_1 \delta_1 - \cdots - d_{k-1}\delta_{k-1} \left(\frac{b_{k-1}}{d_{k-1}} + \cdots + \frac{b_h}{d_h} - 1 \right),$$

we obtain that the Newton polygon $P_k(\widetilde{M})$ is above $P_k(\widetilde{f})$ if and only if

$$\frac{\beta_k}{\ell_{k-1}} + \frac{b_k}{d_k} + \cdots + \frac{b_h}{d_h} \geq 1,$$

or equivalenty

$$b_0 \delta_0 + \cdots + b_k \delta_k + d_k \delta_k \left(\frac{b_{k+1}}{d_{k+1}} + \cdots + \frac{b_h}{d_h} \right) \leq d_k \delta_k.$$

Therefore, M belongs to $W_{\delta_0}(K)$, that is, the polygons $P_k(\widetilde{M})$ are above the polygons $P_k(\widetilde{f})$, if and only if

$$b_0 \delta_0 + \cdots + b_k \delta_k \leq d_k \delta_k \left(1 - \frac{b_{k+1}}{d_{k+1}} - \cdots - \frac{b_h}{d_h} \right) \qquad (3.9)$$

for $k = 1, \ldots, h$. In particular, the condition

$$b_0 \delta_0 + \cdots + b_h \delta_h \leq d_h \delta_h \qquad (3.10)$$

implies the others. In fact, let us suppose (3.10) holds, then, since $b_h < d_h$, we have

$$b_0 \delta_0 + \cdots + b_{h-1}\delta_{h-1} \leq \delta_h(d_h - b_h)$$
$$< \frac{d_{h-1}}{d_h}\delta_{h-1}(d_h - b_h)$$
$$= \delta_{h-1}d_{h-1} \left(1 - \frac{b_h}{d_h} \right),$$

and, after $h - k - 1$ iterations of this argument, we reach inequality (3.9). Moreover, (3.10) implies strict inequality in (3.9) for $k = 1, \ldots, h - 1$.

Now, let $H(X_0, X_1, X_2)$ be a polynomial in $k_{\delta_0}[X_0, X_1, X_2]$. After 3-9, we know H can be written in a unique way as

$$H = \sum_{\underline{b} \in B_0} \lambda_{\underline{b}} \, \underline{G}^{\underline{b}} + \sum_{\underline{c} \in B_1} \lambda_{\underline{c}} \, \underline{G}^{\underline{c}},$$

where the λ's are in k, B_0 and B_1 are contained in B', the elemens of B_0 satisfy (3.5) and the elements of B_1 do not. For $\{\underline{G}^{\underline{b}} \mid \underline{b} \in B' \text{ satisfies (3.5)}\}$ to be a basis of $W_{\delta_0}(K)$, the only thing we have to prove is that the assertion

$$\sum_{\underline{c} \in B_1} \lambda_{\underline{c}} \, G^{\underline{c}} \in W_{\delta_0}(K) \tag{3.11}$$

implies all $\lambda_{\underline{c}}$ are zero. It follows from (3.11) that, at least for two of the $G^{\underline{c}}$, say $G^{\underline{c}}$ and $G^{\underline{c}'}$, the numbers β_k (depending on $\underline{c}$) defined in (3.8) are equal (although they may not be possitive). This implies $\underline{c} = \underline{c}'$, contradicting the fact that all $G^{\underline{c}}$ are different. Therefore, the first assertion in 3-10 is proved.

To prove the second assertion, first note that, since 1, 2 and 3 hold, $d_h \delta_h$ can be written in a unique way as

$$d_h \delta_h = a_{h\,0} \delta_0 + a_{h\,1} \delta_1 + \cdots + a_{h\,h-1} \delta_{h-1},$$

where $a_{h0} \geq 0$ and $0 \leq a_{hj} < n_j$ for $j = 1, \ldots h-1$ (see [16], Section 2, Lemma 1) and in this case we have

$$\beta = \delta_0 - a_{h\,0} - a_{h\,1} \frac{\delta_0}{d_1} - \cdots - a_{h\,h-1} \frac{\delta_0}{d_{h-1}} \geq 0.$$

Now, the germ of H at P defines a general element of I_K if and only if the sequence of Newton polygons for $\widetilde{H}$ is exactly the same as for $\widetilde{f}$. Since inequality (3.10) implies strict inequality in (3.9) for $k = 1, \ldots, h-1$, the only $G^{\underline{b}} \in W_{\delta_0}(K)$ different to F (that is, $b_{h+1} = 0$) for which some of the equalities (3.5) hold is $G^{\underline{a}} = G_0^{a_{h0}} G_1^{a_{h1}} \ldots G_{h-1}^{a_{h\,h-1}} X^\beta$. Therefore, for a dense set of $(\alpha, \beta) \in k^* \times k$ and for any linear combination of the $G^{\underline{b}}$ satisfying (3.5) with $G^{\underline{b}} \neq F$ and $G^{\underline{b}} \neq G^{\underline{a}}$, the germ at P of $H = \alpha F + \beta G_{\underline{a}} + H'$ defines a general element of the ideal I_K. $\qquad \square$

Remark 3-11. *The number of $\underline{b}$'s in B' satisfying (3.5) is the dimension of $W_{\delta_0}(K)$. On the other hand, in (3.3) we have given a bound for $\dim W_{\delta_0}(K)$ explicitly in terms of the δ's.*

Now, we will show with some examples that this bound can be reached, but in general is not equal to $\dim W_{\delta_0}(K)$.

First, let us consider the curve given by $f(x, y) = 0$ where $f(x, y) = y^5 - x^3$. For this curve, $\Gamma_1(P) = \langle \delta_0 = 5, \delta_1 = 3 \rangle$ and obviously $\{g_0 = x, g_1 = y\}$ are approximants of $f(x, y)$. The cluster K associated is represented by figure 3.7, where with this figure we mean that the point P_4 corresponding to

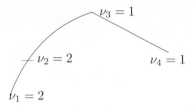

$\nu_3 = 1$

$\nu_2 = 2$ $\nu_4 = 1$

$\nu_1 = 2$

Figure 3.7

ν_4 is proximate to the point P_2 (see [9], libro IV, Chapter I or [8], Section 9). Then, we have

$$\ell(\mathcal{O}_{k^2,P}/I) = \frac{1}{2}(2 \cdot 3 + 2 \cdot 3 + 2 + 2) = 8,$$

and hence, the bound given in (3.3) is

$$\dim, W_{\delta_0}(K) \geq 21 - 8 = 13.$$

On the other hand, proposition 3-10 assures that a basis of $W_{\delta_0}(K)$ is given by $\{X_2^5, X_1X_2^4, X_1^2X_2^3, X_1^3X_2^2, X_1^4X_2, X_0X_2^4, X_0X_1X_2^3, X_0X_1^2X_2^2, X_0X_1^3X_2, X_0^2X_2^3, X_0^2X_1X_2^2, X_0^3X_2^2, F = X_1^5 - X_0^3X_2^2\}$. Therefore $\dim W_{\delta_0}(K) = 13$, which means that in this case the morphism

$$k_{\delta_0}[X_0, X_1, X_2]/W_{\delta_0}(K) \longrightarrow \mathcal{O}_{k^2,P}/I$$

defined in 3-8 is an isomorphism of k-vector spaces.

However, this is not always true. For example, let us consider the curve C given by $f(x,y) = 0$ where $f(x,y) = \left[(y^3 - x)^3 - y^2\right]^5 - y$. The sequence of Newton polygons for the germ of C at P is shown in figure 3.8, and

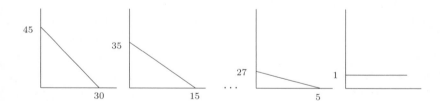

Figure 3.8

$\Gamma_1(P) = \langle \delta_0 = 45, \delta_1 = 15, \delta_2 = 10, \delta_3 = 3 \rangle$ and $\{g_0 = x, g_1 = y, g_2 = y^3 - x, g_3 = (y^3 - x)^3 - y^2\}$ are approximants of $f(x,y)$. Thus, C is a plane curve with only one branch at infinity and having approximants, and the cluster K

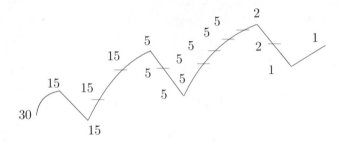

Figure 3.9

associated to C is represented in figure 3.9, where the points proximate to a given Q in the succesive neightbourhoods of a given Q' in the first neighbour-hood of Q (and only these points) are represented on a line segment orthogonal to the segment joing the vertices corresponding to Q and Q' and starting at Q' (see [9] and [8]). Therefore,

$$\ell\left(\mathcal{O}_{k^2,P} \,/\, I\right) = \frac{1}{2}(30 \cdot 31 + 4 \cdot 15 \cdot 16 + 8 \cdot 5 \cdot 6 + 2 \cdot 2 \cdot 3 + 2 \cdot 2) = 1073,$$

and the bound given in (3.3) for this case is

$$\dim W_{\delta_0}(K) \geq 1081 - 1073 = 8.$$

Applying now proposition 3-10, we obtain that a basis of $W_{\delta_0}(K)$ is given by the polynomials $G^{\underline{b}}$ where $G_0 = X_0$, $G_1 = X_1$, $G_2 = X_1^3 - X_2^2$, $G_3 = G_2^3 - X_1^2 X_2^7$, $G_4 = F = G_3^5 - X_1 X_2^{44}$ and $\underline{b} \in \mathbb{N}^6$ satisfies the conditions

1. $45b_0 + 15b_1 + 10b_2 + 3b_3 + 15b_4 \leq 15$.

2. $b_1 < 3$, $b_2 < 3$, $b_3 < 5$.

3. $b_6 = 45 - (b_0 + b_1 + 3b_2 + 9b_3 + 45b_4)$.

$\{X_2^{45}, G_3 X_2^{36}, G_3^2 X_2^{27}, G_3^3 X_2^{18}, G_3^4 X_2^9, G_2 X_2^{42}, G_2 G_3 X_2^{33}, X_1 G_2^{44}, F = G_4\}$ is a basis of $W_{\delta_0}(K)$, and in this case $\dim W_{\delta_0}(K) = 9$ is strictly greater then the bound 8, that is, in this case, the injective morphism

$$k_{\delta_0}[X_0, X_1, X_2]/W_{\delta_0}(K) \longrightarrow \mathcal{O}_{k^2,P}/I$$

is not an isomorphism.

Now, let us reformulate the ideas of this section. Let us consider a field k and the regular surface $\mathbb{P}_k^2$, and let us fix a line L in $\mathbb{P}_k^2$, which will be called line at infinity, a point P in L and a semigroup $\Gamma = \langle \delta_0, \delta_1, \ldots, \delta_h \rangle$ satisfying 1, 2 and 3.

Definition 3-12. *We say that a cluster K with origin at P and satisfying the proximity relations is compatible with Γ if and only if there exists a general element going through K which can be extended to a projective plane curve having only one branch at P and for which the sequence of Newton polygons is exactly the one in figure 3.5.*

Note that we do not care of the transformations from one polygon to the next one. In fact these transformations are the same for all general elements of I_K and different to the transformations of general elements of $I_{K'}$, being $K' \neq K$ another cluster compatible with Γ.

Let K be a cluster with origin at P and compatible with Γ. We define $W_{\delta_0}(K)$ as in (3.2), that is, consisting of all the polynomials in $K_{\delta_0}[X_0, X_1, X_2]$ for which the germ of curve they define at P goes through K. An element $H(X_0, X_1, X_2) \in W_{\delta_0}(K)$ is said to be general if and only if the germ of H at P defines a general element of the ideal I_K.

In this situation we have proved the following result.

Theorem 3-13. *Let K be a cluster with origin at P and compatible with Γ, take coordinates $(X_0\!:\!X_1\!:\!X_2)$ in $\mathbb{P}^2_k$ so that L is defined by $X_2 = 0$ and P is the point $(1\!:\!0\!:\!0)$, and set $x = X_0/X_2$, $y = X_1/X_2$. Let us suppose there is one element F in $W_{\delta_0}(K)$ defining a curve $f(x,y) = 0$ (where $f(x,y) = F(x,y,1)$) with only one branch at infinity and having a sequence of approximants $\{g_i(x,y)\}_{i=0,\ldots,h+1}$. Then, we have:*

(i) *The general elements of $W_{\delta_0}(K)$ define plane curves with only one branch at infinity, having $\{g_i(x,y)\}_{i=0,\ldots,h+1}$ as approximants and such that $\Gamma_1(P) = \langle \delta_0, \delta_1, \ldots, \delta_h \rangle$. Therefore the semigroup of values of any of these curves at P is $S(P) = \langle r_0, \ldots, r_h \rangle$ where*

$$r_0 = \delta_0 \qquad \text{and} \qquad r_i = \frac{\delta_0^2}{d_i} - \delta_i, \qquad 1 \leq i \leq h.$$

(ii) *If one of the general elements of $W_{\delta_0}(K)$ defines a curve smooth in the affine part, then for this curve $\Gamma(P) = \Gamma_1(P) = \langle \delta_0, \delta_1, \ldots, \delta_h \rangle$ and its genus is*

$$g = \frac{1}{2}\left[3 - 2\delta_0 + d_h\delta_h + \sum_{i=0}^{h-1}(n_i\delta_i - \delta_{i+1}) \right].$$

(iii) *A basis of the k-vector space $W_{\delta_0}(K)$ is defined by the polynomials $G^{\underline{b}} = G_0^{b_0} \ldots G_{h+1}^{b_{h+1}} X_2^{b_{h+2}}$ where $G_i \in k_{\delta_0}[X_0, X_1, X_2]$ is the homogenization of $g_i(x,y)$ and*

$$b_0\delta_0 + b_1\delta_1 + \cdots + b_h\delta_h + b_{h+1}d_h\delta_h \;\leq\; d_h\delta_h,$$
$$b_0 + b_1\frac{\delta_0}{d_1} + \cdots + b_h\frac{\delta_0}{d_h} + b_{h+1}\delta_0 + b_{h+2} \;=\; \delta_0,$$

and $0 \leq b_i < n_i$ for $1 \leq i \leq h$. Moreover, if

$$d_h \delta_h = a_{h0} \delta_0 + a_{h1} \delta_1 + \cdots + a_{h\,h-1} \delta_{h-1},$$

where $a_{hj} \in \mathbb{N}$ and $a_{hj} < n_j$ for $1 \leq j \leq h-1$ and we set $G^{\underline{a}} = G_0^{a_{h0}} \ldots G_{h-1}^{a_{h\,h-1}} X_2^{\beta}$ then, for a dense set of $(\alpha, \beta) \in k^* \times k$ and for any linear combination H' of the $G^{\underline{b}}$ above with $G^{\underline{b}} \neq F$ and $G^{\underline{b}} \neq G^{\underline{a}}$, the polynomial $H = \alpha F + \beta G^{\underline{a}} + H'$ is a general element of $\in W_{\delta_0}(K)$.

PROOF. Since the cluster K is compatible with Γ, a general element of $W_{\delta_0}(K)$ defines a projective plane curve C such that the sequence of Newton polygons defined by the germ of C at P is the one in figure 3.5. Therefore C has only one branch at infinity, and (i) and the first assertion in (ii) follow from this sequence of Newton polygons. Now, let us compute the genus equallity in (ii). Each of the general elements of $W_{\delta_0}(K)$ defines a plane curve of degree δ_0. The genus formula applied to any of these curves, say C, gives

$$g = \frac{(\delta_0 - 1)(\delta_0 - 2)}{2} - \sum_{Q \in \operatorname{Sing} C} \delta(Q), \quad \text{where } \delta(Q) = \ell_{\mathcal{O}_{CQ}}(\mathcal{O}_{CQ}/\mathcal{C}_Q),$$

and $\mathcal{C}$ is the conductor sheaf. The cluster defined by $\mathcal{C}_Q$ has the same set of points as the cluster defined by the germ of C at Q and the weights are the weights of C minus one. Therefore, using Result 3-6 and figure 3.5 we can calculate $\delta(P)$ for any of the curves in $W_{\delta_0}(K)$

$$\delta(P) = \frac{1}{2} \left[1 - \delta_0 + (\delta_0^2 - d_h \delta_h) + \sum_{i=0}^{h-1} (n_i \delta_i - \delta_{i+1}) \right],$$

this formula being the same when $\delta_0 - \delta_1$ divides δ_0 and when it does not.

Now, if the curve is smooth in the affine part, then

$$g = \frac{(\delta_0 - 1)(\delta_0 - 2)}{2} - \delta(P) = \frac{1}{2} \left[1 - 2\delta_0 + d_h \delta_h + \sum_{i=0}^{h-1} (n_i \delta_i - \delta_{i+1}) \right],$$

and hence (ii) is proved. Finally, assertion (iii) is nothing but proposition 3-10. $\qquad \square$

When k is an algebraically closed field of characteristic zero, Bertini's theorem assures the existence of curves defined by elements of $W_{\delta_0}(K)$ which are smooth in the affine part. In fact, if $F(X_0, X_1, X_2)$ is an element of $W_{\delta_0}(K)$ and $f(x, y) = F(x, y, 1)$, then the polynomials $\{F + \lambda X_2^{\delta_0}\}_{\lambda \in k}$ are in $W_{\delta_0}(K)$ and hence, for an infinite number of λ's in k the curve $f(x, y) + \lambda = 0$ is smooth in the affine part.

In next section we will prove the existence of curves defined by elements of $W_{\delta_0}(K)$ for specific clusters K compatible with Γ and we will study the smoothness of such curves.

4   A family of examples

Fixed a field k and a semigroup $\Gamma = \langle \delta_0, \delta_1, \ldots, \delta_h \rangle$ satisfying 1, 2 and 3, we are going to construct concrete plane curves with only one branch at infinity P, having a sequence of approximants, and satisfying $\Gamma_1(P) = \Gamma$. Therefore, we will give specific clusters K with origin at P and compatible with Γ satisfying the hypothesis of Theorem 3-13.

Since 1, 2 and 3 hold, for $1 \le i \le h$, $n_i \delta_i$ can be written in a unique way as

$$n_i \delta_i = a_{i0}\delta_0 + a_{i1}\delta_1 + \cdots + a_{i\,i-1}\delta_{i-1}, \tag{4.12}$$

where $a_{i0} \ge 0$ and $0 \le a_{ij} < n_j$ for $j = 1, \ldots, i-1$ (see [16], section 2, lemma 1). Using these a_{ij}, we define elements of $k[x,y]$ as follows: We take $t_1, \ldots, t_h \in k^*$ and define in a recurrent way

$$\begin{cases}
g_0 = x \\
g_1 = y \\
g_2 = g_1^{n_1} - t_1 g_0^{a_{10}} \\
g_3 = g_2^{n_2} - t_2 g_0^{a_{20}} g_1^{a_{21}} \\
\cdots\cdots\cdots\cdots\cdots\cdots\cdots\cdots \\
g_{i+1} = g_i^{n_i} - t_i g_0^{a_{i0}} \cdots g_{i-1}^{a_{i\,i-1}} \\
\cdots\cdots\cdots\cdots\cdots\cdots\cdots\cdots \\
g_{h+1} = g_h^{n_h} - t_h g_0^{a_{h0}} \cdots g_{h-1}^{a_{h\,h-1}}.
\end{cases} \tag{4.13}$$

Let us note that the definition of g_{i+1} depends on $t_1, \ldots, t_i$. We write g_{i+1}^t if we want to refer that dependence. Now, let us determine the degrees of these polynomials.

Proposition 4-1. *Each $g_{i+1} \in k[x,y]$ $(0 \le i \le h)$ is a polynomial whose total degree is equal to its degree in y (that is, the degree of g_{i+1} as a polynomial in y with coefficients in $k[x]$). This degree is*

$$\deg(g_{i+1}) = \deg_y(g_{i+1}) = \frac{\delta_0}{d_{i+1}}, \qquad \text{for } i = 0, \ldots, h.$$

Moreover, the polynomial $g_0^{a_{i0}} \cdots g_{i-1}^{a_{i\,i-1}}$ appearing in the definition of g_{i+1} has a degree strictly less than δ_0/d_{i+1}.

PROOF. Let us prove the result by recurrence on i. For $i = 0$ it is obvious. Let $i > 0$ and define

$$\alpha_1^i(\Gamma) = \frac{\delta_0}{d_{i+1}} - a_{i0} - a_{i1}\frac{\delta_0}{d_1} - a_{i2}\frac{\delta_0}{d_2} - \cdots - a_{i\,i-1}\frac{\delta_0}{d_{i-1}}.$$

By the recurrence hypothesis this is $n_i \deg(g_i) - \deg(g_0^{a_{i0}} \cdots g_{i-1}^{a_{i\,i-1}})$. Let us suppose $\alpha_1^i(\Gamma) > 0$, that is, $\deg(g_0^{a_{i0}} \cdots g_{i-1}^{a_{i\,i-1}}) < \deg(g_i^{n_i})$, then

$$\deg(g_{i+1}) = \deg(g_i^{n_i}) = \frac{\delta_0}{d_{i+1}},$$

and, in the same way,

$$\deg_y(g_{i+1}) = \deg\left(g_i^{n_i}\right) = \frac{\delta_0}{d_{i+1}}.$$

Therefore the proposition is proved once we verify the following Lemma, which will also be systematically used later.   □

Lemma 4-2. *Fixed* $\Gamma = \langle \delta_0, \dots, \delta_h \rangle$ *and* $\{a_{ij}\}_{0 \le j < i \le h}$ *satisfying 1, 2 and 3 and* (4.12) *let us define*

$$\ell_0 = \delta_0 - \delta_1, \quad \ell_i = \frac{d_i \delta_i - d_{i+1}\delta_{i+1}}{d_{i+1}}, \qquad i = 1, \dots, h-1, \tag{4.14}$$

and for $i = 1, \dots, h$ *and* $k = 2, \dots, i$,

$$\begin{cases} \alpha_1^i(\Gamma) = -\delta_0 \left(\dfrac{a_{i0}}{\delta_0} + \dfrac{a_{i1}}{d_1} + \cdots + \dfrac{a_{i\,i-1}}{d_{i-1}} - \dfrac{1}{d_{i+1}} \right), \\[3mm] \alpha_k^i(\Gamma) = n_{k-1} \left[\ell_{k-2} \left(\dfrac{a_{i\,k-1}}{d_{k-1}} + \cdots + \dfrac{a_{i\,i-1}}{d_{i-1}} - \dfrac{1}{d_{i+1}} \right) + \alpha_{k-1}^i(\Gamma) \right]. \end{cases} \tag{4.15}$$

Then $\alpha_k^i(\Gamma)$ *is strictly positive for* $i = 1, \dots, h$ *and* $k = 1, \dots, i$. *Moreover, for* $k = i$, *we have*

$$\alpha_i^i(\Gamma) = \frac{\ell_{i-1}}{d_{i+1}} 5 \tag{4.16}$$

PROOF. The entire $\alpha_k^i = \alpha_k^i(\Gamma)$ is strictly positive if and only if $d_k \alpha_k^i$ is. We have

$$\begin{aligned} d_k \alpha_k^i = {}& d_{k-1}\ell_{k-2} \left(\frac{a_{i\,k-1}}{d_{k-1}} + \cdots + \frac{a_{i\,i-1}}{d_{i-1}} - \frac{1}{d_{i+1}} \right) \\ &+ d_{k-2}\ell_{k-3} \left(\frac{a_{i\,k-2}}{d_{k-2}} + \cdots + \frac{a_{i\,i-1}}{d_{i-1}} - \frac{1}{d_{i+1}} \right) \\ &\cdots\cdots\cdots\cdots\cdots\cdots\cdots\cdots\cdots\cdots\cdots\cdots\cdots \\ &+ d_1 \ell_0 \left(\frac{a_{i1}}{d_1} + \cdots + \frac{a_{i\,i-1}}{d_{i-1}} - \frac{1}{d_{i+1}} \right) \\ &- \delta_0^2 \left(\frac{a_{i0}}{\delta_0} + \cdots + \frac{a_{i\,i-1}}{d_{i-1}} - \frac{1}{d_{i+1}} \right) \\ = {}& \frac{d_{k-1}\delta_{k-1} - d_i \delta_i}{d_{i+1}} + \sum_{k \le r \le i-1} a_{ir} \left(\delta_r - \frac{d_{k-1}}{d_r}\delta_{k-1} \right), \end{aligned}$$

and hence, if $k = i$, then

$$\alpha_i^i = \frac{d_{i-1}\delta_{i-1} - d_i \delta_i}{d_{i+1}} = \frac{\ell_{i-1}}{d_{i+1}}.$$

In the other case, since $d_r \delta_r < d_{k-1} \delta_{k-1}$ for $r \geq k$ and $a_{ij} < n_j$ for $j \geq 1$, we have

$$
\begin{aligned}
d_k \alpha_k^i &> \frac{d_{k-1} \delta_{k-1} - d_i \delta_i}{d_{i+1}} + (n_k - 1)\left(\delta_k - \frac{d_{k-1}}{d_k}\delta_{k-1}\right) + \cdots \\
&\quad + (n_{i-1} - 1)\left(\delta_{i-1} - \frac{d_{k-1}}{d_{i-1}}\delta_{k-1}\right) \\
&= \frac{d_{k-1}\delta_{k-1}}{d_i}(n_i - 1) + \big[\delta_{k-1}n_{k-1} + (n_k - 1)\delta_k + \cdots \\
&\quad + (n_{i-1} - 1)\delta_{i-1} - n_i \delta_i\big] \\
&> \frac{d_{k-1}\delta_{k-1}}{d_i}(n_i - 1) + (1 - n_i)\delta_i = (n_i - 1)\left(\frac{d_{k-1}\delta_{k-1}}{d_i} - \delta_i\right) \\
&> 0.
\end{aligned}
$$

$\square$

Now let us study the polynomials $\{g_i\}_{i=1}^{h+1}$ defined in (4.13). In fact they define curves in $\mathbb{P}_k^2 = \mathrm{Proj}\, k[X_0, X_1, X_2]$ and all these curves pass through the point $P = (1:0:0)$ (due to the second assertion in proposition 4-1, after a recurrent argument). Besides, P is the only point of the intersetion $(g_i = 0) \cap (X_2 = 0)$ for $i = 1, \ldots, h+1$. Let $\{u = X_1/X_0,\ \omega = X_2/X_0\}$ be a coordinate system for the chart $(X_0 \neq 0)$ and let us write equations of those curves in this chart.

$$
\left\{
\begin{aligned}
\tilde{g}_0 &= 1 \\
\tilde{g}_1 &= u \\
\tilde{g}_2 &= \tilde{g}_1^{n_1} - t_1 \omega^{\frac{\delta_0}{d_2} - a_{10}} \\
&\cdots\cdots\cdots\cdots\cdots\cdots\cdots\cdots\cdots\cdots\cdots \\
\tilde{g}_{i+1} &= \tilde{g}_i^{n_i} - t_i\, \tilde{g}_1^{a_{i1}} \cdots \tilde{g}_{i-1}^{a_{i\,i-1}} \omega^{\alpha_1^i} \\
&\cdots\cdots\cdots\cdots\cdots\cdots\cdots\cdots\cdots\cdots\cdots \\
\tilde{g}_{h+1} &= \tilde{g}_h^{n_h} - t_h\, g_1^{a_{h1}} \cdots g_{h-1}^{a_{h\,h-1}} \omega^{\alpha_1^h}.
\end{aligned}
\right.
\tag{4.17}
$$

where the $\alpha_k^i = \alpha_k^i(\Gamma)$ are defined in (4.15).

Proposition 4-3. *Let us suppose $\delta_0 - \delta_1$ does not divide δ_0. Then, there exist transformations $T_1, \ldots, T_h$, each T_i being the composition of a b2-transformation followed by a sequence of b1-transformations, such that, for $j = 2, \ldots, h+1$, the sequence $T_1, \ldots, T_{j-2}$ is the sequence of transformations described in the algorithm of Section 2 for the curve $\tilde{g}_j = 0$, and the Newton polygons so obtained are shown in figure 4.10, where $\gamma_j > \ell_{j-1}/d_j$.*

PROOF. In the course of the proof and for $i = 2, \ldots, h$ (resp. $i = 1$) we call relative i-degree of a monomial $M \in k[u, \omega]$ to the degree of M concerning the graduation for which u has degree d_i (resp. $\delta_0 - \delta_1$) and ω has degree ℓ_{i-1} (resp. δ_0). If $M = u^\alpha \omega^\beta$, then the relative i-degree is greater than $d_i \ell_{i-1}/d_j$ if and only if the point (α, β) is above the i-th Newton polygon P_i^j in the sequence of figure 4.10.

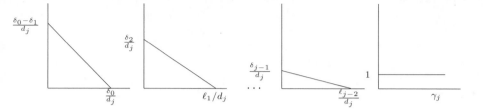

Figure 4.10

Let us prove the proposition by recurrence on j, constructing the suitable transformation in each step. For $j = 2$,

$$\tilde{g}_2 = u^{\frac{\delta_0}{d_2}} - t_1 \omega^{\frac{\delta_0 - \delta_1}{d_2}},$$

and hence the first polygon defined by $\tilde{g}_2$ is the segment with extremes $(\delta_0/d_2, 0)$ and $(0, \delta_0 - \delta_1/d_2)$. Since $\delta_0 - \delta_1$ does not divide δ_0, the transformation obtained from the algorithm 1 is a b2-transformation with parameters (τ, σ) univocally determined by

$$\left| \tau \frac{\delta_0}{d_2} - \sigma \frac{\delta_0 - \delta_1}{d_2} \right| = 1, \qquad 0 \le 2\tau \le \frac{\delta_0 - \delta_1}{d_2}, \qquad 0 \le 2\sigma \le \frac{\delta_0}{d_2}.$$

After this transformation, $\tilde{g}_2$ becomes $w - t_1$ (if $\sigma(\delta_0 - \delta_1) < \tau \delta_0$) or else $1 + t_1 \omega$ (in the other case). Anyway, after a suitable b1-transformation ($\omega' = \omega - \lambda$ for $\lambda \in k^*$) the polygon becomes the point $(0, 1)$. Let T_1 be the composition of the two transformations, then the result is proved for $j = 2$.

Now, let us suppose we have constructed transformations $T_1, T_2, \ldots, T_{i-1}$ in such a way that, for $j \le i$, the sequence of polygons defined by $\tilde{g}_j$ is $\{P_\nu^j\}_{\nu=1}^j$, that is, the one in figure 4.10. Let us denote by $\tilde{g}_j^{(k)}$ the k-transformation of $\tilde{g}_j$ by $T_k \circ \cdots \circ T_2 \circ T_1$ and let

$$Q^{(1)}(u, \omega) = \tilde{g}_1^{a_{i1}} \cdots \tilde{g}_{i-1}^{a_{i\,i-1}} \omega^{\alpha_i^i}$$

in such a way that $\tilde{g}_{i+1} = \tilde{g}_i^{n_i} - Q^{(1)}$.

By recurrence hypothesis, the polygon of $\tilde{g}_i^{n_i}$ is a segment with extremes

$$\left(n_i \frac{\delta_0}{d_i}, 0 \right) = \left(\frac{\delta_0}{d_{i+1}}, 0 \right), \qquad \text{and} \qquad \left(0, n_i \frac{\delta_0 - \delta_1}{d_i} \right) = \left(0, \frac{\delta_0 - \delta_1}{d_{i+1}} \right).$$

Let us show that all the points defined by $Q^{(1)}$ are above this segment, or equivalently, all the monomials appearing in $Q^{(1)}$ have relative i-degree greater than $((\delta_0 - \delta_1)\delta_0) \mid (d_{i+1})$. In fact, let $M = u^c \omega^e$ be one of the monomials appearing in $Q^{(1)}$ with non zero coefficient. By hypothesis, there exist pairs $\{(c_j, e_j)\}_{j=1}^{i-1}$ such that

$$c = a_{i1} + a_{i2} c_2 + \cdots + a_{i\,i-1} c_{i-1},$$
$$e = a_{i2} e_2 + \cdots + a_{i\,i-1} e_{i-1} + \alpha_1,$$

where $\alpha_1 = \alpha_1^i(\Gamma)$ and

$$(\delta_0 - \delta_1)c_j + \delta_0 e_j \geq \frac{\delta_0(\delta_0 - \delta_1)}{d_j}.$$

Then, we have

$$
\begin{aligned}
(\delta_0 - \delta_1)c + \delta_0 e &\geq \left((\delta_0 - \delta_1)a_{i\,1} + \delta_0\alpha_1\right) + \delta_0(\delta_0 - \delta_1)\left(\frac{a_{i2}}{d_2} + \cdots + \frac{a_{i\,i-1}}{d_{i-1}}\right) \\
&= \delta_0\left[(\delta_0 - \delta_1)\left(\frac{a_{i1}}{d_1} + \cdots + \frac{a_{i\,i-1}}{d_{i-1}} - \frac{1}{d_{i+1}}\right) + \alpha_1\right] \\
&\quad + \frac{\delta_0(\delta_0 - \delta_1)}{d_{i+1}} \\
&= \alpha_2 + \frac{\delta_0(\delta_0 - \delta_1)}{d_{i+1}} \\
&> \frac{\delta_0(\delta_0 - \delta_1)}{d_{i+1}}.
\end{aligned}
$$

because $\alpha_2 = \alpha_2^i(\Gamma) \geq 0$ (Lemma 4-2), and therefore the Newton polygon of $\widetilde{g}_{i+1}$ is the segment with extremes $\delta_0/d_{i+1}, 0)$ and $(0, \delta_0 - \delta_1/d_{i+1})$. Since $\delta_0 - \delta_1$ does not divide δ_0, the first transformation defined by the algorithm 1 is exactly T_1.

Now, let us show the Newton polygon of $\widetilde{g}_{i+1}^{(2)}$ is the desired one. A simple calculus shows that if $Q^{(2)}$ is the image of $Q^{(1)}$ by T_1, then

$$
\begin{aligned}
Q^{(2)}(u, \omega) &= \widetilde{g}_2^{(2)\,a_{i2}} \widetilde{g}_3^{(2)\,a_{i3}} \cdots \widetilde{g}_{i-1}^{(2)\,a_{i\,i-1}} u^{\alpha_2} + R_2^1(u, \omega) \\
&= \omega^{a_{i2}} \widetilde{g}_3^{(2)\,a_{i3}} \cdots \widetilde{g}_{i-1}^{(2)\,a_{i\,i-1}} u^{\alpha_2} + R_2(u, \omega)
\end{aligned}
\tag{4.18}
$$

where both R_2 and R_2^1 satisfy the property that all their monomials have relative 2-degree greater than $d_2\ell_1/d_{i+1}$. The same reasoning as in the preceding case shows that if $M = u^c\omega^e$ is a monomial appearing in $P^{(2)} - R_2$ with non zero coefficient, then

$$d_2 c + \ell_1 e \geq \alpha_3^i(\Gamma) + \frac{d_2\ell_1}{d_{i+1}} > \frac{d_2\ell_1}{d_{i+1}}.$$

And hence the Newton polygon for $\widetilde{g}_{i+1}^{(2)}$ is the same as the one for $\widetilde{g}_i^{(2)\,n_i}$, that is, the segment with extremes $(\ell_1/d_{i+1}, 0)$ and $(0, d_2/d_{i+1})$.

Besides,

$$\gcd(d_2, \ell_1) = \gcd(d_2, \delta_2) = d_3 < d_2,$$

and therefore, the following transformation obtained by the algorithm 1 is exactly T_2 (the same as in the preceding cases).

After iterating the calculus of (4.18), for $k \leq i$, we obtain

$$
\begin{aligned}
Q^{(k)}(u, \omega) &= \widetilde{g}_k^{(k)^{a_{ik}}} \, \widetilde{g}_{k+1}^{(k)^{a_{i\,k+1}}} \cdots \widetilde{g}_{i-1}^{(k)^{a_{i\,i-1}}} \, u^{\alpha_k} + R_k^1 \\
&= \omega^{a_{ik}} \, \widetilde{g}_{k+1}^{(k)^{a_{i\,k+1}}} \cdots \widetilde{g}_{i-1}^{(k)^{a_{i\,i-1}}} \, u^{\alpha_k} + R_k,
\end{aligned}
\tag{4.19}
$$

where $\alpha_k = \alpha_k^i(\Gamma)$ and both R_k and R_k^1 satisfy the property that their monomials have relative k-degree greater than $d_k \ell_{k+1}/d_{i+1}$. The same reasoning as in the $k = 2$ case proves that the Newton polygon of $\widetilde{g}_{i+1}^{(k)}$ is the segment with extremes $(\ell_{k-1}/d_{i+1}, 0)$ and $(0, d_k/d_{i+1})$ and the next transformation is T_k.

For $k = i$ we know (Lemma 4-2) that $\alpha_i^i(\Gamma) = (\ell_{i-1}/d_{i+1})$ and, since (4.19) holds, we have

$$
\begin{aligned}
\widetilde{g}_{i+1}^{(i)} &= \widetilde{g}_i^{(i)^{n_i}} - t_i Q^{(i)} \\
&= (\omega - u^{\gamma_i})^{\frac{d_i}{d_{i+1}}} - t_i \, u^{\frac{\ell_{i-1}}{d_{i+1}}} - t_i R_i \\
&= \left(\omega^{\frac{d_i}{d_{i+1}}} - t_i \, u^{\frac{\ell_{i-1}}{d_{i+1}}} \right) + R_i',
\end{aligned}
\tag{4.20}
$$

where both R_i and R_i' satisfy the property that all their monomials have relative i-degree greater than $d_i \, \ell_{i-1}/d_{i+1}$.

Since $\gcd(d_i/d_{i+1}, \ell_{i-1}/d_{i+1}) = 1$, we can carry out a b2-transformation (univocally defined by d_i/d_{i+1} and ℓ_{i-1}/d_{i+1} that will be called T_i'.

Then, we have

$$T_i'\left(\widetilde{g}_{i+1}^{(i)} \right) = \omega - t_1 + R_{i+1} \quad \text{or else} \quad T_i'\left(\widetilde{g}_{i+1}^{(i)} \right) = 1 + t_i \omega + R_{i+1},$$

where the polygon R_{i+1} is a multiple of u, and $t_i \in k^*$. Afterwards, we carry out suitable b1-transformations until the inicial form of $T_i'(g_{i+1}^{(i)})$ is $\omega + c u^{\gamma_{i+1}}$ where $c \in k$ and $\gamma_{i+1} > (\ell_i/d_{i+1})$, and we call T_i the composition of T_i' with these b1-transformations. Then, the sequence of transformations $T_1, T_2, \ldots, T_i$ satisfies the thesis of the proposition for $j \leq i$ and therefore the proposition is proved. $\qquad \square$

The same reasoning applies when $\delta_0 - \delta_1$ divides δ_0 and we obtain the following result.

Proposition 4-4.

a) If $\delta_0 - \delta_1$, does not divide δ_0, then the algorithm in Section 2.1 can be carried out for $\widetilde{f} = \widetilde{g}_{h+1}$ and the sequence of Newton polygons so obtained is shown in figure 4.11.

The polynomials $\{g_i(x, y)\}_{i=0,\ldots,h}$ are approximants of $f(x, y)$.

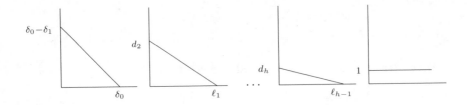

$$\text{Figure 4.11}$$

b) If $\delta_0 - \delta_1$ divides δ_0, then the algorithmin section 2.1 can be carried out for $\tilde{f} = \tilde{g}_{h+1}$ and the sequence of Newton polygons so obtained is shown in figure 4.12.

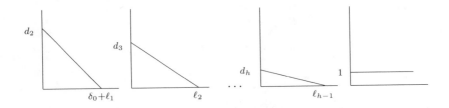

$$\text{Figure 4.12}$$

The polynomials $\{g_i(x,y)\}_{i=0,\dots,h}$ are the approximants of $f(x,y)$.

Corollary 4-5. Let k be an arbitrary field and $\delta_0, \dots, \delta_h$ positive integers such that

(1) $d_1 > d_2 > \dots > d_h > d_{h+1} = 1$, where $d_i = \gcd(\delta_0, \dots, \delta_{i-1})$.

(2) $n_i \delta_i \in \langle \delta_0, \dots, \delta_{i-1} \rangle$ for $1 \le i \le h$, where $n_i = d_i / d_{i+1}$.

(3) $\delta_1 < \delta_0$ and $\delta_i < n_{i-1} \delta_{i-1}$ for $i = 2, \dots, h$.

Then, there exist a plane curve C defined over k satisfying

(i) If we denote by L the line of equation $X_2 = 0$ ($\{X_0, X_1, X_2\}$ being projective coordinates of $\mathbb{P}^2_k$) then $L \cap C$ is an only point that will be called P.

(ii) P is a rational point over k with only one branch and resolution defined over k.

(iii) C is given by $f(x, y) = 0$ where f has a sequence of approximants.

(iv) The semigroup of values of C at P is generated by $r_0, \ldots, r_h$ where

$$r_0 = \delta_0 \quad \text{and} \quad r_i = \frac{\delta_0^2}{d_i} - \delta_i, \quad i = 1 \ldots h.$$

(v) The semigroup $\Gamma_1(P)$ is generated by $\{\delta_0, \ldots, \delta_h\}$.

That is, there is a cluster K compatible with the semigroup $\langle \delta_0, \ldots, \delta_h \rangle$ such that $W_{\delta_0}(K)$ is not empty.

If $C - P$ were smooth, then the Weierstrass semigroup $\Gamma(P)$ of C at P would be equal to $\Gamma_1(P) = \langle \delta_0, \ldots, \delta_h \rangle$ and the genus of C would be

$$g = \frac{(\delta_0 - 1)(\delta_0 - 2)}{2} - \frac{1}{2}\delta(P) = \frac{1}{2}\left[1 - 2\delta_0 + d_h\delta_h + \sum_{i=0}^{h-1}(n_i\delta_i - \delta_{i+1})\right]$$

(see 3-13). Therefore we will study in what conditions the curve C can be taken to be smooth in the affine part.

Proposition 3-10 shows that for any $\gamma \in k$ and for any h-uple $\underline{t} \in k^{*h}$ the curve

$$f_{\overline{\gamma}}^{\underline{t}}(x, y) = g_{\overline{h+1}}^{\underline{t}}(x, y) + \gamma$$

satisfies (i)–(v) in Corollary 4-5. Using only this fact, and when k is a field of characteristic p not dividing both δ_0 and δ_1, we will prove the existence of curves defined over k which are smooth in the affine part and such that they satisfy Corollary 4-5. When $p = 0$ this result is already proved in [17], Section 2, therefore the next theorem generalizes it and shows an alternative proof.

Theorem 4-6. *Let $\delta_0, \ldots, \delta_h$ be positive integers satisfying 1, 2 and 3 and let k be an infinite field of characteristic p not dividing $d_2 = \gcd(\delta_0, \delta_1)$. Then, there exists a plane curve C defined over k such that*

(i) *C has only one point P at infinity line.*

(ii) *P is a rational point over k with only one branch and resolution defined over k.*

(iii) *C is given by $f(x, y) = 0$ where f has a sequence of approximants.*

(iv) *The semigroup of values of C in P is generated by $\{r_0, \ldots, r_h\}$ where $r_0 = \delta_0$ and $r_i = (\delta_0^2/d_i) - \delta_i$ for $1 \leq i \leq h$.*

(v) *C is smooth in the affine part.*

(vi) *The Weierstrass semigroup C at P is generated by $\{\delta_0, \ldots, \delta_h\}$.*

(vii) The genus of C is

$$g(C) = \frac{1}{2} \left[3 - 2\delta_0 + d_h \delta_h + \sum_{i=0}^{h-1} (n_i \delta_i - \delta_{i+1}) \right].$$

PROOF. We denote by $q_i(x,y)$ the polynomial

$$\left(g_0^{\frac{t}{}} \right)^{a_{i0}} \cdot \left(g_1^{\frac{t}{}} \right)^{a_{i1}} \cdots \left(g_{i-1}^{\frac{t}{}} \right)^{a_{i\,i-1}},$$

so that

$$g_{i+1} = \left(g_i^{\frac{t}{}} \right)^{n_i} - t_i q_i^{\frac{t}{}}.$$

Using that $d_i \neq 0$ (because d_i divides d_2 and $d_2 \neq 0$) we will prove inductively that, for a dense set of $\underline{t}$'s and for $1 \leq i \leq h+1$, the polinomials $\partial g_i^{\frac{t}{}}/\partial x$ and $\partial g_i^{\frac{t}{}}/\partial y$ have no common component.

In fact, for $i = 1$ this is obvious. Suppose we have proved it for $1 \leq i \leq r$, then

$$\begin{cases} \dfrac{\partial g_{r+1}^{\frac{t}{}}}{\partial x} = n_r \left(g_r^{\frac{t}{}} \right)^{n_r - 1} \dfrac{\partial g_r^{\frac{t}{}}}{\partial x} - t_h \dfrac{\partial q_r^{\frac{t}{}}}{\partial x} \\[3mm] \dfrac{\partial g_{r+1}^{\frac{t}{}}}{\partial y} = n_r \left(g_r^{\frac{t}{}} \right)^{n_r - 1} \dfrac{\partial g_r^{\frac{t}{}}}{\partial y} - t_h \dfrac{\partial q_r^{\frac{t}{}}}{\partial y}, \end{cases} \tag{4.21}$$

where $n_r \neq 0$ and $\partial q_r^{\frac{t}{}}/\partial x$, $\partial q_r^{\frac{t}{}}/\partial y$ are not zero polymonials (see (4.12)). Moreover, we have

$$\begin{aligned} \deg_y \left(q_r^{\frac{t}{}} \right) &= a_{r\,1} \frac{\delta_0}{d_1} + \cdots + a_{r\,r-1} \frac{\delta_0}{d_{r-1}} \\ &\leq (n_1 - 1) \frac{\delta_0}{d_1} + \cdots + (n_{r-1} - 1) \frac{\delta_0}{d_{r-1}} \\ &= \frac{\delta_0}{d_r} - 1 \\ &= \deg_y \left(g_r^{\frac{t}{}} \right), \end{aligned}$$

and hence $g_r^{\frac{t}{}}$ may not divide $\partial q_r^{\frac{t}{}}/\partial x$. Applying this and the recurrence hypothesis, we obtain that the polymonials

$$(g_r^{\frac{t}{}})^{n_r-1}(\partial g_r^t/\partial x), \quad \partial q_r^{\frac{t}{}}/\partial x, \quad (g_r^{\frac{t}{}})^{n_r-1}(\partial g_r^{\frac{t}{}}/\partial y) \quad \text{and} \quad \partial q_r^t/\partial y$$

have no common factor. Therefore, for a dense set of $\underline{t}$'s $\partial g_{r+1}^{\frac{t}{}}/\partial x$ and $\partial g_{r+1}^{\frac{t}{}}/\partial y$ have no common component.

In particular, when $i = h+1$ we have proved that $\partial f^{\frac{t}{}}/\partial x$ and $\partial f^{\frac{t}{}}/\partial y$ have no common component for some $\underline{t}$. Therefore the intersection of the curves defined by the above polinomials consists of finitely many points. Since k is an infinite field, we can take $\gamma \in k$ such that

$$f_\gamma^t = g_{h+1}^t + \gamma$$

defines a plane curve which is smooth in the affine part. This is a solution to our problem. □

Remark 4-7. *The proof, expressed in geometrical terms, consits exactly in applying Bertini's theorem to the sheaf of projective curves generated by F_γ^t and L, F_γ^t being the projective completion of f_γ^t and L the infinity line. We have chosen the analitic expression of Bertini's argument to have as reference for the case of positive characteristic, that will be studied study next, case in wich Bertini's theorem does not work.*

Theorem 4-8. *Let k be an infinite field of any characteristic and $\delta_0, \ldots, \delta_h$ positive integers satisfying 1, 2 and 3. Let $d_i = \gcd(\delta_0, \ldots, \delta_{i-1})$, $n_i = d_i/d_{i+1}$ and $\{a_{ij}\}_{ij}$ positive integers such that*

$$n_i \delta_i = a_{i0}\delta_0 + a_{i1}\delta_1 + \cdots + a_{i\,i-1}\delta_{i-1},$$

with $a_{ij} < n_j$ for $j \geq 1$. Besides, let us suppose the following conditions are satisfied

a) *If $\delta_0 > d_h\delta_h \geq \delta_1$, one of the two following statements is true:*

   a.1) *$\partial g_h/\partial x \neq 0$ as a polynomial.*

   a.2) *There exists $i \in \{2, \ldots, h\}$ such that $\delta_i < d_h\delta_h$ and $\partial g_i/\partial x \neq 0$.*

b) *If $d_h\delta_h < \delta_1$, one of the following staments is true:*

   b.1) *The polynomials $\partial g_h/\partial x, \partial g_h/\partial y, \partial q/\partial x$ and $\partial q/\partial y$ have no common factor, where $q(x,y) = g_0^{a_{h0}} \cdots g_{h-1}^{a_{hh-1}}$.*

   b.2) *There exist $i \in \{0, \ldots, h-1\}$ such that $\delta_i \leq d_h\delta_h$ and the the polynomials $\partial g_h/\partial x, \partial g_h/\partial y, \partial q/\partial x, \partial q/\partial y, \partial g_i/\partial x$ and $\partial g_i/\partial y$ have no common factor.*

Then, there exists a plane curve C defined over k satisfying

(i) *C has only one point P at the infinity line.*

(ii) *P is a rational point over k with only one branch and resolution defined over k.*

(iii) *C is given by $f(x,y) = 0$ where f has a sequence of approximants.*

(iv) *The semigroup of values of C at P is generated by $\{r_0, \ldots, r_h\}$ where $r_0 = \delta_0$, $r_i = (\delta_0^2/d_i) - \delta_i$ for $1 \leq i \leq h$.*

(v) C is smooth in the affine part.

(vi) The Weierstrass semigroup of C at P is generated by $\{\delta_0, \ldots, \delta_h\}$.

(vii) The genus of C is

$$g(C) = \frac{1}{2}\left[3 - 2\delta_0 + d_h\delta_h + \sum_{i=0}^{h-1}(n_i\delta_i - \delta_{i+1})\right]$$

PROOF. First let us study the case where $d_h\delta_h \geq \delta_0$. Propositions 3-10 and 4-3 state that for any $\underline{t} \in k^{*^h}$, and for any $(\alpha, \beta, \gamma) \in k^3$ with $\alpha \neq t_h$, the algorithm in section 2.1 for the curve defined by

$$f^{\underline{t}}_{\alpha\beta\gamma}(x,y) = f^{\underline{t}}(x,y) + \alpha x + \beta y + \gamma \tag{4.22}$$

around the point P, is the same as the algorithm for $\tilde{f}^{\underline{t}}$, that is, the one described in 4-4. Besides, P is the only point of intersection of the plane curve defined by $f^{\underline{t}}_{\alpha\beta\gamma} = 0$ with the infinity line. Let us fix $\underline{t} = (1, \ldots, 1) = \underline{1}$ and let $f = f^{\underline{1}}$. If we find elements $\alpha \neq 1$, β, γ in k such that $f_{\alpha\beta\gamma}$ is smooth in the affine part, then the theorem is proved for the case $d_h\delta_h \geq \delta_0$. We have

$$\begin{cases} \dfrac{\partial f_{\alpha\beta\gamma}}{\partial x} = \dfrac{\partial f}{\partial x} + \alpha, \\[2mm] \dfrac{\partial f_{\alpha\beta\gamma}}{\partial y} = \dfrac{\partial f}{\partial y} + \beta. \end{cases}$$

Let us suppose $\partial f/\partial x = 0$, then for any $\alpha \neq 0$ the curve $f_{\alpha\beta\gamma} = 0$ is smooth. However if $\partial f/\partial x \neq 0$ we take $\alpha = 0$. Then there must exist an element $\beta \in k$ such that $\partial f/\partial x$ and $(\partial f/\partial y) + \beta$ do not have any common component (since the first polymonial has a finite number of irreducible components and β belongs to an infinite field). Fixed this element $\beta \in k$, the curves $\partial f/\partial x$ and $(\partial f/\partial y) + \beta$ only have a finite number of intersection points. Let us take $\gamma \in k$ such that $f_{\alpha\beta\gamma} = 0$ does not contain any of these points. Then, the curve given by $f_{\alpha\beta\gamma} = 0$ is a solution.

Now, suppose $\delta_0 > d_h\delta_h \geq \delta_1$ and $\partial g_h/\partial x \neq 0$. For any $\underline{\alpha} = (\alpha_1, \ldots, \alpha_{d_h-1}) \in k^{d_h-1}$ set

$$p_{\underline{\alpha}}(z) = z^{d_h} + \alpha_{d_h-1}z^{d_h-1} + \ldots + \alpha_1 z,$$

and let

$$q(x,y) = g_0^{a_{h0}} \cdots g_{h-1}^{a_{h\,h-1}},$$

and for any $(\underline{\alpha}, \beta, \gamma) \in k^{d_h+1}$ set

$$f_{\underline{\alpha}\beta\gamma}(x,y) = p_{\underline{\alpha}}(g_h) - q(x,y) + \beta y + \gamma. \tag{4.23}$$

By Propositions 3-10 and 4-3, if $\beta \neq 1$ then the algorithm in Section 2.1 for the germ of curve defined by f at P is the same as the one for $\tilde{f}_{\underline{\alpha}\beta\gamma}$. Besides

$$
\begin{cases}
\dfrac{\partial f_{\underline{\alpha}\beta\gamma}}{\partial x} = p'_{\underline{\alpha}}(g_h)\dfrac{\partial g_h}{\partial x} - \dfrac{\partial q}{\partial x} \\[3mm]
\dfrac{\partial f_{\underline{\alpha}\beta\gamma}}{\partial y} = p'_{\underline{\alpha}}(g_h)\dfrac{\partial g_h}{\partial y} - \dfrac{\partial q}{\partial y} + \beta
\end{cases}
\tag{4.24}
$$

Let us fix $\underline{\alpha} \in k^{d_h-1}$ such that the first polynomial in (4.24) is non zero (this is possible because $\partial g_h/\partial x \neq 0$). The same reasoning as before shows there exists $\beta \in k$, $\beta \neq 1$ so that the curves defined by the polinomials in (4.24) have no common component. Therefore, there must exist $\gamma \in k$ such that $f_{\underline{\alpha}\beta\gamma} = 0$ is a solution (in fact, γ can be any element in k but a finite number).

Now, suppose $\delta_0 > d_h\delta_h \geq \delta_1$ and $\partial g_h/\partial x = 0$, then, by the hypothesis of the theorem, there exists $i \in \{2,\ldots,h\}$ with $\delta_i \leq d_h\delta_h$ and $\partial g_i/\partial x \neq 0$. That is, for any $(\underline{\alpha},\beta,\gamma,\eta)$ in a dense set in k^{d_h+2}, the curve defined by

$$
g_{\underline{\alpha}\beta\gamma\eta}(x,y) = p_{\underline{\alpha}}(g_h) - q(x,y) + \beta y + \eta g_i + \gamma
\tag{4.25}
$$

around P has the same algorithm than $\tilde{f}$. Besides, we have

$$
\begin{cases}
\dfrac{\partial g_{\underline{\alpha}\beta\gamma\eta}}{\partial x} = -\dfrac{\partial q}{\partial x} + \eta\dfrac{\partial g_i}{\partial x}, \\[3mm]
\dfrac{\partial g_{\underline{\alpha}\beta\gamma\eta}}{\partial y} = -\dfrac{\partial q}{\partial y} + \eta\dfrac{\partial g_i}{\partial y} + \beta.
\end{cases}
\tag{4.26}
$$

Let us take $\eta \in k$ such that the first polynomial in (4.26) is nonzero, afterwards we take β in such a way that the two curves defined by the polinomials in (4.25) have no common component. Then, for almost every $\gamma \in k$, the curve given by $g_{\underline{\alpha}\beta\gamma\eta} = 0$ is solution to our problem.

Let us suppose now that $d_h\delta_h < \delta_1$ and b.1 holds. Then, the curve defined by

$$
g^t_{\underline{\alpha}\gamma}(x,y) = p_{\underline{\alpha}}(g_h) - t_h q(x,y) + \gamma
\tag{4.27}
$$

around P has the same algorithm as $\tilde{f}$, and

$$
\begin{cases}
\dfrac{\partial g^t_{\underline{\alpha}\gamma}}{\partial x} = p'_{\underline{\alpha}}(g_h)\dfrac{\partial g_h}{\partial x} - t_h\dfrac{\partial q}{\partial x}, \\[3mm]
\dfrac{\partial g_{\underline{\alpha}}{}^t{}_\gamma}{\partial y} = p'_{\underline{\alpha}}(g_h)\dfrac{\partial g_h}{\partial y} - t_h\dfrac{\partial q}{\partial y}.
\end{cases}
$$

Therefore, for a suitable choice of $(\underline{\alpha},\gamma,t_h)$ the curve defined by $g^t_{\underline{\alpha}\gamma} = 0$ is solution to our problem.

Finally, suppose b.2 holds and take $i < h$ such that $\delta_i \leq d_h\delta_h$ and the polynomials $\partial g_h/\partial x$, $\partial g_h/\partial y$, $\partial q/\partial x$, $\partial q/\partial y$, $\partial g_i/\partial x$ and $\partial g_i/\partial y$ have no common factor. Then, we consider

$$
g^t_{\underline{\alpha}\eta\gamma}(x,y) = p_{\underline{\alpha}}(g_h) - t_h q(x,y) + \eta g_i + \gamma.
\tag{4.28}
$$

In this case

$$
\begin{cases}
\dfrac{\partial g_{\underline{\alpha}\eta\gamma}^{\,t}}{\partial x} = p'_{\underline{\alpha}}(g_h)\dfrac{\partial g_h}{\partial x} - t_h\dfrac{\partial q}{\partial x} + \eta\dfrac{\partial g_i}{\partial x}, \\[3mm]
\dfrac{\partial g_{\underline{\alpha}\eta\gamma}^{\,t}}{\partial y} = p'_{\underline{\alpha}}(g_h)\dfrac{\partial g_h}{\partial y} - t_h\dfrac{\partial q}{\partial y} + \eta\dfrac{\partial g_i}{\partial y}.
\end{cases}
$$

and the same argument as in preceding cases shows that for a suitable choice of $(\underline{\alpha}, \eta, \gamma, t_h)$ the curve defined by $g_{\underline{\alpha}\eta\gamma}^{\,t} = 0$ is solution to our problem.   □

Note that, for the curves obtined before, we imposed their Newton algorithm to be the one in proposition 4-4, that is, we wanted the minimal number of $b1$-transformations to appear. If Γ does not satisfy the conditions of theorem 4-8 and k is an infinite field of positive characteristic, the preceding process does not guarantee to reach a curve smooth in the affine part. However, allowing more $b1$-transformations to appear in the algorithm, that is, changing the cluster, we can obtain another kind of plane curves defined over k satisfying the conditions (i)–(vii) in theorem 4-8.

Next, we give an example of a semigroup Γ satisfying 1, 2 and 3 for which theorem 4-8 cannot be applied for any infinite field k of characteristic 2 and we show how the last method allows us to find a plane curve defined over k satisfying (i)–(vi) in theorem 4-8.

Example 4-9. Let $\Gamma = \langle \delta_0, \delta_1, \delta_2, \delta_3 \rangle$ where $\delta_0 = 42$, $\delta_1 = 30$, $\delta_2 = 57$, $\delta_3 = 10$ satisfy 1, 2 and 3. If k is an infinite field of characteristic different to 2 or 3 then, by Theorem 4-6 there exists a plane curve defined over k satisfying (i)–(vii). Let us analyze the problem of finding a plane curve defined over an infinite field k of characteristic 2 or 3 and satisfying (i)–(vii). We have $d_3\delta_3 = \delta_1$ and for $\underline{t} \in k^{*3}$

$$
\begin{cases}
g_0 = x \\
g_1 = y \\
g_2 = y^7 - t_1 x^5 \\
g_3 = g_2^2 - t_2 x^2 y \\
g_4 = g_3^3 - t_3 y.
\end{cases}
$$

Therefore, if char $k = 3$ then $\partial g_3/\partial x \neq 0$ and hence Theorem 4-8 guarantees the existence of a plane curve defined over k satisfying (i)–(vii). However, if char $k = 2$ then $\partial g_3/\partial x = 0$ and the hypothesis of Theorem 4-8 are not satisfied.

Let us consider the plane curve C defined over a field k of charactheristic 2 given by $f(x,y) = 0$ where

$$
f(x,y) = \left((y^7 - x^5)^2 - x^2 y - x\right)^3 + y = (g_3^{\frac{1}{3}} - x)^3 + y
$$

The curve C is smooth in the affine part and it has a singularity at the infinity point $P = (1{:}0{:}0)$. The Newton polynomials obtained from the algorithm in Section 2.1 applied to the branch defined by C at P are shown in figure 4.13.

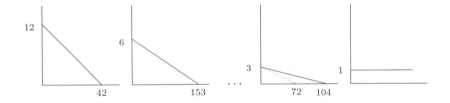

Figure 4.13

That is, to obtain the second diagram from the first one we carry out a b2-transformation followed by a b1-transformation given by $u' = u$ and $w' = w+1$. Again we make the corresponding b2-transformation followed by a translation $u' = u$, $w' = w + 1$ and we reach a segment with extremes in $(72,0)$ and $(0,3)$. Since $72=24.3$, we make the b1-transformation given by $u' = u$, $w' = w - u^{24}$ (this transformation is not allowed by the method followed in Theorem 4-8) and we reach the segment with extremes in $(104,0)$ and $(0,3)$. Finally, after a suitable b2-transformation, the algorithm ends.

Now, we observe that $42 = \delta_0$, $12 = \delta_0 - \delta_1$, $6 = d_2$, $153 = n_1\delta_1 - \delta_2$, $3 = d_3$ and $104 = n_2\delta_2 - \delta_3$, and hence the plane curve C defined over the field k (of characterist 2) satisfies (i)–(vii). However, as it has been said before, this example does not follow the method in theorem 4-8. That means that, besides the curves considered in 4-8, the algorithm in section 2.1 allows us to discuss the existence of other examples for which we have other b1-transformations giving lines as the one in traces in figure 4.13. Moreover, let us note that in this example k could be any finite field of characteristic 2.

References

[1] Abhyankar, S.S. Inversion and invariance of characteristic pairs. Am. J. Math. 89, 363–372 (1967).

[2] Abhyankar, S.S. On the semigroup of a meromorphic curve. Intl. Symp. on Algebraic Geometry, 249–414 (Kyoto, 1977).

[3] Abhyankar, S.S. Lectures on expansion techniques in algebraic geometry. Tata Institute of Fundamental Research, (Bombay 1977).

[4] Abhyankar, S.S. and Sathaye, A. Geometric Theory of Algebraic Space curves. Springer Lecture Notes 423, (1974).

[5] Abhyankar, S.S. and Moh, T.T. Newton-Puiseux expansion and generalised Tschirnhausen transformation. J. Reine Math. 260, 67–83 and 261, 29–54 (1973).

[6] Angermuller, G. Die Wertehalbgruppe einer ebenen irreduziblen algebroiden Kurve. Math. Zeit. 153, 267–282 (1977).

[7] Campillo, A. Algebroid curves in positive characteristic. Springer Lecture Notes 813 (1980).

[8] Casas-Alvero, E. Infinitely near imposed singularities and singularities of polar curves. Math. Ann., 287, 429–454 (1990).

[9] Enriques, F. and Chisini, O. Lezioni sulla teotia geometrica delle equazioni e delle funzioni algebrique. Bologna. Zanichelli, (1915).

[10] Hoskin, M.A. Zero-dimensional valuation ideals associated with plane curve' branch. Proc. London Math. Soc. (3) 6, 70–99 (1956).

[11] Lejeune-Jalabert, M. Linear systems with near base conditions and complete ideals in dimension two. Preprint (1992).

[12] Lipman, J. Rational singularities with applications to algebraic surfaces and unique factorization. Publ. Math. I.H.E.S. Vol 36, 195–279 (1969).

[13] Lipman, J. On complete ideals in regular local rings. Algebraic Geometry and Commutative Algebra in honor of M. Nagata. Kinokuniya, 203–231 (Tokyo, 1987).

[14] Lipman, J. Proximity inequalities for complete ideals in two-dimesional regular local rings. Commutative Algebra week, Mount Holyoke College (1992). Proceedings to appear in Contemporary Mathematics.

[15] Moh, T.T. On characteristic pairs of algebroid plane curves of characteristic p. Bull. Inst. Math. Sinica 1, 75–91 (1973).

[16] Pinkham, H. Séminaire sur les singularités des surfaces (Demazure-Pinkham-Teissier) Cours donné au Centre de Math. de l'Ecole Polytechnique (1977-1978).

[17] Sathaye, A. On planar curves. Amer. J. Math. 99, 1105–1135 (1977).

[18] Zariski, O. and Samuel, P. Commutative Algebra, vol. II, appendix 5. Van Nostrand Princeton (1960).

[19] Zariski, O. Algebraic surfaces. Springer-Verlag (1972).

[20] Zariski, O. Les problèmes des modules pour las branches planes, avec un appendice de B. Teissier. Cours au Centre de Mathématiques de l'Ecole Politechnique. (1973)

Address of author:

Universidad de Valladolid.
Departamento de Álgebra y Geometría.
Facultad de Ciencias, 47005. Valladolid
Spain
Email: areguera@cpd.uva.es

Cubic surfaces with double points in positive characteristic

Marko Roczen

1 Introduction

The classification of singular cubic surfaces, made by Schläfli and Cayley in the last century, and reconsidered by Bruce and Wall [3] from the viewpoint of modern singularity theory (both over the complex numbers) gives rise to the following question: Let k be an algebraically closed field of arbitrary characteristic p, $f = f(x_0, x_1, x_2, x_3)$ an irreducible homogeneous polynomial of degree 3. Let $X \subseteq \mathbb{P}_k^3$ be the set of zeros of f in the projective space.

If X has no triple point (in a way, this is the most general case), it has at most double points. They are seen to be rational singularities from the list of Artin [2], but in general, they do not appear in these normal forms. Hence, it is useful to have a possibility of finding their type. This is given by a "geometric" extension of the "recognition principle" of Bruce and Wall (loc. cit.). An equivalent condition is found via the description of the "local resolution graph" and provides a possibility to avoid some awful coordinate transformations. Now, configurations of double points and the corresponding normal forms can be calculated.

2 Two characterizations of rational double points

Let R be a complete local Cohen Macaulay k-algebra with residue field k of dimension $d \geq 2$. $\operatorname{Spec} R$ is said to be absolutely isolated if there is a resolution of singularities consisting of blowing ups $\varphi_i \colon X_i \to X_{i-1}$ ($i = 1, \ldots, t$), $X_0 = \operatorname{Spec} R$, X_t smooth, $\operatorname{Sing}(X_i)$ finite and φ_i the blowing up of the reduced singular locus $\operatorname{Sing}(X_i)$ of X_i. The set (φ_i) of morphisms is essentially unique and said to be the canonical resolution. We associate to R the "local resolution graph" Γ: This is a directed graph having as vertices the components of the formal scheme

$$\coprod_{i=0}^{t-1} (X_i)_{\operatorname{Sing}(X_i)}^{\wedge};$$

its arrows correspond to the morphisms of complete local rings induced by the φ_i. Thus, e.g. the graph

comes from an isolated singularity which can be resolved by 4 blowing ups as above, the singular locus of X_2 consists of 2 points, and X_1, X_3 both have one singular point.

Now let R be a double point (i.e. of multiplicity 2), then $R \simeq k[\![x]\!]/(f)$, where $x = (x_0, \ldots, x_d)$ are indeterminates, $f \in k[\![x]\!]$ of order 2. Consider any $w = (w_0, \ldots, w_d) \in \mathbb{R}_+^{d+1}$, such that $w_i \leq (1/2)$. f is said to be semiquasihomogeneous (sqh) of weight w if $f = \sum_{\nu} a_\nu x^\nu$ such that

1. $f_1 = \sum_{\nu, w(\nu)=1} a_\nu x^\nu$ defines an isolated singularity,

2. $f - f_1 = \sum_{\nu, w(\nu)>1} a_\nu x^\nu$.

Spec R is said to be sqh of weight w if there exists such an f as above.

2.1   Characterization

For a complete local Cohen Macaulay double point Spec R of dimension $d > 1$, the following conditions are equivalent:

(i) Spec R is absolutely isolated.

(ii) Spec R is sqh of some weight w such that $w_0 + \ldots + w_d > d/2$.

Further, in (ii) the weight is up to permutation one of the following:

$$A_n = \left(\frac{1}{n+1}, \frac{1}{2}, \ldots, \frac{1}{2} \right), \qquad n \geq 1$$

$$D_n = \left(\frac{1}{n-1}, \frac{n-2}{2(n-1)}, \frac{1}{2}, \ldots, \frac{1}{2} \right), \qquad n \geq 4$$

$$E_6 = \left(\frac{1}{3}, \frac{1}{4}, \frac{1}{2}, \ldots, \frac{1}{2} \right),$$

$$E_7 = \left(\frac{1}{3}, \frac{2}{9}, \frac{1}{2}, \ldots, \frac{1}{2} \right),$$

$$E_8 = \left(\frac{1}{5}, \frac{1}{3}, \frac{1}{2}, \ldots, \frac{1}{2} \right).$$

The weight X_n ($X = A$, D or E, respectively) is uniquely determined by R and called the "type" of the singularity. The local resolution graphs are the following ones and correspond to the type as indicated in table 2.1 (m = number of vertices).

The conditions (in brackets) are

S: The exceptional locus of the last blowing up φ_t in the canonical resolution is smooth.

NS: The exceptional locus of φ_t is not smooth.

graph	type (condition)
• ⟶ • ⟶ ... ⟶ • ⟶ •	$A_{2m-1},\ m \geq 1$ (S) $A_{2m},\ \ \ m \geq 1$ (NS) $E_6,\ \ \ \ \ m = 4$ (NI)
• ⟶ • ⟶ ... ⟶ • ⟶ • (with •↑↓• branch above/below)	$D_m,\ m \geq 4,\ m$ even
• ⟶ • ⟶ ... ⟶ • ⟶ • (with ↓• branch below)	$D_{m+1},\ m \geq 4,\ m$ even
• ⟶ • ⟶ • ⟶ • (with •↑, ↓• branches)	$E_7,\ m = 7$
• ⟶ • ⟶ • ⟶ • ⟶ • (with •↑, ↓• branches)	$E_8,\ m = 8$

Table 2.1: Resolution graphs.

NI: For the quadratic suspension of dimension $d + 2$, the exceptional locus of the first blowing up φ_1 has nonisolated singularities (if $R = k[\![x]\!]/(f)$ for any f, then $R' = k[\![x, x_{d+1}, x_{d+2}]\!]/(f + x_{d+1} \cdot x_{d+2})$ is said to be the quadratic suspension of dimension $d + 2$).

PROOF. For the equivalence of (i), (ii) and the uniqueness of w cf. ([6], 3.3). The remaining conditions follow from the proof of ([6], 3.2.). □

Now let $d = 2$. The absolutely isolated double points are known to be rational. Their equations have been computed by Artin ([2], 3.) and can be obtained in the following way:

2.2 Artin's equations of absolutely isolated double points

Choose a quasihomogeneous normal form $f = f(x_1, x_2, x_3)$ of the equation of an isolated singularity for any of the weights A_n, D_n, E_n. Let $X^0 :=$

$\mathrm{Spec}\big(k[x_1, x_2, x_3]/(f)\big)$ be the corresponding singularity. Choose also a monomial k-base of the Tjurina algebra of f and select a monomial $m = m(x_1, x_2, x_3)$
of weight > 1. The corresponding equation $f + m$ is sqh of the same weight.
This procedure for all monomials under consideration gives a finite set X^0, X^1,
$\dots$, X^i, $\dots$ of shq singularities of the given weight, where the order is choosen
such that we have $\tau(X^0) > \tau(X^1) > \cdots > \tau(X^i) > \cdots$ for the Tjurina numbers. In this way we obtain all possible sqh singularities of that weight. Take
e.g., $\mathrm{char}(k) = 2$, $w = E_6$, then $f = x_0^3 + x_1^2 x_2 + x_2^3$ defines the singularity E_6^0.
The only relevant monomial in the Tjurina algebra of f gives the remaining
sqh singularity E_6^1 with equation $f_1 = x_0^3 + x_1^2 x_2 + x_2^3 + x_0 x_1 x_2$.

Remark 2-1.

(i)   The map $X_n^r \mapsto X_n$ gives the type of the singularity.

(ii) The Tjurina number $\tau : \{X_n^r | \text{ all } r\} \to \mathbb{N}$ is injective for a fixed type X_n.

The symbol X_n^r will be used for the corresponding complete local ring and (by
abuse of language) its spectrum, too.

3   Singularities and normal forms

The singularities of the cubic surface X give rise to the possible normal forms
(depending on parameters, in some cases). Though differences from the classical
case can appear only in some characteristics $p \neq 0$, the application of 2.1
simplifies coordinate transformations sometimes.

Let $S = S(X) := X_{\mathrm{Sing}(X)}^\wedge$ be the formal scheme obtained from X by
completion along the singular locus. S will be called the type of the cubic surface
X. The classification can be done via S: If X has only isolated singularities
and contains a triple point, this is the only singularity, and X is the projective
closure of the cone over a smooth plane cubic. In any other case, X contains
at most double points. This is the situation considered here. The following
description extends the list in the paper of Bruce and Wall [3], and some of the
cases (which remain unchanged) are only listed for completeness. Let $P \in X$ be
singular, $P = (0\!:\!0\!:\!0\!:\!1) \in \mathbb{P}^3$ and $(x_0\!:\!x_1\!:\!x_2\!:\!x_3)$ the homogeneous coordinates.
We write

$$f = x_3 f_2 + f_3, \qquad f_i = f_i(x_0, x_1, x_2) \quad \text{homogeneous of degree } i.$$

The classification of quadratic forms (in arbitrary characteristic) gives us the
following possibilities:

A. $f_2 = x_1^2 - x_0 x_2$ B. $f_2 = x_0 x_1$ C. $f_2 = x_0^2$

Let $L := V^+(f_2, f_3) \subseteq \mathbb{P}^2$ be the space of lines in X through P. We identify
the space of all lines through P with $\mathbb{P}^2$ with the coordinates $(x_0 : x_1 : x_2)$.

Case A. Obviously, P is an A_1 singularity of X. Further, $\text{Sing}(X - \{P\})$ is in bijective correspondence with $\text{Sing}(L)$, where a point $Q \in L$ of multiplicity k is mapped to an A_{k-1} singularity of $X - \{P\}$ (cf. [3], Lemma 2). Thus all possibilities for S are

$$S = A_1,\, 2A_1,\, A_1 \perp\!\!\!\perp A_2,\, 3A_1,\, A_1 \perp\!\!\!\perp A_3,\, 2A_1 \perp\!\!\!\perp A_2,$$
$$4A_1,\, A_1 \perp\!\!\!\perp A_4,\, 2A_1 \perp\!\!\!\perp A_3,\, A_1 \perp\!\!\!\perp 2A_2,\, A_1 \perp\!\!\!\perp A_5.$$

Here, the symbol nX always denotes $X \perp\!\!\!\perp \ldots \perp\!\!\!\perp X$ (n disjoint copies).

Case B. (cf. [3], Lemma 3). The singularities of $X - \{P\}$ correspond to the points of $\text{Sing}(L - \{Q\})$, $Q := (0:0:1)$, and under this bijection, a point of multiplicity k is mapped to an A_{k-1} singularity. Further, P is an $A_{k_0+k_1+1}$ singularity if k_i denotes the multiplicity of $L_i = V(x_i, f_3)$ at Q. The only possible k_i are $\{k_0, k_1\} = \{1\}, \{1, 2\}, \{1, 3\}$.

Thus, all possible cases are:

$$S = A_2,\, A_2 \perp\!\!\!\perp A_1,\, 2A_2,\, A_2 \perp\!\!\!\perp 2A_1,\, 2A_2 \perp\!\!\!\perp A_1,\, 3A_2,$$
$$A_3,\, A_3 \perp\!\!\!\perp A_1,\, A_3 \perp\!\!\!\perp 2A_1,\, A_4,\, A_4 \perp\!\!\!\perp A_1,\, A_5,\, A_5 \perp\!\!\!\perp A_1.$$

To see why the singularity at P is of the type claimed above, we may use the local resolutions as follows:

Without loss of generality

$$f = x_3 x_0 x_1 + f_3(x_0, x_1, x_2),$$
$$f_3 = x_0(a_0 x_0^2 + a_1 x_0 x_2 + a_2 x_2^2) + x_1(a_3 x_1^2 + a_4 x_1 x_2 + a_5 x_2^2) + a_6 x_2^3.$$

Putting $x_3 = 1$, we obtain an equation

$$x_0 x_1 + x_0 \cdot a(x_0, x_2) + x_1 \cdot b(x_0, x_2) = 0$$

for X near P corresponding to the origin in $\mathbb{A}^3$.

1. $f_3(Q) \neq 0$, then $a_6 \neq 0$ and P is A_2 (weight $(1/2, 1/2, 1/3)$)

2. $f_3(Q) = 0$, then $a_6 = 0$. We may suppose $k_1 = 1$. Thus $a_2 \neq 0$.

   2.1. $k_0 = 1$ (equivalently $a_5 \neq 0$), P is A_3 (weight $(1/2, 1/2, 1/4)$)

   2.2. $k_0 = 2$ (equivalently $a_5 = 0$ and $a_4 \neq 0$): Blow up P; after an obvious coordinate transformation you obtain a point that is sqh of weight $(1/2, 1/3, 1/2)$ and therefore P is A_4.

   2.3. $k_0 = 3$ (equivalently $a_4 = a_5 = 0$ and $a_3 \neq 0$): One blowing up leads to a point of sqh weight $(1/2, 1/4, 1/2)$ i.e., P is A_5.

Case C. This case will be performed in full detail, including the relevant normal forms.

Let $i := 4 - \#L$, $\#L$ the number of (closed) points of $V^+(x_0, f_3)$. Then $i \in \{1, 2, 3\}$, P is the only singularity of X and has type D_4, D_5, or E_6; for $i = 1, 2, 3$, respectively: Let Ci be the corresponding case, $f = x_3 x_0^2 + x_0 \cdot g_2(x_1, x_2) + g_3(x_1, x_2)$, g_ℓ homogeneous of degree ℓ. Depending on $p = \operatorname{char} k$, we obtain after a linear homogeneous transformation

C1. $g_3 = x_1^3 + x_2^3$ for $p \neq 3$, and $g_3 = x_1^2 x_2 + x_3^3$ for $p = 3$.

C2. $g_3 = x_1^2 x_2$ and $g_2(0, 1) \neq 0$

C3. $g_3 = x_1^3$ and $g_2(0, 1) \neq 0$.

Using (ii) of 2.1, in each case we obtain for P a sqh-singularity of the following type and initial term (i.e. term of weight 1):

C1. $D_4 = (1/2, 1/3, 1/3)$, $x_0^2 + x_1^3 + x_2^3$ for $p \neq 3$, and $x_0^2 + x_1^2 x_2 + x_2^3$ for $p = 3$.

C2. $D_5 = (1/2, 3/8, 1/4)$, $x_0^2 + cx_0 x_2^2 + x_1^2 x_0$, $c \neq 0$.

C3. $E_6 = (1/2, 1/3, 1/4)$, $x_0^2 + cx_0 x_2^2 + x_1^3$, $c \neq 0$

Case C1. $p \neq 3$: For some linear form ℓ, put

$$x_1 := x_1 - \frac{a}{3}x_0, \qquad x_2 := x_2 - \frac{c}{3}x_0, \qquad x_3 := x_3 + \ell(x_0, x_1, x_2)$$

to obtain the normal forms: $f = x_0^2 x_3 + rx_0 x_1 x_2 + x_1^3 + x_2^3$, $r \in \{0, 1\}$, and the singularity at P: D_4 if $p \neq 2, 3$ and D_4^r if $p = 2$. (If $p \neq 2, 3$, the Hessian of f is $x_0^2 x_1 x_2$ and $x_0^2(36x_1 x_2 - x_0^2)$, respectively. Both cases $r = 0, 1$ thus do not provide projectively equivalent surfaces. If $p = 2$, the Tjurina numbers at P are $\tau = 8$ for $r = 0$, $\tau = 6$ for $r = 1$.)

If $p = 3$: Use

$$x_1 := x_1 - (b/2)x_0, \qquad x_2 := x_0 - \alpha x_0, \qquad x_3 := x_3 + \ell(x_0, x_1, x_2).$$

Normal forms: $f = x_0^2 x_3 + rx_0 x_2^2 + x_1^2 x_2 + x_2^3$, $r \in \{0, 1\}$. Singularity at P: D_4 (The Hessians are $x_0^2 x_1^2$ and $x_0^2(x_0 x_2 - x_1^2)$, respectively, thus $r = 0, 1$ provide nonequivalent surfaces).

Case C2. A substitution

$$x_2 := x_2 - ax_0, \qquad x_3 := x_3 + \ell(x_0, x_1, x_2),$$

gives for $p = 2$ normal forms: $f = x_0^2 x_3 + rx_0 x_1 x_2 + x_0 x_2^2 + x_1^2 x_2$, $r \in \{0, 1\}$, and singularity at P: D_5^r. Further, for $p \neq 2$ we choose

$$x_1 := x_1 + \alpha x_0, \qquad x_3 := x_3 + \bar{\ell}(x_0, x_1, x_2)$$

and obtain a single normal form: $f = x_0^2 x_3 + x_0 x_2^2 + x_1^2 x_2$ with singularity at P: D_5.

Case C3. f can be transformed into

$$f = x_0^2 x_3 + a x_0 x_1^2 + b x_0 x_1 x_2 + x_0 x_2^2 + x_1^3.$$

Now we choose a coordinate transformation as before to obtain for $p = 2$, normal forms: $f = x_0^2 x_3 + r x_0 x_1 x_2 + x_0 x_2^2 + x_1^3$, $r \in \{0,1\}$; singularity at P: E_6^r. In the remaining cases, we choose

$$x_1 := x_1 + \alpha x_0, \qquad x_2 := x_2 + \beta x_1, \qquad x_3 = x_3 + \ell(x_0, x_1, x_2).$$

The condition $a = b = 0$ is expressed by

$$b + 2\beta = 0, \qquad a + b\beta + \beta^2 + 3\alpha = 0,$$

which is solvable for $p \neq 3$. Thus we obtain for $p \neq 2, 3$, normal form: $f = x_0^2 x_3 + x_0 x_2^2 + x_1^3$, singularity at P: E_6

And for $p = 3$, normal forms: $f = x_0^2 x_3 + r x_0 x_1 x_2 + x_0 x_2^2 + x_1^3$, $r \in \{0,1\}$ and singularity at P: E_6^r (since for $r = 0$, the Tjurina number is $\tau = 9$, and for $r = 1$ we have $\tau = 7$).

Note, that Schläfli and Cayley mistakenly give only one normal form for surfaces with a singularity of type D_4 ([7], p. 229). This was already remarked in [3]. The given form of Schläfli (loc. cit.) is easily seen to be equivalent to the one above for $r = 1$. In characteristic 2, both cases $r = 0, 1$ give even nonisomorphic singularities of type D_4.

References

[1] Arnold, V.I., Normal forms for functions near degenerate critical points, the Weyl groups A_k, D_k and E_k and Lagrangian singularities, Func. Anal. Appl. 6 (1972) 4, 3–25

[2] Artin, M., Coverings of the rational double points in characteristic p, Complex Analysis and Algebraic Geometry, eds. Baily, W.L. Jr. and Shioda, T., Cambridge 1977, 11–22

[3] Bruce, J.W., Wall, C.T.C., On the classification of cubic surfaces, J. London Math. Soc. 19 (1979), 245–256

[4] Greuel, G.M., Kröning, H., Simple singularities in positive characteristic, Math. Z. 203 (1990), 339–354

[5] Knop, F., Ein neuer Zusammenhang zwischen einfachen Gruppen und einfachen Singularitäten, Invent. Math. 90 (1987), 579–604

[6] Roczen, M., Recognition of simple singularities in positive characteristic, Math. Z. 210 (1992), 641–654

[7] Schläfli, L., On the distribution of surfaces of the third order into species, Phil. Trans. Roy. Soc. 153 (1864), 193–247

Address of author:

Humboldt-Universitt zu Berlin
Institut fr Reine Math.
Postfach 1297
Berlin 1086
Germany
Email: roczen@mathematik.hu-berlin.de

ON THE CLASSIFICATION OF REDUCIBLE
CURVE SINGULARITIES

Jan Stevens

All long lists of singularities are based on the analysis of their defining equations (see e.g. [1, 22, 23]). Milnor number and modality are the basic parameters. For curve singularities one can use other discrete invariants, and base the classification on the parametrisation.

My interest in complex curve singularities started with one-parameter families with smooth general fibre [17]; in this situation the principal invariant is the genus of the Milnor fibre, which is equal to $g = \delta - r + 1$ (δ is the virtual number of nodes and r is the number of branches). This formula defines g also for non smoothable curves. Curves with $g \leq 2$ were classified by Greuel, Steiner and myself [9, 16, 18]. The classification of irreducible curves has been known in principle for some time [7], but the methods used there do not apply to reducible curves. The present paper contains a classification of curves with $g \leq 3$. It takes an inductive approach, based on the fact that the union $(X, 0)$ of two (possibly reducible) curves $(X_1, 0)$ and $(X_2, 0)$ has genus $g(X) \geq g(X_1) + g(X_2)$, with equality if and only if the Zariski tangent spaces of X_1 and X_2 have only the origin in common. This reduces the problem to the determination of the possible configurations of already known curves. The method is to bring X_2 into normal form, using X_1 preserving diffeomorphisms.

In this set-up the embedding dimension of a curve is found as a result of the classification, whereas an approach with equations uses fixed embedding dimensions. A typical example is the curve X, formed by a line and a cusp, intersecting with multiplicity 2; then $g(X) = 2$ and $\delta = 3$. There are two types: besides the plane curve D_5 one has a space curve, where the line lies as parabola in a transversal plane through the tangent line of the cusp; this curve is obtained from E_7 by 'lifting the line out of the plane'. It is not a complete intersection.

In the first section we develop this program for curves with intersection multiplicity $(X_1 \cdot X_2)$ at most 2. It turns out that for $(X_1 \cdot X_2) = 2$ the Zariski tangent spaces of X_1 and X_2 intersect in a line. The classification problem almost reduces to that of a curve X and a smooth branch L with $(X \cdot L) = 2$ (an additional modulus can arise). It is shown how to compute discrete invariants and equations for $X_1 \cup X_2$ from $X_1 \cup L$ and $X_2 \cup L$.

In the second section these results are applied to the classification of curves with $g \leq 3$. The known lists for the case $g \leq 2$ are derived, partly as preparation for the case $g = 3$. I list the possible types of curves. In general the curves depend on moduli. In almost all cases these reduce to the moduli of straight lines through the origin, i.e. moduli of point sets in projective space. This subject is quite extensively covered in [12] and [6]. I include a short discussion, with particular emphasis on the case $g = 2$.

Progress in Mathematics, Vol. 134
© 1996 Birkhäuser Verlag Basel/Switzerland

The final section looks at deformations and smoothings of reducible curve singularities. It contains a general criterion to decide when deformations of sub-curves give rise to a deformation of their union: the intersection multiplicity has to remain constant. With it all curves with $g \leq 3$ are shown to be smoothable, with the well known exception r lines in general position in $\mathbb{C}^{r-3}$, $r > 14$. As example for adjacency computations curves with $g = 2$ of embedding dimension at most 3 are considered. For the singularity of 5 lines through the origin the possible adjacencies depend on the moduli of the curve.

1 Reducible curve singularities

1.1 Definitions and Notations

Let $(X,0) \subset (\mathbb{C}^n,0)$ be a curve singularity, i.e. the germ of a reduced complex space of dimension one with a singular point at the origin. We denote its local ring by $\mathcal{O} = \mathcal{O}_{X,0}$. The normalisation is $n : \widetilde{X} \to X$ with semi-local ring $\widetilde{\mathcal{O}} = \mathcal{O}_{\widetilde{X},n^{-1}(0)}$. Let $r(X)$ be the *number of branches* of X; we simply write r if no confusion can arise. The δ-invariant is $\delta(X) = \dim_{\mathbb{C}}(\widetilde{\mathcal{O}}/\mathcal{O})$. We define the *genus* of X as $g = \delta - r + 1$. If X has a smoothing, then g is just equal to the genus of the Milnor fibre (an open Riemann surface); this follows from the formula $\mu = 2\delta - r + 1$ [4].

Some other invariants of curve singularities also play a role. The *Cohen-Macaulay type* is $t(X) = \dim(\omega/\mathfrak{m}\omega)$, where $\omega = \omega_X$ is the dualising sheaf and $\mathfrak{m}$ is the maximal ideal. The number t is also equal to the rank of the highest syzygy module. This has consequences for the equations defining X. In particular, by the Hilbert-Burch-Schaps theorem the ideal of a space curve (in $\mathbb{C}^3$) is generated by the $(t \times t)$-minors of a $t \times (t+1)$-matrix [14].

The *Tyurina number* τ is the dimension of T_X^1. The dimension e of smoothing components is independent of the component. For quasi-homogeneous curves one has $e = \mu + t - 1$ [9].

(1-1) As the classification of irreducible curve singularities can be handled by the methods of Ebey [7] (see also 3-1), we turn our attention to reducible curves.

Definition 1-2. ([10]) *Let $(X_1,0)$ and $(X_2,0)$ be two curves in $(\mathbb{C}^n,0)$ without common components, defined by ideals I_1 and I_2 in $\mathcal{O}_n = \mathcal{O}_{\mathbb{C}^n,0}$. The intersection multiplicity of $(X_1,0)$ and $(X_2,0)$ is:*

$$(X_1 \cdot X_2) = \dim_{\mathbb{C}} (\mathcal{O}_n/(I_1 + I_2)).$$

Remark 1-3. *The equivalent formula $(X_1 \cdot X_2) = \dim_{\mathbb{C}}(\mathcal{O}_{X_1}/I_2\mathcal{O}_{X_2})$ is sometimes easier to use. One can define $(Y_1 \cdot Y_2)$, if finite, also for general germs, but the interpretation as intersection multiplicity is less clear (a thorough discussion of such problems can be found in the Introduction of [21]). One has, if $X_i \subset Y_i$, that $(X_1 \cdot X_2) \leq (Y_1 \cdot Y_2)$.*

Proposition 1-4. ([10]) *The δ-invariant of the union of two curves X_1 and X_2 is given by:*

$$\delta(X_1 \cup X_2) = \delta(X_1) + \delta(X_2) + (X_1 \cdot X_2).$$

So $g(X_1 \cup X_2) = g(X_1) + g(X_2) + (X_1 \cdot X_2) - 1$. In particular $g \geq 0$.

PROOF. Let $I = I_1 \cap I_2$ be the ideal of $X_1 \cup X_2$; the proposition follows from the exact sequence:

$$0 \longrightarrow \mathcal{O}_n/I \longrightarrow (\mathcal{O}_n/I_1) \oplus (\mathcal{O}_n/I_2) \longrightarrow \mathcal{O}_n/(I_1 + I_2) \longrightarrow 0,$$

once one notices that the integral closure of $\mathcal{O}_n/I$ in its total ring of fractions is equal to the direct sum of the integral closures of $\mathcal{O}_n/I_1$ and $\mathcal{O}_n/I_2$. □

Remark 1-5. *The previous proposition enables an inductive approach to the classification of reducible singularities: one constructs such singularities from known building blocks, and one has to describe how these parts fit together. The next subsections treat this problem for low values of the intersection multiplicity $(X_1 \cdot X_2)$.*

2   Decomposable curves

Definition 2-1. *The curve singularity X is decomposable (into X_1 and X_2), if X is the union of two curves X_1 and X_2, which lie in smooth spaces intersecting each other transversally in the singular point of X. We write $X = X_1 \vee X_2$.*

Lemma 2-2. *Let TX_i be the Zariski tangent space in 0 to the curve $(X_i, 0)$. Then:*

$$(X_1 \cdot X_2) - 1 \geq \dim_{\mathbb{C}}(TX_1 \cap TX_2).$$

PROOF. The dimension of $TX_1 \cap TX_2$ is $\dim(\mathfrak{m}_n/(I_1 + I_2 + \mathfrak{m}_n^2))$. □

Corollary 2-3. ([10]) *One has $(X_1 \cdot X_2) = 1$, if and only if $X_1 \cup X_2 = X_1 \vee X_2$.*

2.1   Some properties of decomposable curves

The analytic type of $X \vee Y$ is completely determined by that of X and Y, so for purposes of classification it suffices to consider indecomposable curves. Numerical invariants can often be computed from the factors in the decomposition.

Let the curve $X \subset \mathbb{C}^m$ be given by equations $f_i(x_1, \ldots, x_m)$, $i = 1, \ldots, s$, and $Y \subset \mathbb{C}^n$ by $g_i(y_1, \ldots, y_n)$, $i = 1, \ldots, t$. Equations for $X \vee Y$ are f_i, g_i and $x_i y_j$, $i = 1, \ldots, m$, $j = 1, \ldots, n$. For the Tyurina number of a wedge of curves, see [3, Appendix].

For the Cohen-Macaulay type one has the following formula:

$$t(X \vee Y) = t(X) + t(Y) + 1,$$

which is also correct if X or Y is smooth, if one puts formally $t(L) = 0$ for a smooth curve L.

PROOF. (OF THE FORMULA) We describe the dualising sheaf ω with Rosenlicht differentials [15, IV.9]: let as before $n: \tilde{X} \to X$ be the normalisation and let $\Omega_{\tilde{X}}(*)$ be the sheaf of differentials on $\tilde{X}$ with arbitrary poles in the points of $n^{-1}(0)$; then:

$$\omega_{X,0} = n_* \left\{ \alpha \in \Gamma(\tilde{X}, \Omega_{\tilde{X}}(*)) \;\middle|\; \sum_{p \in n^{-1}(0)} \mathrm{res}_p f\alpha = 0 \text{ for all } f \in \mathcal{O}_X \right\}.$$

Let s be a local parameter on a branch of X, and t a local parameter on a branch of Y. Then $\omega_{X \vee Y} = \omega_X + \omega_Y + \mathbb{C}[ds/s - dt/t]$, so:

$$\omega/\mathfrak{m}\omega = \mathbb{C}\left[\frac{ds}{s} - \frac{dt}{t}\right] \oplus \omega_X/(\mathfrak{m}_X\omega_X + \mathfrak{m}_X(ds/s)) \oplus \omega_Y/(\mathfrak{m}_Y\omega_Y + \mathfrak{m}_Y(dt/t)).$$

The formula follows because $\mathfrak{m}_X(ds/s) \subset n_*\Omega_{\tilde{X}} = \mathfrak{c}_X\omega_X$ [15], where $\mathfrak{c}$ is the conductor ideal. □

2.2 Curves containing a smooth line

Lemma 2-4. *The embedding dimension of a curve $X = X_1 \cup X_2$ satifies:*

$$\mathrm{ed}(X) \leq \mathrm{ed}(X_1) + \mathrm{corank}(I_1\mathcal{O}_X).$$

In particular, if X_2 is a smooth branch, then $\mathrm{ed}(X) \leq \mathrm{ed}(X_1) + 1$.

PROOF. We recall that $\mathrm{corank}(I_1\mathcal{O}_X)$ is the minimal number of generators of the $\mathcal{O}_X$-ideal $I_1\mathcal{O}_X$; it is equal to the minimal number of generators of the $\mathcal{O}_{X_2}$-ideal $I_1\mathcal{O}_{X_2}$. The lemma follows from the exact sequence:

$$0 \longrightarrow I_1\mathcal{O}_X \longrightarrow \mathfrak{m}_X \longrightarrow \mathfrak{m}_{X_1} \longrightarrow 0. \qquad □$$

Example 2-5. The union of two plane curves can have arbitrary embedding dimension: take two copies of $\mathbb{C}^2$, $Y_1 \cong Y_2 \cong \mathbb{C}^2$, and let $Y = Y_1 \cup Y_2 \subset \mathbb{C}^{n+3}$, with coordinates $(x, y, z_0, \ldots, z_n)$, be given by: $Y_1 = \{z_i = x^{n-i}y^i\}$, $Y_2 = \{z_i = 0\}$. Then $Y_1 \cap Y_2 = \mathcal{O}_2/\mathfrak{m}^n$. Now take curves $X_i \subset Y_i$ with $\mathrm{mult}(X_i) > n$.

(2-6) We first recall some definitions [13]. Let $\mathcal{D}$ be the group of germs of holomorphic diffeomorphisms $h: (\mathbb{C}^n, 0) \to (\mathbb{C}^n, 0)$ and let I be an ideal; then $\mathcal{D}_I = \{h \in \mathcal{D} \mid h^*I = I\}$. We also need k-jets of such diffeomorphisms: let $\mathcal{D}^{(k)} = \{h \in \mathcal{D} \mid h \equiv \mathrm{id} \bmod \mathfrak{m}^{k+1}\}$, then $\mathcal{D}_I^{(k)} = \mathcal{D}_I \cap \mathcal{D}^{(k)}$, $J^k\mathcal{D} = \mathcal{D}/\mathcal{D}^{(k)}$ and $J^k\mathcal{D}_I = \mathcal{D}_I/\mathcal{D}_I^{(k)}$.

Proposition 2-7. *Let the curve X be minimally embedded in $(\mathbb{C}^n, 0)$; let $Y_1 = X \cup L_1$, $Y_2 = X \cup L_2$ with L_i smooth branches given parametrically by $\varphi_i: (\mathbb{C}, 0) \to (\mathbb{C}^{n+1}, 0)$. Suppose that $(X \cdot L_i) = s > 1$. If $j^{s-1}\varphi_1 = j^{s-1}\varphi_2$, then Y_1 is isomorphic to Y_2.*

PROOF. Let t be a coordinate on $\mathbb{C}$ and $(z_0, \ldots, z_n)$ coordinates on $\mathbb{C}^{n+1}$. Because $z_i \circ (\varphi_1 - \varphi_2)(t) \in \mathfrak{m}^s$, there exist functions f_i in the ideal I of X, such that $z_i \circ f_i(t) = z_i \circ (\varphi_1 - \varphi_2)(t)$, so the diffeomorphism $z_i \mapsto z_i - f_i$ is an element of $\mathcal{D}_I$; it gives the required isomorphism. $\qquad\square$

(2-8) The curves $X \cup L_1$ and $X \cup L_2$ are isomorphic under an isomophism which preserves X, if and only if the $(s-1)$-jets $j^{s-1}\varphi_1$ and $j^{s-1}\varphi_2$ are RL-equivalent under the action of $J^{s-1}\mathcal{D}_1 \times J^{s-1}\mathcal{D}_I$. In particular, if $s = 2$, only the tangent direction T of the line L matters.

The branch L of the curve $X \cup L$ can be given in the following normal form: $z_0 = t^s$, $z_i = \varphi_i(t)$, $1 \le i \le n$, with φ_i a polynomial in t with degree less then s. If furthermore $(X \cdot L') = s$, where L' is the projection of L onto the $(z_i, \ldots, z_n)$-space, then the curve is isomorphic to $X \cup L'$, because the projection is a map between curves with the same δ-invariant, which is an isomorphism outside the singular points: such a map is an isomorphism.

Example 2-9. Let X be the curve L_5^3, consisting of 5 lines in general position through the origin in $\mathbb{C}^3$, and let L be a new smooth branch with tangent distinct from the lines in L_5^3. If $(X \cdot L) = 2$, then the isomorphism type of $X \cup L$ is completely determined by the direction of the tangent of L. If L is tangent to the unique quadric cone through X, then the embedding dimension is four, whereas otherwise it is 3: L_6^3 (six lines in $\mathbb{C}^3$) has $\delta = 8$, unless the singularity is the complete intersection of a quadric and a cubic cone (then $\delta = 9$).

(2-10) Consider indecomposable curves of the form $X \cup L$ with $(X \cdot L) = 2$ with X decomposable, so $X = X_1 \vee X_2$. Then L is tangent to the Zariski tangent space TX of X, but not to TX_1 or TX_2: if L is tangent to TX_1, then $(L \cdot X_1) \ge 2$ and:

$$g(X_1) + g(X_2) + 1 = g(X \cup L)$$
$$= (g(X_1) + (L \cdot X_1) - 1) + g(X_2) + ((X_1 \cup L \cdot X_2) - 1),$$

so $X \cup L \cong (X_1 \cup L) \vee X_2$. Conversely, the equations for X (see 2-1) show that $(X \cdot L) = 2$, when L is tangent to TX, but not to TX_1 or TX_2. This argument extends to the case that X is a wedge of several curves.

Corollary 2-11. Let $X = X_1 \vee \ldots \vee X_k$ for some $k > 1$. Consider indecomposable curves $X \cup L$ with $(X \cdot L) = 2$; let T_i be the projection of the tangent line of L onto the Zariski tangent space TX_i of X_i. If at least $k-1$ curves X_i admit a $\mathbb{C}^*$-action, which preserves the direction of T_i, then after a choice of coordinates, in which the curve $T_1 \vee \ldots \vee T_k$ forms the coordinate axes of $(z_1, \ldots, z_k)$-coordinates, the line L can be given by $z_i = t$, $i = 1, \ldots, k$. In particular, the condition on the $\mathbb{C}^*$-action is satisfied for curves $X = X_1 \vee L_k^k$.

PROOF. The argument in 2-10 shows that L is in general position in the vector space spanned by the T_i. Because of the $\mathbb{C}^*$-actions one can put L in the desired position. $\qquad\square$

Example 2-12. Consider $X \cup L$ with $X = D_5 \vee D_5$, and let the projection of L on the tangent space of both factors D_5 be in general position (so the projection forms a $T_{2,4,5}$-singularity). To study moduli, I computed infinitesimal deformations with the computer algebra system *Macaulay* [2]. To get quasi-homogeneous equations, start with the line L in special position, such that the projection is tangent to the cusps in the D_5 singularities. Consider the two parameter family, in which L is given by $z - \lambda x$, $y - xs$ and $w - zs$, so there are six equations:

$$\lambda^2 x(y^2 - x^3) + x^2 z^2 - \lambda x^2 z s^2, \qquad z(w^2 - z^3) + \lambda^2 x^2 z^2 - \lambda x z^2 s^2,$$
$$xz(\lambda x - z), \qquad yw - xzs^2, \qquad xw - xzs, \qquad yz - xzs,$$

where x, y, z and w are variables on $\mathbb{C}^4$. I computed T^1 for specific values of λ, while s was considered to be a constant. For $\lambda = 1$, $s = 0$ the singularity is quasi-homogeneous and $\tau = 19$; for $\lambda = 1$, $s = 1$ it is not quasi-homogeneous, and $\tau = 18$. For $s = 0$ the deformation $[\partial/\partial\lambda]_{\lambda=1}$ is trivial (use the $\mathbb{C}^*$-action on D_5), but not the deformation $[\partial/\partial s]s = 0$. For $s = 1$ the deformation in the s-direction is trivial, but λ gives a modulus.

Lemma 2-13. *Consider the monomial curve $M_{(n)}$ with semigroup $(n, \ldots, 2n - 1)$, and ideal I. Let its osculating spaces be $L_{n-i} : z_{i+1} = \ldots = z_n = 0$, $i = 0, \ldots, n$. The orbits of the action of $J^1\mathcal{D}_I$ on $\mathbb{C}^n$ are the spaces $L_i \setminus L_{i+1}$, $i = 0, \ldots, n - 1$.*

PROOF. Because of the $\mathbb{C}^*$-action of the curve, it suffices to show that the z_j-axis can be moved in the directions in $L_{n-j} \setminus L_{n-j+1}$. Let $\varphi \in \mathcal{D}_I$ have the 1-jet $j^1\varphi = \mathrm{Id} + A$ with A an upper triangular matrix. If $n \colon \mathbb{C} \to M_{(n)}$ is the normalisation, then $z_1 \circ j^1\varphi \circ n = t^n + a_{12}t^{n+1} + \ldots + a_{1n}t^{2n-1}$. Take a new parametrisation $\tau = t(1 + a_{12}t + \ldots + a_{1n}t^{n-1})^{1/n}$ and determine φ such that $z_i \circ \varphi = \tau^{n+i-1}$; one obtains equations for the coefficients of A: they can be expressed in $a_{12}, \ldots, a_{1n}$, with in column j only entering $a_{12}, \ldots, a_{1j}$, and a_{1i} entering linearly in $a_{j-i+1,j}$. Therefore the z_j-axis can be moved in the directions in $L_{n-j} \setminus L_{n-j+1}$.   □

Example 2-14. We describe the possible isomorphism types of indecomposable curves X, consisting of the union of an A_k-singularity (with $g = [k/2]$) and r smooth branches, such that $g(X) = [k/2] + 1$. Let the A_k lie in the (x_1, x_2)-plane: $x_1^2 - x_2^{k+1} = 0$.

Suppose that $r > 1$. Choose one line L and let X' be the union of all other branches, so X' contains the A_k. By the sum formula (1-4) for the genus $g(X) = g(X') + (X' \cdot L) - 1$; because X is indecomposable, $(X' \cdot L) > 1$, so $g(X') = [k/2]$ and $(X' \cdot L) = 2$. Therefore $X' \cong A_k \vee L_{r-1}^{r-1}$. We may take the $r - 1$ smooth branches as coordinate axes in the $x_3, \ldots, x_{r+1}$ directions. The isomorphism type of X depends only on the position of the tangent to the projection of L onto the (x_1, x_2)-plane: if it is in general position, which means that it lies in $L_0 \setminus L_1$, then we can give the smooth branch L by $(x_1, \ldots, x_{r+1}) = (0, t, \ldots, t)$;

we denote X by $S_r^{(0)}(A_k)$. If the projection is tangent to the x_1-axis, then we get the curve $S_r^{(1)}(A_k)$ with L given by $(t, 0, t, \ldots, t)$.

If $r = 1$, then $S_1^{(0)}(A_k) = D_{k+3}$, and $S_1^{(1)}(A_k)$ has embedding dimension 3 with L given by $(t, 0, t^2)$.

2.3   Reducible curves with $(X \cdot Y) = 2$

Consider a reducible curve $X \cup Y$ with $(X \cdot Y) = 2$. Let $T = TX \cap TY$; then $\dim T = 1$ by Lemma 2-2. Essentially we are interested in the curve $X \cup T$; because possibly $(X \cdot T) > 2$, we consider in the following a curve $X \cup L$, where L is a line with $(X \cdot L) = 2$ and T as tangent line; its isomorphism type is uniquely determined. We denote by $Y \cup L$ a similar curve, containing Y.

Lemma 2-15.   *If* $\mathrm{ed}(X \cup L) = \mathrm{ed}(X)$ *or* $\mathrm{ed}(Y \cup L) = \mathrm{ed}(Y)$, *then* $\mathrm{ed}(X \cup Y) = \mathrm{ed}(X) + \mathrm{ed}(Y) - 1$; *otherwise* $\mathrm{ed}(X \cup Y) = \mathrm{ed}(X) + \mathrm{ed}(Y)$.

PROOF. We will bring $X \cup Y$ into a prenormal form. Suppose that $\mathrm{ed}(Y \cup L) > \mathrm{ed}(Y)$ if $\mathrm{ed}(X \cup L) > \mathrm{ed}(X)$. We may suppose that we have coordinates

$$(x_1, \ldots x_{m-1}, y_1, \ldots, y_{n-1}, t, s),$$

where $\mathrm{ed}(X) = m$ and $\mathrm{ed}(Y) = n$; that X lies in $\mathbf{y} = s = 0$ and T is the t-axis, while Y lies in the smooth space $E(Y)$: $s = \varphi_0(\mathbf{y}, t)$, $x_i = \varphi_i(\mathbf{y}, t)$. If $\mathrm{ed}(X \cup L) = \mathrm{ed}(X)$, then $(X \cdot T) = 2$ and (2-7) gives an X preserving transformation of the space $\mathbf{y} = 0$, which brings $E(Y) \cap (\mathbf{y} = 0)$ to the t-axis, and which may be extended to the total space, bringing $E(Y)$ to $\mathbf{x} = s = 0$. Otherwise $\varphi_0|_{y=0}$ has to be quadratic in t, so Y can be brought to lie in $\mathbf{x} = s - t^2 = 0$.   □

Lemma 2-16.   *If X or Y admits a $\mathbb{C}^*$-action, leaving T invariant, then the isomorphism types of $X \cup L$ and $Y \cup L$ determine the type of $X \cup Y$; otherwise there can be an additional modulus.*

PROOF. Suppose that X admits a $\mathbb{C}^*$-action, fixing T. Let $X \cup Y_1$ and $X \cup Y_2$ be two curves in prenormal form and let $\psi \colon Y_1 \cup L \to Y_2 \cup L$ be an isomorphism.

We first consider the case that $\mathrm{ed}(X \cup L) = \mathrm{ed}(X)(= m)$; let $\mathrm{ed}(Y) = n$. Because $(X \cdot T) = 2$, there exist an extension $\widetilde{\psi} \colon \mathbb{C}^{n+m-1} \to \mathbb{C}^n$ of ψ with $\widetilde{\psi}|_{x=0} = \psi$, while the components of $(\widetilde{\psi} - j^1\widetilde{\psi})|_{y=0}$ are in the ideal of X. Choose an X-preserving automorphism $\varphi \colon \mathbb{C}^n \to \mathbb{C}^n$ with $\varphi(T) = T$, $\varphi|_T = j^1\psi|_T$. Let pr_X be the projection on $TX = \{\mathbf{y} = 0\}$, and pr_Y on TY. Then $\widetilde{\psi} + \varphi \circ \mathrm{pr}_X - \varphi|_T \circ \mathrm{pr}_X \circ \mathrm{pr}_Y$ is an isomorphism, which sends $X \cup Y_1$ to $X \cup Y_2$.

In the other case we extend ψ to a map $\widetilde{\psi} \colon \mathbb{C}^{n+m} \to \mathbb{C}^{n+1}$ with $\widetilde{\psi}|_{y=s=0} = j^1\psi|_T$ (if $\psi_i = t^2\alpha_i(t) + \beta_i(\mathbf{y}, t)$ in coordinates $(\mathbf{y}, t)$ on $E(Y)$, we write $\widetilde{\psi}_i = s\alpha_i(t) + \beta_i(\mathbf{y}, t)$) and we take $\varphi \colon \mathbb{C}^{n+m} \to \mathbb{C}^{m+1}$ with $\varphi|_{x=s-t^2=0}(t) = j^1\psi|_T(t)$. Now we use $\widetilde{\psi} + \varphi - j^1\psi|_T \circ \mathrm{pr}_T$.

Without the $\mathbb{C}^*$-action we can do the construction if $j^1\psi|_T = 1$, using $\varphi = \mathrm{id}$; the value of $j^1\psi|_T$ is then a modulus.   □

Example 2-17. Let $X \cong Y \cong D_5$, and let the intersection line T of TX and TY be in general position in both planes. To get quasi-homogeneous equations (to enable *Macaulay* computations) we introduce constants λ and μ. The ideal of X is $(y, s, (x-t)(x^3 - \lambda t^2))$ and that of Y is $(x, s - t^2, (y-t)(y^3 - \mu t^2))$. The dimension of T^1 is 19. Both curves separately are isomorphic to curves with $\lambda = 1$ and $\mu = 1$, but for $X \cup Y$ the ratio $(\lambda : \mu)$ is a modulus: the deformation $\lambda(\partial/\partial\lambda) + \mu(\partial/\partial\mu)$ is trivial, but $\partial/\partial\lambda$ represents a non trivial infinitesimal deformation.

2.4   Equations

We describe equations for $X \cup Y$ in the prenormal form given above. They are closely related to the equations for $X \cup L$ and $Y \cup L$.

Let X be given in $\mathbb{C}^m$, where we have coordinates $(\mathbf{x}, t)$, by equations f_i. Suppose that $\mathrm{ord}_t f_i(\mathbf{0}, t)$ is minimal for $i = 1$; after multiplying with a unit, $f_1(\mathbf{0}, t) = t^p$ for some p, and we can achieve that $f_i(\mathbf{0}, t) \equiv 0$ for $i > 1$, by subtracting a suitable multiple of f_1. Then $(X \cdot T) = 2$ if and only if $p = 2$. Choose similar equations g_i for Y.

If $p_X = p_Y = 2$, then equations for $X \cup Y$ are: $x_i y_j$, f_i with $i > 1$, g_j with $j > 1$ and $f_1 + g_1 - t^2$.

If $2 = p_X < p_Y$, then one has the following equations: $x_i y_j$, f_i with $i > 1$, g_j with $j > 1$, $t^{p_Y - 2} f_1 + g_1 - t^{p_Y}$ and $x_i f_1$. This is not necessarily a minimal set of equations: relations can exist between the equations $x_i f_1$, f_j; as these equations describe $X \cup T$, it is sufficient to look only at this space.

Finally, if $p_X > 2$ and $p_Y > 2$, then $X \cup Y$ is contained in the space given by $x_i y_j$, $x_i s$, $y_j (s - t^2)$ and $s(s - t^2)$. One has to extend the f_i to equations in $\mathbf{x}$, s and t with $f(\mathbf{0}, s, t)$ divisible by $s - t^2$, and the g_i to equations in $\mathbf{y}$, s and t with $g(\mathbf{0}, s, t)$ divisible by s.

Example 2-18. Consider the union of the space curve $X = M_{(3)}$ given by $x = t^3$, $y = t^4$ and $z = t^5$, and a line. The equations of the curve are $f = xz - y^2$, $g = yz - x^3$ and $h = yx^2 - z^2$. If the line L is the z-axis, then h serves as f_1, but f and g generate the ideal of $X \cup L$; we have the curve U_7 [8]. If L is the y-axis, then the equations of $X \cup L$ are g, h and xf. Finally, if L is tangent to the x-axis, then the equations are the (2×2)-minors of the matrix:

$$\begin{pmatrix} x & y & z & s \\ y & z & x^2 - s & 0 \end{pmatrix}.$$

Example 2-19. Let the curve X consist of the union of an A_{k-1} and an A_{l-1} singularity with $(A_{k-1} \cdot A_{l-1}) = 2$. For simplicity we only give parametrisations for even k and l. We look at combinations of the two possible positions of a line in the the plane of A_{k-1} and A_{l-1}. First both positions can be general; we take coordinates (x, y, z), and take x and z-axes as tangent lines: give A_{k-1} by $(s^2, s^k, 0)$ and A_{l-1} by $(0, t^l, t^2)$. We have equations: xz and $y^2 - x^k - z^l$; the curve is $P_{k,l}$ in the list of Wall [22]. If $l = 2$, we have the simple curve S_{k+3} of Giusti [8].

If one position is special, we have A_{k-1} as $(s^2, s^k, 0)$ and A_{l-1} as $(0, t^2, t^l)$. Equations are the (2×2)-minors of the matrix:

$$\begin{pmatrix} z & x & y^{l-2} \\ y^2 - x^k & 0 & z \end{pmatrix}.$$

Finally, if both positions are special, we have a curve of embedding dimension four with parametrisation $(s^2, s^k, 0, 0)$ and $(t^2, 0, t^l, t^4)$.

Lemma 2-20. *The type of $X \cup Y$ with $(X \cdot Y) = 2$ can be computed from that of $X \cup L$ and $Y \cup L$; one has $t(X \cup Y) = t(X \cup L) + t(Y \cup L) - 1$.*

PROOF. As before, we describe ω with Rosenlicht differentials. We first consider $\omega_{X \cup L}$. Because $\omega/n_* \Omega$ is dual to $\tilde{\mathcal{O}}/\mathcal{O}$, and $\delta(X \cup L) = \delta(X) + 2$, we know that $\omega_{X \cup L}$ contains two linear independent differentials with poles on L. Let t be a coordinate on L and $s_1, \ldots, s_r$ local parameters on the branches of X. For the two differentials we take $dt/t - ds_1/s_1$ and $dt/t^2 - \sum \varphi_i ds_i$, with φ_i meromorphic, such that for every extension t of t to a function on the ambient space $\sum \mathrm{res}\, t \varphi_i ds_i = 1$; in particular $ds_1/s_1 - \sum t \varphi_i ds_i \in \omega_X$. So:

$$\omega_{X \cup L} = \omega_X + \mathbb{C}\{t\}dt + \mathbb{C}\left[\frac{dt}{t} - \frac{ds_1}{ds_1}\right] + \mathbb{C}\left[\frac{dt}{t^2} - \sum \varphi_i ds_i\right]$$

and we can describe $\omega_{X \cup L}/\mathfrak{m}\omega$ completely on $\tilde{X}$:

$$\omega_{X \cup L}/\mathfrak{m}\omega = \left(\omega_X + \mathbb{C}\left[\sum \varphi_i ds_i\right]\right) \Big/ \left(\mathfrak{m}\omega_X + \mathfrak{m}\left[\frac{ds_1}{ds_1}\right] + (\mathbf{x})\left[\sum \varphi_i ds_i\right]\right).$$

One has a similar formula for $\omega_{Y \cup L}$; we write s_i' for the parameters on the branches of Y. Then:

$$\omega_{X \cup Y} = \omega_X + \omega_Y + \mathbb{C}[ds_1/s_1 - ds_1'/ds_1'] + \mathbb{C}\left[\sum \varphi_i ds_i - \sum \varphi_i' ds_i'\right].$$

Because $ds_1/s_1 - t \sum \varphi_i ds_i \in \omega_X$, one has $t(\sum \varphi_i ds_i - \sum \varphi_i' ds_i') \equiv ds_1/ds_1 - ds_1'/ds_1'$ modulo $\omega_X + \omega_Y$. One can find an extension t of t, such that t^2 vanishes on Y; therefore $t ds_1/s_1 \in \mathfrak{m}_{X \cup Y}\omega_{X \cup Y}$. Now the formula follows. $\qquad \square$

Corollary 2-21. $t(X \cup Y) \leq t(X) + t(Y) + 1$.

Example 2-22. The type $t(X \cup Y)$ can take any value between 1 and $t(X) + t(Y) + 1$; this is already true for $t(X \cup L)$. Consider the union of a smooth branch L and the monomial curve $M_{(n)}$ (cf.2-13). We parametrise $M_{(n)}$ by $z_j = s^{n+j-1}$. If the tangent of the line lies in the osculating space $L_i \setminus L_{i+1}$, then we assume L to be the z_{n-i}-axis. Let t be a parameter on L. We have the following regular differentials, which are not holomorphic on the normalisation: $ds/s^2, \ldots, ds/s^n$, $dt/t - ds/s$ and $\alpha = dt/t^2 - ds/s^{2n-i}$. One has $z_1 \alpha = ds/s^{n-i}, \ldots, z_{n-i-1}\alpha = ds/s^2$ and $z_{n-i}\alpha = dt/t - ds/s$. So a basis of $\omega/\mathfrak{m}\omega$ is $ds/s^{n-i+1}, \ldots, ds/s^n$, $dt/t^2 - ds/s^{2n-i}$, and therefore the type is $i + 1$.

3 Classification

In this section we apply the methods of the previous section to give a classification of curves with $g = \delta - r + 1 \leq 3$. The list is known in the cases $g = 1$ [9] and $g = 2$ [16, 18], but we include it as preparation for $g = 3$.

A classification scheme gives rise to a system of names for curves; as the number of types grows rapidly with g, one gets complicated, composite names. Different classes are recognised to belong to a same, bigger class. Therefore some curves have more than one name. In general we prefer to use the simpler and more familiar one; e.g., in the notation of Example (2-14) $D_{k+3} = S_1^{(0)}(A_k)$.

3.1 Classification of irreducible curves

The classification problem in this case is solved by Ebey [7]. Let c be the conductor of the semigroup Γ of X, let $n_1 < \ldots < n_k$ be the elements of Γ which are smaller than c, and let $m_{i1} < \ldots < m_{is_i}$ be the gaps between n_i and c for $i = 1, \ldots, k$. Then the local ring of X is isomorphic to a subring of $\mathbb{C}\{t\}$, generated by $t^{n_i} + a_{i1}t^{m_{i1}} + \ldots + a_{is_i}t^{m_{is_i}}$, $i = 1, \ldots, k$. The solvable group G/G_c of transformations $t \mapsto \alpha_1 t + \ldots + \alpha_{c-1}t^{c-1}$ acts on the affine space of the coefficients a_{ij} and two curve singularities are isomorphic if and only if their parameter points are in the same orbit.

Ebey also gives all isomophism classes of irreducible curves with small δ. For $\delta < 5$ one only finds monomial curves; they are given by the following semigroups:

$$\delta = 1: \quad (2,3)$$
$$\delta = 2: \quad (2,5), \ (3,4,5)$$
$$\delta = 3: \quad (2,7), \ (3,4), \ (3,5,7), \ (4,5,6,7)$$
$$\delta = 4: \quad (2,9), \ (3,5), \ (3,7,8), \ (4,5,6), \ (4,5,7), \ (4,6,7,9), \ (5,6,7,8,9).$$

3.2 Classification of curves with $g = 0$

An irreducible curve with $g = 0$ has $\delta = 0$, so is smooth. The genus formula for reducible curves (Proposition 1-4) gives for $X = \bigcup_{i=1}^r X_i$ that $0 = g(X) = g(X_r) + g(\cup_{i<r} X_i) + (X_r \cdot \cup_{i<r} X_i) - 1$, so the r-th branch X_r is smooth and $X \cong X_r \vee \cup_{i<r} X_i$; by induction X is isomorphic to L_r^r, the singularity consisting of the r coordinate axes in $\mathbb{C}^r$.

3.3 Classification of curves with $g = 1$

If $r = 1$, then $X \cong A_2$. Suppose that the reducible curve $X = X_1 \cup \ldots \cup X_r$ is indecomposable. Then $g(X_r) = g(\cup_{i<r} X_i) = 0$ and $(X_r \cdot \cup_{i<r} X_i) - 1 = 1$. So X_r is a smooth branch which intersects a curve of type L_{r-1}^{r-1} with multiplicity two. Therefore, by the results of (2.2), $X \cong A_3$ for $r = 2$ and for $r > 2$: $X \cong L_r^{r-1}$, the singularity consisting of r generic lines through the origin in $\mathbb{C}^{r-1}$.

3.4   Classification of curves with $g = 2$

If the curve X is irreducible, then it is A_4 or $M_{(3)}$. If X is reducible, we suppose that it is indecomposable. We can write $X = X_1 \cup X_2$; such a splitting is not unique, and if necessary, we change it to a more suitable one. The possible values for genus and intersection multiplicity are: $g(X_1) = g(X_2) = 0$ and $(X_1 \cdot X_2) = 3$, or: $g(X_1) = 1$, $g(X_2) = 0$ and $(X_1 \cdot X_2) = 2$. We claim that we can choose the splitting in such a way that X_2 is a line: if $X_2 = L_k^k$ with $k > 1$, we consider a line L in X_2 and the splitting $(X_1 \cup L_{k-1}^{k-1}) \cup L$; then $g(X_1 \cup L_{k-1}^{k-1}) < 2$ by indecomposability.

   If $g(X_1) = g(X_2) = 0$, then $X_1 = L_k^k$ for some k; the line X_2 has $(X_1 \cdot X_2) > 2$ only if X_2 is tangent to a branch of L_k^k, so if $k > 1$ the curve X contains a subcurve of genus 1, which we can use to make a new splitting. Therefore we only have to consider the case $k = 1$; the curve X consists of two smooth branches with intersection multiplicity 3: this is A_5.

   If $g(X_1) = 1$, the classification proceeds by (2-2): we only have to understand the position of a line in the tangent space of an indecomposable curve of genus one: for A_2 and A_3 this question is treated in Example 2-14; in the other cases we deal with L_{n+1}^n, and as the 1-jet of a diffeomorphism fixing L_{n+1}^n is the identity, we get the direction of the tangent to the new branch as modulus; if $n = 2$, the projection of X to the plane of the D_4-singularity is of type $\widetilde{E}_7$, with $\delta = 6$, one too much, so we have a curve $S_4^2(\lambda)$ with embedding dimension 3. For $n \geq 3$ the curve L_{n+1}^n is given by quadratic equations, so X has embedding dimension n: it consists of $n + 2$ lines through the origin in $\mathbb{C}^n$ with $n \geq 3$. We return to the moduli for these curves in (3.8).

   We list the indecomposable curves of genus 2 in Table 3.1. We use the names from [18]: G_{r+1}^c is used for $S_r^{(0)}(A_2)$, $S_{r+1}^c = S_r^{(1)}(A_2)$, $G_{r+2}^t = S_r^{(0)}(A_3)$, and $S_{r+2}^t = S_r^{(1)}(A_3)$ (cf. Example 2-14). In the list of the parametrisations the line $(0, \ldots, 0, t, \ldots, t) \subset \mathbb{C}^{r+s}$ is abbreviated as $(\mathbf{0}_r, t\mathbf{1}_s)$, and $\mathbf{0}_2 \times L_r^r$ stands for the last r coordinate axes in $\mathbb{C}^{r+2}$.

3.5   Classification of curves with $g = 3$

We proceed as in the case $g = 2$. If the curve is irreducible, we have the four types of monomial curves (3.1). If $X = X_1 \cup X_2$ is indecomposable, we have the following possibilities:

	$g(X_1)$	$g(X_2)$	$(X_1 \cdot X_2)$
I	0	0	4
II	1	0	3
III	2	0	2
IV	1	1	2

The cases III and IV fit into the framework of (2.3). We try to find a splitting of type IV. Again we may take X_2 to be a line if $g(X_2) = 0$.

Name	Parametrisation
A_4	(t^2, t^5)
G^c_{r+1}	$(t^2, t^3, \mathbf{0}_{r-1}), \mathbf{0}_2 \times L^{r-1}_{r-1}, (0, t\mathbf{1}_r)$
A_5	$(t, t^3), (t, -t^3)$
G^t_{r+2}	$(t, t^2, \mathbf{0}_{r-1}), (t, -t^2, \mathbf{0}_{r-1}), \mathbf{0}_2 \times L^{r-1}_{r-1}, (0, t\mathbf{1}_r)$
S^c_2	$(t^2, t^3, 0), (t, 0, t^2)$
$S^c_{r+1}, r \geq 2$	$(t^2, t^3, 0, \ldots, 0), \mathbf{0}_2 \times L^{r-1}_{r-1}, (t, 0, t\mathbf{1}_{r-1})$
S^t_3	$(t, t^2, 0), (t, -t^2, 0), (t, 0, t^2)$
$S^t_{r+2}, r \geq 4$	$(t, t^2, \mathbf{0}_{r-1}), (t, -t^2, \mathbf{0}_{r-1}), \mathbf{0}_2 \times L^{r-1}_{r-1}, (t, 0, t\mathbf{1}_{r-1})$
$M_{(3)}$	(t^3, t^4, t^5)
$S^2_4(\lambda)$	$(t, 0, 0), (0, t, 0), (t, t, 0), (t, \lambda t, t^2)$
$L^r_{r+2}(\lambda_p)$	$L^r_r, (t\mathbf{1}_p, \mathbf{0}_{r-p}), (t\lambda_p, t\mathbf{1}_{r-p})$

Remark: $G^c_2 = D_5, G^t_3 = D_6, G^c_3 = S_6$ and $G^c_4 = S_7$ [8].

Table 3.1: Indecomposable curves with $g = 2$.

Case I If X_2 is tangent to a branch of X_1 with intersection multiplicity less than 4, we can find a splitting of a different type, so we may assume that X_1 is irreducible: $X = A_7$.

Case II If X_1 is decomposable, then X_2 is tangent to the tangent space of one of the factors in a decomposition of X_1, for otherwise $(X_1 \cdot X_2) = 2$. This means that we have a subcurve as in case III or IV. So we take X_1 indecomposable with $g(X_1) = 1$.

If $X_1 = A_2$, then $X = E_7$.

Suppose $X_1 = A_3$. Let $\varphi: \mathbb{C} \to X_2$ be a parametrisation of X_2. If the smooth curve $j^2\varphi(\mathbb{C})$, lying in the plane of A_3, has intersection multiplicity at least 3 with a branch $X_{1,1}$ of A_3 (or coincides with $X_{1,1}$), then $X_{1,1} \cup X_2$ is of type A_5 (case III). Otherwise $X_1 \cup j^2\varphi(\mathbb{C})$ is of type $\widetilde{E}_8$, and X depends on one modulus. We call this curve $\widetilde{E}^{(2)}_8$.

If $X_1 = D_4$, then $X = \widetilde{E}_7$ — the cross-ratio of the four lines is a modulus; if this cross-ratio becomes zero, so the new line is tangent to one of the lines of D_4, then we can view this singularity as a D_6 with an extra line.

Finally, if X_1 has more than 3 branches, it is given by quadratic equations; because $(X_1 \cdot X_2) = 3$, X_2 is tangent to one of the branches. We take a new splitting, such that X_1 has a subcurve of type A_5 (case III) or one of type A_3 and $(X_2 \cdot A_3) = 2$; in the last case there exists a second subcurve of X of genus 1 (case IV).

Case III If $g(X_1) = 2$, and $X_1 = X'_1 \vee X''_1$, then X_2 is not tangent to X'_1 or X''_1; if $g(X'_1) = g(X''_1) = 1$, then we can consider X also as the union of the two curves $X_1' \vee X_2$ and X_1'' (case IV). So we may assume that X_1' is indecomposable of genus two. The cases $X_1' = A_4, A_5, M_{(3)}$ are covered by (2-13) and (2-14). For the other cases we need some new arguments; by choosing our splitting carefully we can minimize the computations.

If $X_1 = L_{r-k-1}^{r-k-3} \vee L_k^k$, then the direction of the tangent to the projection of X_2 on X_1' cannot be changed by linear coordinate changes; if this direction is the same as that of a line in X_1', then X has a subcurve of type A_3 or L_n^{n-1} and one of type L_{r-k-2}^{r-k-3} (case IV). Otherwise we get the singularity L_r^{r-3} (r lines through the origin in $\mathbb{C}^{r-3}$), except when $r = 6$ and the lines lie on a quadric cone (cf. Example 2-8). Such curves depend on $2(r-3)$ moduli.

Suppose $X_1' = S_4^2(\lambda)$. We determine the possible tangent directions in $S_4^2(\lambda)$. A tangent to the plane of the reduced tangent cone of S_4^2 gives a new modulus. If X_2 is not tangent to the (x, y)-plane in $\mathbb{C}^3$, then it can be made to be the z-axis: the transformation

$$
\begin{aligned}
x' &= x + az \\
y' &= y + bz + (\lambda a - b)y(x - y)/(1 - \lambda) \\
z' &= z + 2azx + zx^2
\end{aligned}
$$

preserves $S_4^2(\lambda)$, and moves the z-axis to the line with direction $(a : b : 1)$, so its inverse is the required transformation. We obtain the complete intersection U_{10}^* [1] ($= FZ_{0,0}$ [23]).

In the remaining cases X_1 is of the form $L_s^s \cup Y$ with $Y = A_2$ or A_3. If X_2 is tangent to TL_s^s, then we make a new splitting of type IV. If $X_1 = L_{k+m}^{k+m} \cup Y$ is $S_{k+1}^c \vee L_m^m$ or $S_{k+2}^t \vee L_m^m$ with $k \geq 2$, and L a line in L_k^k, then $(X_1 \setminus L) \cup X_2$ has a subcurve of type G_p^c or G_p^t (because X_2 is not tangent to L_{k+m}^{k+m}). So we look at the position of X_2 in X_1 of type G. The intersection of the tangent space TY of Y with L_k^k gives a $(k+1)$-st line in L_k^k, so the direction of the projection T_k of the tangent TX_2 on TL_k^k is fixed (if the projection is non zero), giving $k - 1$ moduli; using the $\mathbb{C}^*$-action we can bring the tangent line in a standard position in the plane spanned by T_k and the kernel of the projection, the tangent line to Y. In case the projection is zero, no moduli appear.

If $X_1 = S_3^t \vee L_m^m$, and the projection T_2 of TX_2 is not the tangent line T of $X_1' = S_3^t$ (if it is, we can make a splitting of type IV), then the three tangent planes to the three subcurves of X_1' of type A_3 and the plane through T_2 and T form four planes in a pencil, having a modulus. We take S_3^t as in Table 3.1, so the third branch lies in $y = z - x^2 = 0$; the transformation $x' = x + az$, $y' = y$ and $z' = z + 2axz + a^2z^2$ preserves S_3^t and shows that all directions in the plane through T_2 and T are equivalent: for the resulting curve $S_1^{(0)}(S_3^t, \lambda)$ we may take L to lie in $x = 0$. For S_2^c one has similar considerations, but no modulus appears; indeed the transformation $z' = z$, $y' = y + a(z - x^2)$, $x' = x - b(z - x^2) + y\psi(x, y)$, where b and ψ are determined by $\tau^3 = t^3 - at^4$, $\tau^2 = t^2 + t^3\psi(t^2, t^3) + bt^4$, is a S_2^c preserving diffeomorphism, which moves the

z-axis out of the (z,x)-plane. We obtain the simple curve U_8 [8]. The case that the line is tangent to the plane of the cusp can also be considered as an D_5 with a line tangent to the cusp, an already treated case.

Case IV We can reduce to the case that one of the curves with $g = 1$ is a cusp or tacnode. For if all branches of X are smooth with distinct tangents, we can construct the curve from a curve of $g = 2$ with distinct tangents and a smooth branch, see case III. If not all tangents are distinct, but X_1 and X_2 do not contain an A_2 or A_3, we choose a new splitting, in which X_1 has a subcurve of type A_3. If we do not obtain an already treated case, we get the situation that $g(X_1) = g(X_2) = 1$ and $X_1 = A_3 \vee L_k^k$. If $X_1 = Y \vee L_k^k$, with $Y = A_2$ or A_3, and $g(X_2) = 1$, then $(Y \cdot (L_k^k \cup X_2)) > 1$ by indecomposability, so $(X_2 \cdot L_k^k) = 1$. Therefore we only have to determine the possible positions of a cusp or tacnode in $X \vee L_k^k$ with X indecomposable of genus 1, and we apply the results of section (2.3).

3.6 Summary

Contrary to the $g = 2$ case we do not list the parametrisations, because such a list is not very illuminating. Instead we only give the discrete invariants multiplicity, embedding dimension, Cohen-Macaulay type and number of components (see Table 3.2, in which $k = 2, 3$, $l = 2, 3$, $m \geq 2$ and $r \geq 0$). In this table the symbol $\lceil p, q \rceil$ denotes the maximum of p and q. We now propose a system of names for curves with $g = 3$.

The symbol L_r^n denotes a curve of r straight lines through the origin in $\mathbb{C}^n$; in (3.8) we discuss the moduli. As we have seen in the case of L_6^3 the δ-invariant may depend on the moduli. We denote by S_r^n a curve with smooth branches, whose tangents form a L_r^n, and $\delta(S_r^n)$ has the (minimal) value of the generic L_r^n; in this case the embedding dimension may depend on the moduli. We allow the case $S_r^n = L_r^n$.

Curves $X \cup L_r^r$ with $(X \cdot L_r^r) = 2$ were treated in (2.2). We denote these by $S_r(X)$. The analytic type depends only of the position of projection of TL_r, the tangent to the last branch, on the first factor of $X \vee L_{r-1}^{r-1}$; if we have an osculating flag as in the case of monomial curves (2-13), we refine this notation to $S_r^{(i)}(X)$, $i \geq 0$, where increasing index signals increasing tangency. In addition moduli may be included. We use a similar notation to denote the position relative to X of an A_k curve with $(A_k \cdot X) = 2$: we write $A_k^{(0)}(X)$ or simply $A_k(X)$ in case of transverse intersection of the tangent line of A_k with TX, and $A_k^{(1)}(X)$ otherwise; if X is itself an A_l singularity, we write e.g. $A_k^{(1)}(A_l^{(1)})$ for the most special position.

3.7 Equations

From the parametrisations equations may be computed. We give them only for low embedding dimensions. For the plane and the complete intersection space

Name	mult	embdim	type	no. cpts
A_6	2	2	1	1
E_6	3	2	1	1
$M_{3,5,7}$	3	3	2	1
$M_{4,5,6,7}$	4	4	3	1
A_7	2	2	1	2
E_7	3	2	1	2
$\widetilde{E}_8^{(2)}$	3	3	2	3
$\widetilde{E}_7$	4	2	1	4
$S_r^{(0)}(A_4)$	$r+2$	$r+1$	1	$r+1$
$S_r^{(0)}(A_5)$	$r+2$	$r+1$	1	$r+2$
$S_r^{(0)}(M_{(3)})$	$r+3$	$r+2$	1	$r+1$
$S_r^{(1)}(A_4)$	$r+2$	$\lceil 3, r+1 \rceil$	2	$r+1$
$S_r^{(1)}(A_5)$	$r+2$	$\lceil 3, r+1 \rceil$	2	$r+1$
$S_r^{(1)}(M_{(3)})$	$r+3$	$r+2$	2	$r+1$
$S_r^{(2)}(M_{(3)})$	$r+3$	$\lceil 4, r+2 \rceil$	3	$r+1$
L_r^{r-3}	r	$r-3$	3	r
S_6^3	6	4	3	6
$S_r^{(0)}(S_4^2(\lambda))$	$r+4$	$r+2$	1	$r+4$
$S_5^2(\lambda)$	5	4	3	5
$S_r^{(1)}(S_4^2(\lambda),\mu), r>1$	$r+4$	$r+2$	3	$r+4$
$S_r^{(0)}(G_{k+1}^c,\lambda_k)$	$k+r+2$	$\lceil 3, k+r \rceil$	2	$k+r+1$
$S_r^{(0)}(G_{k+2}^t,\lambda_k)$	$k+r+2$	$\lceil 3, k+r \rceil$	2	$k+r+2$
$S_r^{(1)}(G_{k+1}^c)$	$k+r+2$	$\lceil 3, k+r \rceil$	2	$k+r+1$
$S_r^{(1)}(G_{k+2}^t)$	$k+r+2$	$\lceil 3, k+r \rceil$	2	$k+r+2$
$S_r^{(0)}(S_3^t,\lambda)$	$r+3$	$r+2$	2	$r+2$
$S_r^{(0)}(S_2^c)$	$r+3$	$r+2$	1	$r+3$
$A_k^{(0)}(A_l^{(0)} \vee L_r^r)$	$4+r$	$3+r$	1	$k+l-2+r$
$A_k^{(1)}(A_l^{(0)} \vee L_r^r)$	$4+r$	$3+r$	2	$k+l-2+r$
$A_k^{(1)}(A_l^{(1)} \vee L_r^r)$	$4+r$	$\lceil 4, r+3 \rceil$	3	$k+l-2+r$
$A_k^{(0)}(L_{m+1}^m \vee L_r^r,\lambda_m)$	$m+r+3$	$m+r+1$	2	$k+m+r$
$A_k^{(1)}(L_{m+1}^m \vee L_r^r,\lambda_m)$	$m+r+3$	$\lceil 4, m+r+1 \rceil$	2	$k+m+r$

Table 3.2: Indecomposable curves with $g = 3$.

Name		Equations
Here	[1, 22]	
$S_1^{(0)}(A_k)$	D_{k+3}	$x(y^2 - x^{k+1})$
$S_2^{(0)}(A_k)$	S_{k+4}	$xz,\; y^2 - x^{k+1} - yz$
$S_1^{(0)}(M_{(3)})$	U_7	$xz - y^2,\; yz - x^3$
$S_1^{(0)}(S_2^c)$	U_8	$yz,\; y^2 - x^3 - xz$
$S_1^{(0)}(S_4^2(\lambda))$	U_{10}^*	$z(y - \lambda x),\; (x - y)(\lambda z - xy)$
$A_k^{(0)}(A_l^{(0)})$	$P_{k+1,l+1}$	$xz,\; y^2 - x^{k+1} - z^{l+1}$

Table 3.3: Equations for complete intersections.

Name	Equations
$\widetilde{E}_8^{(2)}$	$\begin{pmatrix} z & y + \lambda x^2 & (\lambda^2 - 1)x \\ 0 & z - x^3 & y - \lambda x^2 \end{pmatrix}$
$S_r^{(1)}(A_k)$	$\begin{pmatrix} z & y & x^{(k-1)} \\ 0 & x^2 - zx^{(r-1)} & y \end{pmatrix}$
$S_1^{(1)}(M_{(3)})$	$\begin{pmatrix} x & y & z \\ z & x^2 & yx \end{pmatrix}$
$S_r^{(0)}(S_1^{(0)}(A_k))$	$\begin{pmatrix} z & x^k - y & xy \\ 0 & y - x & x^2 - zx^{(r-1)} \end{pmatrix}$
$S_1^{(0)}(S_2^{(0)}(A_k))$	$\begin{pmatrix} xz & x^{k-1} - x & \lambda(z - x - y) \\ 0 & y - x & z - \lambda x \end{pmatrix}$
$S_r^{(1)}(S_1^{(0)}(A_k))$	$\begin{pmatrix} z & x^k & xy \\ 0 & y & x^2 - zx^{(r-1)} \end{pmatrix}$
$S_r^{(0)}(S_3^t, \lambda)$	$\begin{pmatrix} z & \lambda y - x^2 & (\lambda^2 - 1)x \\ 0 & xy & y - \lambda z + \lambda x^2 \end{pmatrix}$
$A_k^{(0)}(A_l^{(1)})$	$\begin{pmatrix} z & x^{(l-1)} & y \\ 0 & y & x^2 - z^{(k+1)} \end{pmatrix}$
$A_k^{(0)}(A_l^{(1)})$	$\begin{pmatrix} z & x^{(l-1)} & y \\ 0 & y & x^2 - z^{(k+1)} \end{pmatrix}$

Table 3.4: Determinantal equations for space curves.

curves in cases III and IV we give in Table 3.3 also the names according to [1] or [23]. The equations are adapted to the chosen parametrisations; the curve S_{k+4} is a special case of the P-series, it is $P_{k,2}$; to see this from the equations one has to replace $y^2 - z^2$ by $y(y - z)$. In Table 3.4 we give equations for the space curves of type 2. The parameter r can have the value 1 or 2.

3.8 Moduli spaces for cones over point sets

The first singularity with moduli, which we encountered in the classification, is $S_4^2(\lambda)$. The moduli problem here is the same as for $\widetilde{E}_7$, four lines in $\mathbb{C}^2$. The moduli space of $\widetilde{E}_7$ singularities is of course the j-line, but we will use the cross ratio as modulus. In general we will use ordered point sets in $\mathbb{P}^n$ instead of unordered ones. This has several reasons. In the first place, we want to consider adjacencies and deformations of the point sets, so we need universal families. When looking at local moduli, there is no harm in ordering the points, which are all distinct. The moduli problem for a curve like $A_k(L_{m+1}^m, \lambda_m)$ reduces to that of L_{m+2}^m, but with one line playing a special role. Furthermore, the moduli spaces for ordered point sets are easier. The concept of association gives a duality, which yields point sets in lower dimensional spaces. As the subject is quite extensively covered in [12] and [6], we will be brief.

Mumford describes the action of $PGL(n + 1)$ on $(\mathbb{P}^n)^r$. He constructs an open subset U_{reg}, consisting of the *pre-stable* points, and proves that $\mathbf{p} = (p_1, \dots, p_r)$ is a geometric point of U_{reg}, if and only if the stabiliser $S(\mathbf{p})$ of $\mathbf{p}$ is zero dimensional, if and only if the cone over $\mathbf{p}$ is an indecomposable curve singularity (we take the cone over the support of $\mathbf{p}$ in case of coinciding points) [12, Prop. 3.3]. A geometric quotient of U_{reg} exists, but is very far from being a scheme.

Definition 3-1. ([12, Def. 3.7]) *The open subset U_{stable} of stable points of $(\mathbb{P}^n)^r$ consists of those points such that for every proper linear subspace $L \subset \mathbb{P}^n$:*

$$\frac{\text{number of points } p_i \text{ in } L}{r} < \frac{\dim L + 1}{n + 1}.$$

For semistability one has this condition with a $\leq$ sign.

The closure of the orbit of a prestable, but not stable point set contains decomposable sets.

Definition 3-2. ([5]) *Let $\mathbf{p} = (p_1, \dots, p_r)$ be a set of r ordered points in $\mathbb{P}^{r-g-1}$, whose coordinates are given by a $(r - g) \times r$-matrix P. A set $\mathbf{q} = (q_1, \dots, q_r)$ of r points in $\mathbb{P}^{g-1}$, given by a $g \times r$-matrix Q is called associated to $\mathbf{p}$ if $PQ^t = 0$.*

Lemma 3-3. *If no subset of $r - 1$ points of $\mathbf{p} = (p_1, \dots, p_r) \in (\mathbb{P}^{r-g-1})^r$ is contained in a hyperplane, then the associated set in $(\mathbb{P}^{g-1})^r$ is uniquely determined up to projective transformations. If the point set $\mathbf{p}$ is stable, then the associated point set is also stable.*

PROOF. Let the matrix P represent $\mathbf{p}$. We may suppose that P is of the form (I_g, A). Write $Q = (Q_1, Q_2)$ with Q_1 a $g \times (r-g)$-matrix. Then $PQ^t = Q_1^t + AQ_2^t = 0$ implies that $Q = (-A^t, I)Q_2$. Because every subset of $r-1$ points spans $\mathbb{P}^{r-g-1}$, no row of A is zero, so $(-A^t, I)$ determines r points in $\mathbb{P}^{g-1}$. This association is independent of the choice of P, because the expression $PQ^t = 0$ is invariant under the action of $Gl(r-g) \times Gl(r) \times Gl(g)$ on $M(r-g,r) \times M(g,r)$, given by $(A, B, C; P, Q) \mapsto (APB^{-1}, CQB^t)$.   □

Suppose that $\mathbf{p}$ is not stable, then there exists a linear subspace L of dimension $l-1$, containing s points with $s/r \geq l/(r-g)$. We can choose A with a $(r-g-l) \times (s-l)$ block of zeroes, so $-A^t$ has a $(s-l) \times (r-g-l)$ block of zeroes, and therefore $r-s$ points lie in a subspace of dimension $g+l-s-1$, while $(r-s)/r \geq (g+l-s)/g$.

3.9   The case $g = 2$

The associated point set of r points in $\mathbb{P}^{r-3}$ is a set of r not necessarily distinct points on the line. The condition that all points in $\mathbb{P}^{r-3}$ are distinct, i.e. no two lie in a zero dimensional subspace, translates in the condition that no $r-2$ points on $\mathbb{P}^1$ coincide.

The associated set of point set $\mathbf{p} = (p_1, \ldots, p_r)$ on the line can be described with rational normal curves [5].

Let $\{q_1, \ldots, q_k\}$ be the set of distinct points in $\mathbf{p}$, and label the points in $\mathbf{p}$ as p_{ij}, $i = 1, \ldots, k$, $j = 1, \ldots, j_i$, such that $p_{ij} = q_i$. Embed $\mathbb{P}^1$ as rational normal curve R_{k-3} of degree $k-3$ in $\mathbb{P}^{k-3} \subset \mathbb{P}^{r-3}$ (if $k = 3$ we get a point $\mathbb{P}^0$). Take in $\mathbb{P}^{r-3}$ for each i a linear space L_i of dimension $j_i - 1$ through $q_i \in R_{k-3}$, such that the L_i span $\mathbb{P}^{r-3}$.

Take j_i points p_{ij} in L_i, such that the points q_i and p_{ij} are in general position. In particular, if $j_i = 1$, then $p_{i1} = q_i$. All choices are projectively equivalent. The points p_{ij} in $\mathbb{P}^{r-3}$ are associated to $\mathbf{p}$.

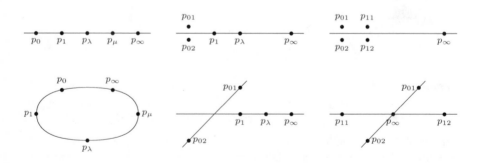

Figure 3.1: Associated point sets with $r = 5$.

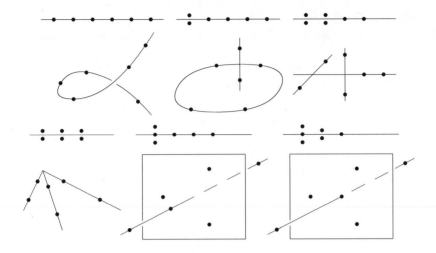

Figure 3.2: Associated point sets with $r = 6$.

For $r = 5$ we obtain only stable point sets. Figure 3.1 shows the three possible configurations; under a picture on the line the associated point set in the plane is shown.

For $r = 6$ we get four stable types and two semistable types (in Figure 3.2 the last two configurations are semi-stable). From $r = 7$ on we also have point sets which are not semistable, the associated point set of

$$(p_0, p_0, p_0, p_0, p_1, p_\infty, p_\lambda)$$

has three points on a line in $\mathbb{P}^4$.

4 Deformations and smoothings

In this section we look at adjacencies among the curves we have classified and consider their smoothability. A very convenient way to produce deformations of curves is by deforming the parametrisation; by a theorem of Teissier this indeed gives a deformation of the curve, if and only if the δ-invariant is constant [20]. More generally, for reducible curves one would like to be able to deform the irreducible components seperately; the following result tells us when this gives a deformation of the curve.

Theorem 4-1. *Let* $X = X_1 \cup X_2$ *be a reducible curve. Let* $\mathcal{X}_1 \to D$ *and* $\mathcal{X}_2 \to D$ *be one parameter deformations with fibres* $\mathcal{X}_{1,t}$ *and* $\mathcal{X}_{2,t}$. *Then* $\mathcal{X} = \mathcal{X}_1 \cup \mathcal{X}_2 \to D$ *defines a deformation of* X *if and only if* $(\mathcal{X}_{1,t} \cdot \mathcal{X}_{2,t})$ *is constant for all* $t \in D$.

PROOF. Suppose first that $(\mathcal{X}_{1,t} \cdot \mathcal{X}_{2,t})$ is constant. We have to show that the special fibre of $\mathcal{X} \to D$ is X. Suppose that the embedding dimension of X is

e, so X can be given by an ideal I in $\mathcal{O}_e$; let I_i be the ideal of X_i. Let $\mathcal{I}$ and $\mathcal{I}_i$ be the ideals of $\mathcal{X}$ and $\mathcal{X}_i$ in $\mathcal{O}_{e+1}$. Now consider the following diagram of exact sequences:

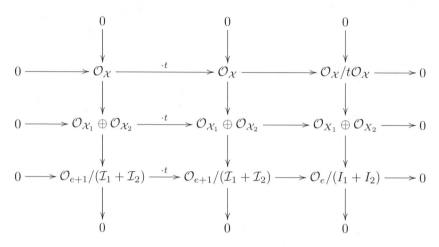

Here the last row is exact because $(X_{1,t} \cdot X_{2,t})$ is constant. Therefore also the last column is exact. From it we conclude that $\mathcal{O}_{\mathcal{X}}/t\mathcal{O}_{\mathcal{X}}$ is equal to $\mathcal{O}_X$. The converse follows by reversing the reasoning.   □

Example 4-2. One can deform the curve singularity $S_r^{(0)}(A_k)$ of Example 2-14, consisting of an A_k-singularity and r smooth branches, into one, containing an A_l-singularity, with $l < k$, and the same number of smooth branches.

Corollary 4-3. A decomposable curve $X \vee Y$ is smoothable if X and Y are smoothable.

PROOF. One can find smoothings X_t and Y_t of X and Y, which still pass through the origin; the curve $X_t \cup Y_t$ is a deformation of $X \cup Y$ and has only one A_1-singularity, so is smoothable.   □

Lemma 4-4. *All curve singularities with $g \leq 2$ are smoothable.*

PROOF. We only have to consider indecomposable curves. Curves with $\delta = r$ are general hyperplane section of simple elliptic singularities of multiplicity r, and therefore smoothable. One can also observe that L_{k+1}^k deforms into L_k^{k-1}: take two lines of L_{k+1}^k and smooth this A_1 singularity in its tangent plane, such that smooth curve is tangent to the hyperplane, spanned by the remaining $k-1$ lines; then the conditions of Theorem 4-1 are satified.   □

In [16] we have shown that all curves with $g = 2$ are smoothable; we give here a direct argument for all types except L_r^{r-2} (a general L_r^{r-2} is hyperplane section of the cone over a hyperelliptic curve of degree r). For the curves of type S_{r+1}^c and S_{r+2}^t we can move the lines into general position (by a δ-constant deformation of the parametrisation), to obtain curves of type G_{r+1}^c and G_{r+2}^t.

We then deform the A_2 or A_3 singularity into A_1, with still $(A_1 \cdot L_r^r) = 2$. The resulting L_{r+2}^{r+1} is smoothable. The curve $S_4^2(\lambda)$ is smoothable, because all space curves are [14]; but also it deforms into S_2^c: D_4 deforms into A_2 with given tangent line (use $x^3 + y^3 + \varepsilon(ax + by)^2$). Finally, the curve $M_{(3)}$ deforms into L_3^3, as shown by the parametrisation $(t^3 - \varepsilon, t(t^3 - \varepsilon), t^2(t^3 - \varepsilon))$.

Proposition 4-5. *All curves of genus 3, different from $L_r^{r-3} \vee L_k^k$, are smoothable. The curve $L_r^{r-3} \vee L_k^k$ is smoothable if and only if its indecomposable factor L_r^{r-3} is; the cone L_r^{r-3} over the points set* **p** *is smoothable, if and only if the associated point set in $\mathbb{P}^2$ lies on a curve of degree 4.*

PROOF. We look at the classification of indecomposable curves in (3.5). The only irreducible curve with embedding dimension greater than three is the curve $M_{(4)}$, which deforms into L_4^4, so it is smoothable. The curves with $g(X_1) \leq 1$ and $g(X_2) = 0$, listed in (3.5), are plane or space curves. If $g(X_1) = 2$ and X_2 is a line, we can deform the line to the most general position. If the resulting curve consists of an A_k and a number of lines with distinct tangents, we deform the A_k into A_1, and obtain a smoothable curve. Except for L_r^{r-3}, the remaining curves are of the type $S_r^{(0)}(Y)$ with Y a non-Gorenstein space curve of genus two, so they all deform into $S_r^{(0)}(M_{(3)})$ (cf. 4.1); the deformation $(t^3 - \varepsilon, t(t^3 - \varepsilon), t^2(t^3 - \varepsilon))$ of $M_{(3)}$ into L_3^3 induces a deformation of $S_r^{(0)}(M_{(3)})$ into L_{r+3}^{r+2}. If $g(X_1) = g(X_2) = 1$, then we have an A_2 or A_3 in $X_1 \vee L_k^k$, which we can deform into A_1 in $X_1 \vee L_k^k$. $\qquad\square$

The cones L_r^{r-3} have no deformations of positive weight, so every smoothing can be obained from a deformation of the projective cone. A set of r points in $\mathbb{P}^{r-4}$ is hyperplane section of a smooth curve C of genus 3, if and only if the associated point set lies on the canonical image of C in $\mathbb{P}^2$ [19, Thm. 9].

So L_r^{r-3} is smoothable if and only if its hyperplane section at infinity is a point set in the closure of the locus of point sets associated to point sets on canonical curves in the plane.

We may assume that decomposable the curve $L_k^k \vee L_r^{r-3}$ is given by an $(k + r - 3) \times (k + r)$ matrix (I, A), with the first rows of A equal to zero: $A^t = (0, A_1^t)$. If this curve is smoothable, then it is a limit of indecomposables, which are cones over hyperplane sections of smooth curves, and can be given as $(I, A(t))$, with $A^t(t) = (A_0^t(t), A_1^t(t))$ and $\lim_{t \to 0} A(t) = A$.

The rows of $A(t)$, and $A_1(t)$ in particular, determine points in $\mathbb{P}^2$ on a canonical curve. Passing to the limit shows that the factor L_r^{r-3} is smoothable.

4.1 Adjacencies

Some adjacencies were already used in the previous proof. The diagrams become complicated. To give a flavor of the results, we include a diagram (Figure 4.3), containing only the plane and space curves of genus 2 without moduli.

The proof of most of these adjacencies is simple and uses the same considerations as used above for showing smoothability. The adjacency $S_2^c \to M_{(3)}$

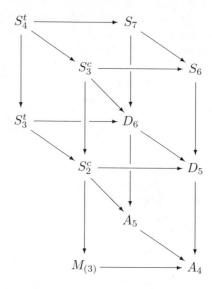

Figure 4.3: Adjacencies of some space curves with $g = 2$.

can be seen from the equations:

$$\begin{pmatrix} x & y & z \\ y & z - x^2 & \varepsilon x^2 \end{pmatrix}.$$

For $M_{(3)} \to A_4$ we can use the deformation of the parametrisation $(t^3 + \varepsilon t^2, t^4, t^5)$; to obtain $S_2^c \to A_5$ we deform the cusp with $(t^2 + \varepsilon t, t^3, \varepsilon^2 t^2)$ to have contact of order 3 with the smooth branch.

To show that our list is complete, we note that $S_3^t \not\to S_6$, and $S_3^c \not\to S_3^t$: such deformations are δ-constant, so would be given by a deformation of the parametrisation; other possible adjacencies are ruled out by semicontinuity of μ, or are well known not to exist (for complete intersections, see [8]).

As for the other space curves, depending on moduli, the curve $S_4^2(\lambda)$ has as plane projection the curve $\widetilde{E}_7$, with the same modulus, and the projection of a deformation of $S_4^2(\lambda)$ defines a deformation of $\widetilde{E}_7$.

Therefore we only have the adjacencies $S_4^2(\lambda) \to S_2^c$ and $S_4^2(\lambda) \to D_6$. The curve S_4^t deforms into curves of the type $S_4^2(\lambda)$, as can be seen by 'rotating' one of the four lines.

4.2   Deformations of L_5^3

The adjacencies of L_5^3 depend on the moduli of this curve. Because T^1 has negative grading, every deformation comes from a deformation of the projective cone over the same five points; furthermore, since the curve is homogeneous, every arithmetically Cohen-Macaulay projective curve with these five points as hyperplane section at infinity occurs in the versal deformation [11, A 2], so to construct deformations of L_5^3 it is sufficient to find curves of degree 5 and $p_a = 2$ with given singularities and given hyperplane section. An irreducible curve of $p_a = 2$ can only have singularities with $\delta \leq 2$, so for other adjacencies we need reducible curves.

$L_5^3 \to S_2^c$, $L_5^3 \to D_6$. Let the curve C consist of two irreducible components, the first one a straight line, say $(s : 0 : 0 : 1)$ with s an affine parameter on $\mathbb{P}^1$, and the second a rational cuspidal curve of degree four, given by $(t^2 : t^3 : t^4 : P_4(t))$ with $P_4(t)$ a polynomial of degree 4 with $P(0) \neq 0$.

Then C has a singularity of type S_2^c, and the hyperplane section at infinity is given by five points on the conic $y^2 = xz$: the point $(1 : 0 : 0)$ and four points $(1 : a_i : a_i^2)$, where the a_i are the roots of P_4. By appropriate choice of the points a_i one finds every isomorphism type of five points in generic position.

The adjacency $L_5^3 \to D_6$ works similar, with the curve consisting of two lines $(s_1 : 0 : 0 : 1)$ and $(s_2 : 0 : 0 : 1)$ and the rational normal curve $(t : t^2 : t^3 : P_3(t))$.

$L_5^3 \to S_6$. To make a degree 5 curve with S_6 we need two lines and a cuspidal curve of degree 3, which is a plane curve. But then the hyperplane section at infinity has three points on a line.

Therefore only the curves of the second type in Fig. 3.1 deforms into S_6. They also deform into S_7 and S_3^c, and for the correct value of the modulus into $S_4^2(\lambda)$.

$L_5^3 \to S_3^t$. To get a S_3^t singularity on a curve of degree 5, one needs two conics and one line, which is the line of intersection of the planes of the two conics. The hyperplane section at infinity is then of the third type in Fig. 3.1. This curve L_5^3 also deforms into S_4^t.

References

[1] V.I. Arnol'd, S.M. Gusein-Zade and A.N. Varchenko, Singularities of Differentiable Maps Vol. I. Basel etc., Birkhäuser 1985

[2] Dave Bayer and Mike Stillman, Macaulay: A system for computation in algebraic geometry and commutative algebra. Source and object code available from zariski.harvard.edu via anonymous ftp.

[3] Kurt Behnke and Jan Arthur Christophersen, Hypersurface sections and obstructions (rational surface singularities). Comp. Math. **77** (1991), 233–268

[4] R.-O. Buchweitz and G.-M. Greuel, The Milnor Number and Deformations of Complex Surface Singularities. Inventiones math. **58** (1980), 241–281

[5] Arthur B. Coble, Point sets and allied Cremona groups I. Trans. A.M.S. **16** (1915), 155–198

[6] Igor Dolgachev and David Ortland, Point sets in projective spaces and theta functions. Astérisque **165** (1988)

[7] Sherwood Ebey, The classification of singular points of algebraic curves. Trans. Am. Math. Soc. **118** (1965), 454–471

[8] Marc Giusti, Classification des singularités isolées simples d'intersections complètes. In: Singularities, Arcata 1981. Proc. Symp. Pure Math. **40**, Part 1 (1983), pp. 457–494

[9] Gert-Martin Greuel, On deformations of curves and a formula of Deligne. In: Algebraic Geometry, La Rábida 1981, pp. 141–168. Berlin etc., Springer 1982. (Lect. Notes in Math.; 961)

[10] H. Hironaka, On the arithmethic genera and the effective genera of algebraic curves. Mem. Coll. Sci. Univ. of Kyoto, Ser. A **30** (1957), 177–195

[11] Eduard Looijenga, The smoothing components of a triangle singularity II. Math. Ann. **269** (1984), 357–387

[12] D. Mumford and J. Fogarty, Geometric Invariant Theory, Second Enlarged Edition. Berlin etc., Springer 1982

[13] Ruud Pellikaan, Finite determinacy of functions with non-isolated singularities. Proc. London Math. Soc. (3) **57** (1988), 357–382

[14] Malka Schaps, Déformations non singulières de courbes gauches. In: Singularités à Cargèse, Astérisque **7** et **8 (1973)**, 121–128

[15] Jean-Pierre Serre, Groupes algébriques et corps de classes. Paris, Hermann, 1959

[16] F. Steiner, Diplomarbeit Bonn 1983

[17] Jan Stevens, Elliptic Surface Singularities and Smoothings of Curves. Math. Ann. **267** (1984), 239–249

[18] Jan Stevens, Kulikov singularities, a study of a class of complex surface singularities with their hyperplane sections. Diss. Leiden 1985

[19] Jan Stevens, On the number of points determining a canonical curve. Indag. Math. **51** (1989), 485–494

[20] Bernard Teissier, The hunting of invariants in the geometry of discriminants. In: Real and Complex Singularities, Oslo 1976, pp. 565–677, Alphen a/d Rijn, Sijthoff & Noordhoff, 1977

[21] W. Vogel, Lectures on Results on Bezout's theorem. Berlin etc., Springer 1984 (Tata Lectures on Mathematics and Physics; 74)

[22] C.T.C. Wall, Classification of unimodal isolated singularities of complete intersections. In: Singularities, Arcata 1981. Proc. Symp. Pure Math. **40**, Part 2 (1983), pp. 625–640

[23] C.T.C. Wall, Notes on the classification of singularities. Proc. London Math. Soc. (3), **48** (1984), 461–513

Address of author:

Mathematisches Seminar der Universität Hamburg
Bundesstr. 55, D 20146 Hamburg
Germany
Email: stevens@math.uni-hamburg.de

Progress in Mathematics

Edited by:

H. Bass
Columbia University
New York
10027
U.S.A.

J. Oesterlé
Dépt. de Mathématiques
Université de Paris VI
4, Place Jussieu
75230 Paris Cedex 05, France

A. Weinstein
Dept. of Mathematics
University of CaliforniaNY
Berkeley, CA 94720
U.S.A.

Progress in Mathematics is a series of books intended for professional mathematicians and scientists, encompassing all areas of pure mathematics. This distinguished series, which began in 1979, includes authored monographs, and edited collections of papers on important research developments as well as expositions of particular subject areas.

We encourage preparation of manuscripts in such form of TeX for delivery in camera-ready copy which leads to rapid publication, or in electronic form for interfacing with laser printers or typesetters.

Proposals should be sent directly to the editors or to: Birkhäuser Boston, 675 Massachusetts Avenue, Cambridge, MA 02139, U.S.A.